Bioinformatics challenges at the interface of biology and computer science

Mind the Gap

Bioinformatics challenges at the interface of biology and computer science:

Mind the Gap

Teresa K. Attwood, Stephen R. Pettifer and David Thorne

The University of Manchester
Manchester, UK

WILEY Blackwell

This edition first published 2016 © 2016 by John Wiley & Sons Ltd.

Registered office: John Wiley & Sons, Ltd., The Atrium, Southern Gate, Chichester, West Sussex, PO19 8SQ, UK

Editorial offices: 9600 Garsington Road, Oxford, OX4 2DQ, UK
The Atrium, Southern Gate, Chichester, West Sussex, PO19 8SQ, UK
111 River Street, Hoboken, NJ 07030-5774, USA

For details of our global editorial offices, for customer services and for information about how to apply for permission to reuse the copyright material in this book please see our website at www.wiley.com/wiley-blackwell.

Library of Congress Cataloging-in-Publication Data
Names: Attwood, Teresa K., author. | Pettifer, Stephen R. (Stephen Robert),
  1970- author. | Thorne, David, 1981-, author.
Title: Bioinformatics challenges at the interface of biology and computer
  science : mind the gap / Teresa K. Attwood, Stephen R. Pettifer and David
  Thorne.
Description: Oxford : John Wiley & Sons Ltd., 2016. | Includes
  bibliographical references and index.
Identifiers: LCCN 2016015332| ISBN 9780470035504 (cloth) | ISBN 9780470035481
  (pbk.)
Subjects: LCSH: Bioinformatics.
Classification: LCC QH324.2 .A87 2016 | DDC 570.285–dc23 LC record available at
  https://lccn.loc.gov/2016015332

A catalogue record for this book is available from the British Library.

Wiley also publishes its books in a variety of electronic formats. Some content that appears in print may not be available in electronic books.

Cover images: Courtesy of the authors, apart from the DNA strand image: Getty/doguhakan.

Set in 10/12pt Sabon LT Std by Aptara Inc., New Delhi, India
Printed and bound in Singapore by Markono Print Media Pte Ltd

1 2016

Table of Contents

PART 2

Preface

0.1 Who this book is for, and why

As you pick up this book and flick through its pages, we'd like to interrupt you for a moment to tell you who we think this book is for, and why. Let's start with the why. One main thought prompted us to tackle this project. There are now many degree courses in **bioinformatics**[1] throughout the world, and many excellent accompanying textbooks. In general, these texts tend to focus either on how to use standard bioinformatics tools, or on how to implement the **algorithms** and get to grips with the **programming languages** that underpin them. The 'how to use' approach is appropriate for students who want to become familiar with the tools of the trade quickly so that they can apply them to their own interests; the 'what's behind them' approach might appeal more to students who want to go on to develop their own **software** and **databases**. Books like this are extremely useful, but there's often something missing – bioinformatics isn't just about writing faster **programs** or creating new databases; it's also about coupling solid **computer science** with an appreciation of how computers can (or can't) solve biological problems.

During the years we've been working together, we realised that this issue is seldom, if ever, tackled head-on – it's an issue rooted in the nature of bioinformatics itself. Bioinformatics is often described as the interface where computer science and biology meet and overlap, as shown in Figure 0.1.

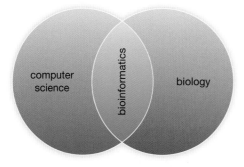

Figure 0.1

The traditional view of bioinformatics.

---

[1] While reading, you'll find that we've **emboldened** some of the more obscure or subject-specific phrases that we've used. These terms are defined in the glossary section.

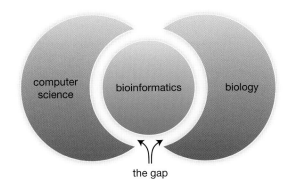

Figure 0.2

Mind the Gap.

the gap

However, the reality is that modern bioinformatics has become a discipline in its own right – bigger, broader and more complex than this seemingly trivial overlap might suggest, and bringing with it many new challenges of its own. The reality, then, tends to look more like Figure 0.2.

A common difficulty for teachers of bioinformatics courses is that they often have to cater for mixed audiences – for students with diverse backgrounds in biology and computer science. As a result, there generally isn't enough time to cover all of the basic biology and all of the basic bioinformatics, and even less time to deal, at an appropriate level, with some of the emerging issues in computer science and how these relate to bioinformatics. It's hard enough to do justice to each of these disciplines on its own, let alone to tackle the problems that emerge when they're brought together.

For teachers and students, this book is therefore an attempt to bridge the interdisciplinary gap, to address some of the challenges for bioinformatics at the interface of biology and computer science. Compared to other textbooks, it offers rather different perspectives on the challenges, and on the relationship between these disciplines necessary for the success of bioinformatics in the future. It will appeal, we hope, to undergraduate and postgraduate students who want to look beyond 'how to use this tool' or 'how to write that program'; it will speak to students who want to discover why the things we often want to do in bioinformatics are not as straightforward as they should be; it will resonate with those who want to understand how we can use computer science to make such tasks more straightforward than they currently are; and it will challenge the thoughtful to reflect on where the limits are in terms of what we want computers to be able to do and what's actually doable.

0.2 Who has written this book

The authors of the book have backgrounds in **biophysics** and computer science. We've worked together successfully for many years, but early in our relationship we often found it harder to communicate than we imagined it either could or should be. It wasn't just about meeting a different discipline head on and having to grapple with new terminology – the real problem was that we often *thought* we understood each other, only to find later that we actually meant quite different things, despite having used what appeared to be the same vocabulary. Learning new terminology is relatively easy; learning to spot that the thing *you* mean and the thing you think *he* means are different is much harder! It's a bit like trying to read the bottom line of an optician's test chart – the scale of the problem only comes into focus when viewed from another perspective: the optician provides the correct lens, and clarity dawns.

The scale of *our* communication problem was thrown into focus when we began developing software together. Initially, there was the usual learning curve to negotiate, as we each had to understand the jargon of the other's discipline. When our new software eventually emerged, we were invited to give a talk about it. It was only when we tried to write the talk, and stood back and looked at the work in a different way, that a kind of fog lifted – thinking we'd been speaking the same language, we'd actually reached this point by understanding different things by using the same words. This revelation provided a **catalyst** for the book: if we could so easily have reached apparent understanding through misunderstanding of basic concepts, could bioinformatics be a similar **house of cards**, waiting to topple? We reasoned that our different scientific perspectives, focused on some straightforward bioinformatics applications, might help to shed a little light on this question. So, let's now briefly introduce our different backgrounds – this should help to explain why the book is written in the way that it is, and why it contains what it does.

0.2.1   The biophysicist

The biophysicist isn't really a biologist and certainly isn't a computer scientist. She was engaged in her early postdoctoral work on protein sequence and structure analysis when the field of bioinformatics was just emerging; some of it rubbed off. The introductory chapters of the book therefore derive mostly from those early experiences, when the term bioinformatics was more or less synonymous with **sequence analysis**. Thus, sequence analysis provides an important backdrop for the book. Of course, it isn't the only theme, either of the book or of bioinformatics in general – it's just a convenient place to start, a place that's both historically and conceptually easy to build from, and one that's arguably even more relevant today than it was when the discipline began. To maintain an appropriate focus, then, we use sequence analysis as our springboard and, in deference to their expertise, we leave other aspects of bioinformatics to the relevant specialists in those particular fields.

0.2.2   The computer scientists

The computer scientists aren't really bioinformaticians and certainly aren't biologists. They began developing software tools in collaboration with bioinformaticians about ten years ago; in the process, a bit of bioinformatics rubbed off on them too. They come at the subject from various perspectives, including those of **distributed systems**, **computer graphics**, **human–computer interaction** and **scientific visualisation**. Especially important for this book, as you'll see if you view the online supplementary materials (which we encourage you to do), is their interest in the design of collaborative and semantically rich software systems for the biosciences, with a particular focus on improving access to scholarly literature and its underlying biochemical/ biomedical data.

0.3 What's in this book

0.3.1 The scope

Rather than offering an authoritative exposition of current hot topics in bioinformatics, this book provides a framework for thinking critically about fundamental concepts, giving new perspectives on, and hence trying to bridge the gap between, where traditional bioinformatics is now and where computer science is preparing to take it in the future. In essence, it's an exploration of the philosophical divide between what we want computers to do for us in the life sciences and what it's actually possible to do, today, with current computer technology. You might wonder why this is interesting – don't all the other books out there deal with this sort of thing? And the answer is, no – this book is different. It isn't about how to do bioinformatics, computer science or biology; it's about what happens when all of these disciplines collide, about how to pick up the pieces afterwards, and about what's likely to happen if we can't put all the bits back together again in the right places.

The problem is, when bioinformatics began to emerge as a new discipline in the 1980s, it was, in a sense, a fix for what the biological sciences needed at the time and what computer science could provide at that time (mostly database technology and fast database-search algorithms). The discipline evolved in a very pragmatic way, addressing local problems with convenient, more-or-less readily available solutions. That, of course, is the very essence of **evolution**: there is no grand plan, no sweeping vision of the future – just a problem that needs to be solved, now, in the most efficient, economical way possible.

The ramifications for systems that evolve are profound: a system that hasn't been designed for the future, but just 'does the job', will almost certainly be an excellent tool for precisely that job, at that time; however, it's unlikely to stand up to uses for which it wasn't originally devised. To be fit for new uses, the system will probably need to be modified in *ad hoc* ways, at each stage of the process gaining further complexity. Eventually, such an evolving system is likely to be confronted by new developments, developments that were not foreseen when the system began its evolutionary journey. At this juncture, we arrive at an all-too-familiar turning point: to move forward, either we must continue to 'make do and mend' the system, to patch it up so that it will continue to 'do the job', or we must take a deep breath and simply throw it away, and start again on a completely new path.

Many systems in bioinformatics have arrived at this point; many more will arrive there soon. So this is a good time to stop and reflect on how best to move forward. But it would be foolish to try to do this without first considering how bioinformatics reached this point in the first place. For this reason, the book begins, unashamedly, by rehearsing a little of the history of where bioinformatics came from; to balance things up, it includes a potted history of computer science too. It then looks at some of the things we want to be able to do now, things that weren't even thought of back then, and discusses why some of these things are so difficult (or impossible) to do with current systems. The book then crosses over to the other side, to look at the issues from a computer science perspective, and to explain what needs to happen in order to make some of the things we want to be able to do doable. Finally, it moves deeper into the emerging technologies, to consider some of the new problems that remain for the future.

0.3.2 The content

As a scientific discipline rooted in the (sequence) data-storage and data-analysis problems of the 1980s and 1990s, bioinformatics evolved hand-in-hand with the burgeoning of **high-throughput biology,** and now underpins many aspects of **genomics, transcriptomics, proteomics, metabolomics** and many other (some rather eccentrically named) 'omics'. Bioinformaticians today must, therefore, not only understand the strengths and limitations of computer technologies, but must also appreciate the basic biology beneath the new high-throughput sciences.

Accordingly, the book is divided into two parts: the first reviews the fundamental concepts, tools and resources at the heart of bioinformatics; the second examines the relationship between these and some of the 'big questions' in current computer science research. To give an idea: biological data may be of many different types, stored in many different formats, using a variety of different technologies, ranging from **flat-files,** to **relational** and **object-oriented databases,** to **data warehouses.** If we're to be able to use these repositories to help us address biological questions, we need to ask to what extent they realistically encapsulate the biological concepts in which we're interested. How accurately do they reflect the dynamic and complex nature of biological systems? To what extent do they allow us to perform seemingly simple tasks, such as comparing **biological sequences** and their 3-dimensional (3D) structures? To what extent do they allow us to reason with the data and to gain confidence that what we're looking at is, in some sense, true or meaningful? In exploring these issues, we examine the concepts involved in database design and organisation, the protocols and mechanisms by which data can be accessed in the modern distributed and networked environment, and we look at the higher-level issues of **data integrity** and **reusability.**

In a nutshell, Chapter 1 introduces the main themes of the book (bioinformatics and computer science) and generally sets the scene; Chapter 2 reviews biology's fundamental building-blocks, both to introduce some basic biological terms and concepts, and to provide sufficient context to start thinking about collecting and archiving biological data; Chapter 3 discusses the emergence of high-throughput sequencing techniques, the consequent data explosion and the spread of biological databases; and Chapter 4 explores the fundamental methods that underpin sequence analysis, with a perspective focused by the need to annotate and add meaning to raw sequence data. At this point, we cross over into Chapter 5 and step into *The Gap*, where some of the challenges at the interface of computer science, bioinformatics and biology arise. Here, we begin to reflect in earnest on how we humans understand the meaning of data, and how much harder it is to teach computers to achieve the same level of comprehension. From here, we move into the second, more technical part of the book: Chapter 6 delves deep into the way in which computers process data, exploring the twin themes of algorithms and complexity, and how understanding these has profound implications for building efficient bioinformatics tools; Chapter 7 considers the thorny issues of data representation and meaning (touching on aspects of **semantic integration, ontologies, provenance,** *etc.*), and examines the implications for building bioinformatics databases and services; Chapter 8 exposes the complexity of linking scientific facts hidden in databases with knowledge embodied in the scientific literature, and includes a case study that draws on many of the concepts, and highlights many of the challenges, outlined in the preceding chapters. We end with an Afterword, in which we try to reflect on how far we've succeeded in making many of our bioinformatics dreams a reality, and what challenges remain for the future. Finally, a comprehensive glossary serves as a reference for some of the biology, bioinformatics and computer science jargon that peppers this text.

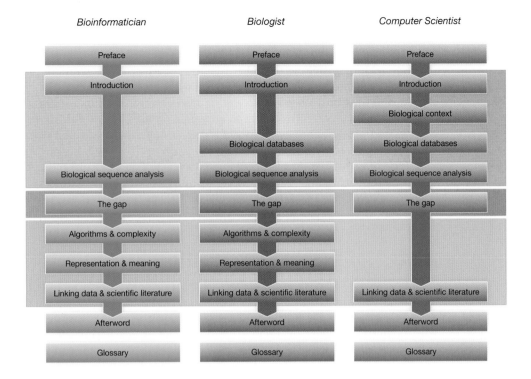

Figure 0.3

Book overview, and possible routes through it for a bioinformatician (left-hand side), a biologist (centre) and a computer scientist (right-hand side). The silver, gold and bronze backdrops represent Part 1 (bioinformatics), Part 2 (computer science) and the Gap respectively.

Figure 0.3 gives an overview of the book, with suggested routes through it for different types of reader: *e.g.*, after the Introduction, a bioinformatician might want to skip directly to 'Biological sequence analysis', moving through 'The Gap' and on to Part 2, referring back to the biological material as appropriate; a computer scientist might want to study Part 1 and 'The Gap' chapters, before moving to the more esoteric discussions in 'Linking data and scientific literature', referring back to the other computer science chapters as necessary; and a biologist might wish to skip the 'Biological context', but otherwise take the chapters as they come.

0.4   What else is new in this book

Most books today come with a variety of online extras, such as image galleries, student tests, instructor websites and so on, to complement the printed work. Alongside this book, one of the chapters is provided online in a form suitable for being read using **Utopia Documents**, a software suite that confers on the digital text all kinds of interactive functionality (*i.e.*, integrated with the chapter directly, not optional extras on a separate website).

To read the chapter online, we recommend that you download the Utopia Documents PDF reader. With this software, as you journey through the chapter, you'll experience new dimensions that aren't possible or accessible from the print version. Of course, you can read the chapter without installing the software; but, for those of you reading the online version, additional functionality will be revealed through the animating lens of Utopia Documents, and the overall experience will be more interesting. You can install the software from getutopia.com. Installation is easy: simply follow the link to the software download, and the guidance notes will talk you through the process for your platform of choice.

Once you've successfully downloaded Utopia Documents, you'll be ready for the new experience. As you read on, the chapter's interactive features will gradually unfold and will grow in complexity. We invite you to explore this functionality at your leisure (for the more adventurous, full documentation is available from the installation site).

0.5 What's not in this book, and why

Having outlined what you can expect to find in this book, we should emphasise again what you shouldn't expect to find, and why. The book doesn't provide a manual on how to use your favourite software or database; it doesn't offer 'teach yourself' **Perl**, **Java**, **machine learning**, probabilistic algorithms, *etc.*; nor does it provide all you need to know about biology – as a quick **Google** or **Amazon** search will reveal, several excellent books of this sort have already been written. Instead, we wanted this book to complement existing texts, to offer a new and thoughtful look at some of the real issues in bioinformatics and the challenges they pose for computer science. This is important, because the future of bioinformatics rests not only on how well we can address the perceived challenges of today, but also on how well we are able to recognise fundamental problems in the first place. And all of this brings us full circle, back to the origins of bioinformatics, when neither the problems ahead, nor the future-proofing steps necessary to make the databases and software of the day sufficiently robust to cope with the needs of scientists now, were fully appreciated.

Of course, it's easy to be wise in hindsight; that's why this book is also not about finding fault with the pioneers of bioinformatics, but rather about learning from their experiences. These scientists were enthusiasts – hard-working, sharp-witted individuals whose new ideas transformed the way we now manipulate biological data; they are directly responsible for how we work today. If we pause to look at what they did then and how they did it, we see that we have a lot to learn, but we also find that the way they tackled their problems has left today's bioinformaticians with new challenges. It is these challenges that form the focus of this book, and that we hope will inspire new cohorts of bioinformatics enthusiasts to take up the gauntlet.

Acknowledgements

We'd like to thank James Marsh for joining us in many of the early discussions that led to the final structure of the book; Robert Stevens and Sean Bechhofer for helping untangle some of the semantics of semantics; and Alex Mitchell for being on hand endlessly to help correct facts and figures whenever we asked. We're indebted also to our anonymous reviewers, whose encouragement and kind criticism helped us fill in some of the book's own gaps – whoever you are, the book is better for your input!

Last, but most certainly not least, two people deserve special mention. First, the long-suffering Nicky McGirr, who commissioned this book, and waited and cajoled and supported us patiently for 15 years until the manuscript was delivered – we'd certainly never have completed this project without her cool, calm presence (and she'd doubtless never have taken it on had she known how much hard work it would be!). Nicky was an enduring model of kind understanding and quiet determination; our debt to her is enormous. Second, Celia Carden, our Development Editor during the last four years, whose effortless charm and natural bonhomie buoyed our flagging spirits, and whooped and cheered us on to the finishing line. Thank you both so much – we'd never have completed this marathon without you!

About the Companion Website

www.wiley.com/go/attwood/bioinformatics

The website includes:

- An interactive PDF of Chapter 8, created in collaboration with Utopia Documents (see http://utopiadocs.com)
- Glossary
- Interactive Multiple-Choice Questions, with answers
- PowerPoint slides of all the figures from the book, to download
- Problems and answers
- References

PART **1**

Chapter 1

Introduction

1.1 Overview

In this chapter, we introduce the main perspectives of the book: **bioinformatics** and **computer science**. In Section 1.2, we offer a working definition of the term 'bioinformatics', we discuss where the discipline came from, and consider the impact of the **genome-sequencing** revolution. In Section 1.3, we discuss the origins of computer science, and note the emerging challenges relating to how to manage and describe biological data in ways that are computationally tractable. Having set the scene, we reflect briefly on some of the gaps that now confront computer science and bioinformatics.

By the end of the chapter, you should have an appreciation of how the field of bioinformatics evolved; you should also have gained insights into the extent to which its future progress is linked to the advances in data management and **knowledge representation** that are engaging computer science today.

1.2 Bioinformatics

1.2.1 What is bioinformatics?

Bioinformatics is a term that means different things to different people, with so many possible interpretations – many of them entirely reasonable – that it can sometimes be difficult to know what bioinformatics actually means, and whether it isn't just **computational biology** by another name. One way of making sense of the bioinformatics landscape is to recognise that it has both service and research components. Its service side primarily involves the routine storage, maintenance and retrieval of biological data. While these may seem like rather humdrum tasks for today's technologies, we'll explore why this is far from being true. By contrast, the research side of bioinformatics largely involves analysis of biological data using a variety of tools and techniques, often in combination, to create complex workflows or pipelines, including components ranging from **pattern recognition** and **statistics**, to visualisation and modelling. As we'll see, a particularly important facet of data analysis also concerns the use and development of prediction tools (later, we'll look at some of the ramifications of our heavy reliance on structure- and function-prediction approaches, especially in light of the emergence of high-throughput biology). The union of all of these capabilities into a broad-based,

Bioinformatics challenges at the interface of biology and computer science: Mind the Gap. First Edition.
Teresa K. Attwood, Stephen R. Pettifer and David Thorne. Published 2016 © 2016 by John Wiley and Sons, Ltd.
Companion website: www.wiley.com/go/attwood/bioinformatics

interdisciplinary science, involving both theoretical, practical and *conceptual* tools for the generation, dissemination, representation, analysis and understanding of biological information sets it apart from computational biology (which, as the name suggests, is perhaps more concerned with the development of mathematical tools for modelling and simulating biological systems).

In this book, we broadly explore issues relating to the computational manipulation and conceptual representation of biological sequences and macromolecular structures. We chose this vantage-point for two reasons: first, as outlined in the Preface, this is our 'home territory', and hence we can discuss many of the challenges from first-hand experience; second, this is where the discipline of bioinformatics has its roots, and it's from these origins that many of its successes, failures and opportunities stem.

1.2.2   The provenance of bioinformatics

The origins of bioinformatics, both as a term and as a scientific discipline, are controversial. The term itself was coined by theoretical biologist **Paulien Hogeweg**. In the early 1970s, she established the first research group specialising in bioinformatics, at the University of Utrecht (Hogeweg, 1978; Hogeweg and Hesper, 1978). Back then, with her colleague Ben Hesper, she defined the term to mean 'the study of informatic processes in **biotic systems**' (Hogeweg, 2011). But the term didn't gain popularity in the community for almost another two decades; and, by the time it did, it had taken on a rather different meaning.

In Europe, a turning point seems to have been around 1990, with the organisation of the *Bioinformatics in the 90s* conference (held in Maastricht in 1991), probably the first conference to include this 'new' term – bioinformatics – in its title. Consider that, during the same period, the National Center for Biotechnology Information[1] (NCBI) had been established in the United States of America (USA) (Benson *et al.*, 1990). But this was a centre for *biotechnology* information, not a bioinformatics centre, and it was established, at least in part, to provide the nation with a long-term 'biology **informatics** strategy' (Smith, 1990), not a 'bioinformatics' strategy.

With a new label to describe itself, a new scientific discipline evolved from the growing needs of researchers to access and analyse (primarily biomedical) data, which was beginning to accumulate, seemingly quite rapidly, in different parts of the world. This sudden data accumulation was the result of a number of technological advances that were yielding, at that time, unprecedented quantities of *biological sequence* information. Hand-in-hand with these developments came the widespread development of the algorithms, and computational tools and resources that were necessary to analyse, manipulate and store the amassing data. Together, these advances created the vibrant new field that we recognise today as bioinformatics.

Looking back, although the full history is convoluted, certain pivotal concepts and milestones stand out in the broadening bioinformatics panorama. In the following pages, we look at two of the major drivers of the evolving story: i) the technological developments that spawned the data deluge and facilitated its world-wide propagation; and ii) the development of **databases** to store the rapidly accumulating data. Before we do so, however, we must first identify a convenient starting point.

---

[1] http://www.ncbi.nlm.nih.gov

1.2.3 The seeds of bioinformatics

It's useful to think about where and when the seeds of bioinformatics were first sown, as this helps to provide a context for the situation we're in today. But where do we start? We could go all the way back to **Franklin** and **Gosling**'s foundational work towards the elucidation of the structure of **DNA** (Franklin and Gosling, 1953a, b, c), or to **Watson** and **Crick**'s opportunistic interpretation of their data (Watson and Crick, 1953). We could focus on the painstaking work of **Sanger**, who, in 1955, determined the **amino acid** sequence of the first **peptide hormone**. We could consider the ground-breaking work of **Kendrew** *et al.* (1958) and of **Muirhead** and **Perutz** (1963) in determining the first 3-dimensional (3D) structures of **proteins**. Or we could fast-forward to the progenitors of the first databases of macromolecular structures and sequences in the mid-1960s and early 1970s. This was clearly a very fertile era, heralding some of the most significant advances in molecular biology, and leading to the award of a series of **Nobel Prizes** (*e.g.*, 1958, Sanger's prize in chemistry; 1962, Watson, Crick and **Wilkins**' shared prize in physiology or medicine, after Franklin's death; and Perutz and Kendrew's prize in chemistry). These advances, any one of which could provide a suitable stepping-off point, each played an important part in the unfolding story.

Because this book focuses on biological sequences, especially protein sequences, we've chosen Fred Sanger's pioneering work on the peptide hormone **insulin** as our reference. Sanger was the first scientist to elucidate the order of amino acids in the **primary structure** of a protein. This was an immensely difficult puzzle, requiring the use of a range of chemical and enzymatic techniques in a variety of different experiments over many years. Each step of this incremental process was sufficiently new and exciting to warrant a separate publication. Eventually, around ten papers detailed the intricate work that led to the elucidation of the sequences, first of **bovine** insulin (*e.g.*, Sanger, 1945; Sanger and Tuppy, 1951a, b; Sanger and Thompson, 1953a, b; Sanger *et al.*, 1955; Ryle *et al.*, 1955), and then of **ovine** and **porcine** insulins (Brown *et al.*, 1955). This monumental achievement had taken a decade to complete. It seems incredible now that such a small protein sequence (containing only ~50 amino acids – see Figure 1.1) could have resisted so many experimental assaults and withheld its secrets for so long; and more so, that its 3D structure would not be known for a further *14 years* (Adams *et al.*, 1969)!

Manual **protein sequencing** was clearly an enormous undertaking, and it was many years before the sequence of the next protein was deduced: this was a small **enzyme** called **ribonuclease**. Work began on ribonuclease in 1955. Following a series of preliminary studies, the first full 'draft sequence' was published in 1960 (Hirs *et al.*, 1960), and a carefully refined final version was published three years later (Smyth *et al.*, 1963). Importantly, this eight-year project offered a stepping-stone towards elucidation of the protein's 3D structure: in fact, knowledge of its amino acid sequence provided a vital piece of a 3D **jigsaw puzzle** that was to take a further four years to solve (Wyckoff *et al.*, 1967). Viewed in the light of today's high-throughput sequence and structure determinations, these time-scales seem unimaginably slow.

Despite the technical and intellectual hurdles, the potential of amino acid sequences to aid our understanding of the functions, structures and evolutionary histories of proteins was compelling. The crucial role of protein sequences was clear to scientists like **Anfinsen**, arguably the originator of the field of **molecular evolution** (Anfinsen, 1959), and later, **Zuckerkandl** and **Pauling**, who helped build the foundations of molecular paleontology, and introduced evolutionary concepts like the **molecular clock**, based on

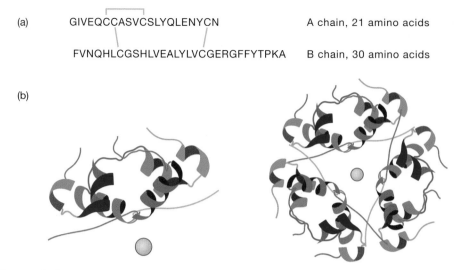

(a)   GIVEQCCASVCSLYQLENYCN   A chain, 21 amino acids

        FVNQHLCGSHLVEALYLVCGERGFFYTPKA   B chain, 30 amino acids

(b)

Figure 1.1

Bovine insulin (INS_BOVIN, P01317[i]): (a) the primary structure, showing intra- and interchain disulphide bonds connecting the A chain and B chain; and (b) its zinc-coordinated tertiary structure (PDB: 2INS[ii]), revealing two molecules in the asymmetric unit, and a hexameric biological assembly.

Source: Protein Data Bank.

[i]http://www.uniprot.org/uniprot/P01317
[ii]http://www.rcsb.org/pdb/explore/explore.do?structureId=2ins

rates of change of nucleotide or **polypeptide** sequences (Zuckerkandl, 1987). In their 1965 paper, Zuckerkandl and Pauling asserted that,

> All the potentialities of an individual may be assumed to be inscribed in polypeptide chains that are actually synthesized, or could be synthesized, by the cells under certain circumstances, and in the structures that control the actual and potential rates of this synthesis (Zuckerkandl and Pauling, 1965).

With great prescience, they envisaged biomolecular sequences as 'documents of evolutionary history'; nevertheless, they recognised that extracting their sequestered histories would require much more sequence information than was then available.

As the field of molecular evolution dawned, our hunger to understand more about the functions and structures of biological macromolecules provided the impetus for further sequencing efforts. **Margaret Dayhoff**, who had a keen interest in discerning evolutionary relationships from **sequence alignments**, was among the first scientists to approach this work systematically. In the early 1960s, to facilitate both her own research and the work of others in the field, she began to collect protein sequence information from scientific papers. This growing compendium of sequences was eventually published as a book – the *Atlas of Protein Sequence and Structure* (Dayhoff *et al.*, 1965), often simply referred to as the *Atlas*. Interestingly, in a letter she wrote in 1967, she observed,

> There is a tremendous amount of information regarding the evolutionary history and biochemical function implicit in each sequence and *the number of known sequences is growing explosively* [our emphasis]. We feel it is important to collect this significant information, correlate it into a unified whole and interpret it (Dayhoff, 1967; Strasser, 2008).

With the creation of the first *Atlas*, that 'explosive growth' amounted to 65 sequences!

During the next decade, the advent of automated processes overtook time-consuming manual peptide sequencing and dramatically increased the rate of sequence determination. In the meantime, another revolution was taking place, spurred on by the elucidation of the first protein atomic structures using the technique of **X-ray crystallography**: those of **myoglobin** and **haemoglobin** (Kendrew *et al.*, 1958; Muirhead and Perutz, 1963). Building on the sequencing work, this advance set the scene for a new era in which structure determination was to take centre stage in our quest to understand the biophysical mechanisms that underpin biochemical and evolutionary processes. So seductive was this approach that many more structural studies were initiated, and the numbers of deduced structures burgeoned.

In parallel with these developments, advances in sequencing technology in the late 1970s meant that, for the first time, DNA sequences could be determined, and their protein products deduced from them, relatively quickly. Incredibly, where it had taken 8–10 years to determine the sequences of the first small proteins (insulin and ribonuclease), dozens of protein sequences could now be rapidly deduced by translation of sequenced DNA.

The technologies that gave rise to manual peptide-sequencing strategies, then to automated peptide and DNA sequencing, and to protein structure determination at atomic resolution, were thus responsible for producing the first waves of sequence and structural data. Key to handling this expanding information was the recruitment of computers to help systematically analyse and store the accumulating data. Initially, newly determined sequences were published in the literature to make them available to the wider community. In this form, any researcher wishing to exploit the information had first to obtain a copy of the original article, and then to type the sequence(s) into a computer by hand, a process that now seems almost unbelievable.

Eventually, it became clear that collating data into electronic repositories would make it more efficient and easier to store and use the data in future. This realisation led to the birth of the first electronic databases. At this time, the idea that molecular information could be managed using electronic repositories was not only very new, but was also very daunting. Consider that technologies we take for granted today (**email**, the **Internet**, the **World-Wide Web** ('the Web')) hadn't yet emerged; there was therefore no simple way to distribute data from a central database, other than posting **computer tapes** and disks to users on request. This model of data distribution was fraught with difficulties: it was cumbersome and slow; it was also relatively costly, and, alarmingly, led some of the first database pioneers to adopt pricing and/or data-sharing policies that threatened to drive away many of their potential users.

From these tentative awakenings, the first biomolecular databases emerged. From the agonisingly slow trickle of determining a single sequence per decade, to a speed of thousands of sequences per second (a rate that will itself seem inconsequential ten years from now), sequencing technology has revolutionised the pace of data acquisition, and has thrown up new challenges for the field of bioinformatics.

1.3 Computer Science

1.3.1 Origins of computer science

The discipline of computer science is associated with many things, from the design of **silicon chips** and the creation of life-style-enhancing gadgets, to the applications and **operating systems** that have become the mainstay of modern life. It's an enormously broad subject, touching on topics as diverse as **electronic engineering, mathematics, aesthetics,** and even human **psychology** and **sociology.** But at its heart, computer science is about one thing: finding mechanisms of representing our universe – both its physical and conceptual nature – such that these can be manipulated and experimented with in ways that are beyond the capacity of the human mind. The process goes something like this: identify an intractable problem in the real world (perhaps one that requires a human to remember too many things at once), or that takes more steps to solve than can be done in a sensible amount of time; devise an abstract representation of that problem and of its constituent parts; and finally, create a device or process that's able to manipulate that representation in an automated way.

This need to manipulate abstract representations in a disciplined and controlled way makes computer science a form of extreme applied mathematics. As we might expect, this anchors its origins much further back in history than bioinformatics. Echoes of the process of abstraction and representation can be seen as far back as 2400 BCE, with the invention of the **abacus.** By drawing lines in the sand, and positioning pebbles in specific patterns, the ancient **Babylonians** were able to represent and manipulate what we would now think of as being positive **integer** numbers – useful for counting concrete physical things such as sheep, slaves and other everyday items of Babylonian life. However, even though it may seem like no enormous conceptual leap to imagine 'half a sheep', the abacus had no way of representing fractional numbers, which did not appear in mathematical systems until several hundred years later, via Egyptian **hieroglyphics.** Neither of these systems coped with negative numbers – what, after all, would it mean to have minus-one **pyramids?** Perhaps more surprisingly, they also lacked a representation for the concept of 'zero': there was no 'thing' that represented 'no thing'. Techniques for capturing and manipulating these more abstract concepts did not appear for at least another two millennia. The pattern of devising increasingly sophisticated and rich representations for abstract concepts, and of building devices that help take the drudge out of manipulating them, has continued to this day. This is the essence of computer science.

Of course, mathematics, and the process of 'computing' answers using mathematical notations, evolved considerably over the centuries, and its history is far too complex and intricate to discuss here; besides, for the most part, its details aren't relevant to the story this book tells. There are, however, a couple of notable exceptions. One was the creation, by Indian mathematician **Brahmagupta** in the 7th century CE, of the idea of an 'algorithm': a formal description of a sequence of steps that, carried out in order, accomplish a specific mathematical task. Another was the realisation, in the 3rd century BCE, by **Pingala,** another Indian mathematician, that numbers could be represented in '**binary notation**' as patterns of *true* or *false* values, *on* or *off* states, or simply by the presence or absence of objects. Centuries later, binary notation became the foundation of computer hardware, with binary numbers being captured, first as the presence or absence of holes in a **punched card,** and later as the presence or absence of a charge in a **valve** or **transistor.** Binary also spawned the idea of 'logic' in a formal, mathematical sense, which, together with the concept of algorithms, became the cornerstone of modern software methods.

Historians will continue to wrangle over exactly what constitutes the first computing device: whether it was the ancient abacus, '**Napier's Bones**' (c. 1610, a contraption for manipulating **logarithms**), the '**Jacquard Loom**' (c. 1800, which used punched cards to control weaving patterns), **Charles Babbage's Difference and Analytical Engines** (1882 and 1837), or one of the many other machines for automating mathematical calculation, remains a topic of enthusiastic (and often heated) academic discussion. To give the recognition deserved to the pioneers of mathematics and computing that steered the path from these early crude machines to the **laptops** and **servers** of modern time is well beyond the scope of this book. Instead, we will leap forward in time, past all the technology-related creativity inspired by the **Second World War**, to the early 1960s, and the coining of the term 'computer science' by numerical analyst **George Forsythe**. By this time, the computer had many of the properties associated with contemporary machines: a **Central Processing Unit** (CPU) able both to manipulate numbers represented in binary and to execute a **program** to choreograph such manipulations; some form of **memory** in which to store data and programs; **back-up devices** to record the contents of memory when the device was switched off; connective infrastructure to allow the machine to communicate with devices other than itself; and various forms of input and output (screens, keyboards, *etc.*) to allow users to interact with the machine.

The first computer programs focused on solving mathematical problems (like finding large **prime numbers**) for which no equation existed that could simply be 'solved' by a human with pen and paper, but for which iterative algorithms could be devised that played to the computer's strength of being able to mindlessly and repetitively 'crunch' numbers according to prescribed rules. Over time, programs were written to perform calculations that were related to more pragmatic ends: calculating payrolls and other business tasks, simulating engineering problems, and recording and searching over data about individuals for all manner of purposes. Today, computers perform innumerable jobs, from sending television programmes across **wireless networks**, to hand-held **mobile devices**, to autonomously guiding the trajectory of space probes visiting the outer reaches of our **solar system**.

Of course, this unashamedly selective, whirlwind tour of the history of computer science overlooks a multitude of incredible developments and technological innovations of the past few decades. On the face of it, many of these appear to be merely incremental improvements to existing ideas, resulting in something that is just a bit faster, smaller, bigger, greener, lighter – insert the adjective of your choice here – than before. Many of these improvements, however, are not just the result of 'trying a bit harder': often, development of the next generation of a particular technology has required considerable research and creative effort to overcome or circumvent what were previously considered unbreachable limits of engineering or physics. Other inventions have resulted in turning points in the way we live and work: the first machines 'cheap' enough to have on one's desk, or efficient enough to run off a battery in one's pocket; the early Internet as a connective infrastructure for commerce and research, and the Web as its more friendly front-end, with much broader appeal. More recently, **social networking**, wireless and mobile technology, and the ability to easily create and distribute audio and video, have changed how society interacts in ways that were inconceivable only a few years ago. Many of these technologies are now so integral to modern society that we fail to recognise they exist, much less the effort and thought that led to their creation.

1.3.2 Computer science meets bioinformatics

So, what has happened to allow today's devices to quietly perform such diverse, incredibly complex tasks since the 'primitive' machines of the 1960s? Although faster, cheaper, smaller, less power-hungry, and in almost every way better, more sophisticated and interconnected than their older counterparts, the modern computer has very much the same basic components as its ancestors: its core remains a device for manipulating binary numbers. What has changed is our ability to use patterns of numbers to represent increasingly complex and sophisticated things. Using a computer to calculate mathematical results is, in some ways, easy – this, after all, is what computers do. Determining which book someone is likely to want to buy next, based on the reading habits of people who have been established as having similar tastes, or – to bring this technological tale back to the life sciences – predicting whether individuals are likely to suffer from particular **diseases**, based on their genetic profiles, are altogether much more complex problems for computers to solve.

Here, then, is a gap: the vast gulf between a computer's ability to manipulate binary numbers, and our desire to use these machines to examine, understand and manipulate concepts – which, ostensibly, have no relationship to binary numbers at all. Bridging this gap, by devising techniques for representing our increasingly sophisticated knowledge in 'computable' form, is a fundamental challenge of modern computer science, one whose solution is poised to transform bioinformatics. Next to the very prominent developments we've just talked about, this is a much quieter revolution, but one that's already shaping how computers deal with data, and particularly **data representation**. In the early days of computing, **flat-file** databases, with their field-based searches, were the norm. In time, however, these were superseded by **relational database** systems that captured some of the structure of the information they modelled; and now we have **graph databases**, which attempt to model the meaning of data through the use of **controlled vocabularies**, allowing 'intelligent' data- and **text-mining** algorithms to scour them for nuggets of knowledge. This trend towards richer, more **semantic** (and thus 'computer friendly') data representation reflects our ongoing quest to make the growing volumes of data accessible to us as knowledge, knowledge that will touch innumerable aspects of our lives, from our relationships with each other and with our planet, to our future medical practice and health.

Albeit very much a thing of the past in the realms of computer science, the flat-file database was the cornerstone of early bioinformatics, playing a pivotal role in housing the gradually accumulating quantities of biomedical information. But modern high-throughput biology changed all this: its data explosion caught many bioinformaticians off-guard, and brought a growing realisation that the technologies underpinning the earliest databases were simply not up to the job. If storing the now-vast quantities of biological data was becoming increasingly demanding, reasoning over the data was becoming more difficult still. Consequently, in the aftershocks that followed, a gulf opened up between what we wanted to be able to do with bioinformatics on the one hand, and what we could actually do with it on the other.

1.4   What did we want to do with bioinformatics?

Because the origins of bioinformatics were rooted in sequence analysis, the earliest analyses aimed to understand what biomolecular sequences could tell us about the functions of **genes** and of their encoded proteins. Ultimately, scientists wanted to discover how amino acid sequences determined 3D **protein folds**, and what their sequences and structures could tell us about their evolutionary histories. Perhaps more importantly, researchers wanted to know how sequence and structure information could be used to elucidate the roles of particular genes and proteins in **pathogenic processes**, and how the assembled data could be used to design better, more efficacious **drugs**.

Later, with the advent of the human **genome-sequencing project**, the goals became increasingly ambitious, and the focus of attention turned more and more towards using bioinformatics to revolutionise molecular medicine. Researchers wanted to identify the genetic determinants both of rare **syndromes** and of common, pervasive diseases like **cancer**; bioinformatics, it was claimed, would play a major role in the development of new approaches to eradicate such diseases, and would pave the way to **personalised therapies**, where an individual's **genome** could be used to determine which drug regime would offer maximal benefit with the minimum of **side-effects**. Ultimately, the goal was to integrate all molecular and cellular data in such a way as to be able to model **biochemical pathways** and, indeed, whole **cells**, and to understand not just how individual cells work, but how complete assemblies of cells work in whole **organs**, including the brain.

To put some of these aspirations in context, in the run up to the publication of the human genome, huge expectations were placed on what bioinformatics should or would be able to deliver. One commentator predicted that a bioinformatics revolution was afoot from which the next step in man's **evolution** would be

> our acquisition of the power to control the evolution of our own **species** and all others on the planet…to create plants that walk[2], animals that can carry out **photosynthesis** and other unlikely **chimeras**…, [ultimately to] go beyond, in human–computer communication, anything we can remotely conceive of at present (Cantor, 2000).

Other predictions weren't quite so far-fetched. Yet there was a general belief that this new discipline would transform research in fields from **genomics** to **pharmacology**, and would probably 'reverse the long-standing **reductionist** paradigm that has held sway in molecular biology' for more than 50 years;

> In addition, bioinformatics will likely provide the methodology finally to make highly accurate predictions about protein **tertiary structure** based on amino acid sequences and a viable means to design drugs based on computer simulation of the **docking** of small molecules to the predicted protein architecture…New computational methods will likely transform taxonomic and phylogenic [sic] studies as well as our ability to understand and predict the results of complex **signal transduction** cascades and the **kinetics** of intricate **metabolic pathways** (Wallace, 2001).

Such views were not uncommon in this exciting new era. Genomics research was making it possible to investigate biological phenomena on a hitherto-impossible scale, amongst other things, generating masses of **gene expression**, gene and protein sequence, protein structure, and protein–protein interaction data.

> How to handle these data, make sense of them, and render them accessible to biologists working on a wide variety of problems is the challenge facing bioinformatics…The 'post-genomic era' holds phenomenal promise for identifying the mechanistic bases of organismal development, metabolic processes, and disease, and we can confidently predict that bioinformatics research will have a dramatic impact on improving our understanding of such diverse areas as the regulation of gene expression, protein structure determination, comparative evolution, and **drug discovery** (Roos, 2001).

As this book unfolds, we'll touch on some of the predictions that have been made for the bioinformatics revolution, and consider how realistic they are in the context of the challenges that still lie ahead. The reality is that *in silico* approaches won't transport us to Star Trek[3] futures quickly. Indeed, more than a decade ago, a rather prescient

---

[2] http://en.wikipedia.org/wiki/Triffid
[3] http://en.wikipedia.org/wiki/Star_Trek

Nature Biotechnology editorial suggested that the transformation we could expect genomics and *in silico* tools to have on traditional **empirical** medicine,

> should take about 10 to 15 years (with a following wind)...Thus, genomics will not rapidly improve the efficiency of drug development. In fact, it may make it even more complicated (Editorial, 2001).

In the chapters that follow, we will explore some of the complexities. We'll look at the transition from numeric to symbolic algorithms that was necessary to allow bioinformatics to move beyond the computation, say, of sequence comparison scores, to manipulation of concepts or entities, such as 'prion', 'promoter', 'helix', and so on (Attwood and Miller, 2001). We'll touch on many of the emerging techniques that are being used to transform data into knowledge, exploring how **ontologies** can be used to give meaning to raw information, how **semantic integration** is beginning to make it possible to join up disparate data-sets, and how visualisation techniques provide a way of harnessing human intuition in situations where computational techniques fall short. As we navigate the Grand Canyon[4] interface between bioinformatics and computer science, we'll see why, despite the power of computers and progress in **information technology**, bioinformatics is still not as straightforward as it perhaps could or should be. In particular, we'll examine the gap between what we wanted to do with bioinformatics and what we can actually do with it, and why the 'following wind' will need to be very much stronger if we're to make substantial progress even in the next 10–15 years.

Before addressing the technical and philosophical issues that arise when we use computers to try to tackle biological problems, the next chapter will take a brief look at the biological context that provides the foundation for molecular sequence- and structure-based bioinformatics today.

1.5 Summary

This chapter explored the nature and roots of bioinformatics, and the origins of computer science. In particular, we saw that:

1 Bioinformatics has both service and research components;
2 The service side of bioinformatics involves storage, maintenance and retrieval of biological data;
3 The research side of bioinformatics largely involves analysis and conceptual modelling of biological data;
4 The term 'bioinformatics' originally had a different meaning, and pre-dates the discipline we recognise today;
5 The discipline evolved from labour-intensive manual technologies that aimed to deduce molecular sequences and structures, and from largely descriptive manual approaches to catalogue this information;
6 The first such catalogues were books;
7 Manual approaches for deriving molecular sequences were gradually superseded by powerful automatic processes;

---

[4] http://en.wikipedia.org/wiki/Grand_Canyon

8 Automation of sequencing technologies catalysed both the spread of databases to store, maintain and disseminate the growing quantities of data, and the development of algorithms and programs to analyse them;

9 Automation of DNA sequencing technologies generated data on a scale that was inconceivable 60 years ago, and even now is almost unimaginable;

10 Computer science involves finding ways of representing real-world problems such that they can be manipulated by machines;

11 The scale of modern bio-data production is demanding new computational approaches for data storage and knowledge representation;

12 There is currently a gap between what computers do (manipulate binary numbers) and what we want them to do (examine and manipulate concepts);

13 The impact of bioinformatics on drug discovery and personalised medicine has been slower to emerge than predicted;

14 Bridging the knowledge-representation gap will help to advance bioinformatics in future.

1.6 References

Adams, M.J., Blundell, T.L., Dodson, E.J. *et al.* (1969) Structure of rhombohedral 2 zinc insulin crystals. *Nature*, **224**, 491– 495.

Anfinsen, C. (1959) *The Molecular Basis of Evolution*. John Wiley & Sons, Inc., New York.

Attwood, T.K. and Miller, C. (2001) Which craft is best in bioinformatics? *Computers in Chemistry*, **25**(4), 329–339.

Benson, D., Boguski, M., Lipman, D.J. and Ostell, J. (1990) The National Center for Biotechnology Information. *Genomics*, **6**, 389–391.

Brown, H., Sanger, F. and Kitai, R. (1955), The structure of pig and sheep insulins. *Biochemical Journal*, **60**(4), 556–565.

Cantor, C. (2000) Biotechnology in the 21st century. *Trends in Biotechnology*, **18**, 6–7.

Dayhoff, M.O., Eck, R.V., Chang, M.A. and Sochard, M.R. (eds) (1965) *Atlas of Protein Sequence and Structure*. National Biomedical Research Foundation, Silver Spring, MD, USA.

Dayhoff, M.O. to Berkley, C. (1967) Margaret O. Dayhoff Papers, Archives of the National Biomedical Research Foundation, Washington, DC.

Editorial. (2001) A cold dose of medicine. *Nature Biotechnology*, **19**(3), 181.

Franklin, R.E. and Gosling, R.G. (1953) **a)** The structure of sodium thymonucleate fibres. I. The influence of water content. *Acta Crystallographica*, **6**, 673–677; **b)** The structure of sodium thymonucleate fibres. II. The cylindrically symmetrical Patterson function. *Acta Crystallographica*, **6**, 678–685; **c)** Molecular configuration in sodium thymonucleate. *Nature*, **171**, 740–741.

Hirs, C.H.W., Moore, S. and Stein, W.H. (1960) the sequence of the amino acid residues in performic acid-oxidized ribonuclease. *Journal of Biological Chemistry*, **235**, 633–647.

Hogeweg, P. (1978) Simulating the growth of cellular forms. *Simulation*, **31**, 90–96.

Hogeweg, P. (2011) The roots of bioinformatics in theoretical biology. *PLoS Computational Biology*, **7**(3), e1002021.

Hogeweg, P. and Hesper, B. (1978) Interactive instruction on population interactions. *Computers in Biology and Medicine*, **8**, 319–327.

Kendrew, J.C., Bodo, G., Dintzis, H.M. *et al.* (1958) A three-dimensional model of the myoglobin molecule obtained by x-ray analysis. *Nature*, **181**, 662–666.

Muirhead, H. and Perutz, M. (1963) Structure of hemoglobin. A three-dimensional Fourier synthesis of reduced human hemoglobin at 5.5 Å resolution. *Nature*, **199**, 633–638.

Roos, D.S. (2001) Bioinformatics – Trying to swim in a sea of data. *Science*, **291**, 1260–1261.

Ryle, A.P., Sanger, F., Smith, L.F. and Kitai, R. (1955) The disulphide bonds of insulin. *Biochemical Journal*, **60**(4), 541–556.

Sanger, F. (1945) The free amino groups of insulin. *Biochemical Journal*, **39**, 507–515.

Sanger, F. and Tuppy, H. (1951) **a)** The amino-acid sequence in the phenylalanyl chain of insulin. 1. The identification of lower peptides from partial hydrolysates. *Biochem. J.*, **49**, 463–481; **b)** The amino-acid sequence in the phenylalanyl chain of insulin. 2. The investigation of peptides from enzymic hydrolysates. *Biochemical Journal*, **49**, 481–490.

Sanger, F. and Thompson, E.O.P. (1953) **a)** The amino-acid sequence in the glycyl chain of insulin. 1. The identification of lower peptides from partial hydrolysates. *Biochem. J.*, **53**, 353–366; **b)** The amino-acid sequence in the glycyl chain of insulin. 2. The investigation of peptides from enzymic hydrolysates. *Biochemical Journal*, **53**, 366–374.

Sanger, F., Thompson, E.O.P. and Kitai, R. (1955) The amide groups of insulin. *Biochemical Journal*, **59**(3), 509–518.

Smith, T.F. (1990) The history of the genetic sequence databases. *Genomics*, **6**, 701–707.

Smyth, D.G., Stein, W.H. and Moore, S. (1963) The sequence of amino acid residues in bovine pancreatic ribonuclease: revisions and confirmations. *Journal of Biological Chemistry*, **238**, 227–234.

Strasser, B. (2008) GenBank – Natural history in the 21st century? *Science*, **322**, 537–538.

Wallace, W. (2001) Bioinformatics: key to 21st century biology. *BioMedNet*, issue **99**, 30 March 2001.

Watson, J.D. and Crick, F.H.C. (1953) Molecular structure of nucleic acids. *Nature*, **171**, 737–738.

Wyckoff, H.W., Hardman, K.D., Allewell, N.M. *et al.* (1967) The structure of ribonuclease-S at 3.5 Å resolution. *Journal of Biological Chemistry*, **242**, 3984–3988.

Zuckerkandl, E. (1987) On the molecular evolutionary clock. *Journal of Molecular Evolution*, **26**, 34–46.

Zuckerkandl, E. and Pauling, L. (1965) Molecules as documents of evolutionary history. *Journal of Theoretical Biology*, **8**, 357–366.

1.7   Quiz

The following multiple-choice quiz will help you to check how much you've remembered of the origins of bioinformatics and computer science described in this chapter. Be mindful that in this and other quizzes throughout the book – just to keep you on your toes – there may be more than one answer!

1   Who first introduced the term bioinformatics?
   A   Fred Sanger
   B   Linus Pauling
   C   Paulien Hogeweg
   D   Margaret Dayhoff

2   Who first sequenced a protein?
   A   Fred Sanger
   B   Linus Pauling
   C   Paulien Hogeweg
   D   Margaret Dayhoff

3   How long did the determination of the sequence of insulin take?
   A   Five months
   B   Five years
   C   Eight years
   D   Ten years

4 Which was the first enzyme whose amino acid sequence was determined?
  A Insulin
  B Ribonuclease
  C Myoglobin
  D Haemoglobin

5 Which was the first protein whose structure was determined?
  A Insulin
  B Ribonuclease
  C Myoglobin
  D Haemoglobin

6 Which of the following statements is true?
  A The first collection of protein sequences was at the National Center for Biotechnology Information
  B The first collection of protein sequences was made by Fred Sanger
  C The first collection of protein sequences was the *Atlas of Protein Sequence and Structure*
  D None of the above

7 Who was responsible for the invention of the Difference Engine?
  A John Napier
  B Charles Babbage
  C Joseph Marie Jacquard
  D George Forsythe

8 What is the smallest number that could be represented using the Babylonian counting scheme?
  A One
  B Two
  C 'Several'
  D Zero

9 Binary representation was first conceived by:
  A George Boole
  B Pingala
  C Alan Turing
  D John Napier

10 Which of the following statements is true?
  A There is a gap between the ability of computers to manipulate concepts and our desire to use them to manipulate binary numbers
  B There is a gap between the ability of computers to manipulate primary numbers and our desire to use them to manipulate concepts
  C There is a gap between the ability of computers to manipulate binary numbers and our desire to use them to manipulate concepts
  D None of the above

1.8 Problems

1 Margaret Dayhoff was one of the pioneers of bioinformatics, actively working in the field in the 1960s, long before the discipline that we know today had even been named. In Section 1.2.3, we described how she was involved in producing the *Atlas of Protein Sequence and Structure*, the first published compendium of protein sequences, one that went on to give life to one of the first protein sequence databases. In addition, she is known for two other pivotal contributions to bioinformatics. What were they?

2 The NCBI, established in 1988, became the new home of the USA's first national nucleotide sequence database in October 1992. What was that database? How many sequences did the database contain when it was first released, and how many were contained in its first release under the auspices of the NCBI?[5] How many sequences does the database contain today?

3 This book is about the gaps we encounter when we explore the interface of bioinformatics and computer science. The nature and types of gap we'll discuss are many and varied: some are subtle and small; others are large and frightening. What is the gap described by Fraser and Dunstan in their 2010 article (*The BMJ*, **342**, 1314–15), and why is it especially disturbing?

---

[5] Hints: http://www.youtube.com/watch?v=mn98mokVAXI; http://www.ncbi.nlm.nih.gov/genbank/statistics

Chapter 2

The biological context

2.1 Overview

In this chapter, we look at some of the data-types and biological concepts that provide the foundations of the field of bioinformatics. We review the fundamental molecules of biology, their molecular building-blocks, the alphabets we use to describe them, and their **physicochemical properties** and structural characteristics. We also provide a reminder of the processes of **transcription** and **translation**, and their relevance to genome-sequencing projects.

By the end of the chapter, you should be able to list the nucleotide and amino acid building-blocks of nucleic acids and proteins respectively, and you should have an appreciation of how to translate a nucleotide sequence in six reading frames; you should also have an insight into the current scale of biological data production.

2.2 Biological data-types and concepts

2.2.1 Diversity of biological data-types

Bioinformaticians deal with a large and incredibly diverse range of inter-related data types. To give an idea, we can begin by thinking about the fundamental macromolecules of life, the cornerstones of molecular biology: **nucleic acids** (**deoxyribonucleic acid** (DNA) and **ribonucleic acid** (RNA)) and proteins. If we just consider types of data related to nucleic acids, then we can count complete genomes, the genes they encode, the **nucleotide bases** that make up their sequences, their genetic **mutations**, and so on. Thinking about protein-related data, on the other hand, we can enumerate their component amino acids, the linear sequences and 3D structures they form, and the molecules or substrates with which they must interact in order to express their native functions.

At another level, we can think of protein or nucleic acid sequences as belonging to related families that share, for example, evolutionarily conserved regions suggestive of possible structural or functional roles. We can compare such sequences by creating **multiple alignments**, we can create diagnostic models of their conserved features, and we can chart their evolutionary histories by building **phylogenetic trees**. If we analyse the

Bioinformatics challenges at the interface of biology and computer science: Mind the Gap. First Edition.
Teresa K. Attwood, Stephen R. Pettifer and David Thorne. Published 2016 © 2016 by John Wiley and Sons, Ltd.
Companion website: www.wiley.com/go/attwood/bioinformatics

similarities and differences between closely related amino acid or nucleotide sequences, we may find that some changes lead to disease states. We may then gather information on the genes and proteins that might be involved in those pathogenic processes, and the drugs that might be created to combat them – this would involve collating data relating to their therapeutic **efficacy** (details of their **Absorption**, **Distribution**, **Metabolism**, **Excretion**, **Toxicity** (**ADME-Tox**) and so on), and to the outcomes of their use in **clinical trials**.

In terms of the possible data-types encountered in bioinformatics, the above is just the tip of an enormous iceberg. A great deal of information relating to this diverse array of data is presented in the biomedical literature. In addition, through systematic efforts of researchers and biocurators during the last 40 years, chunks of this information have also been gathered up and stored in databases. We'll discuss some of these databases in detail later, in Chapter 3. For now, though, to set the scene for those discussions, we'll begin by introducing some of the fundamental concepts and building-blocks of molecular biology.

2.2.2   The central dogma

Today, we tend to take the role of DNA for granted. Yet the identity and nature of the cell's hereditary material were not known before the 1950s: at that time, some scientists believed that *proteins* carried the information responsible for inheritance. Subsequent experiments with a group of **viruses** that parasitise **bacteria** (so-called **bacteriophages** – simple organisms that comprise packets of nucleic acid wrapped in protein shells) went on to show that the infective material was largely DNA (Hershey and Chase, 1952). Nevertheless, just how DNA might mediate inheritance wasn't obvious, not least because no plausible replicative mechanism was known. The first hint came in 1953 when, in proposing their radical new idea for the structure of DNA, Watson and Crick commented,

> It has not escaped our notice that the specific [base] pairing we have postulated immediately suggests a possible copying mechanism for the genetic material (Watson and Crick, 1953).

From this rather understated beginning, Francis Crick went on to propose how the genetic information carried in nucleic acids could be transferred to proteins. In the late 1960s, he formulated his ideas in what he called the **central dogma of molecular biology**, essentially stating that the direction of transfer of the information encoded in DNA is via the sequence information of RNA, and thence to the sequence of proteins (and not from protein to protein, or from protein to nucleic acid). Despite being a major (and, at the time, contentious) over-simplification (Crick, 1970), the central dogma[1] is nevertheless still a useful concept for understanding the basic principles of the genesis of proteins.

The molecular sequences of DNA, RNA and protein each have different chemical components. So, in practice, for this 'simple' genetic transfer to work, the information in a sequence of DNA must first be 'transcribed' into the appropriate chemical sequence of RNA, which must then be 'translated' into the relevant chemical sequence of proteins. This process is illustrated schematically in Figure 2.1. We use the term 'transcription' to describe the first step, because it involves re-writing one sequence of nucleotides

---

[1]https://www.youtube.com/watch?v=41_Ne5mS2ls

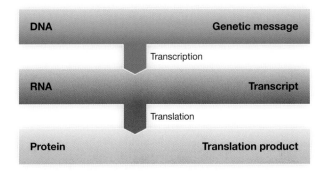

Figure 2.1

The central dogma of molecular biology[i], showing the flow of information from the genetic message in DNA, to the intermediate RNA transcript, and ultimately to the translated protein product.

[i]https://www.dnalc.org/resources/3d/central-dogma.html

into a highly similar sequence of nucleotides, but using a slightly different alphabet. By contrast, we use the term 'translation' to describe the second step, because it involves translating a nucleotide sequence into a sequence that's written in a different language, with a completely different alphabet – that of amino acids. So, when we talk of a transcript, we mean the genetic message encoded as RNA (hence the term **messenger RNA**, or mRNA); and when we speak about a translation product, we mean the transcribed message re-coded as a protein sequence.

As an aside, it's fascinating (and perhaps humbling) to note the similarities between these biological interactions and the internal workings of modern computers: the processes involved in creating proteins from DNA templates bear more than a passing resemblance to the execution of code on a silicon chip. As much as we may recognise the genius of the human pioneers of the silicon age for discovering how to store information in digital form so that it can be copied without error, and how to represent instructions such that they can be interpreted reliably and repeatedly by machines, it's clear that analogous mechanisms evolved naturally millions of years ago.

The biological processes of transcription and translation are extremely complex, involving the interactions of a multitude of protein and nucleic acid molecules. We will not describe these processes in any great detail here, as they are dealt with beautifully in many biology textbooks – some information is, however, given in Section 2.2.5 and Boxes 2.4–2.7. For now, let's explore the building-blocks and alphabets of the key molecular players in the central dogma.

2.2.3 Fundamental building-blocks and alphabets

We start by looking at DNA – deoxyribonucleic acid. We can think of DNA as having a four-letter alphabet, each letter of which denotes one of its constituent nucleotide bases: two of these are known as **purines** (**adenine** (A) and **guanine** (G)), and two are termed **pyrimidines** (**thymine** (T) and **cytosine** (C)). DNA is usually found as a double-stranded, helical molecule – the famous double helix – its two strands winding around each other and held together by chemical bonds between opposing A and T, and C and G bases (for details of the structures of these molecules, please refer to Box 2.1). The component strands run in opposite directions and have complementary sequences, so that, where one has G, the other must have C, and where one has A, the other must have T, as shown here:

```
5'-T-G-A-C-C-G-A-T-3'
3'-A-C-T-G-G-C-T-A-5'
```

Box 2.1 The structures of nucleotide bases and of DNA

The four bases of DNA are Adenine, Thymine, Guanine and Cytosine, more familiarly known as A, T, G and C[a]. Their structures are shown in Figure 2.2.

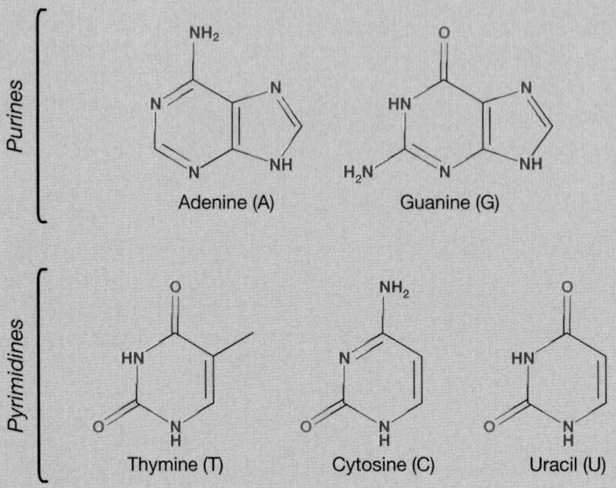

Figure 2.2
Structures of nucleotide bases.

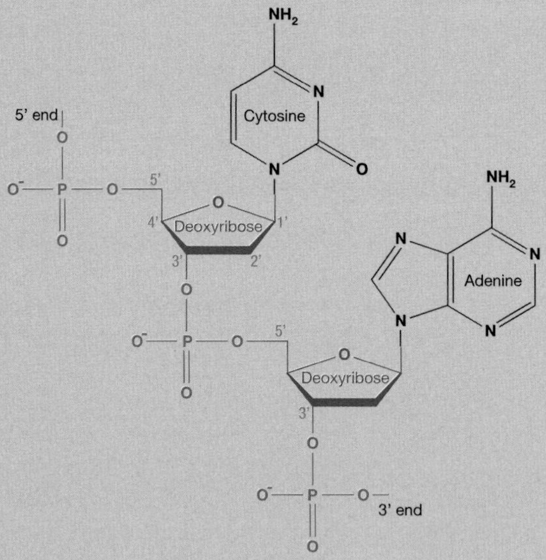

Figure 2.3
The DNA sugar–phosphate backbone.

[a] https://www.dnalc.org/resources/3d/09-5-dna-has-4-units.html

In RNA, Uracil (U) replaces thymine: the structures of uracil and thymine are similar, but U lacks the methyl (CH_3) group linked to the ring of the T molecule. A and G, whose chemical structures contain a double ring, are termed purines; C, T and U, whose chemical structures contain only a single ring, are termed pyrimidines.

The basic building-blocks of nucleic acid chains (the units in which their lengths are measured) are nucleotides: these comprise one base, one sugar molecule, and one phosphate moiety. In DNA, the deoxyribose sugar molecules are linked together via phosphate moieties attached to carbon atoms 3 and 5 of their five-membered carbon ring structures. This arrangement is shown in Figure 2.3. The final carbons at the end of each DNA strand are hence termed 5′ and 3′, and denote the direction in which genetic information is read.

DNA is famous for its double-stranded, helical structure (which forms the basis for numerous logos and is an essential eye-candy component of science articles and science-fiction programmes!). Although its physical structure plays a vital role in the processes that lead to the creation of proteins, its familiar intertwined strands are of a relatively mundane construction: its sugar and phosphate groups essentially form a scaffold on which its bases are anchored in sequence and held in place by chemical bonds between adjacent A and T, and C and G bases on opposing strands – this is sometimes referred to as **Watson–Crick base-pairing**, as shown in Figure 2.4.

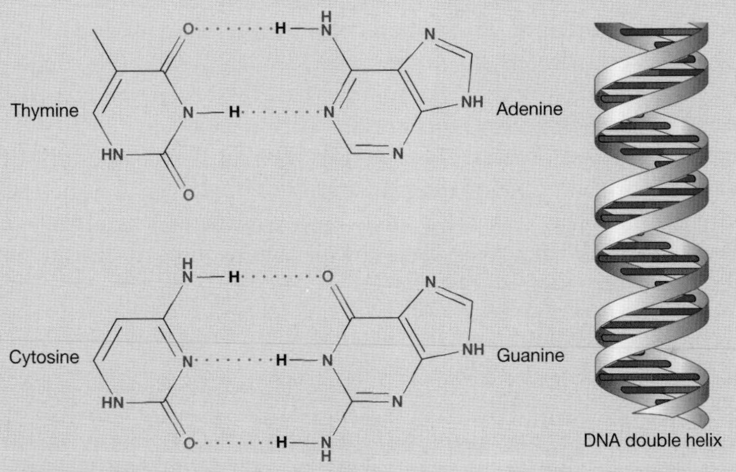

Figure 2.4

Watson–Crick base-pairing gives rise to the characteristic DNA double helix[i].

[i]https://www.youtube.com/watch?v=d0S9qsglg3E

You will see that the termini of these strands have been tagged 5′ and 3′. This notation derives from the fact that the bases that make up DNA molecules are linked to **sugar** molecules, which are, in turn, linked together via **phosphate** groups. Each sugar molecule (termed **deoxyribose**) contains five carbon atoms (numbered 1 to 5) in a ring-like structure; attached to carbon atoms 3 and 5 are the phosphate groups, by means of which adjacent sugars are connected. By convention, the final carbons at the end of each DNA strand are denoted 5′ and 3′, and the sequence is written from the 5′ to the 3′ end, as this is the direction in which genetic information is read.

RNA (ribonucleic acid) is chemically very similar to DNA, but what seem like a couple of relatively minor changes give rise to quite different structures. These seemingly minor differences are, first, the use of **ribose** in place of deoxyribose sugars, and second, use of the chemically similar **uracil** (U) base in place of thymine. RNA's four-letter alphabet is thus A, U, G and C.

As we've just seen, for convenience, single-letter codes are used to denote individual bases – when it comes to storing complete DNA sequences in files and databases, this saves a lot of space (and a lot of typing), as even the following toy sequence reveals:

ThymineGuanineAdenineCytosineCytosineGuanineAdenineThymine
TGACCGAT

For real DNA sequences, with many thousands, or hundreds of thousands of bases, this saving is immensely valuable.

As mentioned earlier, much of the sequence data stored in databases today comes from sequencing DNA (most from genome-sequencing projects). But the process of sequencing DNA is error-prone, and ambiguities sometimes arise. It isn't in the scope of this book to discuss how DNA is sequenced, but it is worth mentioning one of the sources of error. During the process of sequencing, the concentrations of individual bases are read out in the form of **chromatogram** traces, where each base is denoted by a particular fluorescent dye, as shown in Figure 2.5. Computerised robots attempt to interpret or 'call' the sequence of bases from these traces, a step that involves digital signal processing and decoding using appropriate software tools (*e.g.,* **Phred base-calling**). Base-calling isn't a perfect process: at each position in the chromatogram, although one base is usually more clearly abundant than another, sometimes the trace peaks cannot be resolved unambiguously. When no single base (represented by a single colour dye) can be called with confidence, it is useful to be able to indicate this in some kind of simple, but meaningful way.

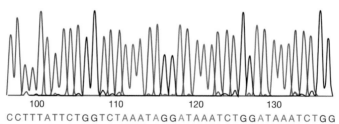

```
            100           110           120           130
          CCTTTATTCTGGTCTAAATAGGATAAATCTGGATAAATCTGG
```

Figure 2.5

Part of a sequencing trace file. Each base is represented by a coloured dye: at each position, base-calling software reads off the most abundant base; where peak heights can't be differentiated and no single base can be called, the software marks the position with an N (adapted from a Wikimedia Commons image[i]).
[i]https://commons.wikimedia.org/wiki/File:DNA_sequence.svg

In practice, to flag ambiguities in the final sequence, a standard set of codes is used – these are detailed in Table 2.1. Thus, for example, if it isn't possible to resolve the abundance of A or T at a given position, the code W can be used; or, if no base can be assigned confidently at a particular position, we can indicate the ambiguity with the letter N. These codes are particularly helpful when we compare nucleotide sequences by aligning them and need a shorthand to denote substitution patterns observed between nucleotide bases: *e.g.*, at positions where purines may interchange, the code R can be used; or, where pyrimidines may interchange, we can use Y. The codes also find use when we wish to encapsulate patterns of conservation in particular regions of an alignment in the form of **consensus expressions** – we'll discuss these in more detail in Chapter 4. For now, it's sufficient to note that when we depict nucleotide sequences, we use a standard single-letter code to denote individual bases, and another standard set of characters to denote either base ambiguities or substitution alternatives at particular positions in sequence alignments.

Table 2.1

IUPAC standard alphabet and ambiguity codes for nucleotides.

Code	Base(s)	Code	Base(s)
A	Adenine	G	Guanine
C	Cytosine	T	Thymine
U	Uracil		
M	A or C	R	A or G
W	A or T	S	C or G
Y	C or T	K	G or T
V	A or C or G	H	A or C or T
D	A or G or T	B	C or G or T
N	A or G or C or T		

These same characters are used in the world of proteins, but with quite different meanings, so it's important not to confuse them. By contrast with the four-letter alphabet of nucleic acids, proteins have a much larger, 20-letter alphabet of amino acids. These, and their ambiguities, may be described using the standard single- or three-letter codes shown in Table 2.2.

As before, these codes are often used in consensus expressions derived from conserved regions of sequence alignments (see, for example, Box 3.3 and Chapter 4 for more details), and they are also used to store sequences in files and databases more compactly, as the following toy sequence illustrates:

```
MethionineLeucineGlutamicAcidLeucineLeucineProlineThreonineAlanine
                        MLELLPTA
```

Although symbolic notations like this are convenient for the purposes of storage or computational manipulation, the danger in using them is that they are essentially free of biological meaning (**semantics**): to a computer, sequences like STEVE or DAVID are no more obviously components of proteins than they are the names of people – they are simply strings of characters. If we are to compute with them, we must therefore supply the necessary protein-relevant contextual information, say to compare them meaningfully in multiple sequence alignments, or to quantify their similarity in database searches.

Table 2.2

IUPAC standard alphabet and ambiguity codes for amino acids.

1-letter code	Amino acid(s)	3-letter code	1-letter code	Amino acid(s)	3-letter code
G	Glycine	Gly	P	Proline	Pro
A	Alanine	Ala	V	Valine	Val
L	Leucine	Leu	I	Isoleucine	Ile
M	Methionine	Met	C	Cysteine	Cys
F	Phenylalanine	Phe	Y	Tyrosine	Tyr
W	Tryptophan	Trp	H	Histidine	His
K	Lysine	Lys	R	Arginine	Arg
Q	Glutamine	Gln	N	Asparagine	Asn
E	Glutamic acid	Glu	D	Aspartic Acid	Asp
S	Serine	Ser	T	Threonine	Thr
B	Asp or Asn	Asx	Z	Glu or Gln	Glx
X	Anything / Unknown	XXX / Unk			

Let's examine this point in a little more detail. The chemical structure of an amino acid is shown in Figure 2.6: it has an **amine** group (NH_2) on one side and a **carboxylic acid** (COOH) group on the other side of a central (or α) carbon. In solution, these groups may be ionised to give NH_3^+ and COO^- (*i.e.*, the groups may carry positive or negative charges). To the **α-carbon** is attached a **side-chain** (denoted R): it is the composition of this side-chain that distinguishes the 20 amino acids from each other (these are listed in Box 2.2).

Linear chains of amino acids are linked together via so-called **peptide bonds**, which are formed via a chemical reaction that removes a **hydroxyl** group (OH) from the **carboxyl** group of one amino acid and a hydrogen from the amine group of the next, to form **water** (H_2O). A chain of two amino acids (a **dipeptide**) joined in this way is shown in Figure 2.8.

Proteins usually comprise chains of hundreds of amino acids. The ends of these chains are termed the **N terminus** and **C terminus**, as the former has an unlinked NH_3^+ group and the latter an unlinked COO^- group. By convention, protein sequences are written from the N to the C terminus, corresponding to the direction in which they are synthesised.

The chemical nature of the peptide bond means that the participating atoms lie in a plane (denoted by boxed areas in Figure 2.9) and can't rotate freely with respect to one another. What flexibility there is derives mostly from rotation about the N–Cα and Cα–C bonds at each α-carbon, as illustrated in Figure 2.9. This torsional freedom confers a wide variety of 3D structures on the polypeptide backbone. The final structure adopted by a protein is then a direct consequence of the specific sequence of its amino acid side-chains ($R_1, R_2 \ldots R_n$), as their sizes or chemical nature either permit or prohibit particular backbone rotations. A protein reaches its native structure through a

Figure 2.6

Chemical structure of an amino acid.

Box 2.2 The chemical structures of amino acid side-chains

The 20 amino acid side-chains may be broadly grouped according to their physicochemical properties: those that are largely hydrophobic are shown in the top tier of Figure 2.7; those that are largely hydrophilic are shown in the middle tier; and those that are ambivalent, or **amphipathic**, are shown at the bottom.

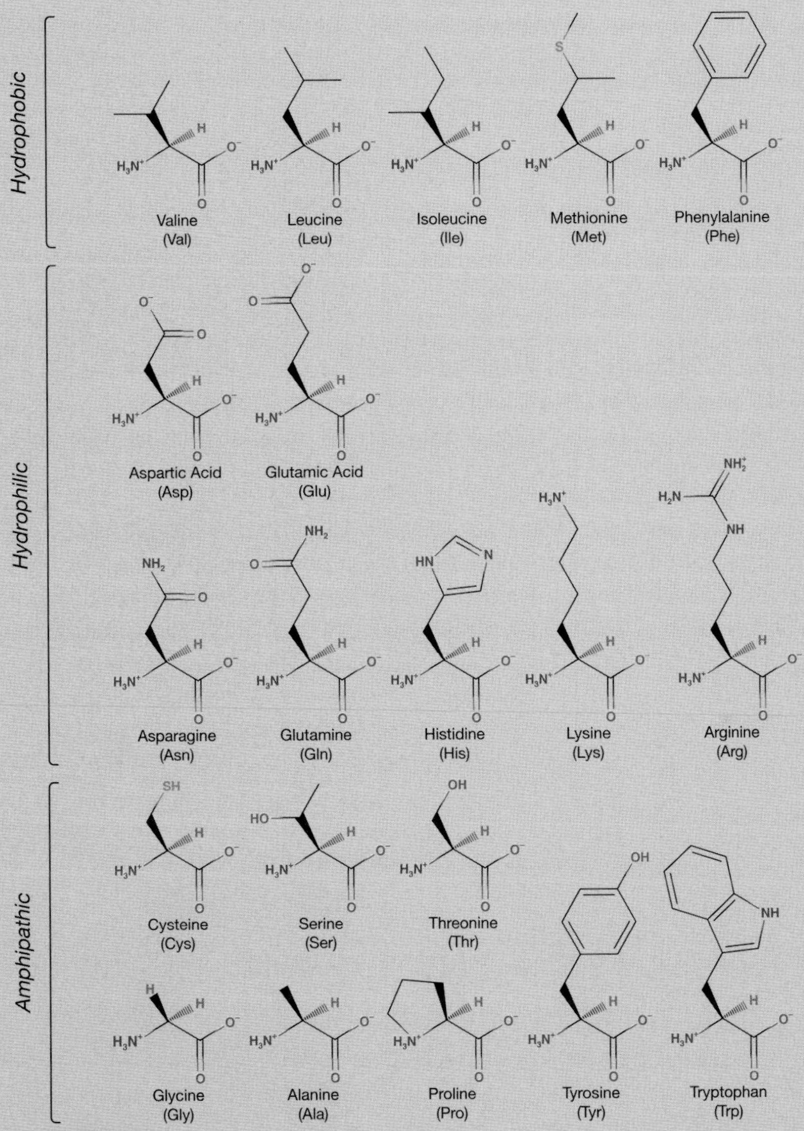

Figure 2.7

Chemical structures of the 20 amino acids[i].

[i]http://chemwiki.ucdavis.edu/Wikitexts/University_of_California_Davis/UCD_Chem_115_Lab_Manual/Lab_7%3A_Electrospray_Mass_Spectrometry

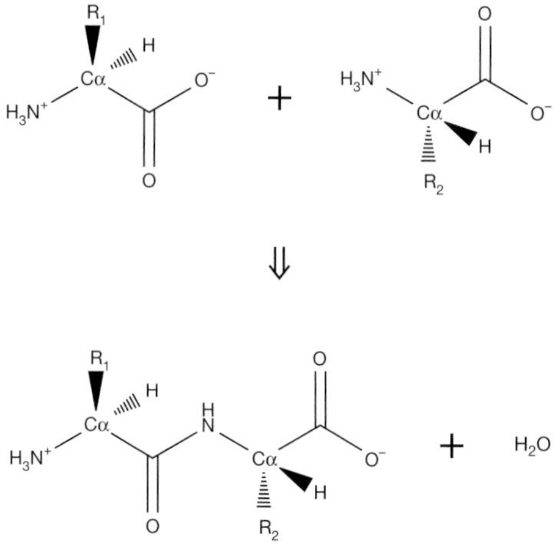

Figure 2.8

Formation of the peptide bond.

highly complex, dynamic folding pathway created by the combination of repulsive and attractive interactions between its different amino acids, and between these amino acids and water. This process is still far from being completely understood.

What we do know is that the properties of the amino acids vary considerably. We can think of these properties at different levels of granularity, depending on the view we're interested in. At the most superficial level, we could perhaps divide the 20 building-blocks into those that have water-attractive (**hydrophilic**) and those that have water-repulsive (**hydrophobic**) properties. But, for most purposes, this would be far too simplistic, as many of the amino acids are both hydrophilic and hydrophobic to some degree, because they contain both polar and non-polar chemical groups: for example,

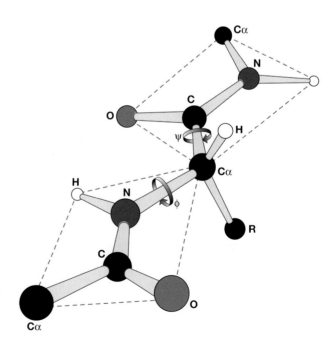

Figure 2.9

Owing to the planarity of peptide bonds, chain flexibility in polypeptides is conferred by rotation about the N–Cα and Cα–C bonds (defined by the **dihedral angles** Φ and Ψ) at each α-carbon. The sequence $R_1, R_2 \ldots R_n$ then determines the final 3D structure of the folded polypeptide chain.

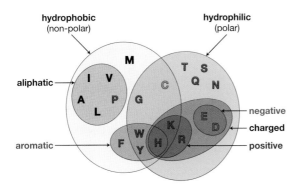

Figure 2.10

Amino acid overlapping properties. Non-polar: **Ala** (A), **Val** (V), **Leu** (L), **Ile** (I), **Met** (M) – black; aromatic: **Phe** (F), **Tyr** (Y), **Trp** (W) – purple; structural properties: **Pro** (P), **Gly** (G) – brown, and **Cys** (C) – yellow; polar negative (acidic): **Glu** (E), **Asp** (D) – red; polar positive (basic): **Lys** (K), **Arg** (R), **His** (H) – blue; polar neutral: **Ser** (S), **Thr** (T), **Gln** (Q), **Asn** (N) – green.

the side-chain of **lysine** (K) contains a highly polar NH_3^+ group at the end of a non-polar carbon chain; the side-chain of **histidine** (H) is polar and may carry a positive charge, but may also be considered non-polar, having partial **aromatic** character; and the side-chain of **tyrosine** (Y) contains a non-polar, aromatic 6-carbon ring structure, attached to which is a polar hydroxyl group.

The extent to which these properties are exhibited in practice depends on the environment of the side-chains: they may be buried in the interior of the protein, which is likely to be highly hydrophobic; they may be freely accessible to the solvent, which is likely to be hydrophilic; they may be partly buried and partly accessible. Thus, the properties of the amino acids are context dependent and overlapping, as illustrated in the **Venn diagram** shown in Figure 2.10 (further details are discussed in Box 2.3).

Box 2.3 Properties of amino acids

As we saw in Box 2.2, the amino acid side-chains have different structures, and these confer on each amino acid different, but nevertheless overlapping, properties; they also play a key role in determining how protein chains fold.

We will look at some of these properties in turn. First, we consider the relevance of amino acid size. Proteins have compact structures: their chains pack tightly, excluding water as they do so. Substituting one amino acid for another without accounting for the size of their side-chains can cause problems: exchanging a large amino acid for a small one, for example, is likely to disrupt the structure. There is significant variation in size or volume between the amino acids: the largest, Trp, has ~3.4 times the volume of the smallest, Gly (measured in units of $Å^3$ (cubic **Ångstroms**)), as shown in Table 2.3 (data taken from Higgs and Attwood, 2005).

An alternative measure of amino acid size is to consider its 'bulkiness' (*i.e.*, the ratio of the side-chain volume to its length), which provides a measure of the average cross-sectional area of the side-chain (in $Å^2$). We can also consider the polarity of the amino acids, calculated in terms of the **electrostatic force** of the amino acid acting on its surroundings at a distance of 10 Å. The total force (in units scaled for convenience) gives a polarity index, also shown in the table. The polarity index (Zimmerman *et al.*, 1968) effectively distinguishes charged and uncharged amino acids.

Another important property is the pI, defined as the pH of the amino acid at its isoelectric point. Acidic amino acids (Asp and Glu) have pI in the range 2–3: these amino acids would be negatively charged at neutral pH owing to ioinisation of the COOH group to COO^-. By contrast, the basic amino acids (Arg, Lys and His) have pI greater than 7. All the others tend to have uncharged side-chains, with pI in the range 5–6. Thus, unlike the polarity index, the pI distinguishes between positive, negative and uncharged side-chains.

A key factor that drives **protein folding** is the so-called **hydrophobic effect**, by which, as mentioned above, protein chains are compelled to pack tightly, to exclude water from their hydrophobic

(continued)

Table 2.3

Selected physicochemical properties of the amino acids.

Amino Acid	Vol.	Bulk.	Polarity	pI	Hyd.1	Hyd.2	S.Area	F.Area
Alanine	67	11.50	0.00	6.00	1.8	1.6	113	0.74
Arginine	148	14.28	52.00	10.76	−4.5	−12.3	241	0.64
Asparagine	96	12.28	3.38	5.41	−3.5	−4.8	158	0.63
Aspartic Acid	91	11.68	49.70	2.77	−3.5	−9.2	151	0.62
Cysteine	86	13.46	1.48	5.05	2.5	2.0	140	0.91
Glutamine	114	14.45	3.53	5.65	−3.5	−4.1	189	0.62
Glutamic Acid	109	13.57	49.90	3.22	−3.5	−8.2	183	0.62
Glycine	48	3.40	0.00	5.97	−0.4	1.0	85	0.72
Histidine	118	13.69	51.60	7.59	−3.2	−3.0	194	0.78
Isoleucine	124	21.40	0.13	6.02	4.5	3.1	182	0.88
Leucine	124	21.40	0.13	5.98	3.8	2.8	180	0.85
Lysine	135	15.71	49.50	9.74	−3.9	−8.8	211	0.52
Methionine	124	16.25	1.43	5.74	1.9	3.4	204	0.85
Phenylalanine	135	19.80	0.35	5.48	2.8	3.7	218	0.88
Proline	90	17.43	1.58	6.30	−1.6	−0.2	143	0.64
Serine	73	9.47	1.67	5.68	−0.8	0.6	122	0.66
Threonine	93	15.77	1.66	5.66	−0.7	1.2	146	0.70
Tryptophan	163	21.67	2.10	5.89	−0.9	1.9	259	0.85
Tyrosine	141	18.03	1.61	5.66	−1.3	−0.7	229	0.76
Valine	105	21.57	0.13	5.96	4.2	2.6	160	0.86
Mean	109	15.35	13.59	6.03	−0.5	−1.4	175	0.74
Std. Dev.	28	4.53	21.36	1.72	2.9	4.8	44	0.11

Key
Vol.:   volume calculated from *van der Waals radii* (Creighton, 1993)
Bulk.:   bulkiness index (Zimmerman *et al.,*1968)
Pol.:   polarity index (Zimmerman *et al.,*1968)
pI: *pH* of the *isoelectric point* (Zimmerman *et al.,* 1968)
Hyd.1:   hydrophobicity scale (Kyte and Doolittle, 1982)
Hyd.2:   hydrophobicity scale (Engelman *et al.,* 1986)
S.Area: surface area accessible to water in unfolded peptide (Miller *et al.,* 1987)
F.Area: fraction of accessible area lost when a protein folds (Rose *et al.,* 1985)

interiors. Several hydrophobicity (sometimes referred to by the rather broader term, **hydropathy**) scales have been defined. The Kyte and Doolittle scale estimates the difference in free energy (in kcal/mol) of an amino acid when buried in the hydrophobic protein interior and when it is in solution in water. Positive values in this scale denote that the residue is hydrophobic. By contrast, the Engelman scale, devised for membrane proteins (whose **lipid bilayer** interior is hydrophobic) estimates the free energy cost for removal of an amino acid from the bilayer to water. These scales are similar; it is interesting to explore the differences shown in the table.

Another important property is the surface area of an amino acid that, accessible to water in an unfolded state, becomes buried when the chain folds. The table shows the accessible surface areas of residues when they occur in a Gly-X-Gly tripeptide (Miller *et al.,* 1987; Creighton, 1993). A different perspective was taken by Rose *et al.* (1985), who calculated the average fraction of the accessible surface area that's buried in the interiors of known crystal structures. The hydrophobic residues have a larger fraction of their surface area buried, supporting the notion that the hydrophobic effect is a driver of protein folding.

2.2.4 The protein structure hierarchy

It is important to understand the properties of the amino acids, and their overlaps, because, as we mentioned earlier, these ultimately determine the structures and functions of their parent proteins. Traditionally, we consider protein structures to be built in a hierarchical way. The **primary structure**, the simplest level, is just the linear sequence of amino acids in the protein:

MLELLPTAVEGVSQAQITGRPEWIWLALGTALMGLGTLYFLVKGMGVSDPDAKKFYAITTLVPAIA…

As a chain of amino acids is synthesised, the amino acids begin a complex dance[2], as their mutual repulsive and attractive interactions play out their folding pathway, maximising and minimising the hydrophilic and hydrophobic interplay with surrounding water molecules. During this process, regions of local regularity are formed in the sequence, and these are referred to as **secondary structures**. There are several types of secondary structure, the most common of which are the so-called α-helix, β-strand, β-sheet and β-turn – these are illustrated in Figure 2.11.

Sometimes, several elements of secondary structure fold into higher-level stable units, which have been given an assortment of descriptive names, such as the **β-barrel, β-propeller, α-solenoid, Jelly roll, Greek key**, and so on. These are often referred to as **super-secondary structures** or **super-folds**.

At the next level of the hierarchy is the **tertiary structure**, which defines the overall chain fold that results from packing of secondary (and super-secondary) structural elements. Often, proteins contain more than one chain. When proteins have several subunits (**haemoglobin**, for example, has four), the relationship between them is described by their **quaternary structure**. Examples of some of these higher-order structures are illustrated in Figure 2.12.

Up to this point, we've been thinking about nucleic acids and proteins in terms of alphabets and sequences of simple building-blocks. Earlier, we mentioned that proteins

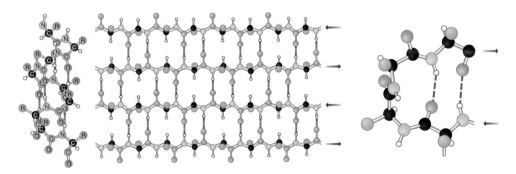

Figure 2.11

Illustration of some of the main forms of protein secondary structure: an α-helix; a series of adjacent β-strands running in opposite directions, bonds between them forming an anti-parallel β-sheet; and a β-turn, which allows the protein chain to turn back upon itself.

[2] https://www.youtube.com/watch?v=sD6vyfTtE4U

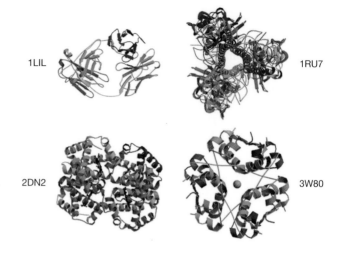

Figure 2.12

Examples of higher-order protein structures: α-helices are shown as coloured coil-like structures, and β-strands as extended, slightly twisted arrows.

Source: Protein Data Bank.

are synthesised from DNA, but indirectly via a messenger molecule – mRNA. Let's now take a brief look at some of the structural components of mRNA sequences and how these are processed within different types of organism to create usable or functional transcripts.

2.2.5 RNA processing in prokaryotes and eukaryotes

The two types of organism that we'll consider are termed **prokaryotes** and **eukaryotes**. Prokaryotes are single-celled organisms that lack a **nucleus**. Evolutionary studies have shed light on two main types of prokaryotic organism: the **archaea** and bacteria (archaeal cells can usually be distinguished from bacteria based on their gene sequence and gene content, many being specialised to conditions of extreme heat or salinity). By contrast, eukaryotic cells do have nuclei; they also usually possess a **cytoskeleton** and contain many other highly specialised **organelles (mitochondria, Golgi apparatus, endoplasmic reticulum** and so on). Eukaryotes are thus more complex than prokaryotes, and their genomes are generally much larger.

One aspect of this increased complexity is seen in the way that RNA transcripts are processed within these different organisms. As we've seen, RNA transcribed from a protein-coding region of DNA is termed messenger RNA, or mRNA (an overview of the molecular events that effect transcription is given in Box 2.4, Figure 2.13). In prokaryotes, mRNA consists of a central sequence that encodes the protein product, flanked by short **untranslated regions** (UTRs) – so-called because they are not destined for translation – at its 5′ and 3′ ends. This simple structure is shown schematically in Figure 2.14 (a). Eukaryotic RNA transcripts have a more complicated structure, as shown in Figure 2.14 (b). Here, RNA is first synthesised in a precursor form (pre-mRNA), which must then be processed before becoming a functional mRNA.

One form of processing occurs at the 3′ end, where a string of ~200 adenine nucleotides is added – this is the **poly(A) tail**. As part of the control of the degradation of mRNA, several different proteins bind to this tail; this helps to limit the number of times an mRNA can be used in translation, and hence controls the amount of protein synthesised. Another form of processing, which is of more direct interest to us, takes place in the central region of the pre-mRNA. Eukaryotic gene sequences contain

Box 2.4 The process of transcription

Until now, we have talked only in very abstract terms about transcription, the process by which linear strands of RNA are synthesised from single-stranded DNA templates (Figure 2.13 (a)):

The biological event is mediated by an enzyme known as **RNA polymerase**. The enzyme binds to one of the two strands of the DNA template, referred to as the **template strand**, which is temporarily separated from the **non-template strand** during the process, as shown schematically in Figure 2.13 (b):

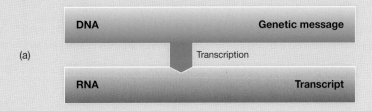

(a)

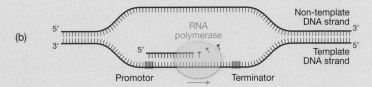

(b)

Figure 2.13

(a) DNA transcription[i].
(b) Synthesis of RNA[ii] from a DNA template.

[i] https://www.youtube.com/watch?v=WsofH466lqk
[ii] https://www.youtube.com/watch?v=ovc8nXObxmQ

The polymerase enzyme catalyses the assembly of individual ribonucleotides into a strand of RNA, which complements its DNA template. Initially, **base pairs** (bps) are formed between the template and the growing RNA strand; however, as the polymerase progresses, the RNA separates from the template, and the two DNA strands re-seal.

RNA polymerase proceeds along the template in the 3′ to 5′ direction; therefore, the RNA strand is synthesised from its 5′ end to its 3′ end. For the process to work, the polymerase needs instructions on where to start and stop; it gets this information from the DNA template, as shown in (b). The start signal is a short sequence of DNA bases (typically quite variable in different genes), termed a promoter. The stop signal, the terminator, often takes the form of a specific base sequence that can form a **hairpin loop** in the RNA strand. This structure inhibits progression of the polymerase and causes it to dissociate from the template.

Because the new RNA sequence is complementary to its DNA template, it is in fact the same as the non-template strand, but with one significant difference. When the template base is a C, T or G, the base added to the growing RNA strand is a G, A or C, as we might expect. However, when the template base is A, a U is added to the RNA in place of a T. Hence, in RNA, we find Us rather than Ts. So, when we think about the DNA sequence of a gene, we usually mean the non-template strand, because it is this sequence that is mimicked by the RNA, and it is this sequence that is subsequently translated into protein.

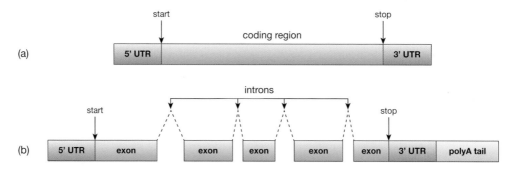

Figure 2.14

Schematic illustration of the structure of mRNA: (a) in prokaryotes, the central coding region is flanked by 5′ and 3′ untranslated (UTR) regions; (b) in eukaryotes, the central region comprises exons and introns – introns must be excised to create a functional transcript.

alternating elements termed **exons** and **introns**: exons are parts of a gene sequence that are expressed – *i.e.*, it is these that are translated; introns are intergenic regions that are destined to be removed from the pre-mRNA, as illustrated in Figure 2.14. As introns are excised, the ends of their flanking exons are joined together to form a continuous strand in a process known as **splicing** (more details of this process are given in Box 2.5). To add to the complexity, for a given pre-mRNA, not all of the exons need be used in the final transcript. As a result, a single pre-mRNA can give rise to several different transcripts, or **splice forms**.

Incredibly, introns weren't known before 1977 (Chow *et al.*, 1977; Berget *et al.*, 1977); their discovery was such a breakthrough that it led to the award of the Nobel Prize in Physiology or Medicine to **Phillip Sharp** and **Richard Roberts** in 1993. Without doubt, the genome still holds many secrets, and there is still much to discover about how its rather monotonous four-base code is transformed into the majestic poetry of proteins. Among the many remaining mysteries is how splice sites are identified at intron–exon boundaries by the cell's molecular machinery, dictating which bits of sequence to remove. While aspects of the process are known, like which signals are recognised (*e.g.*, characteristic **dinucleotides** at the ends of introns: guanine-uracil (GU) at the 5′ end and adenine-guanine (AG) at the 3′ end), the recognition sites are so small (and variable), and the precision of the process so exquisite, that it's breath-taking to think how the process is orchestrated in practice; and if it's hard to imagine how nature does it, think how hard it is to deduce the intron–exon structure of a gene computationally – not surprisingly, this is one of the main sources of error in gene prediction. This problem is compounded by the large numbers of introns in many eukaryotic genes (sometimes up to 20 in a single gene), which further increases the scope for error. Prokaryotes tend to present fewer problems: most of their genes don't contain introns, and their genomes are smaller, making sequencing and gene finding much easier; hence, there are many more prokaryotic than eukaryotic genomes available today.

In eukaryotes, DNA is contained in the nucleus, and transcription and RNA processing occur here. The processed mRNA is then transported from the nucleus through the **nuclear membrane**, into the **cytoplasm**, where translation then takes place. So, let's now think about how the chemical code of mRNA is translated into the corresponding protein code.

Box 2.5 RNA processing

We mentioned that eukaryotic gene sequences are generally more complex than their prokaryotic counterparts. As we saw, one of the principal differences between them is that eukaryotic gene sequences contain alternating coding and non-coding regions, termed exons and introns. Therefore, in order to create a contiguous sequence of coding nucleotides, prior to protein synthesis the molecular machinery must excise the non-coding introns and zip together the coding exons. This RNA-processing step, known as splicing[a], is illustrated in Figure 2.15.

The process of splicing is mediated by a molecular 'machine' termed a **spliceosome**, a complex of proteins and various types of RNA. Just as during the process of transcription, the RNA polymerase makes use of sequence signals to tell it where to stop and start its work, so the spliceosome exploits signals in the pre-mRNA sequence that flag the locations of intron–exon boundaries, and hence which bits of the sequence to remove. Like promoter sequences, splice-site signals are typically relatively short and variable, and the spliceosome recognises them with far greater confidence than we can manage programmatically.

Some introns can excise themselves from an RNA strand without requiring the spliceosome, but such self-splicing introns are rare. The majority of introns in most organisms do require the splicing machinery of the spliceosome, and are hence referred to as spliceosomal introns. Eukaryotic genes can contain large numbers of introns (*e.g.*, up to 20 in one gene is possible). By contrast, many prokaryotic genes are intronless, making them much more tractable to computational analysis.

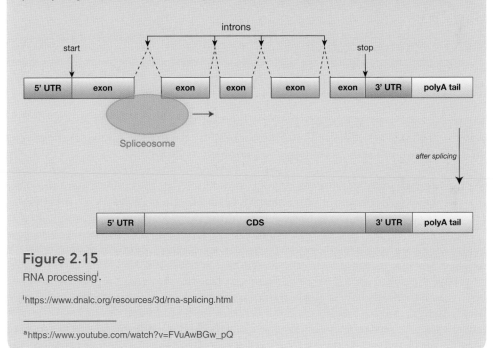

Figure 2.15
RNA processing[i].

[i]https://www.dnalc.org/resources/3d/rna-splicing.html

[a]https://www.youtube.com/watch?v=FVuAwBGw_pQ

2.2.6 The genetic code

The sequence of nucleotide bases in mRNA is read in groups of three: these triplet units are termed **codons**. As there are three base positions and four bases to choose from, there are 64 (4^3) codon permutations. Each codon encodes one amino acid; hence, as there are only 20 naturally occurring amino acids, most have more than one codon. The

First position	Second position				Third position
	U	C	A	G	
U	Phe (F)	Ser (S)	Tyr (Y)	Cys (C)	U
	Phe (F)	Ser (S)	Tyr (Y)	Cys (C)	C
	Leu (L)	Ser (S)	**Stop**	**Stop**	A
	Leu (L)	Ser (S)	**Stop**	Trp (W)	G
C	Leu (L)	Pro (P)	His (H)	Arg (R)	U
	Leu (L)	Pro (P)	His (H)	Arg (R)	C
	Leu (L)	Pro (P)	Gln (Q)	Arg (R)	A
	Leu (L)	Pro (P)	Gln (Q)	Arg (R)	G
A	Ile (I)	Thr (T)	Asn (N)	Ser (S)	U
	Ile (I)	Thr (T)	Asn (N)	Ser (S)	C
	Ile (I)	Thr (T)	Lys (K)	Arg (R)	A
	Met (M)	Thr (T)	Lys (K)	Arg (R)	G
G	Val (V)	Ala (A)	Asp (D)	Gly (G)	U
	Val (V)	Ala (A)	Asp (D)	Gly (G)	C
	Val (V)	Ala (A)	Glu (E)	Gly (G)	A
	Val (V)	Ala (A)	Glu (E)	Gly (G)	G

Figure 2.16

Representations of the standard genetic code used in most prokaryotic genomes and in the nuclear genomes of most eukaryotes. AUG serves the dual purpose of encoding methionine and of indicating the start of the coding region; UAA, UAG and UGA function only as stop codons. The colours are the same as those in Figure 2.10, showing the clustering of amino acid properties.

way in which the 64 codons map to amino acids – an assignment known as the **genetic code** – is shown in Figure 2.16 (left- and right-hand panes).

The heart of the table depicted in Figure 2.16 is divided into 16 sections, each including the products of four different codons; in each section, the codons share the same first two bases (*e.g.*, the codons for **valine** all begin GU). In many cases, all four codons encode the same amino acid – *i.e.*, the base in the third position makes no difference. Other sections contain two different amino acids, depending on whether the third base is a pyrimidine or a purine: *e.g.*, AAY encodes **asparagine**, and AAR **lysine** (where Y and R, remember from Table 2.1, are the ambiguity codes for pyrimidines and purines, respectively). **Isoleucine** has three codons, and **leucine**, **serine** and **arginine** each have six. Two amino acids have a single codon: **tryptophan** (UGG) and **methionine** (AUG). Methionine is unusual because its codon serves both to denote the inclusion of this amino acid when it occurs within a gene sequence, and to indicate the specific starting point of the gene's coding region. Three codons behave as stop signals to terminate the coding region; unlike the **start codon**, however, they don't encode amino acids.

Fifty years ago, the genetic code was believed to be the same for all species. However, although widespread, it turns out that it's not quite universal. The code depicted in Figure 2.16 applies to most prokaryotic genomes (bacterial and archaeal) and to the nuclear genomes of virtually all eukaryotes. Mitochondrial genomes, the nuclear genomes of some unicellular eukaryotes, and the genomes of some bacteria use slightly different codes; but the differences are small, and the bottom line is that the code is generally shared between the three domains of life: archaea, bacteria and eukaryotes.

2.2.7 Conceptual translation and gene finding

Armed with a near-universal genetic code, we can use this information to understand more about the process of translation. However, at this point, we're going to make a brief context switch from the realms of 'natural' translation (what nature does) to the world of 'conceptual' translation (what computers do). There are various reasons for this. First, most protein sequences are now inferred from DNA; it's hence the products of **conceptual translation** that mostly populate our protein sequence databases – we'll therefore not describe the biological translation process in great detail here, but encourage you to look at the overviews in Boxes 2.6 and 2.7 instead. Second, most gene sequences are also inferred from genomic DNA; 'genes' delineated in nucleotide sequence databases are therefore mostly predictions. Outlining the process of computational translation, then, readily leads us into considering some of the issues of gene finding; this, in turn, prepares us to confront the fact that the apparently tangible biological entities housed in databases – the genes and protein sequences – might not be the definitive objects we perhaps expect them to be. Introducing these concepts here provides a useful foundation for the sequence analysis discussions in Chapter 4.

Box 2.6 The process of translation

As with the process of transcription, until now, we have talked only in abstract terms about translation, the process by which protein sequences are synthesised from mRNA templates (Figure 2.17 (a)):

The biological event, rather like splicing, is mediated by another molecular machine, termed a ribosome. Ribosomes comprise a large and a small subunit (represented by the ellipses in Figure 2.17 (b)). In bacterial ribosomes, the small subunit contains so-called small subunit **ribosomal RNA** (SSU rRNA) in complex with ~20 proteins; the large subunit contains large subunit ribosomal RNA (LSU rRNA), a smaller ribosomal RNA (**5S rRNA**), and around 30 proteins. In eukaryotes, the ribosomes are larger: their two major rRNA molecules are much longer, and many more proteins are contained in each subunit.

The mechanism of protein synthesis is illustrated schematically in Figure 2.17 (c). The ribosome binds to the mRNA, inching along its length one codon at a time. Adaptor molecules known as **transfer RNA**, or tRNA (described in Box 2.7), loaded with an amino acid, bind to the mRNA within the ribosome. The amino acid, having been removed from the tRNA, is then attached to the end of a growing protein chain. The unloaded tRNA is freed from the complex and may now be recharged with another, identical, amino acid molecule. A new, charged, tRNA then binds to the next codon in mRNA, and the ribosome moves one position further along.

As we saw with transcription and splicing, the translation machinery also requires signals to tell it where to start and stop its work. Specific codons are employed to do this. One codon (AUG) plays a dual role, on the one hand encoding methionine, and on the other, signifying the start of protein synthesis – this is the start codon. The ribosome begins its work at the first AUG codon it finds; this is downstream of where it initially binds to the mRNA. In bacteria, mRNAs contain a Shine–Dalgarno sequence, a conserved region of about eight nucleotides close to their 5' end; this sequence is complementary to part of the SSU rRNA in the small subunit. The interaction triggers ribosomal binding to the mRNA. The first tRNA recruit is what's known as an **fMet initiator** tRNA. This is a modified methionyl-tRNA, in which a **formyl group** has been added to the Met on the loaded tRNA. Modified Mets are only used when AUG codons behave as initiators; those occurring elsewhere in a sequence lead to the normal recruitment and incorporation of Met into the growing protein chain. Stop codons are more straightforward to grasp – they are simply those that have no associated tRNA. On reaching such a codon, a protein known as a **release factor** enters the ribosome instead of a tRNA, triggering the release of the complete protein chain.

(continued)

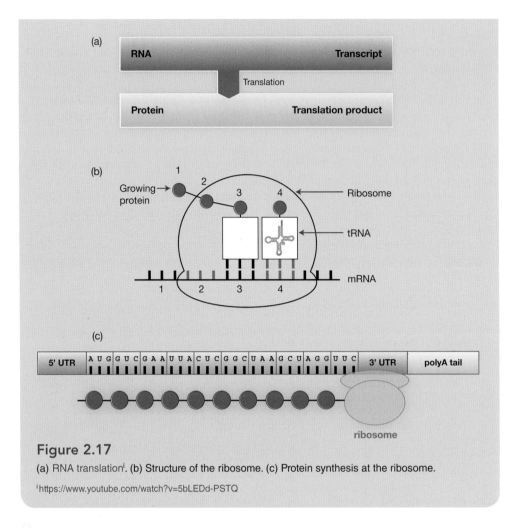

Figure 2.17

(a) RNA translation[i]. (b) Structure of the ribosome. (c) Protein synthesis at the ribosome.

[i]https://www.youtube.com/watch?v=5bLEDd-PSTQ

Before proceeding, note that the code depicted in Figure 2.16 is based on natural translation from RNA; for conceptual translation from DNA, we must therefore replace U with T. Let's suppose that we have a piece of genomic sequence. The first problem we encounter is that we don't know which bits are coding. Possible coding regions are found in what we call **Open Reading Frames** (ORFs). So, what we have to do is look for ORFs: these begin with a start codon (ATG), and end with a **stop codon** (TGA, TAA or TAG). The longer the ORF, the more likely it is to be a **coding region,** as non-coding DNA tends to contain relatively high densities of random stop codons.

The next problem is that we don't know where to start reading. Remember, the sequence is read in triplets. As we traverse the DNA strand from the 5′ to the 3′ end, the triplet code could start at the first, second or third base, and the **coding sequence** (CDS) could lie in any one of these three different frames, as shown in Figure 2.19 (a). But there's another complication – DNA sequences contain two strands, either of which may encode a gene sequence. So, in fact, there are six frames in which to look for ORFs: three forward and three reverse. As we only have one piece of genomic sequence, we must therefore also look in the reverse complement – Figure 2.19 (b): here, note that we find start codons both in the first forward frame and in the third frame of the reverse complement.

Box 2.7 The role of tRNA in translation

A key molecule in the process of translation is transfer RNA (tRNA), a single-stranded molecule in which base-pairing between short regions of complementary sequence within the strand give rise to stem- or hairpin-loop-like structures. These are most familiar to us in tRNA's characteristic **clover-leaf structure**, as depicted in Figure 2.18.

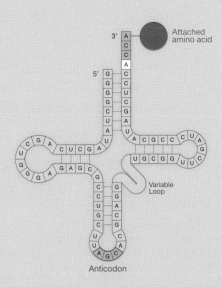

Figure 2.18

Structure of tRNA.

The three bases at the heart of the central hairpin loop of the clover-leaf are termed the **anti-codon**. tRNA acts as an adaptor molecule. The anticodon connects to the mRNA, and the other end of the molecule connects to the growing protein chain. The sequence shown here, whose anticodon is AGC (reading from 5′ to 3′), is that of the tRNA for alanine, otherwise referred to as tRNA-Ala (or tRNA^{Ala}). This can form complementary base pairs with the codon sequence GCU (reading from 5′ to 3′ in the mRNA), which is a codon for alanine in the standard genetic code, as shown below (remember, reading codons in mRNA (5′ to 3′) requires anticodons to bind in the opposite direction):

```
  |   |
 C G A    tRNA
-G C U-   mRNA
```

Organisms possess populations of tRNAs that can base-pair with the 61 codons that encode amino acid molecules. While these tRNAs have different anticodons, and a variety of other sequence differences, they all exhibit the same clover-leaf structure.

Many tRNAs can pair with more than one codon, because of the flexibility of the pairing rules at the third position in the codon, which has been dubbed **wobble**. The number of tRNAs varies widely between organisms. In eukaryotes, the wobble rules tend to be less flexible than those in prokaryotes, and hence a greater number of distinct tRNAs are required.

As we saw earlier, in prokaryotes, translation is normally just about finding the start codon and decoding the longest bit of sequence uninterrupted by a stop codon, because most prokaryotes don't have introns. This is a simplification, of course, and there are

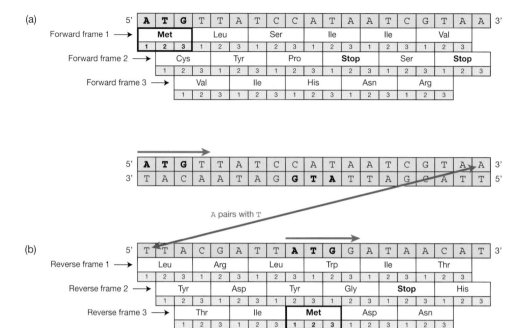

Figure 2.19

Conceptual translation of DNA proceeds in three forward frames (a) and in three reverse frames, which is equivalent to three forward frames of the reverse complement (b). The red arrow denotes formation of the reverse complement; the green arrows denote the direction of translation from the start codons in the first forward frame and in the third frame of the reverse complement.

many likely sources of error: *e.g.*, DNA sequencing errors may result in incorrectly assigned or missed start and/or stop codons, leading to truncation of a gene, extension of a gene, or missing a gene completely. Sometimes there are overlapping ORFs (whether on the same or opposite DNA strand), and sometimes the shorter of the two is the real gene! Figure 2.20 illustrates the sort of complexity that arises with overlapping ORFs in different reading frames.

As a result of these complications, evidence from a variety of sources needs to be used to confirm whether the sequence encoded in an ORF really is a protein. One trusted method of confirmation is to search a protein sequence database to discover whether something like this sequence already exists (a step that wouldn't be possible, of course, if comprehensive sequence databases weren't readily available). Another method is to

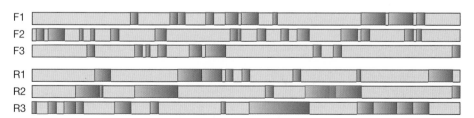

Figure 2.20

Distribution of ORFs in three forward (F) and three reverse (R) reading frames of a sequence of DNA. The horizontal grey bars show reading frames; the blue blocks denote ORFs. Here, there are numerous ORFs in each frame, and in overlapping positions. Such complexity makes gene finding difficult.

Table 2.4

Example prokaryotic genomes[i], their genome sizes and predicted gene counts.

Prokaryotes	# of genes	Genome size (base pairs)	Reference
Mycoplasma genitalium	476	580,076	Fraser *et al.*, 1995
Rickettsia prowazekii	834	1,111,523	Andersson *et al.*, 1998
Helicobacter pylori 26695	1,566	1,667,867	Tomb *et al.*, 1997
Neisseria meningitidis A	2,121	2,184,406	Parkhill *et al.*, 2000
Yersinia pestis Antiqua	4,167	4,702,289	Chain *et al.*, 2006
Streptomyces coelicolor	7,825	8,667,507	Redenbach *et al.*, 1996
Bradyrhizobium japonicum	8,317	9,105,828	Kaneko *et al.*, 2002

[i] http://en.wikipedia.org/wiki/List_of_sequenced_bacterial_genomes

look at the codon usage or other statistical features of the sequence: the **guanine–cytosine (GC)-content**, codon frequency and **oligonucleotide** composition characterise ORFs – hence, these features can be analysed to determine whether they are consistent with known protein-coding genes from the same organism. Other signals to look out for include the presence of a ribosome-binding site (also known as a **Shine–Dalgarno sequence**) prior to the coding sequence (the **ribosome** is the molecular machine responsible for natural translation), or the presence of typical regulatory (or **promoter**) regions upstream of the coding sequence. The best computational methods today use a combination of these approaches.

The genomes of prokaryotes typically contain hundreds or a few thousands of genes. By contrast, the genomes of eukaryotes, which range from simple unicellular organisms to multicellular organisms, vary in size from a few thousand genes to tens of thousands. Finding genes in eukaryotic genome sequences is thus considerably more complicated than it is in prokaryotes. Tables 2.4 and 2.5 compare the predicted gene counts from some example prokaryotic (bacterial) and eukaryotic organisms. Incidentally, note the near-linear relationship between the number of ORFs and the genome size for prokaryotes, and how this breaks down in eukaryotes; note, in particular, the smaller numbers of genes in the human and mouse genomes relative to those of plants, and yet their massively bloated genome sizes.

Table 2.5

Example eukaryotic genomes[i], their genome sizes and predicted gene counts.

Eukaryotes	# of genes	Genome size	Reference
Saccharomyces cerevisiae	6,294	12.1Mb	Goffeau *et al.*, 1996
Drosophila melanogaster	13,600	165Mb	Adams *et al.*, 2000
Caenorhabditis elegans	19,000	100Mb	*C. elegans* sequencing consortium, 1998
Homo sapiens	20,251	3.2Gb	Lander *et al.*, 2001
Mus musculus	22,011	2.5Gb	Mouse genome sequencing consortium, 2002
Arabidopsis thaliana	25,498 27,400 31,670	119Mb	*Arabidopsis* genome initiative, 2000
Zea mays	39,656	2,300Mb	Schnable *et al.*, 2009

[i] http://en.wikipedia.org/wiki/List_of_sequenced_eukaryotic_genomes

So why this discrepancy in eukaryotes? As we saw earlier, coding regions in eukaryotes are interrupted by introns, which must be excised to form a continuous transcript for translation. It turns out that introns can be many hundreds of bases long, and genes can be hundreds of bases apart. Obviously, the smaller the intergenic regions, and the fewer the introns, the easier it is to find genes. For some of the smaller eukaryotes, which are more prokaryote-like in terms of their genome organisation, gene identification is a much more tractable problem; however, for those with intron-rich genomes, there is no single, reliable method for predicting protein-coding genes – for most such organisms, it is therefore difficult or impossible to know how many genes they really contain (*e.g.*, note the three different figures in Table 2.5 for *Arabidopsis thaliana*, and see Box 2.8). For multicellular eukaryotes, then, with numerous introns and large

Box 2.8 How many genes in the human genome?

The level of uncertainty around the number of human genes can hardly be overstated. During preparation of the first 'working draft' of the human genome, the number of genes was estimated to lie between 35,000 and 150,000 (numbers that were likely to have been, at least in part, extrapolated from then currently sequenced genomes, like *Drosophila* and *Arabidopsis*, which had ~13,000 and ~25,000 genes respectively, organisms that were considered much less complex than *Homo sapiens*).

Opinion was divided. In fact, it became so contentious that a 'gene sweepstake'[a] was organised between 2000 and 2003 to try to resolve the debate. The issue was not so much the number of genes itself, but rather, how to define what genes are and how to count them. During the three years of the 'genesweep', betting was open to all (including children!), with the caveat that only bets registered at Cold Spring Harbor could be considered. But placing a bet wasn't free: the cost in the first year was $1.00, rising to $5.00 and $20.00 in the second and third years – the increased fees in years two and three reflected the expectation that advances in gene counting would provide greater evidence on which to base guesses in later years.

By the time the 'book' was closed, 281 bets had been placed: the lowest guess was 27,462 genes, the highest 312,278 (see Table 2.6). While it's possible that some of the outliers resulted from bets of non-scientists, the spread of results shown in the table below indicates just how much doubt and ambiguity there was in the scientific community itself during this period; and all of these figures are larger than the best predictions available today, one of which is as low as 18,877 – what's clear is that estimates of this fundamental component of the human genome are likely to continue to fluctuate for years to come. (To find out more, we recommend reading *Between a chicken and a grape: estimating the number of human genes,* by Pertea and Salzberg, 2010).

Table 2.6
Summary of Genesweep results.

Genesweep	
Bets	281
Mean	67,006
Median	61,302
Lowest	27,462
Highest	312,278

[a] http://web.archive.org/web/20000816213907/http:/www.ensembl.org/genesweep.html

intergenic regions, gene identification still poses major problems. To put this in perspective, only about 3 per cent of the three billion bases in the human genome are coding – gene finding is thus like the proverbial needle-in-a-haystack problem.

As we saw in Section 2.2.5, a major difficulty in eukaryotic gene identification is in correctly recognising intron–exon boundaries. Boundary identification relies on correct identification of **splice sites**, *i.e.* the positions in the nucleotide sequence between which to remove non-coding regions. Most splice sites have similar sequences, and many are characterised by two particular dinucleotides at the ends of each intron; but there are variants of these sequences, making reliable prediction of intron–exon boundaries very tricky.

Figure 2.21 illustrates some of the problems that can arise when gene-prediction software goes wrong. If intron–exon boundaries are not recognised accurately, incorrect gene structures are generated, with missing exons or introns, extended exons, or additional exons (within or beyond the correct structure). Note that the figure shows a highly simplified gene model: it doesn't account for the complications that arise from gene nesting, interleaved genes, and fusion or splicing of genes. Overall, then, gene prediction in higher eukaryotes poses major challenges, and best results are obtained from combined approaches that use several different programs in parallel, that integrate extrinsic and intrinsic data, and train the prediction algorithms on organism-specific training sets.

In 2000 (when the first draft of the human genome was being assembled), a comparison of gene-prediction software against the *Drosophila* genome revealed that the best tools could correctly predict only ~40 per cent of genes (Reese *et al.*, 2000). At the time, estimates for the numbers of genes encoded in the human genome varied dramatically (from tens of thousands to hundreds of thousands – Box 2.8); the inaccuracy of the gene-prediction tools available at the time might be one reason for such uncertainty. Even today, precise numbers of genes for most organisms are unknown, and this is why the numbers quoted in Tables 2.4 and 2.5 are given as 'predicted' gene counts. The numbers should therefore be considered as snapshots of our current thinking, tempered by the limits of the technology of the day. In time, many of these numbers will probably change (indeed, some may already have done so), as improvements are made both in sequencing technology and in gene-prediction software.

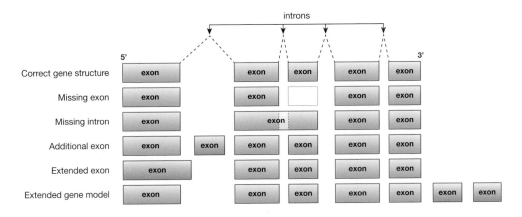

Figure 2.21

Simplified schematic illustration of eukaryotic gene modelling, showing the effects of addition and loss of exons, loss of introns, extension of exons, and inclusion of exons beyond the correct gene structure.

2.3   Access to whole genomes

The late 1980s and early 1990s were fertile years, giving rise to a flourishing number of new molecular structures and sequences, and to new databases in which to store them. Looking back at this period of prolific activity, it's incredible to reflect that two major developments had yet to take place: together, these would not only seed an explosion of biological data but would also spur their global dissemination – they were the advent of the Web and the arrival of **high-throughput DNA sequencing**.

The significance of the development of high-throughput technologies is that it made whole-genome sequencing a practical reality for the first time. Seizing this opportunity, there followed an unprecedented burst of sequencing activity, yielding in rapid succession, for example, the genomes of *Haemophilus influenzae* and *Mycoplasma genitalium* in 1995 (Fleischmann *et al.*, 1995; Fraser *et al.*, 1995), of *Methanocaldococcus jannachii* and *Saccharomyces cerevisiae* in 1996 (Bult *et al.*, 1996; Goffeau *et al.*, 1996), of *Caenorhabditis elegans* in 1998 (*C. elegans* sequencing consortium, 1998), of *Drosophila melanogaster* in 2000 (Adams *et al.*, 2000) and, the ultimate prize, of *Homo sapiens* in 2001 (Lander *et al.*, 2001; Venter *et al.*, 2001; International Human Genome Sequencing Consortium, 2004; Gregory *et al.*, 2006). Thousands of genomes have been sequenced since this fruitful dawn – Table 2.7 provides an overview of some of the major landmarks.

Hand-in-hand with these activities came the development of numerous organism-specific databases to store the emerging genomic data: *e.g.*, **FlyBase** (Ashburner and Drysdale, 1994), **ACeDB** (Eeckman and Durbin, 1995), **SGD** (Cherry *et al.*, 1998), **TAIR** (Huala *et al.*, 2001), **Ensembl** (Hubbard *et al.*, 2002), and many more. For some, the value of this genomic 'gold rush' was not entirely clear: with much of the amassed data seemingly impossible to characterise, and vast amounts of it non-coding, the hoped-for treasure troves were beginning to seem to some commentators about as inspiring as large-scale butterfly collections (Strasser, 2008), and suggested to others that molecular biology had entered a rather vacuous era of 'high-tech stamp collecting' (Hunter, 2006). Interestingly, arguments like this were used by some of the early opponents of the first publicly funded nucleotide sequence databases, and characterised much of the substantial resistance to the **Human Genome Project** (HGP) a few years later (Strasser, 2008).

Table 2.7

Landmark events in whole genome sequencing (adapted from Higgs and Attwood, 2005).

Year	Organism	Landmark
1995	*Haemophilus influenzae*	First bacterial genome
1996	*Methanocaldococcus jannaschii*	First archaeal genome
1997	*Saccharomyces cerevisiae*	First unicellular eukaryote
1998	*Caenorhabditis elegans*	First multicellular eukaryote
2000	*Arabidopsis thaliana*	First plant genome
2001	*Homo sapiens*	First draft of the human genome
2004	*Homo sapiens*	'Finished' draft of the human genome

Whether or not such views are still held today, the onslaught of genome sequencing has continued, unabated. The scale of data gathering from world-wide sequencing projects is daunting, and compels us to think in more disciplined ways about how to manage, store and annotate the information, so that future generations of researchers can exploit it more efficiently and effectively. This situation provides much of the context for the computer science chapters in this book, and sets the scene for the next chapter, in which we introduce some of the databases that have been created over the years to capture and manage the prodigious quantities of accumulating biological data.

2.4 Summary

This chapter introduced some of the basic biological concepts and data-types that provide the foundations for bioinformatics. Specifically, we saw that:

1   The central dogma states that the flow of genetic information is in the direction of DNA to RNA to protein;
2   Nucleic acids have a four-letter alphabet: DNA is built from the bases A, T, G and C; RNA from A, U, G and C;
3   mRNA is produced from DNA via the process of transcription;
4   Proteins are produced from mRNA via the process of translation;
5   Before translation of eukaryotic mRNA, non-coding regions – introns – are spliced out;
6   The molecular machine responsible for splicing is called a spliceosome;
7   The molecular machine that mediates translation of mRNA is the ribosome;
8   Proteins are built from a 20-letter alphabet of 20 amino acids;
9   The physicochemical properties of the amino acids overlap; their structures and properties induce unique local and global folding patterns in protein chains;
10   Protein structure is hierarchical: the primary structure is the linear amino acid sequence; secondary structures are regions of local structural regularity within the chain; the tertiary structure is the overall protein fold;
11   Most protein sequences today are produced by conceptual translation from DNA sequences generated from whole-genome sequencing projects;
12   Translation requires use of the genetic code to convert the language of nucleic acids to the language of proteins;
13   Translation of DNA must be performed in six reading frames: three forward and three reverse; coding regions may be found in forward and reverse frames;
14   Identifying and building genes requires the correct recognition of translation start and stop signals, and of intron–exon boundaries, a computational process that can be highly error-prone;
15   World-wide genome projects are producing sequence data on a vast scale.

2.5 References

Adams, M.D., Celniker, S.E., Holt, R.A. *et al.* (2000) The genome sequence of *Drosophila melanogaster*. *Science*, **287**, 2185–2195.

Andersson, S.G., Zomorodipour, A., Andersson, J.O. *et al.* (1998) The genome sequence of *Rickettsia prowazekii* and the origin of mitochondria. *Nature*, **396**, 133–140.

Arabidopsis Genome Initiative (2000) Analysis of the genome sequence of the flowering plant *Arabidopsis thaliana*. *Nature*, **408**, 796–815.

Ashburner, M. and Drysdale, R. (1994) FlyBase – the *Drosophila* genetic database. *Development*, **120**(7), 2077–2079.

Berget, S.M., Moore, C. and Sharp, P.A. (1977) Spliced segments at the 5' terminus of adenovirus 2 late mRNA. *Proceedings of the National Academy of Sciences of the U.S.A.*, **74**(8), 3171–3175.

Bult, C.J., White, O., Olsen, G.J. *et al.* (1996) Complete genome sequence of the methanogenic archaeon, *Methanococcus jannaschii*. *Science*, **273**, 1058–1073.

C. elegans Sequencing Consortium (1998) Genome sequence of the nematode *C. elegans*: a platform for investigating biology. *Science*, **282**, 2012–2018.

Chain, P.S., Hu, P., Malfatti, S.A. *et al.* (2006) Complete genome sequence of *Yersinia pestis* strains Antiqua and Nepal516: evidence of gene reduction in an emerging pathogen. *Journal of Bacteriology*, **188**(12), 4453–4463.

Cherry, J.M.; Adler, C. *et al.* (1998) SGD: Saccharomyces Genome Database. *Nucleic Acids Research*, **26**(1), 73–79.

Chow, L.T., Gelinas, R.E., Broker, T.R and Roberts, R.J. (1977) An amazing sequence arrangement at the 5' ends of adenovirus 2 messenger RNA. *Cell*, **12**(1), 1–8.

Creighton, T.E. (1993) *Proteins: structures and molecular properties*, 2nd edn, W.H. Freeman, San Francisco.

Crick, F. (1970) Central dogma of molecular biology. *Nature*, **227**, 561–563.

Eeckman, F.H. and Durbin, R. (1995) ACeDB and macace. *Methods in Cell Biology*, **48**, 583–605.

Engelman, D.A., Steitz, T.A. and Goldman, A. (1986) Identifying nonpolar transbilayer helices in amino acid sequences of membrane proteins. *Annual Review of Biophysics and Biophysical Chemistry*, **15**, 321–353.

Fleischmann, R.D., Adams, M.D., White, O. *et al.* (1995) Whole-genome random sequencing and assembly of *Haemophilus influenzae* Rd. *Science*, **269**, 496–512.

Fraser, C.M., Gocayne, J.D., White, O. *et al.* (1995) The minimal gene complement of *Mycoplasma genitalium*. *Science*, **270**, 397–403.

Goffeau, A., Barrell, B.G., Bussey, H. *et al.* (1996) Life with 6000 genes. *Science*, **274**, 546–567.

Hershey, A. and Chase, M. (1952) Independent functions of viral protein and nucleic acid in growth of bacteriophage. *Journal of General Physiology*, **36**(1), 39–56.

Higgs, P. and Attwood, T.K. (2005) *Bioinformatics and Molecular Evolution*. Blackwell, Oxford.

Huala, E., Dickerman, A.W., Garcia-Hernandez, M. *et al.* (2001) The Arabidopsis Information Resource (TAIR): a comprehensive database and web-based information retrieval, analysis, and visualization system for a model plant. *Nucleic Acids Research*, **29**(1), 102–105.

Hubbard, T., Barker, D., Birney, E. *et al.* (2002) The Ensembl genome database project. *Nucleic Acids Research*, **30**(1), 38–41.

Hunter, D.J. (2006) Genomics and proteomics in epidemiology: treasure trove or 'high-tech stamp collecting'? *Epidemiology*, **17**(5), 487–489.

International Human Genome Sequencing Consortium (2004) Finishing the euchromatic sequence of the human genome. *Nature*, **431**, 931–945.

Kaneko, T., Nakamura, Y., Sato, S. *et al.* (2002) Complete genomic sequence of nitrogen-fixing symbiotic bacterium *Bradyrhizobium japonicum* USDA110. *DNA Res.*, **9**(6), 189–197.

Kyte, J. and Doolittle, R.F. (1982) A simple method for displaying the hydropathic character of a protein. *Journal of Molecular Biology*, **157**, 105–132.

Lander, E.S., Linton, L.M., Birren, B. *et al.* (2001) Initial sequencing and analysis of the human genome. *Nature*, **409**, 860–921.

Miller, S., Janin, J., Lesk, A.M. and Chothia, C. (1987) Interior and surface of monomeric proteins. *Journal of Molecular Biology*, **196**, 641–57.

Mouse Genome Sequencing Consortium (2002) Initial sequencing and comparative analysis of the mouse genome. *Nature*, **420**, 520–562.

Parkhill, J., Achtman, M., James, K.D. *et al.* (2000) Complete DNA sequence of a serogroup A strain of *Neisseria meningitidis* Z2491. *Nature*, **404**, 502–506.

Pertea, M. and Salzberg, S.L. (2010) Between a chicken and a grape: estimating the number of human genes. *Genome Biology*, **11**(5), 206. doi:10.1186/gb-2010-11-5-206

Redenbach, M., Kieser, H.M., Denapaite, D. *et al.* (1996) A set of ordered cosmids and a detailed genetic and physical map for the 8 Mb *Streptomyces coelicolor* A3(2) chromosome. *Molecular Microbiology*, **21**(1), 77–96.

Reese, M.G., Hartzell, G., Harris, N.L. *et al.* (2000) Genome annotation assessment in *Drosophila melanogaster*. *Genome Research*, **10**, 483–501.

Rose, G.D., Geselowitz, A.R., Lesser, G.J. *et al.* (1985) Hydrophobicity of amino acid residues in globular proteins. *Science*, **228**, 834–838.

Schnable, P.S., Ware, D., Fulton, R.S. *et al.* (2009) The B73 maize genome: complexity, diversity, and dynamics. *Science*, **326**, 1112–1115.

Service, R.F. (2006) The race for the $1000 genome. *Science*, **311**, 1544–1546.

Strasser, B. (2008) GenBank – Natural history in the 21st century? *Science*, **322**, 537–538.

Tomb, J.F., White, O., Kerlavage, A.R. *et al.* (1997) The complete genome sequence of the gastric pathogen *Helicobacter pylori*. *Nature*, **388**, 539–547.

Venter, J.C., Adams, M.D., Myers, E.W. *et al.* (2001) The sequence of the human genome. *Science*, **291**, 1304–1351.

Watson, J.D. and Crick, F. (1953) Molecular structure of nucleic acids: a structure for deoxyribose nucleic acid. *Nature*, **171**, 737–738.

Zimmerman, J.M., Eliezer, N. and Simha, R. (1968) The characterization of amino acid sequences in proteins by statistical methods. *Journal of Theoretical Biology*, **21**, 170–201.

2.6 Quiz

This multiple-choice quiz will help you to check how much you've remembered of the basic background material introduced in this chapter.

1 Which of the following bases are found in RNA but not DNA?
    A Adenine
    B Thymine
    C Uracil
    D Guanine

2 3′ and 5′ describe the termini of which of the following molecular sequences?
    A Proteins
    B DNA
    C RNA
    D All of the above

3 Which of the following contain sulphur atoms?
    A Arginine
    B Lysine
    C Methionine
    D Tryptophan

4 Which of the following are not valid amino acid sequences?
    A ATTWOOD
    B PETTIFER
    C MARSH
    D THORNE

5 Gln-Trp-Tyr-Lys-Glu-Asn-Asp corresponds to which of the following amino acid sequences?
    A NFWKDQE
    B EYFLQDN
    C DYWLNEQ
    C QWYKEND

6 Which two of the following DNA sequences (written in the 5′ to 3′ direction) are complementary to one another?
    1) CTTACGG   2) TCGGATT   3) GAATGCC   4) ACCGTAA
    A 1 and 2
    B 2 and 3
    C 2 and 4
    D 1 and 3

7 Which of the following statements about the standard genetic code is correct?
    A Phenylalanine and glycine each have only one codon.
    B Isoleucine and serine each have six codons.
    C There are three possible stop codons.
    D There are three possible start codons.

8 Which of the following statements about translation is correct?
  A Translation is initiated at a start codon.
  B Translation is carried out by polymerases.
  C Translation involves complementary pairing of DNA and RNA strands.
  D None of the above.

9 The primary structure of a protein is:
  A the regions of local folding in a polypeptide chain.
  B the alpha helices and beta strands in a polypeptide chain.
  C the sequence of amino acids in a polypeptide chain.
  D the fully folded structure of a polypeptide chain.

10 The spliceosome is the molecular machine responsible for:
  A DNA replication.
  B DNA synthesis.
  C RNA processing.
  D protein synthesis.

2.7 Problems

1 *Sphingopyxis alaskensis*, a marine bacterium, has a genome size of 3,345,170 bases. Using the data in Table 2.4, estimate the number of genes encoded in the genome. *Vitis vinifera*, a popular fruit plant, contains 30,434 genes. Give an estimate for its likely genome size. How accurate were your estimations? If there was a discrepancy, can you provide an explanation?

2 Why is it harder to determine the protein products of eukaryotic gene sequences than those of prokaryotic gene sequences? Why is this a difficult problem to tackle computationally? Mention some of the issues that can arise when this process goes wrong.

3 The DNA sequence of bacteriorhodopsin from *Halobacterium salinarium*[3] is shown:

```
gggtgcaaccgtgaagtccgccacgaccgcgtcacgacaggagccgaccagcgacacccagaaggtgc
gaacggttgagtgccgcaacgatcacgagtttttcgtgcgcttcgagtggtaacacgcgtgcacgcat
cgacttcaccgcgggtgtttcgacgccagccggccgttgaaccagcaggcagcgggcatttacagccg
ctgtggcccaaatggtggggtgcgctattttggtatggtttggaatccgcgtgtcggctccgtgtctg
acggttcatcggttctaaattccgtcacgagcgtaccatactgattgggtcgtagagttacacacata
tcctcgttaggtactgttgcatgttggagttattgccaacagcagtggagggggtatcgcaggcccag
atcaccggacgtccggagtggatctggctagcgctcggtacggcgctaatgggactcgggacgctcta
tttcctcgtgaaagggatgggcgtctcggacccagatgcaaagaaattctacgccatcacgacgctcg
tcccagccatcgcgttcacgatgtacctctcgatgctgctggggtatggcctcacaatggtaccgttc
ggtggggagcagaaccccatctactgggcgcggtacgctgactggctgttcaccacgccgctgttgtt
gttagacctcgcgttgctcgttgacgcggatcagggaacgatccttgcgctcgtcggtgccgacggca
tcatgatcgggaccggcctggtcggcgcactgacgaaggtctactcgtaccgcttcgtgtggtgggcg
```

[3] http://www.ncbi.nlm.nih.gov/nuccore/43536?report=fasta

```
atcagcaccgcagcgatgctgtacatcctgtacgtgctgttcttcgggttcacctcgaaggccgaaag
catgcgccccgaggtcgcatccacgttcaaagtactgcgtaacgttaccgttgtgttgtggtccgcgt
atcccgtcgtgtggctgatcggcagcgaaggtgcgggaatcgtgccgctgaacatcgagacgctgctg
ttcatggtgcttgacgtgagcgcgaaggtcggcttcgggctcatcctcctgcgcagtcgtgcgatctt
cggcgaagccgaagcgccggagccgtccgccggcgacggcgcggccgcgaccagcgactgatcgcaca
cgcaggacagccccacaaccggcgcggctgtgttcaacgacacacgatgagtcccccactcggtcttg
tactc
```

Use an online translation tool[4] to deduce the sequence of the encoded protein. Now look at the DNA sequence[5] of human **rhodopsin**. What are the translations of the 3rd and 5th exons? To automate the process of DNA translation, what issues must be tackled, and what extra evidence can help to reduce likely errors?

4   Are the amino acids Trp, Pro, His, Cys, Arg and Lys hydrophobic or hydrophilic (refer to Boxes 2.2 and 2.3)? According to Table 2.3 in Box 2.3, which is the most hydrophobic and which the most hydrophilic amino acid?

There are many amino acid hydropathic rankings; some are shown here:

Zimmerman	LYFPIVKCMHRAEDTWSGNQ
von Heijne	FILVWAMGTSYQCNPHKEDR
Efremov	IFLCMVWYHAGTNRDQSPEK
Eisenberg	IFVLWMAGCYPTSHENQDKR
Cornette	FILVMHYCWAQRTGESPNKD
Kyte	IVLFCMAGTSWYPHDNEQKR
Rose	CFIVMLWHYAGTSRPNDEQK
Sweet	FYILMVWCTAPSRHGKQNED

Do these rankings support your answers above? If not, what were the discrepancies and how can you explain them?

---

[4] http://web.expasy.org/translate
[5] http://www.ncbi.nlm.nih.gov/nuccore/U49742

Chapter 3

Biological databases

3.1 Overview

In this chapter, we outline some of the principal databases that are in common use in bioinformatics today. We broadly introduce some of the main kinds of repository, in terms of the kinds of data they store (nucleotide and protein sequences, protein structures, and so on); we also look at some of the challenges that have arisen from advances in high-throughput data-generation technologies, in terms of data storage, management and, especially, annotation. We build the story largely from a historical perspective, in order to illustrate how individual databases evolved in response to perceived data 'crises', rather than being designed in line with some visionary grand plan.

By the end of the chapter, you should be able to list some of the main databases, you should appreciate the types of data they contain, and you should have insights into some of the key issues thrown up by high-throughput science.

3.2 What kinds of database are there?

In Chapters 1 and 2, we saw that massive amounts of data have become available as a consequence of the advent of genome sequencing; and we hinted that genomes are now just one of the many different outputs of, or data-types relevant to, high-throughput biological methods. The ocean of data accumulating from this 'new biology' has created an urgent need for better, more efficient means of storage: the greater the level of organisation, the more amenable the data become to computational manipulation and algorithmic attack.

Given the range of biological data-types, many different kinds of repository have been developed over the years to capture and manage them. To get an idea of the number of databases now used in the field of bioinformatics, take a look at the Database Issue[1] of *Nucleic Acids Research*, an entire journal volume annually dedicated to the description of new and existing resources (*e.g.*, the January 2015 issue listed more than 1,500 repositories in its online compendium[2]). It's clearly impossible to do justice to the

---

[1] http://nar.oxfordjournals.org/content/43/D1.toc
[2] http://www.oxfordjournals.org/nar/database/a

Bioinformatics challenges at the interface of biology and computer science: Mind the Gap. First Edition.
Teresa K. Attwood, Stephen R. Pettifer and David Thorne. Published 2016 © 2016 by John Wiley and Sons, Ltd.
Companion website: www.wiley.com/go/attwood/bioinformatics

Table 3.1

Overview of some of the biological databases mentioned in this chapter.

Database type	Example resource	Content
Primary (nucleic acid)	EMBL, GenBank, DDBJ	Nucleotide sequence data and annotation
Primary (protein)	PIR, Swiss-Prot, TrEMBL	Amino acid sequence data and annotation
Composite primary	NRDB, UniProtKB	'Non-redundant' sequence data and annotation
Signature	PROSITE, PRINTS, Pfam	Protein family 'signatures', sometimes with annotation
Composite signature	InterPro	Amalgamated signatures, rationalising different family views
Structure	PDB, PDBj, ePDB	3D coordinates of biological macromolecules
Structure classification	SCOP, CATH	Classification of protein folds into related groups
Organism	FlyBase, SGD, TAIR	Organism-specific genetic and genomic information
Genome	Ensembl	Genomic data and annotation for a variety of organisms

full spectrum of resources available today; in this chapter, therefore, we'll give a flavour of just a few of them, but with a specific focus on those commonly used in sequence analysis and that form part of the particular story we tell in later chapters – the snapshot we offer here is therefore necessarily selective. A broad overview is given in Table 3.1, and we refer you to the various boxes throughout the chapter for further information.

As can be seen from Table 3.1, the databases range from those that house raw sequence data and their **annotations**, to those that store **diagnostic family signatures** and 3D macromolecular structures, those that store genomic and genetic information pertinent to specific organisms, and so on. To try to put the development of resources like this into perspective, in the sections that follow we chart their evolution, from the inception of one of the oldest scientific databases to the emergence of some of the newer repositories that have emerged in recent years. We offer, here, high-level descriptions of the selected databases, primarily to set the scene for discussions in Chapter 4 about how the data are stored in practice, and hence how they may be interrogated and analysed.

3.3   The Protein Data Bank (PDB)

One of the earliest, and hence oldest, scientific databases was established in 1965 at the **Cambridge Crystallographic Data Centre** (CCDC), under the direction of **Olga Kennard** (Kennard *et al.*, 1972; Allen *et al.*, 1991). This was a collection of **small-molecule** structures derived using **X-ray crystallography** – it was called the Cambridge Structural Database[3] (or CSD). The CSD began life in printed form. Nevertheless, it was eventually produced electronically, in essence because Kennard had a vision: she wanted to be able to use the collected data to discover new knowledge, above and beyond the results yielded by individual experiments (Kennard, 1997) – archiving data electronically was an essential step towards realising this dream (it was also quite a radical step, given that computers weren't then the familiar tools they are today).

Some years after the creation of the CSD, Walter Hamilton and colleagues discussed the possibility of creating a similar kind of 'bank' for housing the 3D coordinate data of proteins – the idea was to **mirror** this archive at sites in the UK and the USA

---

[3] http://www.ccdc.cam.ac.uk/solutions/csd-system/components/csd

(Berman, 2008). Hamilton himself volunteered to set up the 'master copy' of the American bank at the **Brookhaven National Laboratory**; meanwhile, Kennard agreed both to host the European copy in Cambridge and to extend the CCDC small-molecule format to accommodate protein structural data (Kennard *et al.*, 1972; Meyer, 1997). These events gave birth to the **Protein Data Bank**[4] (PDB), a joint venture of the CCDC and Brookhaven.

The new resource was announced in a short news bulletin in October 1971 (Protein Data Bank, 1971); its first public release held a mere seven structures (Berman *et al.*, 2000). Interestingly, Kennard envisaged the PDB as a prototype for Europe's nucleotide sequence data library, although this resource was not to materialise for a further decade (Smith, 1990). The PDB became fully operational in 1973 (Protein Data Bank, 1973). By then, the corpus of data it had been established to house still only amounted to nine structures (see Table 3.2).

Kennard knew that, for the resource to be successful, it had to have buy-in from the entire crystallography community – it was dependent on their support; but gaining sufficient community momentum to back the initiative was a slow process (*e.g.*, note the absence from its 1973 holdings of ribonuclease, whose structure, as we saw in Chapter 1, had been determined six years earlier).

In the years that followed, as the number of structures submitted to the PDB grew, the archive included its first example of a transfer RNA (tRNA) structure. This stimulated a debate about whether 'Protein Data Bank' was still an appropriate name (Bernstein *et al.*, 1977). Despite the reservations, however, the name stuck: in November 2015, still known as the PDB, the resource included more than 8,000 nucleic acid and protein–nucleic acid complexes, and is a testament to the pioneers of structural biology[5].

By 1977, the database contained 77 sets of coordinates relating to only 47 macromolecules, highlighting a rather sluggish growth. Perhaps this explains why Berstein *et al.* tried to get news of the resource out by publishing the same PDB article verbatim in May and November of 1977, and again in January 1978, in three different journals (Bernstein *et al.*, 1977a, b; 1978). Whatever the real reasons, the PDB expanded very gradually compared to the CSD (Kennard, 1997), and the number of unique structures remained small – to put this in context, by 1996, there were ~6,000 structures in the PDB

Table 3.2

PDB holdings, August 1973.

Protein structures	PDB ID
Cyanide methaemoglobin V from sea lamprey	2DHB
Cytochrome b_5	1CYO
Basic pancreatic trypsin inhibitor	4PTI
Subtilisin BPN (Novo)	1SBT
Tosyl α-chymotrypsin	2CHA
Bovine carboxypeptidase Aα	3CPA
L-Lactate dehydrogenase	6LDH
Myoglobin	1MBN
Rubredoxin	4RXN

[4] http://www.rcsb.org
[5] http://pdb101.rcsb.org/motm/142

compared to ~150,000 structures in the CSD (although this is an unfair comparison, because small-molecule structures present fewer obstacles to structure determination than do proteins and protein–nucleic acid complexes). Perhaps a more pressing concern highlighted by the 1977 holdings was that the database was beginning to accrete a significant amount of redundancy – by 1992, this was calculated to be ~7-fold (Berman, 2008; Hobohm *et al.*, 1992). This redundancy was an inevitable consequence of the nature of structural biology research. Let's take a moment to think about why this is so.

Structural studies are immensely valuable. For example, they've shown that the physical similarities shared by some proteins reflect **common descent** through **divergent evolution**, during which mutations (substitutions, insertions and/or deletions) have accumulated in their underlying sequences. For distantly related proteins, such mutations can be extensive, yielding folds in which the numbers and orientations of secondary structures vary considerably, but whose basic scaffolds nevertheless act to preserve the structural environments of critical functional or active-site residues. The diversity and elegance of such protein scaffolds has been captured in a number of **fold-classification databases**, which are described in more detail in Box 3.1.

Box 3.1 Structure classification databases

The PDB is a highly redundant resource. This is largely because of the nature of scientific research, which increases our knowledge incrementally by repeating detailed analyses of the same system until its intricacies are better understood. Some of the earliest protein structures to have been deduced (those of myoglobin and haemoglobin, for example) have therefore been determined repeatedly, in order to better resolve their atomic details, and to learn more about the likely biophysical and biochemical mechanisms they use to effect their molecular functions.

By rationalising the structural data we have at our disposal, we can begin to put in place suitable foundations from which to try to further our understanding of nature's structural principles – it turns out that nature also uses only a limited repertoire of basic folds from which to build its portfolio of protein architectures (*i.e.*, the natural structure universe is highly redundant too). Some of the systematic approaches to rationalise the available structural data have yielded fold libraries (for use in structure prediction and/or data-mining studies) and more comprehensive fold-classification databases, like SCOP and CATH. These resources group proteins that share common folds and domains; however, in doing so, they use slightly different hierarchical approaches. Let's take a closer look.

The SCOP (**Structural Classification Of Proteins**[a]) hierarchy attempts to classify proteins according to evolutionary and structural relationships. This task is complicated by the fact that protein structures are highly diverse: some comprise only single small domains, while others are vast multi-domain assemblies (hence, some are classified as individual domains and others as multi-domain proteins). This isn't an exact science, of course, and so algorithms for comparing structures automatically can't always unravel the complexities completely reliably; SCOP therefore combines both manual and automated methods.

SCOP's hierarchy has three levels: family, superfamily, fold. The boundaries between these may be subjective, but the higher echelons generally reflect the clearest structural similarities:
1. at the family level are proteins that share >30 per cent sequence identity;
2. at the superfamily level are proteins that are believed to share a common evolutionary origin based on shared functional characteristics, but retain only moderate sequence identity;
3. at the fold level are proteins that have the same organisation of secondary structures, irrespective of their evolutionary origins (*i.e.*, regardless of the mechanisms of divergence or convergence).

---

[a] http://scop2.mrc-lmb.cam.ac.uk

CATH stands for **Class, Architecture, Topology, Homology**[b]. Its classification strategy differs in several ways from that used in SCOP: most notably, its hierarchy derives from a domain-based classification and the database is derived largely automatically (with the caveat that manual inspection is necessary when the automatic processes fail). Despite its name, CATH's hierarchy has five levels:

1. *Class* describes the secondary-structure content in terms of four broad classes: mainly-α, mainly-β, α-β (alternating α/β or $\alpha+\beta$ elements), and low secondary-structure content;

2. *Architecture* describes the overall shape of the tertiary fold using simple descriptive terms like barrel, roll, sandwich, *etc.*, but doesn't consider the connectivity of the constituent secondary-structure elements (hence, similar architectures may have different topologies);

3. *Topology* organises architectures according to the connectivity of secondary structures (hence, proteins at the topology level have the same overall fold, with both a similar number and arrangement of secondary structures and the same connectivity);

4. *Homology* then groups those domains that share >35 per cent sequence identity, and whose high level of structural and/or functional similarity suggests evolutionary ancestry; and

5. *Sequence* further clusters sequences in those groups that have significant similarity or are identical (*e.g.*, the same protein from different species), to produce one or more families per homologous group.

By analogy with the **EC (Enzyme Classification) system**, CATH notation assigns a 4-number (not 5!) code to each protein domain, which simplifies parsing of the database and renders it more attractive for computational manipulation.

Interestingly, SCOP and CATH have been used to estimate the number of unique folds represented by protein structures housed in the PDB. As shown in Figure 3.1 for CATH, the RCSB provides year-on-year statistics to illustrate the growth of the PDB in terms of the annual number of unique folds[c] contributed to the database (red bars) – cumulative figures are shown in blue. As is apparent from Figure 3.1, no new folds have been released for structures published since 2009, and the cumulative total is now ~1,300.

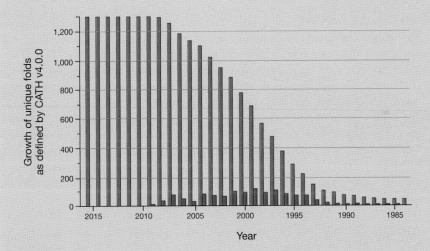

Figure 3.1

PDB growth statistics.

Source: Protein Data Bank.

This is particularly fascinating from the perspective of the Protein Structure Initiative (PSI), which partly sought to tackle PDB's redundancy issues by targeting novel structures. Nevertheless, the number of unique folds entering the database didn't significantly increase in the 8–10 years following the start of the PSI in 2000, and, as we can see, has remained at zero since 2009.

[b] http://www.cathdb.info
[c] http://www.rcsb.org/pdb/statistics/contentGrowthChart.do?content=fold-cath

Structural studies have also revealed that the similarities shared by some proteins may reflect **convergent evolution** from unrelated ancestors, as a result, say, of physical principles that favour particular packing arrangements and fold topologies. Over the years, investigations of these phenomena have driven structural biologists to study the minutiae of protein architectures from many different perspectives. Once elucidated, a structure thus becomes a kind of starting point from which its atomic details may be refined through repeated structure determinations, and new molecular insights gained by replicating the analysis with the protein bound to its cognate ligand, or with the introduction of a particular mutation, and so on. Consequently, numerous coordinate sets, with only minor variations, have been deposited in the PDB. Therefore, as the resource has evolved, its contents have been skewed by a handful of well-studied proteins, like myoglobin, haemoglobin, immunoglobulins, lysozymes and so on.

Returning to the evolution of the PDB, in 1996, a new database of macromolecular structures was created, the E-MSD (Boutselakis *et al.*, 2003), at the newly established **European Bioinformatics Institute** (EBI) in the UK. Building directly on PDB data, the E-MSD was originally conceived as a pilot study to explore the feasibility of using relational database technologies to manage structural data more effectively. As it turned out, the prototype database emerging from the pilot was successful in its own right, and the E-MSD consequently became established as a major EBI resource.

During this period, a concerted effort was made to hasten the pace of knowledge acquisition from structural studies. This was partly motivated by the desire to augment the still-limited number of structures available in the PDB, but partly also to temper its redundancy. The idea was to establish a program of high-throughput X-ray crystallography in a high-profile venture known as the Protein Structure Initiative, or PSI (Burley *et al.*, 1999). Several feasibility studies had already been launched; these, coupled with the broad-sweeping vision of the PSI (which was gearing up to produce 10,000 new structures over the next ten years), provided the first indications that coping with high-throughput structure-determination pipelines would require new ways of gathering, storing, distributing and 'serving' the data to end users.

One of the PDB's responses to this, and to the many challenges that lay ahead, was to form a new management structure. This was to be embodied in a three-membered **Research Collaboratory for Structural Bioinformatics** (RCSB), a consortium including **Rutgers**, The State University of New Jersey; the **San Diego Supercomputer Center** at the University of California; and the Center for Advanced Research in Biotechnology of the **National Institute of Standards and Technology** (Berman *et al.*, 2000; Berman *et al.*, 2003). With the establishment of this consortium, the Brookhaven PDB ceased operations, and the RCSB formally took the helm on 1 July, 1999.

With the RCSB PDB in the USA, the E-MSD established in Europe, and a sister resource (PDBj[6]) subsequently announced in Japan (Nakamura *et al.*, 2002), structure-collection efforts had taken on an international dimension. As a result, the three repositories were brought together beneath an umbrella organisation known as the world-wide Protein Data Bank (wwPDB)[7], to harmonise their activities and maintain a single, global, publicly available archive of macromolecular structural data (Berman *et al.*, 2003). By 2009, perhaps to align its nomenclature in a more obvious way with its consortium partners, E-MSD was re-named PDBe[8] (Velankar *et al.*, 2009). Today, the

---

[6] http://www.pdbj.org
[7] http://www.wwpdb.org
[8] http://www.ebi.ac.uk/pdbe

RCSB remains the 'archive keeper', with sole write-access to the PDB, controlling its contents, and distributing new PDB **identifiers** (IDs) to all deposition sites.

In November 2015, the PDB holdings[9] amounted to 114,080 structures: of these, ~106,000 were protein structures deduced primarily using **X-ray diffraction, Nuclear Magnetic Resonance** (NMR) spectroscopy and **electron microscopy**; the rest were mostly structures of nucleic acids and nucleic acid–protein complexes, and of some **carbohydrates**. This figure is obviously much larger than the seven structures deposited in the PDB 44 years earlier; nevertheless, growth of the resource has been slower than might have been predicted in the wake of high-throughput structure determination. In fact, it became apparent that high-throughput studies were fraught with difficulties, and the experimental pipelines had high attrition rates (Yon and Jhoti, 2003): for example, some proteins couldn't be purified in sufficient quantities to be useful; of those that could be purified, some wouldn't readily crystallise; and, of those that did crystallise, some produced only poor-quality crystals that wouldn't produce usable **diffraction patterns**. Consequently, the rate of high-throughput structure determination didn't match the original projections, and the initial eight years of the PSI produced fewer than 3,000 structures[10].

As you'll now appreciate, the total figure of ~106,000 protein structures mentioned above includes a considerable amount of redundancy. We can get an idea of just how much redundancy by clustering PDB's contents according to different levels of sequence identity. Table 3.3 shows the result of doing this for the November 2015 holdings. At 30 per cent identity, there are ~24,000 clusters; but even at this level, many of the proteins will still have the same fold, because protein structures are more conserved than their underlying sequences. More rigorous studies, using the fold-classification databases as benchmarks, suggest that the number of unique folds in the PDB is currently somewhere between 1,300 and 1,400 (see Box 3.1 for further details).

Even if we consider the full redundant figure, the number of available structures is very small relative to the number of known protein sequences – or, to put this another way, the sequence–structure deficit is large (see Figure 3.3). This means that, at the time of writing, there is still no 3D information available for most proteins.

Table 3.3

Sequence-based redundancy in the PDB[a], November 2015, calculated using the commonly used pairwise alignment algorithm, **BLAST** (Altschul *et al.*, 1990).

% identity	# of clusters
100	63,995
90	42,812
70	37,915
50	32,610
40	28,889
30	24,601

[a] http://www.rcsb.org/pdb/statistics/clusterStatistics.do

---

[9] http://www.rcsb.org/pdb/statistics/holdings.do
[10] https://commons.wikimedia.org/wiki/File:PSI_chart_numStructures-vs-time.PNG

In practice, then, 3D information for the majority of newly determined sequences has to be inferred by comparison with proteins whose structures have already been determined. For the millions of sequences now pouring from genome-sequencing projects, high-throughput structural initiatives will therefore remain a vital part of the continued efforts to augment the contents of the PDB and provide structural insights on a large scale.

The PDB was created in the same era that interest in collecting protein sequences was growing. Like the sequence repositories, the database evolved long before the advent of high-throughput techniques; its format is therefore a reflection of the technology current at that time. As a result, the last 40 years have borne witness to a patchwork of developments to try to make the data more self-consistent, and to try to make the format sufficiently robust to sustain its usefulness in the 21st century.

3.4   The EMBL nucleotide sequence data library

Despite the advances in protein sequence- and structure-determination technologies between the mid-1940s and -1970s, sequencing nucleic acids had remained problematic. The key issues related to size and ease of molecular purification. It had proved possible to sequence tRNAs, mostly because they're short (typically less than 100 bases long), and hence individual molecules could be purified; but chromosomal DNA molecules are in a completely different ball-park, containing many millions of bases. Even if such molecules could be broken down into smaller chunks, their purification was still a major challenge. The longest fragment that could then be sequenced in a single experiment was ~500bp; and yields of potentially around half a million fragments per **chromosome** were simply beyond the technology of the day to handle.

Progress on these problems came in the mid 1970s, when Sanger developed a technology (which was to become known as the **Sanger method**) that made it possible to work with much longer nucleotide fragments. The new technique allowed completion of the sequencing of the 5,386 bases of the single-stranded **bacteriophage φX174** (Sanger *et al.*, 1978), and subsequently permitted rapid and accurate sequencing of even longer fragments (an achievement of sufficient magnitude, as we saw in Chapter 1, to earn him his second Nobel Prize in Chemistry). With this approach, Sanger went on to sequence human mitochondrial DNA (Anderson *et al.*, 1981) and **bacteriophage λ** (Sanger *et al.*, 1982). Landmark achievements in their own right (see Table 3.4), these projects also provided the first direct evidence of the phenomenon of overlapping gene sequences and of the non-universality of the genetic code, mentioned in Chapter 2 (Sanger, 1988; Dodson, 2005). But it was automation of DNA sequencing techniques from the mid-1980s that catalysed a major leap forward: this not only significantly increased productivity, but also made much larger genomes (especially the human genome) realistic sequencing targets.

These advances prepared the way for a revolution that was to rock the foundations of molecular biology and make the gathered fruits of all sequencing efforts before it appear utterly inconsequential. Together, they created a dramatic turning point: for the first time, it dawned on scientists that the new sequencing machines were shunting the bottlenecks away from data production *per se* and onto the requirements of data management:

> the rate limiting step in the process of nucleic acid sequencing is now shifting from data acquisition towards the organization and analysis of that data (Gingeras and Roberts, 1980).

Table 3.4

Sequencing landmarks.

Year	Protein	RNA	DNA	No. of residues
1935	Insulin			1
1945	Insulin			2
1947	**Gramicidin S**			5
1949	Insulin			9
1955	Insulin			51
1960	Ribonuclease			120
1965		$tRNA_{Ala}$		75
1967		5S rRNA		120
1968			Bacteriophage λ	12
1977			Bacteriophage φX 174	5,375
1978			Bacteriophage φX 174	5,386
1981			Mitochondria	16,569
1982			Bacteriophage λ	48,502
1984			**Epstein–Barr virus**	172,282
2004			*Homo sapiens*	2.85 billion

This realisation had profound consequences across the globe, as it was generally agreed that a centralised data bank was now inescapable as a tool for managing nucleic acid sequence information efficiently.

From this point, the clock was ticking, and the race was on to establish the first public nucleotide sequence repository. First past the post was the **European Molecular Biology Laboratory** (EMBL) in Heidelberg, which set up the EMBL data library. After an initial pilot period, the first release of 568 sequences was made in June 1982. The aim of this resource was not only to make nucleic acid sequence data publicly available, and to encourage standardisation and free exchange of data, but also to provide a European focus for computational and biological data services (Hamm and Cameron, 1986).

Nevertheless, from the outset, it was clear that maintenance of such a centralised database, and of its attendant services, would require international collaboration. In the UK, copies of the EMBL data library were being maintained at Cambridge University, together with the user-manual, indices and associated sequence analysis, search and retrieval software. This integrated system also provided access to the **GenBank** library of sequences that was being developed in parallel at **Los Alamos National Laboratory** (Kanehisa *et al.*, 1984). At this stage, it's fascinating to read that,

> this system is presently being used by over 30 researchers in eight departments in the University and in local research institutes. These users can keep in touch with each other via the MAIL command[!]

Not only do such comments give a flavour of the novelty of electronic facilities like email (now the scourge of modern life!), but they also provide an insight into the limits of the then available computer networks. Today, we take our ease of network communication for granted, and would consider centralised database access for 30 users rather trivial; however, more than a quarter of a century ago, these developments were clearly something to be excited about.

These services were eventually extended beyond the bounds of Cambridge to the wider UK community via **JANET**, the Joint Academic NETwork (Kneale and Kennard, 1984). As with the PDB before it, it was important not only to push the data out to researchers, but also to pull their data in. Hence, one of the planned developments was to centralise collection of nucleic acid data from UK research groups, and to periodically transfer the information to the EMBL data library. It was hoped that this would minimise both data-entry errors and the workload of EMBL staff at a time when the number of sequence determinations was predicted to 'increase greatly' (Kneale and Kennard, 1984). Of course, more than 30 years ago, the size of this 'great increase' could hardly have been predicted: release 125, in September 2015, contained >620 million entries.

3.5 GenBank

Six months after the first release of the EMBL data library came the first public release of GenBank[11], with 606 sequences. Three years earlier, at a scientific meeting at **Rockefeller University** in New York, a consensus had emerged on the necessity of creating an international nucleic acid sequence repository. Several groups expressed an interest in being part of this effort, including those led by Dayhoff at the **National Biomedical Research Foundation** (NBRF); **Walter Goad** at Los Alamos National Laboratories; Doug Brutlag at Stanford; Olga Kennard and Fred Sanger at the **MRC** Laboratory in Cambridge; and Ken Murray and Hans Lehrach at the EMBL (Smith, 1990). At that time, each of these groups had created its own nucleotide sequence collection; all therefore had a vested interest in hosting the proposed international database, and each understood the enormity of the effort that would be involved in scaling their projects up to an international dimension.

Politically, the situation in the USA was complicated. It took the best part of three years for an appropriate database-funding model to emerge from the US **National Institutes of Health** (NIH), by which time the EMBL data library was already up and running under the direction of Greg Hamm. By then, three proposals remained on the table for NIH support: two of these were from Los Alamos (one with **Bolt, Beranek and Newman** (BBN), the other with IntelliGenetics), and the third from NBRF. In the end, remarkably, it was agreed to establish the new GenBank resource at Los Alamos (with BBN, Inc.) rather than at the NBRF (Smith, 1990; Strasser, 2008).

Although there was a general sense of relief that a decision had finally been made, the outcome surprised some members of the community (including, no doubt, Dayhoff herself). After all, the NBRF seemed like a more appropriate home for GenBank, given Dayhoff's successful track record as a curator of protein sequence data (Smith, 1990). More to the point, although Los Alamos undoubtedly offered excellent computer facilities, it was probably best known for its role in the creation of atomic weapons – this was hardly an obvious environment in which to establish the nation's first public

---

[11] http://www.ncbi.nlm.nih.gov/genbank

nucleotide sequence database! The crux of the matter lay in the different philosophical approaches embodied in the NBRF and Los Alamos proposals; in particular, they each espoused very different policies with regard to scientific priority, data sharing/privacy and intellectual property.

Dayhoff had planned to continue gathering sequences directly from literature sources and from bench scientists, and wasn't particularly interested in matters of provenance (Eck and Dayhoff, 1966); the Los Alamos team, by contrast, advocated collaboration with journal editors to oblige authors to deposit their sequences into the database as part of their routine publication process. This latter approach was especially compelling, as it would allow scientists to assert priority, and to keep their research results private until formally published and their provenance established; perhaps more importantly, it was unencumbered by proprietary interest in the data. Unfortunately, Dayhoff had prevented redistribution of NBRF's protein sequence library and, as we will see in the next section, sought revenues from its sales (albeit only to cover costs); this was held against her, because allowing the data to become 'the private hunting grounds' of any one group of researchers was considered contrary to the spirit of open access (Strasser, 2008). At the time, it was considered paramount that the data and associated software tools should be free and open – that the highly secured area of what may have appeared to some as *The Atomic City*[12] should have been chosen as the home for the new international database was therefore somewhat ironic.

As an aside, it's interesting that the vision of free data and software was advocated so strongly at this time, not least because there was no funding model to support it. Thirty years on, the same arguments are being vehemently propounded with regard to free databases, free software and free literature (*e.g.*, Lathrop *et al.*, 2011); but even now, database funding remains an unsolved and controversial issue. As Olga Kennard put it around 20 years ago,

> Free access to validated and enhanced data worldwide is a beautiful dream. The reality, however, is more complex (Kennard, 1997).

Returning to our theme, perhaps the final nail in the coffin of Dayhoff's proposal was that the NBRF had only limited means of data distribution (via **modems**). The Los Alamos outfit, on the other hand, had the enormous benefit of being able to distribute their data via **ARPANET**, the computer network of the **US Department of Defense**. Together, these advantages were sufficient to swing the pendulum in favour of the Los Alamos team, who hence become the official custodians of the new GenBank.

In spite of this major leap forward in creating a home for GenBank, the database didn't (indeed, couldn't) function in isolation. From its inception, it evolved in close collaboration with the EMBL data library and, from 1986 onwards, also with the DNA Data Bank of Japan (DDBJ)[13], a tripartite alliance managed by the International Nucleotide Sequence Database Collaboration (INSDC)[14]. Although the databases were not identical (each had its own format, naming convention, and so on), the teams adopted common data-entry standards and data-exchange protocols in order to improve data

---

[12] http://en.wikipedia.org/wiki/The_Atomic_City
[13] http://www.ddbj.nig.ac.jp
[14] http://www.insdc.org

quality, and to manage the growth and annotation of the data more effectively. Reflecting on this collaborative process in 1990, Temple Smith commented,

> By working out a division of labor with the EMBL and newer Japanese database efforts, and by involving the authors and journal editors, GenBank and the EMBL databases are currently keeping pace with the literature.

Today, the boot is very much on the other foot, as the literature can no longer keep pace with the data: by October 2015, GenBank contained >188 million entries, presenting almost insurmountable management and annotation hurdles (note that this total appears smaller than the size of the EMBL data library because it doesn't conflate sequences from **Whole Genome Shotgun** (WGS) projects). GenBank and EMBL are now so large that they're split into divisions, both to help manage their data and to make searching them more efficient. For example, GenBank has 12 taxonomic divisions (including primate, rodent, invertebrate, plant, bacteria); five high-throughput data divisions (**Expressed Sequence Tags** (ESTs), **Genome Survey Sequences** (GSS), *etc.*); a patent division containing records supplied by patent offices; and a WGS division containing sequences from WGS projects. Overall, the database contains entries from >300,000 species, the bulk of its data deriving from the major model organisms (human, mouse, rat, *etc.*, as shown in Table 3.5). The pace of electronic data submission is such that GenBank

Table 3.5

Number of entries and bases of DNA/RNA for the 20 most sequenced organisms in GenBank Release 210.0 (October 2015), excluding chloroplast, mitochondrial, metagenomic, WGS, 'constructed' CON-division and uncultured sequences.

Species	Entries	Bases
Homo sapiens	22,701,156	17,829,049,608
Mus musculus	9,730,521	10,006,508,725
Rattus norvegicus	2,195,488	6,527,104,159
Bos taurus	2,227,648	5,412,641,459
Zea mays	4,177,686	5,204,532,318
Sus scrofa	3,297,508	4,895,700,592
Hordeum vulgare subsp. vulgare	1,346,739	3,230,965,051
Danio rerio	1,728,327	3,165,947,207
Ovis Canadensis canadensis	62	2,590,569,059
Triticum aestivum	1,808,894	1,938,085,894
Cyprinus carpio	204,950	1,835,949,253
Solanum lycopersicum	744,762	1,744,786,361
Oryza sativa Japonica Group	1,373,106	1,639,633,769
Apteryx australis mantelli	326,939	1,595,384,171
Strongylocentrotus purpuratus	258,062	1,435,471,103
Macaca mulatta	458,084	1,297,938,863
Spirometra erinaceieuropaei	490,068	1,264,189,828
Xenopus (Silurana) tropicalis	1,588,003	1,249,276,612
Arabidopsis thaliana	2,579,492	1,203,018,566
Nicotiana tabacum	1,779,678	1,201,354,674

staff are providing authors with accession numbers at a rate of ~3,500 per day – allowing for sleep, that's more than one accession number every eight seconds (Benson *et al.*, 2015). Like all sequence databases, then, GenBank is growing[15] at an alarming rate.

Perhaps not surprisingly, the initial funding for GenBank was insufficient to adequately maintain this mass of rapidly accumulating data; responsibility for its maintenance, with increased funding under a new contract, therefore passed to IntelliGenetics in 1987; then, in 1992, it became the responsibility of the NCBI, where it remains today (Benson *et al.*, 1993; Benson *et al.*, 2015; Smith, 1990).

3.6 The PIR-PSD

To some extent, the gathering momentum of nucleic acid sequence-collection efforts were beginning to overshadow the steady progress that was being made in the world of protein sequences, most notably with Dayhoff's *Atlas*. By October 1981, the *Atlas* had run into its fifth volume, a large book with three supplements, listing more than 1,600 proteins. As with all data collections, this information required constant updating and revision in light of new knowledge and of new data appearing in the literature – and all of this was becoming increasingly difficult to harvest and maintain by hand.

Perhaps more importantly, the community had become increasingly keen to harness the efficiency gains of central, electronic data repositories. Inevitably, as more databases began appearing on the horizon, making and maintaining cross-references between them had to become part of annotation and data-update processes, to allow scientists to exploit new and existing sequence data to the full. Under the circumstances, continued publication of the *Atlas* in paper form simply became untenable: the time was ripe to exploit the advances in computer technology that had given rise to the CSD, the PDB, the EMBL data library and GenBank. Consequently, in 1984, the *Atlas* was made available for the first time on computer tape as the Protein Sequence Database (PSD)[16].

Later, to facilitate protein sequence analysis more broadly, the NBRF established the **Protein Identification Resource** (PIR) (George *et al.*, 1986). This new online system included the PSD, several bespoke query and analysis tools (*e.g.*, the Protein Sequence Query (PSQ), SEARCH and ALIGN programs), and a new, efficient search program, FASTP. The latter was a modification of an earlier algorithm for searching protein and nucleic acid sequences (Wilbur and Lipman, 1983). Interestingly, given that the number of deduced sequences had, by then, grown into the thousands, the great advantage of Wilbur and Lipman's method was considered to be its speed; their paper reported a, 'substantial reduction in the time required to search a data bank'. Improving on this further, the new FASTP algorithm could compare a 200-amino-acid sequence to the 2,677 sequences of the PSD in 'less than 2 minutes on a minicomputer, and less than 10 minutes on a microcomputer (IBM PC)' (Lipman and Pearson, 1985). Looking back, such search times on such small numbers of sequences seem incredibly slow; at the time (when a contemporary algorithm required eight hours for the same search), they were revolutionary.

The PIR was built on NBRF's existing resources. This means that, in addition to the PSD, it also made available its DNA databank (Dayhoff *et al.*, 1981a) and associated software tools, together with copies of GenBank and the EMBL data library. In an

---

[15] ftp://ftp.ncbi.nih.gov/genbank/gbrel.txt
[16] http://pir.georgetown.edu/pirwww/dbinfo/pir_psd.shtml

attempt to cover the cost of providing such comprehensive services, the PIR exploited the NBRF's cost-recovery model, levying a charge for distributing copies of its databases on magnetic tape and an annual subscription fee for use of its online services: in 1988, these amounted to $200 per tape release and $350 per annum respectively (Dayhoff *et al.*, 1981b; Sidman *et al.*, 1988) – that would be something like $400 and $700 today.

By 1992, the **PIR-PSD** had shown steady growth, with increasing contributions from European and Asian protein sequence centres, most notably from the **Munich Information Center for Protein Sequences** (MIPS) and from the **Japan International Protein Information Database** (JIPID). To reflect this situation, a tripartite collaboration was established, termed PIR-International, to formalise these relationships and to establish, maintain and disseminate a comprehensive set of protein sequences (Barker *et al.*, 1992). By this time, charging for access to the resource was no longer mentioned, possibly because of this more formal distribution arrangement. Doubtless, however, the recent advent of browsers like **Mosaic** also had a significant impact, suddenly and dramatically changing the way in which information could be broadcast and received over the newly established Web, literally at the click of a mouse button. In 1997, PIR changed its name to the Protein Information Resource[17] (George *et al.*, 1997) and, by 2003, with 283,000 sequences (Wu *et al.*, 2003), the PSD was the most comprehensive protein sequence database in the world.

3.7 Swiss-Prot

With these events unfolding in the background, a newly qualified Swiss student (who, as a teenager, had had an enthusiasm for space exploration and the search for extra-terrestrial life) signed up to undertake a Masters project – this was Amos Bairoch[18]. Bairoch had specifically chosen a **mass spectrometry** project with both 'wet' and 'dry' components, but immediately hit problems when he discovered that the new mass spectrometer that was to have been used for the experimental side of his project didn't work properly. Not to lose time, he set to work instead developing protein sequence-analysis programs on the spectrometer's computer system. These were the first steps towards creating the software suite that was later to be known as PC/Gene, and was to become the most widely used PC-based sequence analysis package of its day (Bairoch, 2000).

Part of the uniqueness of this software was its focus on proteins at a time when the analysis of nucleotide sequences was much more in vogue. In the process of writing the software, Bairoch had to manually enter >1,000 protein sequences into his computer: some of these he gleaned from the literature; most were taken from the *Atlas*, which had not then been released in electronic form. This task was clearly both immensely tedious and highly error-prone. Realising this, and anxious to help alleviate the burdens for other scientists in future, he wrote a letter to the *Biochemical Journal* recommending that researchers publishing protein and peptide sequences should compute **checksums** to 'facilitate the detection of typographical and keyboard errors' (Bairoch, 1982). Within the letter, Bairoch explained how such a 'checking number' might be computed for an imaginary peptide, as illustrated in Figure 3.2. This recommendation was never widely adopted by publishers, but he was at least able to ensure that it was implemented in the database that he went on to create.

---

[17] http://pir.georgetown.edu
[18] http://en.wikipedia.org/wiki/Amos_Bairoch

Peptide: H E L P I H A T E M A T H

CN **computation:** $CN = 1\cdot9 + 2\cdot7 + 3\cdot11 + 4\cdot15 + 5\cdot10 + 6\cdot9 + 7\cdot1 + 8\cdot17 + 9\cdot7 + 10\cdot13 + 11\cdot1 + 12\cdot17 + 13\cdot9 = 788$

$COMP = A_2R_0N_0D_0C_0Q_0E_2G_0H_3I_1L_1K_0M_1F_0P_1S_0T_2W_0Y_0V_0$

$NR = 13 \qquad MMP = 1186.66 \qquad CN = 788$

Figure 3.2

Computation of a 'checking number' (CN) for an imaginary peptide, published by Bairoch in a letter to the *Biochemical Journal* in 1982. The journal editors either didn't notice, or perhaps chose to ignore, the hidden message in the peptide.

Source: Bairoch (1982). Reproduced with permission from Portland Press Ltd.

Several other major developments were to emerge from the work of this industrious student, essentially as by-products of his PC/Gene software package. To maximise the usefulness of the software, he needed to bundle with it both a nucleotide- and a protein sequence database. In 1983, he acquired a computer tape containing 811 sequences in version 2 of the EMBL data library; for his protein database, he initially used sequences he'd typed in for his Masters project. However, the following year, he received the first electronic copy of the *Atlas*, and was quick to appreciate the relative merits of the different database formats. The main strength of the PIR-PSD was its protein annotations, while the strength of the EMBL data library was its semi-structured format. Coupling the two, he realised, would give him access to a unique and powerful resource. His innovation was thus to convert the PSD's manually annotated data into something like EMBL's semi-structured format, producing a resource with the strengths of both – he called this PIR+, and released it side-by-side with PC/Gene, which by that time he'd commercialised through IntelliGenetics (Bairoch, 2000).

Although many researchers welcomed these advances, use of the public PIR data-set in this way caused problems. Setting aside commercial and proprietary issues, there were other hurdles: for example, it was difficult to parse PIR-PSD files to extract specific information (*e.g.*, relating to **post-translational modifications** (PTMs), *etc.*); some of the newer database entries lacked functional annotations; protein sequences lacked cross-references to their parent DNA; and so on. A little ironically, in light of what he went on to achieve, Bairoch has written of this period,

> As I was not interested in building up databases I kept sending letters to PIR to ask them to remedy this situation.

Nevertheless, his pleas went unheeded. Faced with increasing demands for easier access to PIR+, decoupled from PC/Gene, he decided to make the database freely available to the entire research community. The new, public version of the resource was released on 21 July 1986 and contained around 3,900 sequences (the precise number is unknown as Bairoch lost the original floppy disks!) This was **Swiss-Prot** (Bairoch and Boeckmann, 1991), which was to become the world's pre-eminent manually annotated protein-sequence database.

Since its inception, the guiding philosophy of Swiss-Prot has been to provide high-level annotations describing, amongst other things, protein functions, their domain

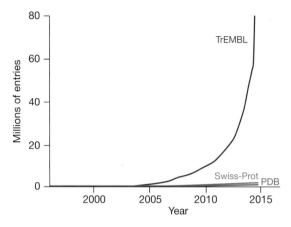

Figure 3.3

Illustration of the growth of TrEMBL[i] (purple line), Swiss-Prot[ii] (red line) and PDB (green line). The gulf between manual-annotation and computer-annotation strategies is clear, as is the size of the current sequence–structure deficit.

[i]http://www.ebi.ac.uk/uniprot/TrEMBLstats
[ii]http://web.expasy.org/docs/relnotes/relstat.html

organisms, their family and disease relationships, their sequence variants, structural information, relevant supporting literature, and so on. The extent and quality of this information set Swiss-Prot apart from other sequence databases and ensured its popularity, but there was a price to pay: providing such comprehensive annotation for the ever growing volume of sequences was demanding the input of more and more curators. Yet it would take an entire army to keep pace with the rate of data acquisition from the sequencing centres. The unfortunate truth is, manual annotation processes simply don't scale, as illustrated in Figure 3.3. By October 2015, Swiss-Prot contained 549,832 entries (derived from >13,000 species, mostly from model organisms – human, mouse, *Arabidopsis*, rat, yeast, *etc.*). This figure is an order of magnitude smaller than the holdings of its companion database, **TrEMBL**, as we shall see in Section 3.9.

3.8 PROSITE

An almost inevitable consequence of the systematic collection of protein and nucleotide sequences into databases was the desire to organise and classify these molecular entities in meaningful ways. The first attempt to categorise protein sequences into evolutionarily related families, and to provide diagnostic tools to help detect potential new family members, again arose as a by-product of Bairoch's PC/Gene software package (more detailed information about protein families and their broad diagnostic approaches is given in Box 3.2).

Bairoch had read **Russell Doolittle**'s newly published sequence analysis primer, *Of URFs and ORFs* (Doolittle, 1986). Inspired, he began creating a compendium of short sequences (so-called **consensus expressions** or **patterns** – see Box 3.3) that were, according to literature and other sources, known to be characteristic of particular protein families or of specific binding and/or active sites. At the same time, he developed a program

[19] http://prosite.expasy.org

Box 3.2 Protein family signatures

Many proteins show evidence of evolutionary kinship; those with highly similar sequences are likely to have descended from a common ancestor. We can use these similarities to group sequences into families: this is useful, because members of such families often (not always) have the same or related functions.

Classifying sequences into families isn't just a stamp-collecting exercise. An enormous challenge facing bioinformatics today is to provide at least some level of annotation (including, especially, evidence of their potential functions) to the millions of sequences billowing from world-wide sequencing projects into our databases. The idea is that if we can distil the essence of particular protein families into 'signatures' of some sort, then these could provide the means to diagnose the family membership, and likely functional attributes, of hundreds, thousands or even millions of uncharacterised sequences.

Over the years, efforts to create such **diagnostic signatures** have produced several different databases. The underlying methodologies are different, but the central principle is the same. Essentially, when related proteins are compared by aligning their sequences, blocks of conservation become visible, separated by non-conserved regions. The conserved blocks, usually referred to as **motifs**, can be thought of as islands of evolutionary stability in a sea of mutational change. In Figure 3.4, the sequences being compared (aligned) are denoted by horizontal bars. The sequences have different lengths – for the shorter sequences, this necessitates the addition of **gaps** to allow downstream regions to be brought into the correct register, which is why some of the bars are broken. Figure 3.4 shows three islands of conservation, delineating three motifs (the yellow boxes), emerging in this way.

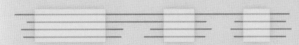

Figure 3.4
Three motifs in a sequence alignment.

Here, the motifs are clear because the aligned sequences are very similar. For more dissimilar or divergent sequences, alignment is more difficult, the regions of conservation shrink, and it becomes harder to spot motifs, as shown in Figure 3.5.

Figure 3.5
Alignment of increasingly divergent sequences.

In this example, an initial comparison of two highly similar sequences highlights two well-defined conserved regions (the green blocks). As more dissimilar sequences are included in the alignment, however, the patterns of conservation change, and grow gradually smaller, until eventually only four tiny motifs remain (the small pink blocks).

Motifs like this, which lie in different parts of the alignment, may nevertheless come close to each other in 3D space as the sequence adopts its tertiary structure. Typically, such motifs include key amino acids at particular positions, reflecting some vital structural or functional role that has been preserved through the ravages of evolution (a catalytic or binding site, for example, or some kind of folding knot). As a result, they can provide powerful diagnostic signatures that can, in principle, allow us to make highly specific functional and structural inferences, say, for uncharacterised sequences found to share the same motifs.

There are several ways of encoding the conserved parts of alignments, which have different diagnostic consequences. The three main approaches exploit single motifs (the method originally used in PROSITE), multiple motifs (as used in the PRINTS and Blocks databases) or complete domains (used in the Profiles and Pfam databases). More details of these methods are given in Chapter 4.

to scan this collection of sequence patterns, so that users could test or diagnose the presence of such characteristic sites in their own query sequences. This part-program, part-database chimera he called **PROSITE**[19] (Bairoch, 1991). Its first release, in March 1988, as part of the PC/Gene suite, contained 58 entries.

As with Swiss-Prot before it, PROSITE swiftly became popular. In the beginning, the database was very small, which obviously restricted its general utility. To expand its coverage, users therefore began to suggest additional patterns that could be included in the database. Perhaps more importantly, its growing user community urged Bairoch to liberate PROSITE from the commercial shackles of PC/Gene. He bowed to this pressure the following year, announcing a new public version in October 1989, and making a formal release the following month with 202 entries (version 4.0).

Box 3.3 Consensus sequences and sequence patterns

When related sequences are aligned, regions that have survived the mutational pressures of evolution often stand out as more highly conserved than others, because they share higher levels of similarity than surrounding parts of the alignment. Such regions include key conserved residues at particular aligned positions, and these can serve as diagnostic landmarks against which other sequences can be compared.

A simple way to visualise regions of conservation is to read sequences into an alignment editor – an editor is a software tool that allows sequences to be aligned automatically and the resulting alignment to be refined manually. Often, editors will display some sort of **consensus sequence**, providing a rapid visual overview of residues that are highly or completely conserved, as shown in Figure 3.6.

OPRD_HUMAN	A L Y S A V C A V G L L G N V L V M F G I V R Y T K M
OPRD_RAT	A L Y S A V C A V G L L G N V L V M F G I V R Y T K L
OPRD_MOUSE	A L Y S A V C A V G L L G N V L V M F G I V R Y T K L
OPRK_HUMAN	A V Y S V V F V V G L V G N S L V M F V I I R Y T K M
OPRK_RAT	A V Y S V V F V V G L V G N S L V M F V I I R Y T K M
OPRK_MOUSE	A V Y S V V F V V G L V G N S L V M F V I I R Y T K M
OPRM_HUMAN	A L Y S I V C V V G L F G N F L V M Y V I V R Y T K M
OPRM_RAT	A L Y S I V C V V G L F G N F L V M Y V I V R Y T K M
OPRM_MOUSE	A L Y S I V C V V G L F G N F L V M Y V I V R Y T K M
Consensus	A Y S V V G L G N L V M I R Y T K

Figure 3.6
An alignment with its consensus sequence.

Figure 3.6 shows part of an alignment of nine sequences (human, rat and mouse δ, μ and κ opioid receptors), in which 17 of the 27 residues are completely conserved, as denoted by the amino acid single-letter codes in the Consensus row beneath the alignment. Where alignment positions contain several different residues, a darker shade is used: the darker the shade, the greater the number of different residues at that position. Here, the largest number of different residues at a position is 3; hence, positions 5, 12 and 15 have the darkest colour.

In this alignment, residues are coloured according to the physicochemical properties described in Section 2.2.3. The uniformity of colour in some columns shows that, even if a particular residue isn't conserved, its properties have been preserved at this position. Consider column 5: the consensus indicates this to be one of the most variable positions; nevertheless, the colour scheme suggests that all of the residues at this position share the same, or similar, physicochemical properties – all are hydrophobic. This is important, because it allows us to see additional information that's lost in the simple consensus. One approach to capture this kind of information more explicitly is to list the types of residue that occur at each alignment position, using a kind of **shorthand notation**[a]. For Figure 3.6, focusing on the first six residues, and using the PROSITE syntax, gives the following:

A-[LV]-Y-S-[AVI]-V

Here, in addition to the four residues that are completely conserved, some residue groups appear in square brackets – this notation is used to denote residues that broadly share similar properties. For example, in the second alignment column, leucine and valine occur: according to Figure 2.10 (Chapter 2), these occupy the same region of the Venn diagram, and hence we can group them together. Similarly, in column 5, all three residues sit in the same region of the Venn diagram and can hence be grouped.

Such **consensus expressions** are often referred to as **patterns**, because they neatly summarise patterns of residue conservation within alignments. Being simple to create, patterns like this were used to encapsulate conserved residues of protein active sites, binding sites, *etc.*, and this collection provided the foundations of the first protein family database, PROSITE. This was an important development because, for the first time, a collection of expressions was available to the community for users to search with their own query sequences. Being able to characterise newly determined sequences using tools like this has been one of the major goals of bioinformatics since the first sequence databases were created. The appeal of this particular method was, and still is, its simplicity; its simplicity is also its drawback and led to the development of other, more sophisticated, approaches. These issues are explored in Chapter 4.

---

[a] http://prosite.expasy.org/scanprosite/scanprosite_doc.html

Diagnostically, sequence patterns suffer certain limitations. Specifically, matching a pattern is a binary 'match/no-match' event: hence, even the most trivial difference (a single amino acid) results in a mis-match. As Swiss-Prot grew and accommodated more and more divergent sequences, the more evident and problematic this weakness became. Indeed, it generated a kind of circularity in which the more diverse the sequences in Swiss-Prot became, the more likely particular PROSITE patterns were to fail to diagnose or characterise them correctly, and the more work curators had to do to upgrade or revise the sequence patterns in order to better reflect the changing contents of Swiss-Prot. This intimate dependence on both the size and composition of Swiss-Prot was to become one of the toughest challenges for PROSITE curators (and indeed for the curators of all such protein family-based databases in future).

Many researchers were aware of the diagnostic limitations of sequence patterns, and worked to develop more powerful 'descriptors' of protein families and functional sites. One approach was to create so-called **position-specific weight matrices**, or **profiles** (see Box 3.2). These are built from comprehensive sequence alignments, and are much more tolerant of amino acid changes or substitutions than are patterns; they're also more flexible in how they handle differences in the lengths of sequences that arise from insertions or deletions. In principle, therefore, profiles allow the relationships between families of sequences to be modelled more 'realistically'. Philipp Bucher was one of the first to research and develop profile-based protein sequence-analysis methods, and he therefore began work with Bairoch to augment PROSITE with a range of these new sequence profiles – the first release to include them was version 12.0, in June 1994 (Bairoch and Bucher, 1994).

A rather different approach, which evolved independently from PROSITE (at least in terms of the underlying analysis method), was that of protein family 'fingerprinting'. By contrast with sequence patterns, which focus on unique sites or **motifs** that are found in multiple sequence alignments, **fingerprints** are created from *groups* of conserved motifs. It is the unique inter-relationships between these motifs that provide distinctive signatures for particular protein families or structural/functional domains (see Box 3.4). The fingerprint method was originally devised to analyse and characterise a superfamily of proteins known as **G protein-coupled receptors** (GPCRs), whose distinguishing trait is the presence of seven well-conserved **transmembrane (TM) domains** – the first fingerprint therefore encoded these seven domains as sequence motifs; but the method had more general utility, and was hence used to characterise several other families. The results were used to populate a database that was initially known as the Features Database, a component of the SERPENT information storage and analysis resource for protein sequences (Akrigg *et al.*, 1992). The Features Database (later to be re-born as PRINTS[20]) saw its first release in October 1991, with 29 entries. Most of these had equivalent entries in PROSITE, which by then held a much more substantial 441 family descriptions.

Although disparate in size, the Features and PROSITE databases shared certain attributes. The most important of these was the principle of adding value to the database contents by manually annotating them. In 1992, given the similarities between the databases, at least in terms of the manual-annotation burdens they incurred, Bairoch and Attwood discussed the possibility of unifying the PROSITE and Features databases into the world's first integrated, annotated protein family resource. Together with colleagues at the University of Leeds, they submitted a bid for European funding; however, the proposal was not successful, and they had to wait a further seven years before their goal was finally realised.

In the meantime, inspired by the value of PROSITE, on the one hand, and by its diagnostic limitations, on the other, a range of other signature databases began to emerge. One of the earliest of these, Blocks[21], was developed by Steve and Jorja Henikoff (Henikoff and Henikoff, 1991). The Blocks approach was very similar to fingerprinting in so far as it used multiple motifs; however, the method used a different scoring system and didn't exploit the relationships between motifs, and hence its diagnostic behaviour was rather different. A few years later came ProDom[22] (Sonnhammer and Kahn, 1994), which adopted a new approach, essentially using a clustering algorithm to sort Swiss-Prot contents into apparent 'families'; later still came Pfam[23] (Sonnhammer *et al.*, 1997), which used a probabilistic method very similar to profiles. Initially linked closely to the annotation of predicted proteins from genomic sequencing of *Caenorhabditis elegans*, Pfam was to become one of the most widely used protein family databases in Europe and the USA.

---

[20] http://www.bioinf.manchester.ac.uk/dbbrowser/PRINTS
[21] http://blocks.fhcrc.org
[22] http://prodom.prabi.fr/prodom/current/html/home.php
[23] http://pfam.xfam.org

Box 3.4 Fingerprints

The concept of a protein family fingerprint is illustrated below. The main feature that distinguishes a fingerprint from a consensus pattern (Box 3.3) is that it draws its diagnostic power from groups of conserved regions, rather than from individual motifs. The potency of fingerprints derives from the fact that not only do they use multiple motifs, but they also inherently encapsulate information relating to the number of motifs they contain, to their order in the alignment, and to the distance or interval between them (as shown in Figure 3.7); the resulting discriminator is thus a powerful signature for the given protein family.

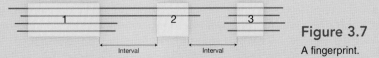

Figure 3.7
A fingerprint.

Of course, different families are characterised by different patterns of conservation, and their fingerprints differ accordingly. Figure 3.8 shows a schematic illustration of fingerprints for three different families: the first two fingerprints each contain four motifs, but their sizes and the distances between them are different; the third fingerprint contains five motifs, again with different sizes and inter-motif distances.

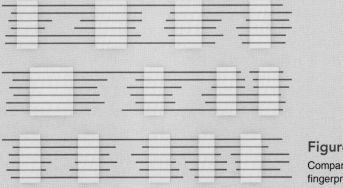

Figure 3.8
Comparison of three different fingerprints.

An especially important feature of protein fingerprints is that a query sequence need not match all of its constituent motifs equally well, nor need it match *all* of the motifs to be able to diagnose it as a potential family member – in other words, fingerprints can tolerate mismatches both within individual motifs and within the fingerprint as a whole.

In principle, any protein family that's sufficiently well conserved can be uniquely fingerprinted. The flexibility of the approach, with regard to how well particular motifs match the fingerprint, and how many motifs match, lends great analytical potential, and hence offers powerful diagnostic opportunities beyond those afforded by consensus patterns.

3.9 TrEMBL

By 1996, the first quakes that followed the explosion of whole-genome sequencing projects were being felt. The impact was greatest for databases whose maintenance involved significant amounts of manual annotation; some were not able to adapt in time; many did not survive. Swiss-Prot was a survivor of the onslaught, but this required both radically new philosophical approaches to be embraced and new processes to be adopted.

At this time, Swiss-Prot had the highest standard of annotation of any publicly available protein sequence database: from the outset, one of its leading goals had been to provide critical analyses and annotation for all of its constituent sequences. To this end, each entry was accompanied by a significant amount of manually gleaned information: this was derived primarily from original publications and review articles by an expanding group of curators, sometimes with input from an international panel of experts. But this high degree of meticulous manual annotation was creating a bigger and bigger bottleneck, inhibiting the release rate of the resource. Eventually, faced with the relentless data flow from the growing number of genome projects, this hugely labour-intensive process became untenable.

To keep pace with the flood of data, a new approach was needed. The imperative was to release the products of genomic sequences into the public domain as swiftly as possible; the problem was how this could be done without either compromising the high quality of the existing Swiss-Prot data, or eroding the editorial standards of the database. The solution was to create a computer-generated supplement, with a Swiss-Prot-like format, in which entries were derived by translation of coding sequences in the EMBL data library – this was TrEMBL, first released in October 1996 (Bairoch and Apweiler, 1996). TrEMBL 1.0 contained almost 105,000 entries, not far off twice the size of Swiss-Prot 34.0, with which it was released in parallel.

Initially, TrEMBL contained no annotations – it was essentially just the result of an automatic pipeline that extracted coding sequences from the EMBL database and reformatted them. The idea behind its creation was really twofold. In addition to giving Swiss-Prot users access to the protein products of genomic data more quickly, the new resource was intended to act as a corral in which to hold 'raw' sequences prior to their annotation by curators, in readiness for accession to Swiss-Prot. As such, it had two main sections: SP-TrEMBL, containing entries that would eventually go into Swiss-Prot, once suitably annotated; and REM-TrEMBL, containing sequences that would never be used in Swiss-Prot (fragments of less than eight residues, synthetic sequences, *etc.*).

The raw state of TrEMBL entries meant that a lot of work was necessary to upgrade them to the necessary Swiss-Prot standard. To try to accelerate this process, various tools and protocols were therefore developed over the years to try to provide some level of annotation automatically: *e.g.*, including information about the potential functions of proteins, their subcellular locations, the metabolic pathways in which they're involved, active sites they may contain, cofactors, binding sites, domains they may possess, and so on. Initially, binding-site, domain- and family-based information was derived from searches of databases like PROSITE, PRINTS and Pfam; later, information from the amalgamated protein family resource, **InterPro**, was also used. By February 2015, with >90 million entries, TrEMBL[24] was 163 times larger than Swiss-Prot, illustrating the vast gulf that has opened between manual and computer-assisted annotation strategies.

The scale of the problem is illustrated in Figure 3.3, which contrasts the growth of Swiss-Prot and TrEMBL during the last 20 years or so, up to February 2015, and

---

[24]Redundancy was becoming increasingly problematic – UniProtKB doubled in size in 2014 alone – as submissions of multiple genomes for the same or similar organisms (especially for strains of bacterial species) poured into the resource. The high levels of redundancy were skewing the database contents, and increasing the amount of data to be processed. Consequently, from April 2015, redundant proteomes within species groups were removed, and redundant sequences stored in UniParc, resulting in a dramatic decrease in the size of the resource: in November 2015, UniProtKB reported ~54.5 million entries, a drop of something like 40 million sequences (http://www.ebi.ac.uk/uniprot/TrEMBLstats)!

effectively demonstrates why computer annotation has become vital. It should be clear from this figure why some manually annotated databases were unable to survive the genome-sequencing revolution. Incidentally, the figure also shows the growth of the PDB, barely discernible above the base-line, illustrating the widening gulf between the volume of sequence data and the number of proteins for which 3D structural information is now available. The impact of whole-genome sequencing has been profound; but the next chapter in the history of sequencing will make these vast quantities of data seem utterly trivial, as we will see in Section 3.12.

3.10 InterPro

The development of TrEMBL was spearheaded at the EBI by **Rolf Apweiler** in collaboration with Bairoch at the Swiss Institute of Bioinformatics (SIB). In 1997, Apweiler, Bairoch and Attwood again discussed the feasibility of uniting PROSITE and PRINTS. This time, however, the plans were much more ambitious: the goal was to create an instrument to analyse and functionally annotate the growing numbers of uncharacterised genomic sequences, essentially by developing a tool to help annotate TrEMBL. The case for this proposal was much stronger, especially as there was now also ProDom to bring into the picture, and the first release of Pfam had just been announced. A new proposal was therefore submitted to the **European Commission**, and the vision of an integrated protein family database was finally funded.

A beta release of the unified resource was made in October 1999, characterising 2,423 families and domains in Swiss-Prot 38.0 and TrEMBL 11.0 – this was InterPro[25] (Apweiler *et al.*, 2001). By that time, PROSITE and the Features Database had undergone major changes: PROSITE had grown threefold to 1,370 entries, while the Features Database had grown 40-fold to 1,157 entries, and had been re-named 'PRINTS' (Attwood *et al.*, 1994). InterPro's first release thus combined the contents of PROSITE 16.0 and PRINTS 23.1; it also included 241 descriptors from the Profiles database and 1,465 entries from Pfam 4.0.

Although ProDom was part of the original InterPro consortium (as shown in Figure 3.9), it wasn't included in the first release, initially because there was no obvious way of doing so. ProDom is built automatically by clustering Swiss-Prot. As such, it isn't a true signature database, because it doesn't involve the creation and refinement of diagnostic discriminators; moreover, its sequence clusters need not have precise biological correlations, but can change between database releases. This is because, as successive Swiss-Prot releases expand, the composition of the database fluctuates, for example because one release may include more data from a particular sequencing project, or projects, relative to another – such compositional fluctuations necessarily yield different results when the contents of particular releases are clustered. Assigning stable **accession numbers** to ProDom entries was therefore impossible, and this issue had to be addressed before it could be meaningfully included in InterPro.

Other factors rendered a step-wise approach to the development of InterPro desirable. The scale of amalgamating just PROSITE, PRINTS and Pfam was immense. Trying to sensibly merge what appeared to be equivalent database entries that, in fact, defined specific families, or domains within those families, or even repeats within those domains, presented enormous challenges. In the beginning, InterPro therefore focused

---

[25] http://www.ebi.ac.uk/interpro

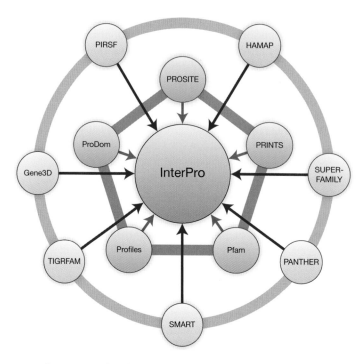

Figure 3.9

Schematic illustration of the relationship between InterPro, the integrating hub (blue), its founding databases (green) and its later partners (yellow), all of which contribute diagnostic signatures and, in some cases, protein family annotation. The arrows and connectors indicate that information is shared between satellite databases, and between satellites and the central hub. See Table 3.6 for more details.

on amalgamating databases that offered some level of hand-crafted annotation, because it was reasoned that this would facilitate the integration process.

Over the years, further partners joined the InterPro consortium (see Figure 3.9). At the time of writing, with 12 sources, the integration challenges are legion, and give InterPro a degree a complexity that wasn't envisaged 20 years ago, when combining PROSITE and PRINTS was first discussed. A flavour of this complexity can be discerned from Table 3.6: this lists the versions of the partner databases contributing to the October 2015 release, their sizes and the numbers of their signatures included in the resource.

Table 3.6

Composition of InterPro release 54.0, October 2015

Signature Database	Version	Signatures	Integrated Signatures
GENE3D	3.5.0	2,626	1,716
HAMAP	201502.04	2,001	1,994
PANTHER	10.0	95,118	4,626
PIRSF	3.01	3,285	3,224
PRINTS	42.0	2,106	2,007
PROSITE patterns	20.113	1,308	1,289
PROSITE profiles	20.113	1,112	1,085
Pfam	28.0	16,230	15,374
ProDom	2006.1	1,894	1,125
SMART	6.2	1,008	996
SUPERFAMILY	1.75	2,019	1,381
TIGRFAMs	15.0	4,488	4,461

For most partners, the bulk of their contents are incorporated; however, in some cases, errors or discrepancies found in or between the data from the different partners have caused integration of some of their signatures to be deferred. Other difficulties include the relative complexity of the data sources (*e.g.*, reconciling sequence-based families with the structure-based families in Gene3D), and the relative volumes of their data (such as the incredibly large number of entries in PANTHER)[26]. Issues like this make efficient processing extremely difficult, especially as InterPro includes a significant amount of manual annotation. In consequence, note the discrepancies in the figures shown in columns 3 and 4 of Table 3.6. Nevertheless, despite the challenges, with ~28,500 entries in release 54.0, InterPro is the most comprehensive integrated protein family database in the world (Hunter *et al.*, 2009; Mitchell *et al.*, 2015).

3.11 UniProt

The year 2004 marked a turning point for the way in which protein-sequence data were to be collected and disseminated globally. As we saw in Section 3.6, the PIR-PSD had been available online since 1986, and Swiss-Prot became available in the same year. In the years that followed, problems arose concerning how to maintain such similar, but disparate and independent resources when, world-wide, funds for database maintenance were dwindling. The situation was exacerbated by the appearance of TrEMBL in 1996. For PIR, some of the difficulties were mitigated, at least in the early years, by charging for copies of their databases and for online access to their software; later, this funding model changed, as the international collaboration with MIPS and JIPID, supported by **National Science Foundation** (NSF) and European grants, helped to sustain the resource.

Meanwhile, despite its huge popularity, Swiss-Prot was having a rocky ride and had had to be rescued from the brink of closure. The database had fallen foul of a 'Catch-22' scenario: the Swiss government had declined to provide further financial support unless the database also gained additional or matching funds from a **European Union** (EU) grant, because they now viewed Swiss-Prot as an international resource; however, a joint proposal with the EBI for an EU infrastructure grant had been declined because Swiss-Prot was not being supported by the Swiss government! In May 1996, with only two months of salary remaining for the Swiss-Prot entourage, Bairoch launched an Internet appeal in which he announced that Swiss-Prot and its associated databases and software tools were to be terminated owing to lack of funds. His appeal stimulated a storm of protest of such force (not just on the Internet, but also in high-profile academic journals and the media) that the Swiss government had to reconsider: their compromise was to offer interim funding until the end of the year. In the negotiations that followed, it was clear that a more stable vehicle for long-term funding of Swiss-Prot was needed; in the ensuing debate, draft plans emerged to establish what was to become the **Swiss Institute of Bioinformatics** (Bairoch, 2000).

Eventually, the continued difficulties of supporting such pivotal protein sequence databases in Europe and the USA culminated in a joint bid for multinational funding: from the NIH, the NSF, the Swiss federal government and the EU. Thus, in 2002, Swiss-Prot, TrEMBL and the PIR-PSD joined forces as the **UniProt consortium**. Aside

---

[26] http://www.pantherdb.org

from ameliorating their combined funding problems, the idea of forming a Universal Protein Resource (UniProt) was to build on the partners' many years of foundational work, to streamline their efforts by providing a stable, high-quality, unified database. This would serve as the world's most comprehensive protein sequence **knowledgebase**, the ultimate protein catalogue, replete with accurate annotations and extensive cross-references, and accompanied by freely-available, easy-to-use query interfaces.

In fact, UniProt[27] provides an umbrella for several different resources. Initially, it housed three separate database layers, tailored for specific uses (Apweiler *et al.*, 2004):

1   the UniProt Archive (**UniParc**)[28], which aims to provide a complete, non-redundant repository of all publicly available protein sequence data, accurately reflecting the histories of all sequences in the resource;

2   the UniProt Knowledgebase (**UniProtKB**)[29], which consists of Swiss-Prot and TrEMBL, acting as the central access point for protein sequences, with accurate, consistent and rich annotation (relating to function, family classifications, disease associations, sequence features, *etc.*); and

3   the UniProt Reference Clusters, or NREF databases (**UniRef**)[30], which provide non-redundant subsets of UniProtKB, for efficient searching.

From a user perspective, despite including three different primary sequence sources, UniProtKB appears to reflect the contents only of Swiss-Prot and TrEMBL. There are probably several reasons for this. As we have seen, Swiss-Prot had become the *de facto* gold standard protein sequence database, because of the quality of its manually-derived annotations; but it also had its origins in PIR-PSD data, and was largely redundant with it. Consequently, the redundant PIR-PSD sequences were retired, although they can still be accessed via UniParc. Some of the issues involved in amalgamating databases from different sources to produce non-redundant composites are outlined in Box 3.5.

By 2011, UniProt also included a Metagenomic and Environmental Sequence component, termed **UniMES** (The UniProt Consortium, 2011), and offered UniMES clusters to provide comprehensive coverage of sequence space at different resolutions. By February 2015, UniProt's combined UniProtKB/Swiss-Prot and UniProtKB/TrEMBL layer contained >91 million entries (although this number had dropped to ~55 million entries by November 2015, following the implementation of stricter redundancy criteria earlier that year). The collaborative effort embodied in UniProt, allowing several of the world's major databases to be united, marked a significant step forward in standardising protein sequence data, and made access to it, and use of it, much more straightforward for users across the globe. Nevertheless, the amassing quantities of protein sequence data, and the management, analysis and annotation headaches this was creating, were as nothing compared to the revolution waiting in the wings. An unprecedented torrent of nucleotide sequence data was beginning to be unleashed by a new breed of sequencing technology that heralded the world-wide initiation of what are now known as **Next-Generation Sequencing** (NGS) projects.

---

[27] http://www.ebi.ac.uk/uniprot
[28] http://www.uniprot.org/uniparc
[29] http://www.uniprot.org/uniprot
[30] http://www.uniprot.org/uniref

Box 3.5 Composite protein sequence databases

Although choice is usually a good thing, the abundance of databases can also complicate matters. Faced with a range of resources that contain similar information and do similar things, a user will want to know which is the most comprehensive or the most up-to-date – in short, which is the best to use.

One solution to the proliferation of databases is to amalgamate related resources into a composite, preferably removing some of the redundancy in order to make data storage and searching more efficient. Over the years, several researchers saw the advantages of amalgamating similar databases, which in turn gave rise to several composites, each with different format and redundancy criteria. Today, the main composite protein sequence resources are the NCBI's non-redundant database, **NRDB**, and the EBI's UniProtKB; their data sources are listed in the Table 3.7.

Table 3.7

Comparison of the contents of the NRDB and UniProtKB databases.

NRDB	UniProtKB
GenBank CDS translations	Swiss-Prot
PDB	TrEMBL
Swiss-Prot	PIR-PSD
PIR-PSD	
PRF	

The availability of different composites begs the same questions: which is best, which is the most comprehensive, which the most up-to-date, which should be used? In practice, the choice is determined largely by the task at hand, and whether we care, say, more about performing the most comprehensive search, or whether we're more interested in performing the most rapid search.

Although **composite databases** offer various advantages, such as efficiency of searching, more intelligible search outputs (uncomplicated by multiple matches to the same sequence) and so on, they also have pitfalls. Much depends on how a composite is made, the level of redundancy it contains and its error rate. A database built by eliminating identical sequences within and between its sources will guarantee that copies of the same sequence aren't included (and hence interrogated or searched twice when users query the resource). This scenario is shown in Figure 3.10 for two source databases (left-hand side and centre). The resulting non-identical composite (right) is smaller than the sum of the contents of its sources, because it has eliminated duplicate sequences (those denoted by the black, white and primary colours).

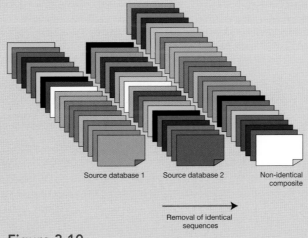

Source database 1   Source database 2   Non-identical composite

Removal of identical sequences

Figure 3.10

Building a non-identical database.

(continued)

If, in addition, entries are eliminated on the basis of some defined level of sequence similarity, the level of internal redundancy can be reduced still further, rendering searches much more efficient. This situation is illustrated in Figure 3.11, again for two source databases (on the left-hand side and centre). The resulting non-redundant composite (right) is significantly smaller than the sum of the contents of its sources, because it has eliminated both duplicate sequences (as before, those denoted by the black, white and primary colours) *and* highly similar sequences (denoted by the different shades of the primary colours).

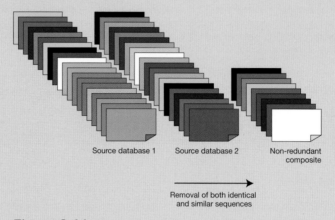

Source database 1   Source database 2   Non-redundant composite

→

Removal of both identical
and similar sequences

Figure 3.11

Building a non-redundant database.

In practice, producing a **non-redundant database** in this way is more challenging, because it can be hard to decide where exactly to draw the line: should we eliminate sequences that are 99 per cent similar, on the assumption that they're probably the same thing (the only difference, perhaps, being a sequencing error, or retention of the initial methionine residue)? Or is this too conservative, and should we choose a 95 per cent similarity cut-off instead? No matter which approach we use, the likelihood is that a composite database will still contain some level of redundancy and some amount of error – this is because the source databases contain non-systematic (usually human) errors, which are very hard to trace and can't be corrected automatically.

Consider NRDB, the default database of the NCBI's BLAST[a] service (see Chapter 4). This resource includes weekly updates of Swiss-Prot[b] and daily updates of GenBank, so it is up-to-date and comprehensive; however, its construction is fairly simplistic. This means that multiple copies of the same protein may be retained as a result of polymorphisms or sequencing errors; errors that have been corrected in Swiss-Prot are re-introduced when sequences are re-translated back from the nucleotide sequences in GenBank; and numerous sequences are duplicates of existing fragments. The contents of the database are hence error-prone and redundant (strictly speaking, it's more appropriate to consider NRDB as a **non-identical database**).

Redundancy issues are tackled more systematically in UniProt (which amalgamates Swiss-Prot, TrEMBL and PIR). In UniProt, closely similar sequences are clustered to yield representative sequence subsets using three different sequence identity cut-offs: the resulting resources are termed UniRef100, UniRef90 and UniRef50. In UniRef90 and UniRef50, no pair of sequences in the representative set has >90 per cent or >50 per cent mutual sequence identity; in UniRef100, identical sequences and their sub-fragments are presented as a single entry. The creation of UniRef[c] not only facilitates sequence merging in UniProt, but also allows faster and more informative similarity searches.

---

[a] http://blast.ncbi.nlm.nih.gov
[b] http://web.expasy.org/docs/swiss-prot_guideline.html
[c] http://www.uniprot.org/help/uniref

3.12 The European Nucleotide Archive (ENA)

The method of sequencing pioneered by Sanger (Section 3.4) was the predominant approach for around 30 years. Although automation of the process made whole-genome sequencing possible, the time-scales were long and the costs were high – the Human Genome Project[31], for example, took more than ten years to complete and was estimated to have cost around $3 billion. While the time-spans and prices did come down, and scientists had access to more and more completed genomes, their hunger for yet more data, at more affordable prices, couldn't be satisfied.

In 2005, this drive to access more data fuelled the development of a new technology, one that was capable of an approximate 100-fold increase in raw sequence throughput over the Sanger method – this technology was able to sequence 25 million bases, with high fidelity, in a single 4-hour run (Marguiles *et al.*, 2005). Perhaps because the accuracy of the new technology had not actually been proven in the field, and perhaps also because of the considerable financial investment already made in Sanger technology, take-up by the community was initially quite slow, despite the fact that the new machines could generate 50-times the throughput for around a sixth of the cost (Schuster, 2008). Nevertheless, inspired by some of the early NGS studies, including landmarks such as analysis of the genomes of the **mammoth** (Poinar *et al.*, 2006) and **Neanderthal** (Green *et al.*, 2006; Noonan *et al.*, 2006), recent years have witnessed the blossoming of NGS sequencing projects. Today, the cost is such that NGS projects can now be conducted outside national sequencing centres, in the laboratories of small academic groups – effectively, 'desk-top' sequencing has become a reality.

The real motivation driving the sequencing revolution was to bring down the cost of sequencing a human genome far enough to enable sequencing to become a feasible biomedical and clinical diagnostic tool (*e.g.*, in early diagnosis of diseases like cancer). Significant progress has been made, and the cost of human genome sequencing is now much lower than it was ten years ago. At the time of writing, sequencing can be completed with a single machine, processing more than 100 million samples per run, in about a week, for ~$20,000. But this isn't enough, and the goal is now to develop **third-generation (3Gen) sequencing** technologies that are able to provide high-quality human genome sequencing for $1,000. One thing is clear: when this target is reached, the data-generation landscape will be radically different from the one we see today.

Meanwhile, a mounting tsunami of nucleotide sequence data has been gaining force as NGS technologies have swept the globe. In 2013, there were estimated to be 2,000 sequencing instruments operating in laboratories world-wide, collectively able to yield 15 quadrillion nucleotides per annum (*i.e.*, ~15 petabytes of compressed data). It has been argued that such quantities of data aren't unprecedented, but are familiar, say, in fields such as astronomy and high-energy physics – *e.g.*, detectors at the Large Hadron Collider[32] produced ~13 petabytes of data in 2010. But there's a difference – in these latter fields, the data are generated by just a few centralised instruments; genomic data, on the other hand, are spewing out of thousands – and soon, possibly, tens of thousands – of instruments distributed around the globe[33] (Schatz and Langmead, 2013).

Several important developments occurred in the wake of this data onslaught. For example, it became clear that providing access not only to the most recent versions of sequences, but also to their historical artefacts was imperative – the rush to patent genetic

[31] http://www.ornl.gov/sci/techresources/Human_Genome
[32] http://en.wikipedia.org/wiki/Large_Hadron_Collider
[33] http://omicsmaps.com

information made issues of priority, and the need to see sequence entries exactly as they appeared in the past, increasingly important. As a result, the EBI established a Sequence Version Archive (Leinonen *et al.*, 2003), to store both current and earlier versions of entries in the EMBL data library (which, by then, had changed its name to **EMBL-Bank**). By September 2004, EMBL-Bank had grown prodigiously, to >42 million entries (Kanz *et al.*, 2005); by 2007, it was accompanied by the Ensembl Trace Archive (ETA) – the ETA was set up to provide a permanent archive for single-pass DNA sequencing reads (from WGS, EST and other large-scale sequencing projects) and associated traces and quality values. Together, EMBL-Bank and the ETA became known as ENA, the **European Nucleotide Archive**[34], Europe's primary nucleotide-sequence repository (Cochrane *et al.*, 2008).

The volume and nature of the data contained in the ENA continued to grow. By 2008, a third component had been added to the resource to allow archiving of raw NGS read data – this was the **Sequence Read Archive**, or SRA (Cochrane *et al.*, 2009). Overall, by October 2010, ENA contained a phenomenal ~500 billion raw and assembled sequences, comprising 50×10^{12} bps. During this period, NGS reads held in the SRA had become the largest and fastest growing source of new data, and accounted for ~95 per cent of all base pairs made available by the ENA (Leinonen *et al.*, 2011). Just three years later, ENA included data from >20,000 sequencing studies and >18,000 assembled genomes, and contained more than a colossal 570×10^{12} bps (Pakseresht *et al.*, 2014). Information relating to these genomes is housed in a host of organism-specific databases. These are, of course, incredibly valuable resources; however, as our focus is primarily on protein sequences (as explained in Chapter 1), we'll not discuss them further here – instead, we provide a taste in Box 3.6, and encourage interested readers to explore a little further.

Box 3.6 Organism-specific databases and Ensembl

The development of high-throughput technologies made whole-genome sequencing a practical reality in the mid-1990s. The sequencing frenzy that followed, producing the first bacterial genome in 1995 and the first draft of the human genome in 2001, has continued unabated, and has created in its wake numerous organism-specific databases. It is not in the scope of this chapter to try to describe these resources in detail. Some of the notable databases include **FlyBase**[a] (the repository of *Drosophila* genes and genomes), **WormBase**[b] (the repository of mapping, sequencing and phenotypic information relating to *C. elegans* and other nematodes), **SGD**[c] (a database dedicated to the molecular biology and genetics of the yeast *S. cerevisiae*), **TAIR**[d] (a database of genetic and molecular biology data for the model higher plant *Arabidopsis thaliana*), and **Ensembl**[e] (the system housing data from the Human Genome Project (HGP)).

The Ensembl project (a joint operation of the EBI and the Wellcome Trust Sanger Institute in the UK) began in 1999, before the human genome was available in draft form. Early on, it was obvious that manual annotation of the information encoded in the sequence's three billion bps wouldn't be practicable. The Ensembl system was therefore conceived as a means of storing and automatically annotating the genome, augmented with data from other sources. It was first launched on the Web in July 2000.

---

[a] http://flybase.org
[b] http://www.wormbase.org
[c] http://www.yeastgenome.org
[d] http://www.arabidopsis.org
[e] http://www.ensembl.org

---

[34] http://www.ebi.ac.uk/ena

Up to the summer of 2014, Ensembl provided data based on the December 2013 *Homo sapiens* high-coverage assembly GRCh38 from the **Genome Reference Consortium**[f] (GRC), which places sequences into a chromosomal context. The GRC attempts to provide the best possible reference assembly for *H. sapiens* by generating multiple representations (alternate loci) for complex regions, and releasing regional 'fixes', termed patches, to improve the representation of specific chromosomal regions. Figure 3.12 shows an ideogram of human genome[g] assembly, GRCh38 (September 2015 saw the release of GRCh38.p5, the fifth patch release for the GRCh38 reference assembly).

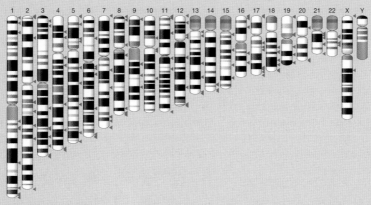

◄ Region containing alternate loci

Figure 3.12

Human assembly GRCh38.

Source: Genome Reference Consortium[i] (GRC).

[i]www.genomereference.org

But Ensembl is particularly noteworthy because it is much more than just a repository for storing information relating to the human genome. Its wider vision was to provide a portable bioinformatics framework for orchestrating biological investigations, able to facilitate the storage, analysis and visualisation of very large genomes (Hubbard *et al.*, 2002). More than a decade on, the system houses the genomes of numerous vertebrate organisms, including those of our close cousins, the chimpanzee and gorilla, and those of more exotic species, such as the megabat and Tasmanian devil. Release 73, in September 2013, included data from 77 species, with greatest support given to human, mouse, rat, a variety of farm animals, and zebrafish (Flicek *et al.*, 2014) – a full current list[h] is accessible from the Ensembl home-page.

Various non-vertebrate model species are also supported by Ensembl, including yeast, fly and worm. Annotation for their genomes is imported from their respective organism databases in a collaborative project known as Ensembl Genomes. Overall, Ensembl Genomes provides access to a battery of genomes of non-vertebrate species, from bacteria, protists and fungi, to plants and invertebrate metazoa (Kersey *et al.*, 2014) – again, a full current list[i] can be accessed from the Ensembl Genomes home-page[j]. Still more information is available from PreEnsembl[k], which offers early access to genomes that are in the process of annotation – here, tools for BLAST searches and genome visualisation are accessible, but not full gene builds.

The enormous wealth of available information, coupled with powerful visualisation tools that allow users to drill their analyses down through different views of the data, has rendered Ensembl an invaluable system for comparative genomics studies. It is an impressive complement both to the many organism-specific resources and to the more general-purpose nucleotide databases.

[f]http://www.ncbi.nlm.nih.gov/projects/genome/assembly/grc
[g]http://www.ncbi.nlm.nih.gov/projects/genome/assembly/grc/human
[h]http://www.ensembl.org/info/about/species.html
[i]http://www.ensemblgenomes.org/info/species
[j]http://ensemblgenomes.org
[k]http://pre.ensembl.org/index.html

The daunting march of progress just described comes at a cost, challenging current IT infrastructures to the limit. Some of the oldest data in ENA date back to the early 1980s, when the EMBL data library was first created. As an aside, it is ironic that, even in those days, there were distribution headaches. Bairoch, for example, relates how difficult it was to transfer version 2 of the EMBL data library from computer tape to a mainframe computer and thence to his microcomputer, because the mainframe had no communication protocol to talk to a microcomputer – he therefore had to spend the night transferring the data, screen by screen, using a 300 **baud acoustic modem** (Bairoch, 2000). To put this in perspective, this version of the database contained 811 nucleotide sequences (with more than 1 million bps) – this is about the same amount of data that currently enters ENA every 1 or 2 seconds.

If ENA housed >20 terabases of raw NGS reads[35] in 2010, and, combined with EMBL-Bank's annotation information, *etc.*, then occupied 230 terabytes of disk space, the amount of space required to accommodate its current >1,000 terabases of nucleotide data is virtually unimaginable. The infrastructure required to store, maintain and service such a vast archive, and the cost of doing so, is beyond anything that either the originators of the first databases, or the developers of the new sequencing technologies could have conceived (as an aside, with a global investment of $105 billion in 2012, energy consumption of digital data centres was estimated to be at the level of 38 GW, rising to $120 billion in 2013 and consumption >40 GW (Venkatraman, 2012), a truly staggering figure – alleged to be equivalent to the output of ~30 nuclear power plants (Glanz, 2012)). The strain on funding agencies has already become apparent, and how they and the community will react to the predicted further onslaught from 3Gen technologies is unclear. In a hint of what might lie ahead, in February 2011, owing to budget constraints, the NCBI announced that it would be discontinuing its Sequence Read and Trace Archives for high-throughput sequence data in a phased closure over 12 months. Shortly afterwards, further negotiations with the NIH yielded interim funding for the archives until October 2011 (the resource was still accessible in February 2015, but the last software update was in June 2013 – an upgrade to the SRA BLAST service to make searching its 700 trillion bases more efficient). What is clear from this incredible trajectory is that new funding and service strategies will be necessary both to maintain these resources and to provide continued access to the existing data archives, now and in the future. The European **ELIXIR**[36] framework for a bioinformatics infrastructure for the life sciences and the trans-US-NIH initiative – Big Data to Knowledge (BD2K) – are likely to play key roles as this drama continues to unfold.

A summary of some of the key sequencing milestones and the databases that grew up in response to the various data crises of the last 40 years is given in Figure 3.13, from the 'explosion' of 65 sequences that gave rise to Dayhoff's *Atlas*, on which the PIR-PSD was built; the flurry of seven protein structures that gave birth to the PDB; the flood of 568 nucleotide sequences that spurred the development of the EMBL data library; the deluge of 105,000 protein sequences that inaugurated TrEMBL; the continued onslaught of nucleotide data that bloated EMBL-Bank with 630 million sequences; to the barrage of ENA's >500 billion reads and their hundreds of trillions of base pairs. The impact of genomic sequencing in the mid 1990s and of subsequent NGS technologies is clear; the shock-waves of 3Gen sequencing are yet to be felt.

---

[35] http://www.ebi.ac.uk/ena/about/statistics
[36] https://www.elixir-europe.org

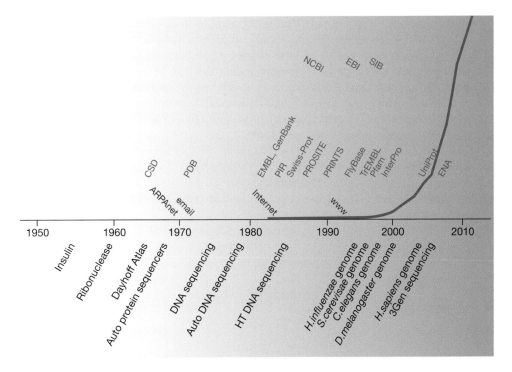

Figure 3.13

Milestones in the history of sequencing, from the determination of the first amino acid sequence, to the creation of an archive of 500 billion nucleotide sequences. Selected milestones are denoted in black; key computing innovations are purple; example databases are blue; bioinformatics organisations and institutions are green; the growing volume of data is shown in the red curve and in the background gradient.

3.13 Summary

This chapter introduced some of the main biological sequence and structure databases in common use today. Specifically, we saw that:

1 The PDB stores the 3D coordinates of macromolecular structures, including those of proteins, protein–nucleic acid complexes and carbohydrates;

2 The SCOP and CATH databases classify protein structures according to their evolutionary and structural relationships;

3 EMBL-Bank, GenBank and DDBJ are the repositories of nucleotide sequence data held in Europe, the USA and Japan;

4 The UniProt Knowledgebase (UniProtKB) is the principal source of protein sequence data; its Swiss-Prot and TrEMBL components contain high-quality manual annotation and computer-generated annotation, respectively;

5 Composite databases provide efficiency gains for database searching, but the levels of redundancy and error rates of different composites differ;

6 UniProt offers three subsets, clustered at different levels of sequence identity for efficiency of searching: UniRef100, UniRef90 and UniRef50;

7 Automated, high-throughput DNA sequencing technologies in the mid-1980s made whole-genome sequencing feasible in realistic time-frames;

8 High-throughput technologies have created a gulf between the accumulating mass of raw sequence data and the amount of data that can be annotated by hand;

9 High-throughput technologies have also widened the gap between the number of available protein sequences and their deduced 3D structures;

10 InterPro is the world's principal unified protein family resource, offering comprehensive hand-crafted annotation;

11 The founding partners of InterPro (PROSITE, PRINTS, Profiles, ProDom and Pfam) make use of different diagnostic techniques;

12 The main diagnostic methods encode conserved regions of sequence alignments as individual motifs (consensus expressions/patterns), groups of motifs (fingerprints) and profiles/HMMs;

13 ENA consists of three main databases: the Sequence Read Archive (SRA), the Ensembl Trace Archive (ETA) and EMBL-Bank;

14 3Gen technologies are being developed to bring the speed and cost of sequencing human genomes down to $1,000;

15 The ultimate goal of 3Gen sequencing is to be able to feasibly deploy sequencing technologies in clinical settings.

3.14 References

Akrigg, D.A., Attwood, T.K., Bleasby, A.J. *et al.* (1992) SERPENT – An information storage and analysis resource for protein sequences. *CABIOS*, **8**(3), 295–296.

Allen, F. H., Davies, J.E., Galloy, J.J. *et al.* (1991) The development of versions 3 and 4 of the Cambridge Structural Database System. *Journal of Chemical Information and Computer Sciences*, **31**, 187–204.

Altschul, S.F., Gish, W., Miller, W. *et al.* (1990) Basic local alignment search tool. *Journal of Molecular Biology*, **215**(3), 403–10.

Anderson, S., Bankier, A.T., Barrell, B.G. *et al.* (1981) Sequence and organization of the human mitochondrial genome. *Nature*, **290**, 457–465.

Apweiler, R., Attwood, T.K., Bairoch, A. *et al.* (2001) The InterPro database, an integrated documentation resource for protein families, domains and functional sites. *Nucleic Acids Research*, **29**(1), 37–40.

Apweiler, R., Bairoch, A., Wu, C.H. *et al.* (2004) UniProt: the Universal Protein knowledgebase. *Nucleic Acids Research.*, **32**(Database issue), D115–119.

Attwood, T.K., Beck, M.E., Bleasby, A.J. and Parry-Smith, D.J. (1994) PRINTS – a database of protein motif fingerprints. *Nucleic Acids Research*, **22**(17), 3590–3596.

Bairoch, A. (1982) Suggestion to research groups working on protein and peptide sequence. *Biochemical Journal*, **203**(2), 527–528.

Bairoch A. (1991) PROSITE: a dictionary of sites and patterns in proteins. *Nucleic Acids Research*, **19** Suppl., 2241–2245.

Bairoch, A. (2000) Serendipity in bioinformatics, the tribulations of a Swiss bioinformatician through exciting times! *Bioinformatics*, **16**(1), 48–64.

Bairoch, A. and Apweiler, R. (1996) The SWISS-PROT protein sequence data bank and its new supplement TREMBL. *Nucleic Acids Research*, **24**(1), 21–25.

Bairoch, A. and Boeckmann, B. (1991) The SWISS-PROT protein sequence data bank. *Nucleic Acids Research*, **19** Suppl., 2247–2249.

Bairoch, A. and Bucher, P. (1994) PROSITE: recent developments. *Nucleic Acids Research,* **22**(17), 3583–3589.

Barker, W.C., George, D.G., Mewes, H.W. and Tsugita, A. (1992) The PIR-International Protein Sequence Database. *Nucleic Acids Research,* **20** Suppl., 2023–2026.

Benson, D., Lipman, D.J. and Ostell, J. (1993) GenBank. *Nucleic Acids Research,* **21**(13), 2963–2965.

Benson, D.A., Clark, K., Karsch-Mizrachi, I. *et al.* (2015) GenBank. *Nucleic Acids Research,* **43**(Database issue), D30–D35.

Berman, H.M., Westbrook, J., Feng, Z., *et al.* (2000) The Protein Data Bank. *Nucleic Acids Research,* **28**(1), 235–242.

Berman, H., Henrick, K. and Nakamura, H. (2003) Announcing the worldwide Protein Data Bank. *Nature Structural Biology,* **10**, 980.

Berman, H. (2008) The Protein Data Bank: a historical perspective. *Foundations of Crystallography,* **64**(1), 88–95.

Bernstein, F.C., Koetzle, T.F., Williams, G.J. *et al.* (1977) The Protein Data Bank. A computer-based archival file for macromolecular structures. *Journal of Molecular Biology,* **112**(3), 535–42. Reprinted in *European Journal of Biochemistry,* **80**(2), 319–324 (1977), and *Archives of Biochemistry and Biophysics,* **185**(2), 584–591 (1978).

Boutselakis, H., Dimitropoulos, D., Fillon, J. *et al.* (2003) E-MSD: the European Bioinformatics Institute Macromolecular Structure Database. *Nucleic Acids Research,* **31**(1), 458–462.

Burley, S.K., Almo, S.C., Bonanno, J.B. *et al.* (1999) Structural genomics: beyond the human genome project. *Nature Genetics,* **23**(2), 151–157.

Cochrane, G., Akhtar, R., Aldebert, P. *et al.* (2008) Priorities for nucleotide trace, sequence and annotation data capture at the Ensembl Trace Archive and the EMBL Nucleotide Sequence Database. *Nucleic Acids Research,* **36**(Database issue), D5–D12.

Cochrane, G., Akhtar, R., Bonfield, J. *et al.* (2009) Petabyte-scale innovations at the European Nucleotide Archive. *Nucleic Acids Research,* **37**(Database issue), D19–D25.

Dayhoff, M.O., Schwartz, R.M., Chen, H.R. *et al.* (1981) Nucleic Acid Sequence Database. *DNA,* **1**, 51–58.

Dayhoff, M.O., Schwartz, R.M., Chen, H.R. *et al.* (1981) Data Bank. *Nature,* **290**, 8.

Dodson, G. (2005) Fred Sanger: sequencing pioneer. *The Biochemist* (December issue).

Doolittle, R.F. (1986) *Of Urfs and Orfs: a primer on how to analyze derived amino acid sequences.* University Science Books, Mill Valley, CA.

Eck, R.V. and Dayhoff, M.O. (1966) *Atlas of Protein Sequence and Structure.* National Biomedical Research Foundation, Silver Spring, MD.

Flicek, P., Amode, M.R., Barrell, D. *et al.* (2014) Ensembl 2014. *Nucleic Acids Research,* **42**(Database issue), D749–D755.

George, D.G., Barker, W.C. and Hunt, L.T. (1986) The protein identification resource (PIR). *Nucleic Acids Research,* **14**(1), 11–15.

George, D.G., Dodson, R.J., Garavelli, J.S. *et al.* (1997) The Protein Information Resource (PIR) and the PIR-International Protein Sequence Database. *Nucleic Acids Research,* **25**(1), 24–28.

Gingeras, T.R. and Roberts, R.J. (1980) Steps towards computer analysis of nucleotide sequences. *Science,* **209**, 1322–1328.

Glanz, J. (2012) Power, pollution and the internet. *New York Times,* 22 September 2012.

Green, R.E., Krause, J., Ptak, S.E. *et al.* (2006) Analysis of one million base pairs of Neanderthal DNA. *Nature,* **444**, 330–336.

Hamm, G.H. and Cameron, G.N. (1986) The EMBL data library. *Nucleic Acids Research,* **14**(1), 5–9.

Henikoff, S. and Henikoff, J.G. (1991) Automated assembly of protein blocks for database searching. *Nucleic Acids Research,* **19**(23), 6565–6572.

Hobohm, U., Scharf, M., Schneider, R. and Sander, C. (1992) Selection of representative protein data sets. *Protein Science,* **1**(3), 409–417.

Hubbard, T., Barker, D., Birney, E. *et al.* (2002).The Ensembl genome database project. *Nucleic Acids Research*, **30**(1), 38–41.

Hunter, S., Apweiler, R., Attwood, T.K. *et al.* (2009) InterPro: the integrative protein signature database. *Nucleic Acids Research*, **37**(Database Issue), D211–D215.

Kanehisa, M., Fickett, J.W. and Goad, W.B. (1984) A relational database system for the maintenance and verification of the Los Alamos sequence library. *Nucleic Acids Research*, **12**(1), 149–158.

Kanz, C., Aldebert, P., Althorpe, N. *et al.* (2005) The EMBL Nucleotide Sequence Database. *Nucleic Acids Research*, **33**(Database issue), D29–D33.

Kennard, O. (1997) From private data to public knowledge. In *The Impact of Electronic Publishing on the Academic Community*, an International Workshop organised by the Academia Europaea and the Wenner–Gren Foundation, Wenner–Gren Center, Stockholm, 16–20 April, 1997 (ed. Ian Butterworth), Portland Press, London.

Kennard, O., Watson, D. G. and Town, W. G. (1972) Cambridge Crystallographic Data Centre. I. Bibliographic File. *Journal of Chemical Documentation*, **12**(1), 14–19.

Kersey, P.J., Allen, J.E., Christensen, M. *et al.* (2014) Ensembl Genomes 2013: scaling up access to genome-wide data. *Nucleic Acids Research*, **42**(Database issue), D546–D552.

Kneale, G.G. and Kennard, O. (1984) The EMBL nucleotide sequence data library. *Biochemical Society Transactions*, **12**, 1011–1014.

Lathrop, R.H., Rost, B., ISCB Membership, ISCB Executive Committee, ISCB Board of Directors and ISCB Public Affairs Committee (2011) ISCB Public Policy Statement on Open Access to Scientific and Technical Research Literature. *PLoS Computational Biology*, **7**(2), e1002014.

Leinonen, R., Nardone, F., Oyewole, O. *et al.* (2003) The EMBL sequence version archive. *Bioinformatics*, **19**(14), 1861–1862.

Leinonen, R., Akhtar, R., Birney, E. *et al.* (2011) The European Nucleotide Archive. *Nucleic Acids Research*, **39**(Database issue), D28–D31.

Lipman, D.J. and Pearson, W.R. (1985) Rapid and sensitive protein similarity searches. *Science*, **227**, 1435–1441.

Margulies, M., Egholm, M., Altman, W.E. *et al.* (2005) Genome sequencing in microfabricated high-density picolitre reactors. *Nature*, **437**, 376–80.

Meyer, E.F. (1997) The first years of the Protein Data Bank. *Protein Science*, **6**, 1591–1597.

Mitchell, A., Chang, H.Y., Daugherty, L. *et al.* (2015) The InterPro protein families database: the classification resource after 15 years. *Nucleic Acids Research*, **43**(Database issue), D213–D221. doi:10.1093/nar/gku1243

Nakamura, H., Ito, N. and Kusunoki, M. (2002) Development of PDBj: advanced database for protein structures. *Tanpakushitsu Kakusan Koso. Protein, Nucleic Acid, Enzyme*, **47**(8 Suppl), 1097–1101.

Noonan, J.P., Coop, G., Kudaravalli, S. *et al.* (2006) Sequencing and analysis of Neanderthal genomic DNA. *Science*, **314**, 1113–1118.

Pakseresht, N., Alako, B., Amid, C. *et al.* (2014) Assembly information services in the European Nucleotide Archive. *Nucleic Acids Research*, **42**(Database issue), D38–D43. doi:10.1093/nar/gkt1082

Poinar, H.N., Schwarz, C., Qi, J. *et al.* (2006) Metagenomics to paleogenomics: large-scale sequencing of mammoth DNA. *Science*, **311**, 392–394.

Protein Data Bank (1971) *Nature New Biology*, **233**, 223.

Protein Data Bank (1973) *Acta Crystallographica, sect. B*, **29**, 1746.

Sanger, F. (1988) Sequences, sequences, and sequences. *Annual Reviews in Biochemistry*, **57**, 1–28.

Sanger, F., Coulson, A.R., Friedmann, T. *et al.* (1978) The nucleotide sequence of bacteriophage phiX174. *Journal of Molecular Biology*, **125**(2), 225–246.

Sanger, F., Coulson, A.R., Hong, G.F. *et al.* (1982) Nucleotide sequence of bacteriophage lambda DNA. *Journal of Molecular Biology*, **162**(4), 729–773.

Schatz, M.C. and Langmead, B. (2013) The DNA Data Deluge. *IEEE Spectrum*, 27 June 2013.

Schuster, S.C. (2008) Next-generation sequencing transforms today's biology. *Nature Methods*, **5**, 16–18.

Sidman, K.E., George, D.G., Barker, W.C. and Hunt, L.T. (1988) The protein identification resource (PIR). *Nucleic Acids Research*, **16**(5), 1869–1871.

Smith, T.F. (1990) The history of the genetic sequence databases. *Genomics*, **6**, 701–707.

Sonnhammer, E.L., Eddy, S.R. and Durbin, R. (1997) Pfam: a comprehensive database of protein domain families based on seed alignments. *Proteins*, **28**(3), 405–420.

Sonnhammer, E.L. and Kahn, D. (1994) Modular arrangement of proteins as inferred from analysis of homology. *Protein Science*, **3**(3), 482–492.

Strasser, B. (2008) GenBank – Natural history in the 21st century? *Science*, **322**, 537–538.

UniProt Consortium. (2011) Ongoing and future developments at the Universal Protein Resource. *Nucleic Acids Research*, **39**(Database issue), D214–D219.

Velankar, S., Best, C., Beuth, B. *et al.* (2010) PDBe: Protein Data Bank in Europe. *Nucleic Acids Research*, **38**, D308–D317.

Venkatraman, A. (2012) Global census shows datacentre power demand grew 63% in 2012. ComputerWeekly.com, 8 October 2012. http://www.computerweekly.com/news/2240164589/Datacentre-power-demand-grew-63-in-2012-Global-datacentre-census (accessed 4 May 2016)

Wilbur, W.J. and Lipman, D.J. (1983) Rapid similarity searches of nucleic acid and protein data banks. *Proceedings of the National Academy of Sciences of the U.S.A.*, **80**(3), 726–730.

Wu, C.H., Yeh, L.S., Huang, H. *et al.* (2003) The Protein Information Resource. *Nucleic Acids Research*, **31**(1), 345–347.

Yon, J. and Jhoti, H. (2003) High-throughput structural genomics and proteomics: where are we now? *TARGETS*, **2**(5), 201–207.

3.15 Quiz

This multiple-choice quiz will help you to see how much you've remembered about the principal biological databases described in this chapter.

1 Which of the following databases house nucleotide sequences?
   A PROSITE
   B ENA
   C TrEMBL
   D NRDB

2 Which of the following databases house diagnostic protein signatures?
   A DDBJ
   B PRINTS
   C InterPro
   D PIR-PSD

3 The PDB, PDBe and PDBj are sources of:
   A nucleotide sequence data.
   B protein family data.
   C macromolecular structure data.
   D protein sequence data.

4 UniRef100, UniRef90 and UniRef50 are sources of:
  A nucleotide sequences clustered at different levels of sequence identity.
  B protein sequences clustered at different levels of sequence identity.
  C protein families clustered at different levels of sequence identity.
  D protein structures clustered at different levels of sequence identity.

5 Which of the following statements are correct?
  A Swiss-Prot initially took data from Dayhoff's *Atlas*.
  B Swiss-Prot was inspired by the format of entries in the EMBL data library.
  C Swiss-Prot was built on data in the PIR-PSD.
  D None of the above.

6 Which of the following statements are correct?
  A TrEMBL is a source of high-quality manual annotation.
  B UniProt was created to integrate protein family information.
  C ENA was created as the world-wide archive of protein sequence data.
  D None of the above.

7 Signature databases were developed to:
  A classify structures into evolutionarily related families.
  B help analyse and functionally annotate uncharacterised sequences.
  C classify sequences into evolutionarily related families.
  D None of the above.

8 Which of the following statements are correct?
  A The sequence–structure deficit refers to over-abundance of protein structures relative to the number of known protein sequences.
  B The sequence–structure deficit is shrinking owing to advances in high-throughput sequencing technology.
  C The sequence–structure deficit refers to over-abundance of protein sequences relative to the number of known protein structures.
  D The sequence–structure deficit is shrinking owing to advances in high-throughput structure determination.

9 Which of the following statements are correct?
  A NGS technologies reduce the need for manual annotation.
  B NGS technologies make fully manual annotation strategies unrealistic.
  C NGS technologies reduce the need for computer-generated annotation.
  D NGS technologies make automatic-annotation strategies unrealistic.

10 Which of the following statements are correct?
  A The goal of 3Gen sequencing technology is to produce more sequence data than NGS technologies.
  B The goal of 3Gen sequencing technology is to make whole-genome sequencing cheaper and faster.
  C The goal of 3Gen sequencing technology is to make sequencing technology feasible as a diagnostic biomedical and clinical tool.
  D None of the above.

3.16 Problems

1 Consensus expressions use a kind of shorthand notation to summarise the conserved residues in parts of sequence alignments. Consider the expression shown in Box 3.3:

```
A-[LV]-Y-S-[AVI]-V-[CF]-[AV]-V-G-L-[LVF]-G-N-X-L-V-M
```

This sequence, AVYSVVCVVGLLGNYLVM, matches the pattern of conserved residues, but this one, AVYSIVCVVGLIGNYLVM, doesn't. Why? Based on your answer, can you write a version of the latter sequence that would match the expression?

Fingerprints are groups of motifs that encode conserved regions of sequence alignments. Consider the fingerprints illustrated in Box 3.4, reproduced in Figure 3.14.

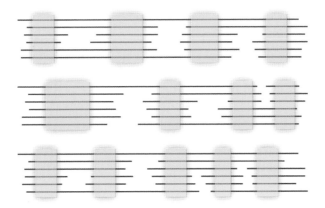

Figure 3.14

Three different fingerprints.

Let's assume that we've searched these fingerprints with a query sequence; the result, is an imperfect match to one of them, as shown in Figure 3.15:

Figure 3.15

A partial fingerprint match.

The gapped (missing) region in the centre shows that the query is shorter than some of the sequences and shares most similarity at its N- and C-termini. Which fingerprint does it most resemble, and why? How many (and which) motifs does it lack?

2 According to the information available from the RCSB, the number of unique protein folds housed in the PDB in November 2015 was 1,375 according to CATH and 1,393 according to SCOP. What are CATH and SCOP, and why might their estimates of the number of unique folds differ? In November 2015, the total number of protein structures available from the PDB was 105,951. Why does this number differ so much from the fold estimates given by SCOP and CATH?

3   Which is likely to be larger: a non-identical composite database or a non-redundant composite database? Why?

There are two widely used composite protein sequence databases. What are they, and in what ways do they differ (comment on the likely nature and sources of error)?

4   InterPro combines data from more than a dozen source databases. What is InterPro? What are the principal diagnostic methods and which of InterPro's partner databases do they underpin? What are some of the main integration challenges that must be tackled in creating InterPro releases?

Chapter 4

Biological sequence analysis

4.1 Overview

In this chapter, we outline a variety of sequence analysis methods, some of which under-pin databases we discussed in Chapter 3. We begin by looking at the need to annotate raw sequence data, and show how the database flat-file format evolved as a necess-ary part of making biological data reusable and searchable. In the central part of the chapter, we introduce some of the tools that are used routinely to compare pairs of sequences and groups of sequences, and discuss how these are used to assess evolution-ary relationships. To complete the picture, we then show how the information latent in multiple alignments can be used for the purpose of sequence annotation.

By the end of the chapter, you should be able to list some of the principal elements of database flat-file formats, and understand how some of these are used by database index-ing software. You should also have an appreciation of the strengths and weaknesses not only of the main scoring matrices used to quantify sequence similarity, but also of the key pattern-recognition methods used to characterise and annotate sequence data.

4.2 Adding meaning to raw sequence data

In the previous chapter, we very broadly described the range of information that's avail-able in sequence- and structure-based databases. We now want to take a step back to look more closely at raw sequence data, and to consider what we need to add to bio-logical sequences in order to make them both useful and reusable. We'll use this dis-cussion to build a picture of how we ultimately store the data in databases.

As we've seen, for convenience, we use single-letter notations to denote protein and nucleotide sequences. Figure 4.1(a) shows a typical fragment of genomic DNA – a sim-ple, continuous, monotonous string of As, Gs, Cs and Ts.

On its own, of course, a string of hundreds of A, C, T and G characters (or many thou-sands of characters, if we're dealing with a completely sequenced genome) is meaningless. Just like natural language text, it requires punctuation for it to be comprehensible. By way of illustration, consider how difficult is it to read the tract of English prose shown

Bioinformatics challenges at the interface of biology and computer science: Mind the Gap. First Edition.
Teresa K. Attwood, Stephen R. Pettifer and David Thorne. Published 2016 © 2016 by John Wiley and Sons, Ltd.
Companion website: www.wiley.com/go/attwood/bioinformatics

(a)
```
gggtgcaaccgtgaagtccgccacgaccgcgtcacgacaggagccgaccagcgacacccagaaggtgcgaacggttgagtgccgcaacg
atcacgagtttttcgtgcgcttcgagtggtaacacgcgtgcacgcatcgacttcaccgcgggtgtttcgacgccagccggccgttgaac
cagcaggcagcgggcatttacagccgctgtggcccaaatggtggggtgcgctattttggtatggttttggaatccgcgtgtcggctccgt
gtctgacggttcatcggttctaaattccgtcacgagcgtaccatactgattgggtcgtagagttacacacatatcctcgttaggtactg
ttgcatgttggagttattgccaacagcagtggaggggggtatcgcaggcccagatcaccggacgtccggagtggatctggctagcgctcg
gtacggcgctaatgggactcgggacgctctatttcctcgtgaaagggatgggcgtctcggacccagatgcaaagaaattctacgccatc
acgacgctcgtcccagccatcgcgttcacgatgtacctctcgatgctgctggggtatggcctcacaatggtaccgttcggtggggagca
gaaccccatctactgggcgcggtacgctgactggctgttcaccacgccgctgttgttgttagacctcgcgttgctcgttgacgcggatc
agggaacgatccttgcgctcgtcggtgccgacggcatcatgatcgggaccggcctggtcggcgcactgacgaaggtctactcgtaccgc
ttcgtgtggtgggcgatcagcaccgcagcgatgctgtacatcctgtacgtgctgtcttcggggttcacctcgaaggccgaaagcatgcg
ccccgaggtcgcatccacgttcaaagtactgcgtaacgttaccgttgtgttgtggtccgcgtatcccgtcgtgtggctgatcggcagcg
aaggtgcgggaatcgtgccgctgaacatcgagacgctgctgttcatggtgcttgacgtgagcgcgaaggtcggcttcgggctcatcctc
ctgcgcagtcgtgcgatcttcggcgaagccgaagcgccggagccgtccgccggcgacggcgcggccgcgaccagcgactgatcgcacac
gcaggacagccccacaaccggcgcggctgtgttcaacgacacacgatgagtcccccactcggtcttgtactc
```

(b)
```
thisisabitofpaddingtomakethestartofthemessagehardtofinditisusualtofindlotsofpaddingbefore
atruemessageislocatedthedifficultyisindistinguishingtruemessagesfromthingsthatlooklikemes
sagesbutarentbecausethesamealphabetisusedthroughoutthesequenceandhowthealphabetisusedchan
gesdependingonitslocationetcetcetcchoweverafilelikethiscontainingastringofhundredsoreventh
ousandsofactandgshere1229ismeaninglessonitsownthesequenceoflettersneedstobeinterpretedsom
ehowlikenaturallanguagetexttobecomprehensibleitneedspunctuationaddingcorrectpunctuationto
nucleotidesequencesisthehardparttoaddonelevelofmeaningtoourgenesequenceoneofthethingswene
edtodoisfindoutwhichpartsofthesequencearelikelytoencodeproteininwesawearlierthatthecodonfor
methionineatgalsofunctionsasastartcodonsowithinthefilewecouldindicatethelocationsofallatg
codonsetcetcetcmorepaddingoccursattheendofthemessageandthispaddingusesthesamealphabetandc
ontainsmorestartandstopcodonsmakingtheprocessofgenefindingseemfairlydifficult...
```

Figure 4.1

(a) Fragment of raw genomic DNA – a continuous string of text; (b) fragment of English prose without punctuation.

in Figure 4.1(b). Tricky, isn't it? Even if we know the language in which a piece of text is written, it can still be difficult to comprehend without punctuation (if there are also errors of grammar and spelling, or idiomatic usage, understanding the text can be quite a challenge). Genomic DNA is much the same – but adding correct punctuation is harder.

To add one level of meaning to a raw string of nucleotide sequences, one thing we might want to do is find out which parts are likely to encode proteins. We saw earlier that the codon for methionine, ATG, also functions as a start codon. So, for the example shown in Figure 4.1, we could begin by indicating the locations of all ATG codons, as illustrated in Figure 4.2(a).

It should be immediately obvious that this isn't enough: knowing where a message might start doesn't tell us where it *actually* starts, nor does it shed light on where the message stops. In Chapter 2, we saw that there are three stop codons: TGA, TAA and TAG. As shown in Figure 4.2(b), if we now add these to the sequence, the picture appears to become even more complicated.

Clearly, just knowing the positions of start and stop codons isn't the full story. What we need to do is look for start codons that are followed by stop codons at a sufficient distance for the intervening sequence to be able to encode a full-length protein sequence. But there are numerous potential starts and stops, and several of them overlap; so what does this mean?

An essential part of the process of reading text is to separate the constituent characters into words. We saw earlier that there are several frames in which a genomic message can be read, depending on which of the three bases in a given codon triplet we start reading from. Suppose we start arbitrarily from the first nucleotide and delineate all the triplet words – this transformation is shown in Figure 4.3(a).

Quickly scanning through the array of three-letter words, we can see that the number of in-frame start and stop codons is much smaller than the 'wordless' sequence in Figure 4.2

(a)
```
gggtgcaaccgtgaagtccgccacgaccgcgtcacgacaggagccgaccagcgacacccagaaggtgcgaacggttgagtgccgcaacga
tcacgagttttttcgtgcgcttcgagtggtaacacgcgtgcacgcatcgacttcaccgcgggtgtttcgacgccagccggccgttgaacca
gcaggcagcgggcatttacagccgctgtggcccaaatggtggggtgcgctattttggtatggtttggaatccgcgtgtcggctccgtgtc
tgacggttcatcggttctaaattccgtcacgagcgtaccatactgattgggtcgtagagttacacacatatcctcgttaggtactgttgc
atgttggagttattgccaacagcagtggaggggtatcgcaggcccagatcaccggacgtccggagtggatctggctagcgctcggtacg
gcgctaatgggactcgggacgctctatttcctcgtgaaagggatgggcgtctcggacccagatgcaaagaaattctacgccatcacgacg
ctcgtcccagccatcgcgttcacgatgtacctctcgatgctgctggggtatggcctcacaatggtaccgttcggtggggagcagaacccc
atctactgggcgcggtacgctgactggctgttcaccacgccgctgttgttgttagacctcgcgttgctcgttgacgcggatcagggaacg
atccttgcgctcgtcggtgccgacggcatcatgatcgggaccggcctggtcggcgcactgacgaaggtctactcgtaccgcttcgtgtgg
tgggcgatcagcaccgcagcgatgctgtacatcctgtacgtgctgttcttcgggttcacctcgaaggccgaaagcatgcgccccgaggtc
gcatccacgttcaaagtactgcgtaacgttaccgttgtgttgtggtccgcgtatcccgtcgtgtggctgatcggcagcgaaggtgcggga
atcgtgccgctgaacatcgagacgctgctgttcatggtgcttgacgtgagcgcgaaggtcggcttcgggctcatcctcctgcgcagtcgt
gcgatcttcggcgaagccgaagcgccggagccgtccgccggcgacggcgcggccgcgaccagcgactgatcgcacacgcaggacagcccc
acaaccggcgcggctgtgttcaacgacacacgatgagtcccccactcggtcttgtactc
```

(b)
```
ggggtgcaaccgtgaagtccgccacgaccgcgtcacgacaggagccgaccagcgacacccagaaggtgcgaacggttgagtgccgcaacg
atcacgagttttttcgtgcgcttcgagtggtaacacgcgtgcacgcatcgacttcaccgcgggtgtttcgacgccagccggccgttgaacc
agcaggcagcgggcatttacagccgctgtggcccaaatggtggggtgcgctattttggtatggtttggaatccgcgtgtcggctccgtgt
ctgacggttcatcggttctaaattccgtcacgagcgtaccatactgattgggtcgtagagttacacacatatcctcgttaggtactgttg
catgttggagttattgccaacagcagtggaggggtatcgcaggcccagatcaccggacgtccggagtggatctggctagcgctcggtac
ggcgctaatgggactcgggacgctctatttcctcgtgaaagggatgggcgtctcggacccagatgcaaagaaattctacgccatcacgac
gctcgtcccagccatcgcgttcacgatgtacctctcgatgctgctggggtatggcctcacaatggtaccgttcggtggggagcagaaccc
catctactgggcgcggtacgctgactggctgttcaccacgccgctgttgttgttagacctcgcgttgctcgttgacgcggatcagggaac
gatccttgcgctcgtcggtgccgacggcatcatgatcgggaccggcctggtcggcgcactgacgaaggtctactcgtaccgcttcgtgtg
tgggcgatcagcaccgcagcgatgctgtacatcctgtacgtgctgttcttcgggttcacctcgaaggccgaaagcatgcgccccgaggt
cgcatccacgttcaaagtactgcgtaacgttaccgttgtgttgtggtccgcgtatcccgtcgtgtggctgatcggcagcgaaggtgcggg
aatcgtgccgctgaacatcgagacgctgctgttcatggtgcttgacgtgagcgcgaaggtcggcttcgggctcatcctcctgcgcagtcg
tgcgatcttcggcgaagccgaagcgccggagccgtccgccggcgacggcgcggccgcgaccagcgactgatcgcacacgcaggacagccc
cacaaccggcgcggctgtgttcaacgacacacgatgagtcccccactcggtcttgtactc
```

Figure 4.2

(a) Fragment of raw genomic DNA with start codons highlighted; (b) same sequence fragment with stop codons also highlighted.

(a)
```
ggg tgc aac cgt gaa gtc cgc cac gac cgc gtc acg aca gga gcc gac cag cga cac cca gaa
ggt gcg aac ggt tga gtg ccg caa cga tca cga gtt ttt cgt gcg ctt cga gtg gta aca cgc
gtg cac gca tcg act tca ccg cgg gtg ttt cga cgc cag ccg gcc gtt gaa cca gca ggc agc
ggg cat tta cag ccg ctg tgg ccc aaa tgg tgg ggt gcg cta ttt tgg tat ggt ttg gaa tcc
gcg tgt cgg ctc cgt gtc tga cgg ttc atc ggt tct aaa ttc cgt cac gag cgt acc ata ctg
att ggg tcg tag agt tac aca cat atc ctc gtt agg tac tgt tgc atg ttg gag tta ttg cca
aca gca gtg gag ggg gta tcg cag gcc cag atc acc gga cgt ccg gag tgg atc tgg cta gcg
ctc ggt acg gcg cta atg gga ctc ggg acg ctc tat ttc ctc gtg aaa ggg atg ggc gtc tcg
gac cca gat gca aag aaa ttc tac gcc atc acg acg ctc gtc cca gcc atc gcg ttc acg atg
tac ctc tcg atg ctg ctg ggg tat ggc ctc aca atg gta ccg ttc ggt ggg gag cag aac ccc
atc tac tgg gcg cgg tac gct gac tgg ctg ttc acc acg ccg ctg ttg ttg tta gac ctc gcg
ttg ctc gtt gac gcg gat cag gga acg atc ctt gcg ctc gtc ggt gcc gac ggc atc atg atc
ggg acc ggc ctg gtc ggc gca ctg acg aag gtc tac tcg tac cgc ttc gtg tgg tgg gcg atc
agc acc gca gcg atg ctg tac atc ctg tac gtg ctg ttc ttc ggg ttc acc tcg aag gcc gaa
agc atg cgc ccc gag gtc gca tcc acg ttc aaa gta ctg cgt aac gtt acc gtt gtg ttg tgg
tcc gcg tat ccc gtc gtg tgg ctg atc ggc agc gaa ggt gcg gga atc gtg ccg ctg aac atc
gag acg ctg ctg ttc atg gtg ctt gac gtg agc gcg aag gtc ggc ttc ggg ctc atc ctc ctg
cgc agt cgt gcg atc ttc ggc gaa gcc gaa gcg ccg gag ccg tcc gcc ggc gac ggc gcg gcc
gcg acc agc gac tga tcg cac acg cag gac agc ccc aca acc ggc gcg gct gtg ttc aac gac
aca cga tga gtc ccc cac tcg tc ttg tac tc
```

(b)
```
MLELLPTAVEGVSQAQITGRPEWIWLALGTALMGLGTLYFLVKGMGVSDPDAKKFYAITTLVPAIAFTMYLSMLLGYGLTMVPF
GGEQNPIYWARYADWLFTTPLLLLDLALLVDADQGTILALVGADGIMIGTGLVGALTKVYSYRFVWWAISTAAMLYILYVLFFG
FTSKAESMRPEVASTFKVLRNVTVVLWSAYPVVWLIGSEGAGIVPLNIETLLFMVLDVSAKVGFGLILLRSRAIFGEAEAPEPS
AGDGAAATSD
```

Figure 4.3

(a) Same sequence fragment as shown in Figure 4.2, with codons delineated from the first forward frame; the highlighted region indicates an Open Reading Frame (ORF); (b) ORF translation product.

led us to believe. The first in-frame start codon is encountered about a third of the way through the sequence, and the first in-frame stop codon appears towards the end, as shown by the highlighted region.

So now we've made progress: we've identified what we believe to be a likely message region, or Open Reading Frame (ORF) – the longest piece of sequence between an in-frame start and an in-frame stop codon. We can therefore now translate this sequence using the genetic code to discover the putative encoded protein (Figure 4.3(b)) – see Box 4.1 for more details.

Box 4.1 Using the genetic code to translate an ORF

When a reading frame has been correctly established, its codons must be translated into amino acids via the genetic code in order to reveal the encoded message. This is a simple stepwise process of mapping each codon to its position in the look-up table and assigning the relevant amino acid. As shown in Figure 4.4, this sequence begins with the start codon, which also encodes methionine, and continues with codons for leucine, glutamic acid, leucine, and so on, terminating with the stop codon (which doesn't encode an amino acid).

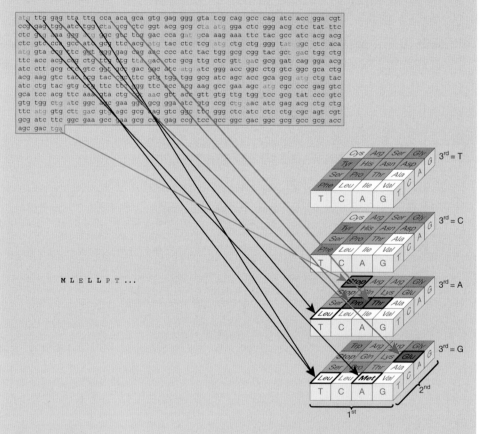

Figure 4.4
Conceptual translation of a nucleotide sequence.

And there it is – a piece of cake! Or is it? To get to this point, we've assumed that there were no spelling or grammatical mistakes in the nucleotide sequence – if there were, they will have been inherited by the translation. Remember also that we haven't yet verified that this is a real protein – to do so, we would need to establish whether a sequence like this had been seen before (*e.g.*, by searching a protein sequence database), or perhaps better, we would need to characterise the sequence experimentally. It's also important to understand that this particular scenario resulted simply from reading the sequence in a forward direction from the first nucleotide – *i.e.*, in the first forward reading frame. It's usual to perform this analysis in all six reading frames in order not to miss any potential ORFs. Doing this often adds to the complexity of the result.

The main reason for the relative simplicity of this example is that the sequence came from a prokaryotic organism – recall that prokaryotic genes don't usually have introns. So let's now consider a eukaryotic sequence in which a number of introns do interrupt the coding sequence (CDS). As discussed in Section 2.2.5, there are various ways to distinguish introns from exons in eukaryotic gene sequences; however, locating their boundaries correctly is extremely difficult because the signals marking them are extremely short, and because the same alphabet and the same words are used in coding and non-coding regions. To give an idea of the difficulty, consider again the English tract we saw in Figure 4.1(b). This time, for clarity, we've highlighted the real message. As you can see in Figure 4.5, this is flanked by padding words, the beginning and ends of which are marked by the familiar shorthand device, '*etc., etc., etc.*'. Knowing the language makes it a tiny bit easier to read the message.

Inevitably, this is a much harder problem for DNA, whose language we still don't fully understand. Fortunately, statistical features of the sequence, such as the GC-content and word-usage, differ sufficiently between coding and non-coding regions to make intron–exon boundary-detection possible. We'll now put this into practice with our example eukaryotic sequence. To save time, we've already identified the start and stop codons, and have delineated the triplet words. Having done this, the interruption of the CDS (highlighted) by non-coding regions is striking, as shown in Figure 4.6(a).

Once the introns have been recognised in this way, they may be excised and the exons sewn together to form a continuous gene transcript, as shown in Figure 4.6(b);

```
thisisabitofpaddingtomakethestartofthemessagehardtofinditisusualtofindlotsofpadd
ingbeforeatruemessageislocatedthedifficultyisindistinguishingtruemessagesfromthi
ngsthatlooklikemessagesbutarentbecausethesamealphabetisusedthroughoutthesequence
andhowthealphabetisusedchangesdependingonitslocationetcetcetchoweverafilelikethi
scontainingastringofhundredsoreventhousandsofactandgshere1229ismeaninglessonitso
wnthesequenceoflettersneedstobeinterpretedsomehowlikenaturallanguagetexttobecomp
rehensibleitneedspunctuationaddingcorrectpunctuationtonucleotidesequencesistheha
rdparttoaddonelevelofmeaningtoourgenesequenceoneofthethingsweneedtodoisfindoutwh
ichpartsofthesequencearelikelytoencodeproteinwesawearlierthatthecodonformethioni
neatgalsofunctionsasastartcodonsowithinthefilewecouldindicatethelocationsofallat
gcodonsetcetcetcmorepaddingoccursattheendofthemessageandthispaddingusesthesameal
phabetandcontainsmorestartandstopcodonsmakingtheprocessofgenefindingseemfairlydi
fficult…
```

Figure 4.5

Message, surrounded by padding text, hidden in the English tract depicted in Figure 4.1(b).

(a)
```
aggccgcgccccgggctccgcgccagccaatgagcgccgcccggccgggcgtgcccccgcgccccaagcataaaccctggcgc
gctcgcggcccggcactcttctggtccccacagactcagagagaacccacc atg gtg ctg tct cct gcc gac aag
acc aac gtc aag gcc gcc tgg ggt aag gtc ggc gcg cac gct ggc gag tat ggt gcg gag gcc
ctg gag agg tgaggctccctcccctgctccgacccgggctcctcgcccgcccggacccacaggccaccctcaaccgtcct
ggccccggacccaaaccccacccctcactctgcttctccccgcagg atg ttc ctg tcc ttc ccc acc acc aag
acc tac ttc ccg cac ttc gac ctg agc cac ggc tct gcc caa gtt aag ggc cac ggc aag aag
gtg gcc gac gcg ctg acc aac gcc gtg gcg cac gtg gac gac atg ccc aac gcg ctg tcc gcc
ctg agc gac ctg cac gcg cac aag ctt cgg gtg gac ccg gtc aac ttc aag gtgagcggcgggccg
ggagcgatctgggtcgaggggcgagatggcgccttcctctcagggcagaggatcacgcgggttgcgggaggtgtagcgcaggc
ggcggcgcggcttgggccgcactgacccctcttctctgcacag ctc cta agc cac tgc ctg ctg gtg acc ctg
gcc gcc cac ctc ccc gcc gag ttc acc cct gcg gtg cac gct tcc ctg gac aag ttc ctg gct
tct gtg agc acc gtg ctg acc tcc aaa tac cgt taa gctggagcctcggtagccgttcctcctgcccgctg
ggcctcccaacgggccctcctcccctccttgcaccggcccttcctggtctttgaataaagtctgagtgggcggcagcctgtgt
gtgcctgggttctctctgtcccggaatgtgccaacaatgcgaggtgtttacctgtctcagaccaaggacctctctgcagctgca
tggggctggggagggagaactgcagggagtatgggagggggaagctgaggtgggcctgctcaagagaaggtgctgaaccatccc
ctgtcctgagaggtgccagcctgcaggcagtggc
```

(b)
```
atg gtg ctg tct cct gcc gac aag acc aac gtc aag gcc gcc tgg ggt aag gtc ggc gcg cac
gct ggc gag tat ggt gcg gag gcc ctg gag agg atg ttc ctg tcc ttc ccc acc acc aag acc
tac ttc ccg cac ttc gac ctg agc cac ggc tct gcc caa gtt aag ggc cac ggc aag aag gtg
gcc gac gcg ctg acc aac gcc gtg gcg cac gtg gac gac atg ccc aac gcg ctg tcc gcc ctg
agc gac ctg cac gcg cac aag ctt cgg gtg gac ccg gtc aac ttc aag ctc cta agc cac tgc
ctg ctg gtg acc ctg gcc gcc cac ctc ccc gcc gag ttc acc cct gcg gtg cac gct tcc ctg
gac aag ttc ctg gct tct gtg agc acc gtg ctg acc tcc aaa tac cgt taa
```

(c)
```
MVLSPADKTNVKAAWGKVGAHAGEYGAEALERMFLSFPTTKTYFPHFDLSHGSAQVKGHGKKVADALTNAVAHVDDMPNALSA
LSDLHAHKLRVDPVNFKLLSHCLLVTLAAHLPAEFTPAVHASLDKFLASVSTVLTSKYR
```

Figure 4.6

(a) Eukaryotic DNA sequence with codons delineated, and start and stop codons shown in green and red; the highlighted regions indicate exons; (b) contiguous coding sequence, with introns removed; (c) translation product of the coding sequence.

and, as before, this may be translated to give the putative protein sequence, shown in Figure 4.6(c). At this point, just as before, we've assumed that the original nucleotide sequence was error free, that we've used the correct reading frame, that the intron–exon boundaries were predicted correctly, and consequently that this is likely to be a real protein – follow-up searches now need to be performed to validate the product.

4.2.1 Annotating raw sequence data

Annotation can mean different things to different people. For the purposes of this discussion, we'll be using the term in its broadest sense to mean any kind of extra information that can add value to raw data in order to make it useful and reusable. In the previous section, we explored the process of reading DNA sequences: adding start and stop codons, delineating introns and exons, identifying and translating coding sequences, *etc.* We now want to consider how we might archive this information so that it can be used again. Assuming our gene product is a real protein, we could store the raw nucleotide sequence and its translation in a file, as shown in Figure 4.7.

However, without labels to say where the nucleotide sequence came from, and what it is, and without any clues to indicate how to derive its product, we would have to go through the entire process again to determine which part, or parts, of the sequence gave rise to the deduced protein. Aside from locating protein-coding sequences, we might also want to know if there are other important regions sequestered in a piece of genomic sequence: there might be regions that encode RNAs (*e.g.*, transfer or ribosomal

(a)

```
aggccgcgccccgggctccgcgccagccaatgagcgccgcccggccgggcgtgccccgcgccccaagcataaaccctggcgcgctcgcggcccggcactcttctggtcccc
acagactcagagagaacccaccatggtgctgtctcctgccgacaagaccaacgtcaaggccgcctggggtaaggtcggcgcgcacgctggcgagtatggtgcggaggccctg
gagaggtgaggctccctccccctgctccgacccgggctcctcgcccgcccggacccacaggccaccctcaaccgtcctggcccccggacccaaaccccaccccctcactctgctt
ctccccgcaggatgttcctgtccttccccaccaccaagacctacttcccgcacttcgacctgagccacggctctgcccaagttaagggccacggcaagaaggtggccgacgc
gctgaccaacgccgtggcgcacgtggacgacatgcccaacgcgctgtccgccctgagcgacctgcacgcgcacaagcttcgggtggacccggtcaacttcaaggtgagcggc
gggcggggagcgatctgggtcgaggggcgagatggcgccttcctctcagggcagaggatcacgcgggttgcgggaggtgtagcgcaggcggcggcgcggcttgggccgcact
gaccctcttctctgcacagctcctaagccactgcctgctggtgaccctggccgcccacctccccgccgagttcacccctgcggtgcacgcttcctggacaagttcctggct
tctgtgagcaccgtgctgacctccaaataccgttaagctggagcctcggtagccgttcctcctgcccgctgggcctcccaacgggccctcctccccctccttgcaccggccct
tcctggtctttgaataaagtctgagtgggcggcagcctgtgtgtgcctgggttctctctgtcccggaatgtgccaacaatggaggtgtttacctgtctcagaccaaggacct
ctctgcagctgcatggggctggggaggggagaactgcagggagtatgggaggggaagctgaggtgggcctgctcaagagaaggtgctgaaccatcccctgtcctgagaggtgc
cagcctgcaggcagtggc
```

(b)

```
MVLSPADKTNVKAAWGKVGAHAGEYGAEALERMFLSFPTTKTYFPHFDLSHGSAQVKGHGKKVADALTNAVAHVDDMPNALSALSDLHAHKLRVDPVNFKLLSHCLLVTLAA
HLPAEFTPAVHASLDKFLASVSTVLTSKYR
```

Figure 4.7

(a) Nucleotide sequence and (b) its protein product. In this naked form, the sequences alone aren't very useful.

RNA), there might be regulatory regions, there could be splice sites, and so on. For a sequence with thousands of nucleotides, directly superposing on to it all the start and stop codons, the correct triplet reading-frames, the intron–exon boundaries, regulatory regions, and so on, is impracticable. It probably makes more sense to leave the sequence in its raw form and add supplementary notes and instructions; but just how much information should we include?

To start, it might be helpful to note what the sequence is and where it came from, in terms both of its location in the genomic sequence and of its source organism. For the example above, we gain useful information from knowing that it is genomic DNA of human origin, and that it is derived from the α-globin germ line gene, which is 1,138 nucleotide bases in length:

```
source: genomic DNA, alpha-globin germ line gene, bases 1-1138
organism: Homo sapiens
```

This is useful, but it hasn't identified where the primary transcript and its exons are located, nor has it indicated the name of the translated protein. We could do this in the following way, to show, at a glance, that there are three exons in the primary transcript, lying between bases 98 and 929:

```
primary transcript: 98-929 (exon 1, 98-230; exon 2, 348-551; exon 3, 92-929)
product: alpha globin
```

By noting features like this, we add increasing value to our raw sequence, such that we can re-use the information quickly and conveniently. The more notes, or annotations, we add, the richer the content. But there's a catch. There are many different ways in which annotations could be arranged in a file. If we plan to use the information again in a particular way, this might determine how we add the notes; or, if we were to think of other potential uses of the information (things that perhaps we don't want to do now, but might want to do in future), this might make us add the notes in a different way, or it might make us consider how to accommodate different kinds of annotation in future.

This is the heart of the matter. As we mentioned in the opening pages of the book, the first biological databases were created to address particular needs at that time, with no idea of the data-generation revolutions ahead. Hence, how the data were stored

```
description: alpha-globin germ line gene
organism: Homo sapiens
source: genomic DNA, bases 1-1138
        primary transcript 98-929
        exon 1 98-230
        exon 2 348-551
        exon 3 692-929
coding sequence: join nucleotide bases (135-230,348-551,692-820)
sequence: 1138 bp
aggccgcgcccgggctccgcgccagccaatgagcgccgcccggccgggcgtgccccgcgcgcccaagcataaaccctggcgcgctcgcggcccggcactcttctggtccccc
acagactcagagagaacccaccatggtgctgtctcctgccgacaagaccaacgtcaaggccgcctgggggtaaggtcggcgcgcacgctggcgagtatggtgcggaggccctg
gagaggtgaggctccctccctgctccgacccgggctcctcgcccgcccggacccacaggccaccctcaaccgtcctggcccccggacccaaacccaccctcactctgctt
ctccccgcaggatgttcctgtccttccccaccaccaagacctacttcccgcacttcgacctgagccacggctctgcccaagttaagggccacggcaagaaggtggccgacgc
gctgaccaacgccgtggcgcacgtggacgacatgcccaacgcgctgtccgccctgagcgacctgcacgcgcacaagcttcgggtggacccggtcaacttcaaggtgagcggc
gggccgggagcgatctgggtcgaggggcgagatggcgcctcctctcagggcagaggatcacgcgggttgcgggaggtgtagcgcaggcggcggcggcggcttgggccgcact
gaccctcttctctgcacagctcctaagccactgcctgctggtgaccctggccgcccacctccccgccgagttcacccctgcggtgcacgcttcctggacaagttcctggct
tctgtgagcaccgtgctgacctccaaataccgttaagctgagcctcggtagccgttcctcctgcccgctgggcctcccaacgggccctcctccctccttgcaccggccct
tcctggtctttgaataaagtctgagtgggcggcagcctgtgtgtgcctgggtctctctgtcccggaatgtgccaacaatggaggtgtttacctgtctcagaccaaggacct
ctctgcagctgcatggggctgggggagggagaactgcagggagtatgggaggggaagctgaggtgggcctgctcaagagaaggtgctgaaccatcccctgtcctgagaggtgc
cagcctgcaggcagtggc
```

```
translation:
MVLSPADKTNVKAAWGKVGAHAGEYGAEALERMFLSFPTTKTYFPHFDLSHGSAQVKGHGKKVADALTNAVAHVDDMPNALSALSDLHAHKLRVDPVNFKLLSHCLLVTLAA
HLPAEFTPAVHASLDKFLASVSTVLTSKYR
product: alpha globin
```

Figure 4.8

Annotated nucleotide sequence, including information on how to derive its protein product. In this form, the sequences are more useful than they are in the raw file shown in Figure 4.7.

reflected those needs, and not the unknown needs of future users. To build a database well, however, requires us not only to know precisely how we want to use the data now, but also to have a good idea of how they might be used at some future point, possibly 30–50 years away. This is virtually impossible to achieve without the gift of second sight, and is largely why the field of bioinformatics is now so littered with legacy data-format problems.

Returning to our globin gene, Figure 4.8 shows a simple way of organising the annotations mentioned above. This layout, for humans at least, is straightforward to read, and the presented information makes it relatively easy to understand how the protein translation was derived. However, without formal structure, how amenable would this information be to computational manipulation?

To add further value to the data, it would also be helpful to add 'database specific' details, such as when the sequence was submitted, and when the database entry was last updated. We could also include literature citations, to indicate when the sequence was first published, and we could add cross-references to information in related databases. However, if we are to keep on adding annotations, and if we want computers to be able to access particular pieces of this information reliably, we need to store the data in a more structured way. This was the challenge faced by the creators of the earliest biological databases, and many of the simple solutions they devised remain part of the currency of bioinformatics data storage and interchange that's still in use today.

4.2.2 Database and sequence formats

Most of the earliest biological databases were originally created as plain **ASCII** text files – flat-files – populated with a collection of **records** containing different fields of

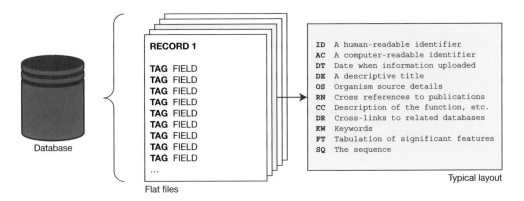

Figure 4.9

Construction of a flat-file database from multiple individual text files. The inset shows the type of detail, denoted by specific tags, that might be held in each field.

data, denoted by a series of specific tags (see Figure 4.9). The tags help to identify the nature of the data being stored in that part of the file; the field contains the actual data specified by the tag. One of the most simple, and hence successful, flat-file formats was that devised to store nucleotide sequences in the EMBL database. This provided the basis of the formats of a number of databases that are still in wide use (*e.g.*, Swiss-Prot, TrEMBL, PROSITE).

The EMBL format uses a series of two-letter tags to describe the data stored on each line of the file, as shown in the inset of Figure 4.9. To illustrate how this works for our α-globin sequence, let's take a look at its EMBL entry, shown in Figure 4.10.

The file begins with an identifying (ID) code (HSAGL1) and an accession (AC) number (V00488): the AC number is designed to be easily read by computers; the ID code is intended to convey slightly more meaning to a human reader – here, with an experienced eye, we might deduce that HSAGL1 was *Homo sapiens* α-globin 1, which is certainly not deducible from V00488. The AC and ID codes are intended to uniquely specify a given entry in a given database release, but of the two, only the AC number (in principle) is invariant. If, for any reason, database entries are merged, the AC numbers are retained, allowing older numbers to be tracked. AC numbers are thus critical to tracking, maintaining and merging data. IDs can and do change if entries are modified. For all operational purposes, the AC number should therefore always be used rather than the ID.

The line following the AC number provides the sequence version (SV). Here, V00488.1 informs us that this is version 1 of the nucleotide sequence. Sequence versioning is useful, as it specifically flags to users when changes have been made to the sequence rather than to the annotation; nevertheless, it took more than ten years to introduce versioning to nucleotide sequence databases, and even longer to introduce the concept to protein sequence databases.

Another important piece of information for tracking changes is the date (DT) of accession of a sequence to the database, and the date on which changes were last made to that database record. This helps to 'future proof' it, should it change in subsequent

```
ID    HSAGL1      standard; genomic DNA; HUM; 1138 BP.
AC    V00488;
SV    V00488.1
DT    09-JUN-1982 (Rel. 01, Created)
DT    24-APR-1993 (Rel. 35, Last updated, Version 5)
DE    Human alpha-globin germ line gene.
KW    alpha-globin; germ line; globin.
OS    Homo sapiens (human)
OC    Eukaryota; Metazoa; Chordata; Craniata; Vertebrata; Euteleostomi; Mammalia;
OC    Eutheria; Euarchontoglires; Primates; Catarrhini; Hominidae; Homo.
RN    [1]
RP    1-1138
RX    PUBMED; 6452630.
RA    Liebhaber S.A., Goossens M.J., Kan Y.W.;
RT    "Cloning and complete nucleotide sequence of human 5'-alpha-globin gene";
RL    Proc. Natl. Acad. Sci. U.S.A. 77(12):7054-7058(1980).
DR    EPD; EP07071; HS_HBA1.
DR    GDB; GDB:119293.
DR    GDB; GDB:119294.
DR    TRANSFAC; R00506; HS$AG_01.
CC    KST HSA.ALPGLOBIN.GL [1138]
FH    Key             Location/Qualifiers
FT    source          1..1138
FT                    /db_xref="taxon:9606"
FT                    /mol_type="genomic DNA"
FT                    /organism="Homo sapiens"
FT    prim_transcript 98..929
FT    exon            98..230
FT                    /number=1
FT    CDS             join(135..230,348..551,692..820)
FT                    /db_xref="GOA:P01922"
FT                    /db_xref="GOA:P69905"
FT                    /db_xref="InterPro:IPR000971"
FT                    /db_xref="InterPro:IPR002338"
FT                    /db_xref="InterPro:IPR002339"
FT                    /db_xref="PDB:1LFL"
FT                    /db_xref="PDB:1LFQ"
FT                    /db_xref="PDB:1LFT"
FT                    /db_xref="PDB:1LFV"
FT                    /db_xref="PDB:1LFY"
FT                    /db_xref="PDB:1LFZ"
FT                    /db_xref="PDB:1YHR"
FT                    /db_xref="PDB:1YIE"
FT                    /db_xref="PDB:1YIH"
FT                    /db_xref="PDB:1YZI"
FT                    /db_xref="UniProt/Swiss-Prot:P69905"
FT                    /product="alpha globin"
FT                    /protein_id="CAA23748.1"
FT                    /translation="MVLSPADKTNVKAAWGKVGAHAGEYGAEALERMFLSFPTTKTYFP
FT                    HFDLSHGSAQVKGHGKKVADALTNAVAHVDDMPNALSALSDLHAHKLRVDPVNFKLLSH
FT                    CLLVTLAAHLPAEFTPAVHASLDKFLASVSTVLTSKYR"
FT    exon            348..551
FT                    /number=2
FT    exon            692..929
FT                    /number=3
SQ    Sequence 1138 BP; 183 A; 412 C; 350 G; 193 T; 0 other;
      aggccgcgcc ccgggctccg cgccagccaa tgagcgccgc ccggccgggc gtgccccgc          60
      gccccaagca taaaccctgg cgcgctcgcg gcccggcact cttctggtcc ccacagactc         120
      agagagaacc caccatggtg ctgtctcctg ccgacaagac caacgtcaag gccgcctggg         180
      gtaaggtcgg cgcgcacgct ggcgagtatg gtgcggaggc cctggagagt tgaggctccc         240
      tcccctgctc cgacccgggc tcctcgcccg cccggaccca caggccaccc tcaaccgtcc         300
      tggccccgga cccaaacccc acccctcact ctgcttctcc ccgcaggatg ttcctgtcct         360
      tccccaccac caagacctac ttcccgcact cgacctgag ccacggctct gcccaagtta         420
      agggccacgg caagaaggtg gccgacgcgc tgaccaacgc cgtggccgac gtggacgaca         480
      tgcccaacgc gctgtccgcc ctgagcgacc tgcacgcgca caagcttcgg gtggacccgg         540
      tcaacttcaa ggtgagcggc gggccgggag cgatctgggt cgaggggcga gatggcgcct         600
      tcctctcagg gcagaggatc acgcgggttg cgggaggtgt agcgcaggcg gcggcgcggc         660
      ttgggccgca ctgaccctct tctctgcaca gctcctaagc cactgcctgc tggtgaccct         720
      ggccgcccac ctcccccgcg agttcacccc tgcggtgcac gcttccctgg acaagttcct         780
      ggcttctgtg agcaccgtgc tgacctccaa ataccgttaa gctggagcct cggtagccgt         840
      tcctcctgcc cgctgggcct cccaacgggc cctcctcccc tccttgcacc ggcccttcct         900
      ggtctttgaa taaagtctga gtgggcggca gcctgtgtgt gcctgggttc tctctgtccc         960
      ggaatgtgcc aacaatggag gtgtttacct gtctcagacc aaggacctct ctgcagctgc        1020
      atggggctgg ggagggagaa ctgcagggag tatgggaggg gaagctgagg tgggcctgct        1080
      caagagaagg tgctgaacca tccctgtcc tgagaggtgc cagcctgcag gcagtggc          1138
//
```

Figure 4.10

The EMBL database entry for human α-globin. The series of tags (ID, AC, DT, DE, *etc.*) identify the type of information held in each database field. Owing to its simplicity, this format was adopted by many other biological databases.

database releases; it also allows users to ascertain the currency of a given entry and hence to determine whether it's out-of-date.

Given that AC and ID codes can be difficult to read (even with the eye of experience!), the file includes a descriptive (DE) title indicating what the entry contains – the title in the example confirms that this is the human alpha-globin germ line gene (to further facilitate swift computational parsing of the file, many of these terms are also included as keywords (KW)). Although the title shows that we're dealing with a human gene, the file also specifies the organism species (OS) more precisely (*i.e.*, Homo sapiens) and its biological classification (OC) (here, Eukaryota, Metazoa, . . . Hominidae, Homo).

In addition, because we might want to know, say, when the sequence was published and other relevant background information (*e.g.*, relating to structural and functional studies), the file includes a number of bibliographic citations – these are stored on the R'x' lines (RN is the reference number; RP gives the subject; RX the PubMed cross-reference; RA the authors; RT the title; RL the location of the publication). Cross-links to related information in other databases is achieved via the DR lines, here with links to EPD (the **Eukaryotic Promoter Database**), GDB (the **Human Genome Database**) and TRANSFAC (the **TRANScription FACtor database**).

Another potentially rich source of annotation is the comment (CC) field. In this example, little further information is provided, but in other databases, the comment field provides the bulk of the useful annotation: *e.g.*, in UniProtKB/Swiss-Prot, the CC field is semi-structured, with different headings to indicate details of function, structure, disease associations, family relationships, *etc.* (this is one of the features that made Swiss-Prot such a popular resource).

To make the entry more informative, characteristics or features of the sequence itself are stored in the **Feature Table** (FT). As mentioned in Section 3.5, the partner databases of the INSDC had adopted common data standards and protocols, partly to streamline data-annotation practices. In particular, they used a common FT format, to help improve data consistency and reliability, and to facilitate database interoperation. Many different features and regions of a sequence may be documented, including those that perform particular functions; affect the expression of function or the replication of a sequence; interact with other molecules; have secondary or tertiary structure; and so on. Regulating the content, vocabulary and syntax of such feature descriptions ensures that the data are released in a form that can be exchanged efficiently and is readily accessible to analysis software. In this example, the FT indicates that the source of the entry is human genomic DNA: within the 1,138 bases of its sequence, the primary transcript resides between bases 98 and 929, where the first exon runs from 98 to 230, the second from 348 to 551, and the third from 692 to 929. To create the coding sequence from these exons, it's necessary to join the three coding segments, from bases 135 to 230, 348 to 551 and 692 to 820.

The FT provides a number of cross-references to related entries in resources such as the PDB, InterPro and the Swiss-Prot component of UniProtKB. It also indicates that the protein product is alpha globin, with ID CAA23748.1, which is followed by the full protein sequence translation. Finally, the nucleotide sequence is stored in the SQ field using the single-letter code, together with attributes such as its length and base composition (here, Sequence 1138 BP; 183 A; 412 C; 350 G; 193 T; 0 other). The entry is completed with the // terminator symbol.

What we've seen here is only a subset of the EMBL format, but it should give a flavour of how we can take a raw sequence of nucleotides and successively add information to it via additional lines of annotation. As noted earlier, in Swiss-Prot, which houses a much smaller collection of protein sequences, the amount of manually-added free-text is more extensive. Consequently, the number of annotation lines, especially in the CC field, is usually much greater – for comparison, compare the actual UniProtKB/Swiss-Prot entry using the AC number in Figure 4.10, P69905[1].

Up to this point, we've discussed the kinds of information or annotation that can be attached to raw sequences in order to make them meaningful. Before considering how we derive such information, there is one further aspect of sequence formats that we should briefly introduce. Figure 4.10 shows the EMBL flat-file database format; but this is a format for *storing* raw sequences and their annotations – it is too bloated for most forms of computational processing. Sequence databases therefore usually provide options to download their sequence data in a more streamlined form, as illustrated in Figure 4.11.

The examples seen in Figure 4.11(a) and (b) depict what's known as '**FASTA** format', which has become the 'standard' text-based format for representing nucleotide and amino acid sequences, largely owing to its simplicity. The file begins with a 'greater than' symbol (>), which is followed by the sequence AC number, and sometimes also the ID, and the sequence description or title. On the next line of the file is

(a)
```
>ENA|V00488|V00488.1 Human alpha-globin germ line gene.
AGGCCGCGCCCCGGGCTCCGCGCCAGCCAATGAGCGCCGCCCGGCCGGGCGTGCCCCCGCGCCCCAAGCATAAACCCTG
GCGCGCTCGCGGCCCGGCACTCTTCTGGTCCCCACAGACTCAGAGAGAACCCACCATGGTGCTGCTCTCCTGCCGACAAG
ACCAACGTCAAGGCCGCCTGGGGTAAGGTCGGCGCGCACGCTGGCGAGTATGGTGCGGAGGCCCTGGAGAGGTGAGGCT
CCCTCCCCTGCTCCGACCCGGGCTCCTCGCCCGCCCGGACCCACAGGCCACCCTCAACCGTCCTGGCCCCGGACCCAAA
CCCCACCCCTCACTCTGCTTCTCCCCGCAGGATGTTCCTGTCCTTCCCCACCACCAAGACCTACTTCCCGCACTTCGAC
CTGAGCCACGGCTCTGCCCAAGTTAAGGGCCACGGCAAGAAGGTGGCCGACGCGCTGACCAACGCCGTGGCGCACGTGG
ACGACATGCCCAACGCGCTGTCCGCCCTGAGCGACCTGCACGCGCACAAGCTTCGGGTGGACCCGGTCAACTTCAAGGT
GAGCGGCGGGCCGGGAGCGATCTGGGTCGAGGGGCGAGATGGCGCCTTCCTCTCAGGGCAGAGGATCACGCGGGTTGCG
GGAGGTGTAGCGCAGGCGGCGGCGGCTTGGGCCGCACTGACCCTCTTCTCTGCACAGCTCCTAAGCCACTGCCTGCT
GGTGACCCTGGCCGCCCACCTCCCCGCCGAGTTCACCCCTGCGGTGCACGCTTCCCTGGACAAGTTCCTGGCTTCTGTG
AGCACCGTGCTGACCTCCAAATACCGTTAAGCTGGAGCCTCGGTAGCCGTTCCTCCTGCCCGCTGGGCCTCCCAACGGG
CCCTCCTCCCCTCCTTGCACCGGCCCTTCCTGGTCTTTGAATAAAGTCTGAGTGGGCGGCAGCCTGTGTGTGCCTGGGT
TCTCTCTGTCCCGGAATGTGCCAACAATGGAGGTGTTTACCTGTCTCAGACCAAGGACCTCTCTGCAGCTGCATGGGGC
TGGGGAGGGAGAACTGCAGGGAGTATGGGAGGGGAAGCTGAGGTGGGCCTGCTCAAGAGAAGGTGCTGAACCATCCCCT
GTCCTGAGAGGTGCCAGCCTGCAGGCAGTGGC
```

(b)
```
>sp|P69905|HBA_HUMAN Hemoglobin subunit alpha OS=Homo sapiens GN=HBA1 PE=1 SV=2
MVLSPADKTNVKAAWGKVGAHAGEYGAEALERMFLSFPTTKTYFPHFDLSHGSAQVKGHGKKVADALTNAVAHVDDMPN
ALSALSDLHAHKLRVDPVNFKLLSHCLLVTLAAHLPAEFTPAVHASLDKFLASVSTVLTSKYR
```

(c)
```
>P1;HBA_HUMAN
Hemoglobin subunit alpha OS=Homo sapiens GN=HBA1 PE=1 SV=2
MVLSPADKTNVKAAWGKVGAHAGEYGAEALERMFLSFPTTKTYFPHFDLSHGSAQVKGHGKKVADALTNAVAHVDDMPN
ALSALSDLHAHKLRVDPVNFKLLSHCLLVTLAAHLPAEFTPAVHASLDKFLASVSTVLTSKYR
*
```

Figure 4.11

(a) FASTA format for the human α-globin germ-line gene; (b) FASTA format for its product, the haemoglobin α subunit; (c) PIR format for the same sequence.

---

[1] http://www.uniprot.org/uniprot/P69905.txt?version=149

the sequence itself, usually in 80 columns (a throw-back to the days when scientists used DEC VT100[2] (or similar) terminals, most of which could display 80 characters per line).

FASTA format is a simplified version of the original NBRF-PIR format, an example of which is shown in Figure 4.11(c). There are three significant differences to note: first, immediately after the '>' symbol is the tag 'P1;' – this means that the file contains a protein sequence (for DNA sequences, the tag is 'D1;', and for sequence fragments, 'F1;'); second, the descriptive title appears on its own line; and third, the file ends with an asterisk.

Although, in principle, simple to parse, there is no standard for the information that appears in the first line of a FASTA file; in practice, therefore, making sense of the information that follows the '>' symbol can be tricky. The PIR format obviates these problems; nevertheless, FASTA format predominates.

FASTA is also the format commonly used for multiple sequences (*e.g.*, for input to multiple alignment programs). In this case, the individual FASTA files are simply concatenated, each sequence ending when the next line commencing with the '>' symbol appears in the file.

There are many other sequence-database formats, and sequence-input and alignment-output formats, far too numerous to describe here; fortunately, excellent guides are available online[3]. Given the variety of disparate data formats, the key challenge is how to get databases and software tools to talk to each other.

4.2.3 Making tools and databases interoperate

Probably the first tool for providing integrated access to related information in disparate databases was SRS, the Sequence Retrieval System (Etzold and Argos, 1993). SRS (now a commercial package) is an information-indexing and -retrieval tool designed specifically to access diverse data-fields in different flat-file databases. Physical integration of all bio-databases world-wide clearly isn't feasible, but integration at the level of interrogation software is. The beauty of SRS is that it allows any flat-file database to be indexed to any other, permitting queries across many different database types via a single interface, irrespective of their underlying data-structures, query languages, and so on (see Figure 4.12). The relative simplicity of the system made SRS an immensely popular tool: at one time, there were >40 SRS servers around the world providing access to >1,200 databases (Etzold *et al.*, 1996).

Figure 4.12(a) illustrates links between several disparate resources: a nucleotide sequence database (EMBL), a protein sequence database (UniProtKB/Swiss-Prot), a protein signature database (PROSITE), a protein structure database (PDB) and a literature database (**MEDLINE**). Note that EMBL, Swiss-Prot and PROSITE use the same basic flat-file format, but PDB and MEDLINE do not. Links between them are made via the AC numbers and IDs unique to each resource – the more cross-references stored internally by a given database, the greater the web of connectivity that's possible from it, as illustrated by the UniProtKB/Swiss-Prot entry at the centre of this example.

---

[2] http://en.wikipedia.org/wiki/VT100
[3] http://toolkit.tuebingen.mpg.de/reformat/help_params

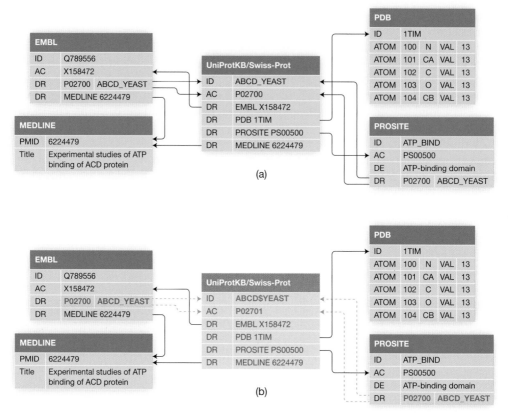

Figure 4.12

(a) Linking database entries via AC numbers and IDs; (b) illustration of how changes to the UniProtKB/ Swiss-Prot ID and AC number shown here result in broken links in the host database (grey arrows), and invisibility of the UniProtKB/Swiss-Prot entry to any database that links to it via its primary AC number or ID (grey box).

Here, the currency of connectivity is the primary database AC number; any change in this number requires all databases around the world to echo the change, otherwise that entry effectively becomes invisible to all of those resources, as illustrated in Figure 4.12(b). Another challenge faced by tools like SRS is that databases have grown to such a size that many institutes struggle to house and maintain local copies – the larger the database (or set of databases), the longer the indexing software takes to run (usually overnight). Hence, typically, two copies of a database are required: one for online use and the other for off-line indexing.

MRS is a query and information-retrieval tool that was developed, at least in part, to try to address this problem, having been optimised for speed and ease-of-use (Hekkelman and Vriend, 2005). MRS-files contain both the data and their indices, which hence remain synchronous, compressing them so that they require less disk space than the raw data. Multiple MRS-files can be integrated, allowing public and private in-house data to be merged. MRS thus offers a fast, robust and freely available alternative to SRS.

Up to now, we've discussed in very simple terms how biological sequences have been annotated and stored in databases, and how database records can be linked together, via unique accession numbers and database identifiers, to allow navigation between them. We now need to think about how annotations, particularly those used to help structurally and functionally characterise sequences, are actually derived using different analysis techniques.

4.3 Tools for deriving sequence annotations

Many analysis tasks require researchers not simply to navigate between related database entries but, rather, to seek relationships at the level of the amino acid (or nucleotide) sequences themselves. Typically, this might involve searching a database with an uncharacterised sequence, or sequence fragment, to which no useful annotation has yet been attached, to try to uncover clues to its likely function, structure or evolutionary history. In the following sections, we explore how the degree of similarity between sequences is typically quantified and the nature of their relationships surmised.

4.3.1 Methods for comparing two sequences

To determine how similar one sequence is to another, the naïve approach is to align them, one below the other, and to introduce additional 'gap' characters, if the sequences differ in length, to bring the character strings into vertical register. In the simple example shown in Figure 4.13(a), two short, highly similar sequence fragments are compared: here, fragment 1 is one amino acid residue shorter than fragment 2. To create the optimal alignment between the sequences requires a single gap to be inserted after the 7th amino acid, as shown in Figure 4.13(b). To deduce the alignment scores, we then simply count the number of identical residues between the sequences: in this example, the unaligned score is 7 and the aligned score is 19.

Usually, we want to compare much longer sequences – often, many hundreds of amino acids in length. The comparison process is consequently now much more complicated, involving the alignment of many non-identical residues and the positioning of gaps, generally of different length, in a variety of locations. Computing the optimal alignment in such scenarios is thus more complex. In principle, we could try to maximise the number of identical matches between the sequences by inserting gaps in an unrestricted way; however, this wouldn't yield a biologically meaningful result. In practice, we introduce scoring penalties (to minimise the number of gaps that are opened) and extension penalties (to inhibit their length, once open): the alignment score is then a function of the number of identical residues and the gap penalties incurred during the alignment process.

```
(a)   Fragment 1:  KTYFPHFDLSHGSAQVKGHG
                   *******
      Fragment 2:  KTYFPHFSDLSHGSAQVKAHG

(b)   Fragment 1:  KTYFPHF-DLSHGSAQVKGHG
                   ******* ********** **
      Fragment 2:  KTYFPHFSDLSHGSAQVKAHG
```

Figure 4.13

Pairwise comparison of sequences of unequal length, (a) unaligned, and (b) aligned by insertion of a gap character to bring the sequences into vertical register. Asterisks denote identical residues: 7 in the unaligned pair, 19 in the aligned pair.

	C	S	T	P	A	G	N	D	E	Q
C	1	0	0	0	0	0	0	0	0	0
S	0	1	0	0	0	0	0	0	0	0
T	0	0	1	0	0	0	0	0	0	0
P	0	0	0	1	0	0	0	0	0	0
A	0	0	0	0	1	0	0	0	0	0
G	0	0	0	0	0	1	0	0	0	0
N	0	0	0	0	0	0	1	0	0	0
D	0	0	0	0	0	0	0	1	0	0
E	0	0	0	0	0	0	0	0	1	0
Q	0	0	0	0	0	0	0	0	0	1

Figure 4.14

Part of the unitary scoring matrix for amino acids.

In this discussion, we've considered only residue identities and have scored them using a **unitary matrix**: *i.e.*, a matrix that assigns a score of 1 to identical residue matches, and 0 to non-identical matches (see Figure 4.14). Unitary matrices are sparse – that is to say, most of their cells are populated with zeros. Consequently, their diagnostic power is poor, because all identical matches carry equal weight and no non-identical matches contribute to the score, even if they're biologically meaningful. Ideally, to improve diagnostic performance, we want to be able to boost the scoring potential of weak but biologically significant signals so that they too can contribute to the overall score, but without the risk of amplifying noise. This tension is the very essence of sequence analysis – the desire to be able to distinguish low-scoring, biologically relevant matches from high-scoring, mathematically significant but biologically irrelevant matches.

To try to inject greater biological insight into the scoring of alignments, amino acid substitution rates have been investigated with respect to different evolutionary distances, and the derived substitution probabilities encapsulated in a variety of **scoring matrices**. These became very popular because they increase the sensitivity of the alignment process, allowing different weights to be assigned both to identical and to non-identical residue matches. However, this increase in sensitivity also brings an increase in noise, because the matrices indiscriminately weight all relationships, whether or not they are ever observed in the sequences being compared (see Box 4.2 for more details) – *i.e.*, random matches are boosted along with those of genuine signals. Unfortunately, this can encourage over-interpretation of results, especially when levels of sequence similarity are low. Care should therefore always be used when trying to understand what alignments mean, because comparison algorithms will always give a mathematically best answer, regardless of its biological significance.

In the next section, we look briefly at some of the characteristics of two of the most popular series of similarity matrices: the **PAM** and **BLOSUM** series.

4.3.2 The PAM and BLOSUM matrices

In Chapter 1, we saw how Margaret Dayhoff's interest in discerning evolutionary relationships from sequence alignments led her to avidly collect protein sequences, and to publish them in the *Atlas of Protein Sequence and Structure*. As part of her research, Dayhoff analysed more than 1,500 mutations in ~70 of the protein families published in the *Atlas*, and used the amino acid substitution data to derive replacement probabilities at various evolutionary distances (Dayhoff *et al.*, 1978).

Box 4.2 Visualising the effects of unitary and similarity scoring

A simple device to help visualise the similarity shared by two sequences is the dotplot. A dotplot is essentially just a matrix (Maizel and Lenk, 1981) in which the sequences being compared are arranged along the *x* and *y* axes, and the matrix cells are populated by the scores of successive pairwise residue comparisons. Using the unitary scoring matrix, all residue identities score 1, and all non-identical matches have zero score. Thus, for identical sequences, the plot is characterised by an unbroken diagonal line; but comparing identical sequences is seldom useful! For the more realistic scenario, where two *similar* sequences are compared, the plot will be characterised by a broken diagonal, where the interrupted regions indicate the locations of the residue mis-matches. The dotplot for the highly similar sequence fragments shown in Figure 4.13 is shown in Figure 4.15.

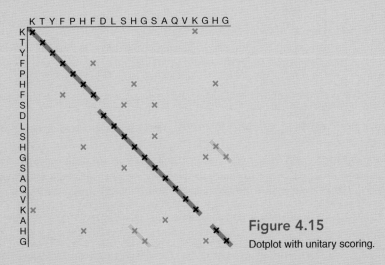

Figure 4.15
Dotplot with unitary scoring.

Within the plot, a number of features are noteworthy. First, the principal diagonal is broken in two places: the initial interruption denotes the insertion of a serine residue in the sequence along the *y*-axis relative to that along the *x*-axis; the second denotes a substitution of glycine for alanine in the *y*-axis sequence. Importantly, the residue insertion causes not just a break in the diagonal, but also an offset – the value of this offset denotes the number of gaps required to bring the two sequences into register. The second feature seen in the plot is a pair of small off-diagonal diagonals. Closer inspection reveals that these denote a repetition at the C terminus of a two-letter 'word', HG – a histidine–glycine pair – that's first seen midway along the sequences. Hence, dotplots are very efficient and powerful tools for pinpointing the positions of internal repeats within sequences. The final point to note is that, aside from the main diagonal and a few off-diagonal internal matches, most elements of this matrix are empty – the matrix is relatively sparse. Overall, the plot gives an at-a-glance summary of the degree of identity between the two sequences, and indicates how they can be lined up – *i.e.*, by joining the dots on the principal (black) diagonal, taking into account mis-matches and offsets resulting from indels.

As already mentioned, the scoring potential of unitary matrices is relatively poor, because only identical matches are scored. This prompted the introduction of a variety of substitution matrices, such as the PAM and BLOSUM series, which allow biologically feasible amino acid replacements also to be counted. To illustrate the effect of similarity scoring, compare the dotplots in Figures 4.15 and 4.16: the latter uses the same pair of sequence fragments as in Figure 4.13, but with residue similarities scored using the BLOSUM 62 substitution matrix. It's immediately apparent that the matrix is no longer sparse, but is densely populated with negative (red) and positive (blue) scores – very few matrix elements are empty, as before (Figure 4.16). The appeal of such matrices

(continued)

is that they increase the sensitivity of sequence comparisons; their inherent disadvantage is that they weight all possible amino acid substitutions, regardless of whether they ever occur in the types of sequence being compared: hence, the background noise they introduce may mask the weak biological signals they were designed to find.

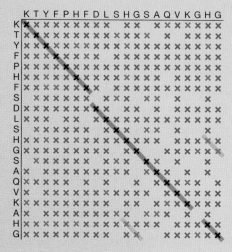

Figure 4.16
Dotplot with BLOSUM 62 scoring.

Here, the dotplot is effectively a signal-to-noise landscape, offering a tangible, graphic expression of the challenge of discriminating signal from noise. When substitution matrices are used in dotplots like this (especially for more realistic, full-length sequences), it becomes essential to filter out the background noise. In practice, this can be done by implementing a sliding-window calculation as a smoothing function, which improves the signal-to-noise ratio.

The pivotal concept behind her work was the 'accepted **point mutation**': *i.e.*, an amino acid replacement (arising from a mutation in the parent gene) that has been accepted by natural selection as the predominant new form within a species. Under such circumstances, the new amino acid is likely to share similar physicochemical properties and hence have a related function to the one it has replaced. As part of Dayhoff's study, not only was the frequency of each amino acid exchange with every other amino acid considered, but also the propensity of each to remain unchanged during a given evolutionary interval – in other words, the amino acid's 'relative mutability' was also evaluated. The information from the different types of mutation observed, coupled with the relative mutabilities of the amino acids, was ultimately combined into a distance-dependent mutation probability matrix. The evolutionary interval was 1 PAM (Point Accepted Mutation), equivalent to 1 accepted point mutation per 100 amino acids.

With this approach, different mutation probability matrices can be derived for any evolutionary distance measured in PAMs, essentially by multiplying the 1 **PAM matrix** by itself. For example, a much larger evolutionary distance is encapsulated in the mutation probability matrix for 250 PAMs. Sequences reflected in this matrix can be considered to have undergone 250 mutation events per 100 residues, leaving 20 per cent of their amino acids unchanged. The latter broadly corresponds to the **Twilight Zone** of sequence similarity (see Box 4.3), and hence led PAM 250 to become the default scoring-matrix in many sequence analysis and alignment packages.

Box 4.3 The Twilight and Midnight Zones

The goal of many sequence analysis tasks is to identify relationships between new, often uncharacterised sequences and those that have been deposited in databases, and hence whose functions (and possibly structures) we hope have already been elucidated. At high levels of sequence identity, where sequence alignment methods perform relatively reliably, this is reasonably straightforward (although, even for highly similar sequences, there are still traps for the unwary). Below about 50 per cent identity, however, it becomes increasingly difficult to establish relationships with confidence, as depicted in Figure 4.17.

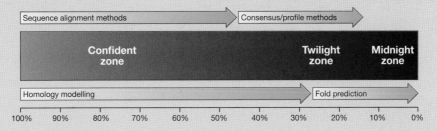

Figure 4.17
Tools for probing the Twilight and Midnight Zones.

Analyses can be pursued with decreasing certainty towards what has been termed the 'Twilight Zone', a region of ~15–25 per cent sequence identity in which alignments may seem plausible to the eye but are not statistically significant – *i.e.*, the same alignment could have arisen by chance (Doolittle, 1986).

Many analytical methods have been devised to penetrate deeper into the Twilight Zone, ranging from pairwise alignment methods to consensus or profile methods. Each of these approaches offers a slightly different perspective, depending on the type of information used in the search (whether implementing pairwise or multiple sequence comparisons, substitution probability matrices or frequency matrices, *etc.*). It is important to appreciate, however, that there's a theoretical limit to the effectiveness of sequence-based comparison techniques. The problem is that, for distantly related proteins, tertiary structures are generally more conserved than are their underlying primary structures. This means that some evolutionary relationships are apparent only at the level of shared structural features, which therefore can't be detected even using the most sensitive sequence comparison methods. The region of identity where sequence comparisons fail completely to detect structural similarity has been termed the Midnight Zone (Rost, 1998). The Midnight Zone also denotes a philosophical limit to our understanding because, in the absence of significant sequence similarity, we cannot know whether structural similarities have arisen as a result of divergent evolution from a common ancestor or as a result of convergent evolution to a particularly favourable or stable packing arrangement.

Exploration of the Midnight Zone requires structural approaches (such as fold recognition), which are outside the scope of this book.

For mathematical convenience, the PAM 250 mutation probability matrix is often expressed as a log-odds matrix, as illustrated in Figure 4.18. Here, positive values denote likely mutations, negative values denote unlikely mutations, and values equal to zero are neutral. The amino acids have been grouped to try to reflect their physicochemical properties. Closer inspection of the matrix reveals that the pattern of substitution probabilities is broadly consistent with our intuition that chemically similar amino acids are more likely to replace each other: the acidic group, the basic group, the acid/amide group, the aliphatic group, and so on. These patterns of replacement reflect the similarity of the roles

C	12																			
P	-3	6																		
G	3	-1	5																	
M	-5	-2	-3	6																
I	-2	-2	-3	2	5															
L	-6	-3	-4	4	2	6														
V	-2	-1	-1	2	4	2	4													
A	-2	1	1	-1	-1	-2	0	2												
S	0	1	1	-2	-1	-3	-1	1	2											
T	-2	0	0	-1	0	-2	0	1	1	3										
N	-4	-1	0	-2	-2	-3	-2	0	1	0	2									
Q	-5	0	-1	-1	-2	-2	-2	0	-1	-1	1	4								
D	-5	-1	1	-3	-2	-4	-2	0	0	0	2	2	4							
E	-5	-1	0	-2	-2	-3	-2	0	0	0	1	2	3	4						
K	-5	-1	-2	0	-2	-3	-2	-1	0	0	1	1	0	0	5					
R	-4	0	-3	0	-2	-3	-2	-2	0	-1	0	1	-1	-1	3	6				
H	-3	0	-2	-2	-2	-2	-2	-1	-1	-1	2	3	1	1	0	2	6			
F	-4	-5	-5	0	1	2	-1	-4	-3	-3	-4	-5	-6	-5	-5	-4	-2	9		
Y	0	-5	-5	-2	-1	-1	-2	-3	-3	-3	-2	-4	-4	-4	-4	-4	0	7	10	
W	-8	-6	-7	-4	-5	-2	-6	-6	-2	-5	-4	-5	-7	-7	-3	2	-3	0	0	17
	C	P	G	M	I	L	V	A	S	T	N	Q	D	E	K	R	H	F	Y	W

Figure 4.18

Log-odds matrix for 250 PAMs, with amino acids broadly grouped according to their physicochemical properties.

of the amino acids in forming protein tertiary structures: *e.g.*, in terms of their size, shape, local charge, **hydrogen-bonding** potential, ability to form salt-bridges.

The PAM matrices saw wide take-up in the bioinformatics community, and, as mentioned above, PAM 250 became a kind of standard for many sequence analysis tasks. Nevertheless, as more sequence data became available during the 1980s, researchers saw opportunities to derive more sensitive and more accurate models of amino acid substitution.

Theoretically, there are two main weaknesses of the PAM model of evolutionary change. First, their substitution rates were derived from alignments of sequences that were at least 85 per cent identical. For most practical purposes, however, sequence analysis tasks are likely to involve the detection of more distant relationships. The problem is, such relationships are only inferred from the PAM model, because matrices corresponding to larger evolutionary intervals are derived by successive multiplication of the original 1 PAM matrix. Thus, any errors or inconsistencies in the 1 PAM matrix are propagated to all others in the series, becoming larger with increasing evolutionary distance.

A number of scientists recognised this; among them were Steve and Jorja Henikoff. They endeavoured to represent distant relationships more explicitly by exploiting information latent in aligned sequences held in their own Blocks database (Blocks is no longer maintained but, in November 2015, was still accessible from the Fred Hutchinson Cancer Research Center). Similar in concept to the PRINTS fingerprint database (see Box 3.4), Blocks contains groups of ungapped motifs – or **blocks** – that characterise particular protein families. The set of BLOcks SUbstitution Matrices they developed is referred to as the BLOSUM series (Henikoff and Henikoff, 1992).

The way in which the matrices were created was simpler than for the PAM matrices, largely because the method didn't involve deriving an evolutionary model, but extrapolated from the alignment data directly to the respective log-odds matrices. Within each block, the constituent sequences were clustered according to a percentage identity greater than some specific cut-off value. Sequences in the same cluster were then weighted as a single sequence, thereby helping to reduce any bias introduced by groups of identical and near-identical sequences (of which there were likely to be many per block). The log-odds matrix resulting from a particular clustering percentage is referred to as the corresponding number BLOSUM matrix: *e.g.*, BLOSUM 62 (see Figure 4.19), which includes substitutions between sequences that are <62 per cent identical; and BLOSUM 80, which includes substitutions between sequences that are <80 per cent identical. Note that, by contrast with PAM matrices, the higher the BLOSUM number, the greater the sequence similarity represented by the matrix.

BLOSUM 62 became a popular scoring matrix, often replacing PAM250 as the default in sequence analysis and alignment programs. This is probably because the Henikoffs published evidence that their matrices performed better in alignments and similarity searches than those based on accepted mutations, and, in particular, that BLOSUM 62 was the single best matrix (Henikoff and Henikoff, 1992; Henikoff and Henikoff, 1993).

Comparing the PAM 250 and BLOSUM 62 matrices shown in Figures 4.18 and 4.19, we can see that key trends are generally preserved in terms of which amino acids are the most and which the least mutable; nevertheless, the numerical details differ. This means that when we use such matrices for database searches or sequence alignment, the details of the program outputs are also likely to differ slightly. When seeking relationships in the Twilight Zone, where the overall resemblance between sequences is slight, such disparities could make the difference between spotting a distant relationship or missing it.

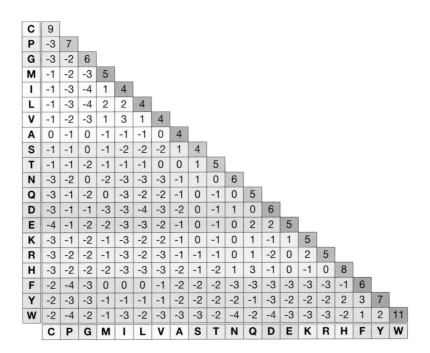

	C	P	G	M	I	L	V	A	S	T	N	Q	D	E	K	R	H	F	Y	W
C	9																			
P	-3	7																		
G	-3	-2	6																	
M	-1	-2	-3	5																
I	-1	-3	-4	1	4															
L	-1	-3	-4	2	2	4														
V	-1	-2	-3	1	3	1	4													
A	0	-1	0	-1	-1	-1	0	4												
S	-1	-1	0	-1	-2	-2	-2	1	4											
T	-1	-1	-2	-1	-1	-1	0	0	1	5										
N	-3	-2	0	-2	-3	-3	-3	-1	1	0	6									
Q	-3	-1	-2	0	-3	-2	-2	-1	0	-1	0	5								
D	-3	-1	-1	-3	-3	-4	-3	-2	0	-1	1	0	6							
E	-4	-1	-2	-2	-3	-3	-2	-1	0	-1	0	2	2	5						
K	-3	-1	-2	-1	-3	-2	-2	-1	0	-1	0	1	-1	1	5					
R	-3	-2	-2	-1	-3	-2	-3	-1	-1	-1	0	1	-2	0	2	5				
H	-3	-2	-2	-2	-3	-3	-3	-2	-1	-2	1	3	-1	0	-1	0	8			
F	-2	-4	-3	0	0	0	-1	-2	-2	-2	-3	-3	-3	-3	-3	-3	-1	6		
Y	-2	-3	-3	-1	-1	-1	-1	-2	-2	-2	-2	-1	-3	-2	-2	-2	2	3	7	
W	-2	-4	-2	-1	-3	-2	-3	-3	-3	-2	-4	-2	-4	-3	-3	-3	-2	1	2	11

Figure 4.19

BLOSUM 62 matrix, with amino acids broadly grouped according to their physicochemical properties.

This is why appropriate matrix selection is important and why it is wise to compare the results from different analysis and alignment tools.

4.3.3 Tools for global and local alignment

By now, it should be evident that when we create and score sequence alignments we are invoking particular mathematical models whose behaviour we can manipulate by modifying a range of parameters (such as penalties for opening and extending gaps, the underlying evolutionary model for scoring amino acid substitutions, structural constraints, and so on). To this extent, there is no right or wrong alignment; rather, there are different models that reflect different biological perspectives.

Two general ways of thinking about alignments involve consideration either of the degree of similarity shared across the full sequence lengths or of the similarity that's confined to specific regions of the sequences: the former results in a global alignment, the latter produces a local alignment. It's important to appreciate these different perspectives: often, it's the features that give rise to local similarity (*e.g.*, binding sites, active sites, modification sites, hydrophobic regions, structural or functional domains) that we're most interested in (see Box 4.4); under these circumstances, algorithms that attempt to optimise local alignment may produce more sensitive and biologically meaningful results than those that attempt to optimise alignment over the full sequence lengths.

It isn't in the scope of this book to describe in detail algorithms for global and local alignment; instead, we will mention some of the most popular local alignment methods that have been optimised specifically for database searching. '**Dynamic programming**' methods, like those of Needleman and Wunsch (1970) and Smith-Waterman (1981), compute the optimal alignment between two sequences, for a given scoring system, in a time that's proportional to the product of the lengths of the sequences being compared. When scaling up to a full database search, the computation time increases in proportion to the database size, making such calculations impractically slow. To address this problem, approximate (**heuristic**) algorithms are used to reduce the time to find good (not necessarily optimal) alignments. Amongst the first implementations of such heuristic algorithms was the FASTP program, developed in 1985 by Lipman and Pearson as an efficient modification of the earlier Wilbur and Lipman algorithm (1983); later, this yielded to their improved FASTA program (Pearson and Lipman, 1988), which was quickly followed by the Basic Local Alignment Search Tool (more commonly known as BLAST), developed by Altschul *et al.* in 1990; this, in turn, yielded to subsequent innovations, such as 'gapped BLAST' and **PSI-BLAST** (Altschul *et al.*, 1997).

The algorithm in FASTA is based on the concept of identifying '**k-tuples**': these are short words, or sequence fragments, that occur in both the sequences being compared, and are detected by sliding a window of length k successively along the query sequence and each database sequence. The value of k is important, because it determines the speed and sensitivity of the search: *e.g.*, a value of 1 will take into account identities between individual amino acids, a value of 2 will consider identical amino acid pairs, and so on; hence, the greater k, the faster the search but the lower its sensitivity. Comparing these words or k-tuples, and their relative offsets, is like focusing on the diagonal matches in a **dotplot** – recall Box 4.2. FASTA locates the ten highest-scoring diagonal regions, based on the number of k-tuple matches and the distance between them; these are then re-scored using a substitution matrix (originally, PAM 250 for proteins, but online implementations also allow the BLOSUM matrices to be

Box 4.4 Identifying local features

As mentioned in Section 4.3.4, in the process of sequence analysis, we're often interested in identifying features that give rise to local similarity, such as binding sites, active sites, hydrophobic regions, and so on. A method that's found wide practical use in the detection of local sequence features is the humble property profile, created using a sliding-window technique. Here, we illustrate how these can be applied to the detection of transmembrane (TM) domains.

We saw in Chapter 2 that amino acids have overlapping, context-dependent properties, which can be measured in a variety of different ways. A property of particular interest is hydrophobicity or hydropathy, because this can be used to ascertain the likelihood of a protein being globular (soluble) or membrane-bound (insoluble). A major driving force in the folding of globular proteins is considered to be the 'hydrophobic effect', the tendency of hydrophobic amino acid residues to sequester within the protein interior, away from surrounding water. In the primary structures of globular proteins, hydrophobic residues may be relatively far apart in sequence, but are nevertheless capable of forming stable hydrophobic pockets during the dynamic dance of protein folding. Sometimes, however, elements of secondary structure may be partly or fully buried within the hydrophobic core. When this happens, characteristic patterns of hydrophobicity in the sequence bear witness to the nature of the buried structure(s). This is particularly noticeable for membrane proteins, whose architectural components span cells' lipid milieux to form, sometimes vast, TM scaffolds, which are predominantly hydrophobic in nature. So how do we identify these computationally?

The starting point is to assign a hydropathy value to each amino acid in the sequence under investigation. The table in Box 2.3 listed two hydropathy scales, which assigned slightly different individual values to the amino acids. Figure 4.20 illustrates the hydropathic rankings arising from several other scales: these show similar trends overall, but nevertheless differ in detail, owing to the different ways in which the hydrophobic properties were measured.

Zimmerman	LYFPIVKCMHRAEDTWSGNQ
von Heijne	FILVWAMGTSYQCNPHKEDR
Efremov	IFLCMVWYHAGTNRDQSPEK
Eisenberg	IFVLWMAGCYPTSHENQDKR
Cornette	FILVMHYCWAQRTGESPNKD
Kyte	IVLFCMAGTSWYPHDNEQKR
Rose	CFIVMLWHYAGTSRPNDEQK
Sweet	FYILMVWCTAPSRHGKQNED

Figure 4.20

Amino acid hydropathic rankings according to different hydropathy scales.

Having chosen a suitable scale, commencing from the N terminus, the hydropathy values of amino acids that fall in a window of defined size are averaged. The window size is determined by the nature of the feature being investigated: *e.g.*, for a TM helix, the window would need to encompass ~20 amino acids, as this is the number of residues in helical conformation needed to span a membrane; features in β-conformation, or in α-conformation in globular proteins are likely to be shorter, and smaller window sizes (*e.g.*, ~9 residues) might be more appropriate (the selected window size will determine the level of smoothing in the final plot). The window is moved on, stepwise through the sequence, and the calculation is performed for each window, until the C terminus is reached, as illustrated in Figure 4.21.

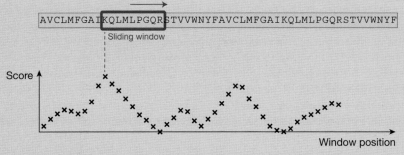

Figure 4.21

Sliding window calculating the hydropathy profile for an amino acid sequence.

(continued)

The resulting profile is characterised by a series of peaks and troughs, as the sliding window delineates the most hydrophobic and most hydrophilic parts of the sequence. A typical hydropathy profile for a TM receptor is shown in Figure 4.22, which was created using the online DAS transmembrane prediction server[a] (Cserzo *et al.*, 2002). To help interpret the plot, two significance cut-offs are provided: peaks above the strict cut-off (solid line) are considered likely to be TM domains; those falling between the strict and loose cut-off (dotted line) are much less confident predictions; peaks below the loose cut-off are highly unlikely to be TM domains.

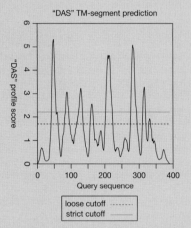

"DAS" TM-segment prediction

loose cutoff -------
strict cutoff ———

Figure 4.22

Typical hydropathy profile.

A weakness of plots like this is that while they are good at highlighting TM domains that are significantly hydrophobic, they struggle to pinpoint hydrophilic TM domains. This is a general problem for TM-domain prediction tools, which has resulted in a number of annotation errors in databases. In the profile shown, how many TM domains is this sequence likely to possess?

As mentioned earlier, sliding-window techniques aren't simply confined to the detection of TM domains, but provide a simple, versatile and effective approach for depicting a range of amino acid properties. Hence, for example, they may also be used to illustrate regions of sequences that are most likely to be buried, those that are likely to be the most flexible, those that are most likely to be solvent exposed, and so on.

---

[a] http://www.sbc.su.se/~miklos/DAS

used) to permit both runs of identities and conservative replacements to contribute to the score. For each of these best regions, a sub-region with maximal score is found (termed the 'initial region') and, depending on their scores and locations, these are joined by applying a 'join penalty' (rather like a gap penalty), to create an optimal alignment of initial regions with maximal score. This score is used to rank sequences during a database search. The top-scoring database matches are then aligned using a modification of the optimisation method described by Needleman and Wunsch (1970) and Smith and Waterman (1981).

FASTA was widely used for several years, but its use was more or less superseded by the development of BLAST (Altschul *et al.*, 1990), which benefitted from greater efficiency and built-in statistics, and later, by implementations of 'gapped-' and PSI-BLAST. The BLAST algorithm is relatively straightforward, having as its central concept 'segment-' or word-pairs (rather like the k-tuples of FASTA) – when comparing two sequences, these are pairs of same-length sub-sequences that form an ungapped alignment. BLAST computes all such word-pairs between a query sequence and those in a database, according to a scoring threshold. When two non-overlapping word-pairs on the same diagonal are found within a defined distance of each other, an extension step is invoked – a gapped alignment algorithm uses dynamic programming to extend the pair of aligned residues in both directions (Altschul *et al.*, 1997) and yield the final gapped alignment.

During this process, for the original implementations of BLAST, the best-scoring of all the extended segments (the Maximal Segment Pair, or MSP) was assigned a probability value (or **p-value**), denoting the probability of achieving at least one score greater than or equal to that of the MSP by chance. Later versions of BLAST introduced the **e-value**, denoting the number of alignments with score greater than or equal to the calculated alignment score likely to be found by chance when comparing random sequences of the same length and composition as the query to a database of the same size and composition. This ability to ascribe a measure of statistical significance to retrieved matches very soon made BLAST the tool of choice for database searching.

In Position-Specific Iterated BLAST, or PSI-BLAST (Altschul *et al.*, 1997), a **Position-Specific Scoring Matrix (PSSM)**, or profile, is created from sequences identified above a specified threshold by an initial BLAST run. Successive BLAST searches are then made, each time adding newly identified sequences to the profile and increasing the **sensitivity** of subsequent searches. The appeal of PSI-BLAST is its speed and ease-of-use. By combining generic pairwise database comparisons with more sensitive, profile-based searches, it appears to provide the best of both worlds. Overall, however, it lacks the power of motif-based methods, and its iterative nature necessitates close supervision – if a false match is incorporated into the profile, subsequent searches will retrieve more false matches, as the profile is successively corrupted. Used wisely, PSI-BLAST is nevertheless an important component in the sequence analyst's armoury.

Today, a wide range of implementations of the BLAST and FASTA database search tools can be found online. These programs are swift and easy to use, but they are not the end of the sequence analysis line. Often, the results they produce will provide broad brush-strokes, indicating a generic family to which a query sequence may belong; however, they are likely to be insensitive to the specific traits that differentiate closely related sequences – the ones that allow curators, for example, to assign possible functions or structural features to newly deposited database sequences. To achieve this kind of resolution usually requires additional analyses, including those that seek to identify particular functional motifs or structural domains, or those designed to elucidate unique functional determinants across extended protein family hierarchies. We will explore some of these methods in the next sections; but first, a warning.

Proteins exhibit rich evolutionary relationships and complex molecular interactions, and hence present many challenges for *in silico* investigations. Problems arise if we lose sight of the underlying biology and fail to understand the limitations of the methods we're using. Tools like BLAST and FASTA, and those we'll discuss next, are essentially biology-unaware, character-string matchers. If we're to make inferences of evolutionary relatedness and of precise biological functions, we must temper our expectations by considering how much light such tools can realistically shed on biology's dynamic soup of molecular and cellular components. When we extrapolate from a string of letters returned by a database search algorithm to the assignment of a piece of functional annotation to a new database sequence, it's imperative to examine evidence from different quarters and, guided by the relevant biological context, then to evaluate its biological significance, irrespective of its statistical significance.

4.3.4   Tools for comparing multiple sequences

As already mentioned, a number of different methods have been developed to help identify conserved functional sites, specific functional determinants that differentiate close family members, structural domains, and other characteristic sequence features. Many of these approaches hinge on the ability to make connections between more than two sequences; most necessitate the creation of multiple sequence alignments, because only in this context can conserved family traits be seen. Multiple alignments provide succinct overviews of the relationships between members of gene families, allowing us to recognise aspects that are similar, to identify features that are different, and to determine whether observed similarities and differences are meaningful.

As in the previous section, it isn't the goal of this book to provide detailed descriptions of the algorithms that underpin the process of multiple sequence alignment; instead, we'll briefly mention some general approaches, before focusing on how alignments are exploited to create diagnostic signatures suitable for identifying generic functional sites and family-specific traits.

Computing multiple sequence alignments is non-trivial. As we saw in Section 4.3.4, dynamic programming algorithms guarantee to find the optimal alignment between two sequences. When extended to multiple sequences, however, these algorithms become computationally prohibitive, and heuristics become necessary to reduce the time to find good (not necessarily optimal) alignments. The majority of alignment packages are based on the 'progressive algorithm' (Hogeweg and Hesper, 1984), a greedy heuristic assembly method that aligns sequences in pairs, following the branching order of a family tree, or **guide tree**. The progressive algorithm can be embedded in an iterative loop, in which both guide tree and alignment are re-estimated until convergence: example implementations of this approach include **ClustalW** (Thompson *et al.*, 1994), **T-Coffee** (Notredame *et al.*, 2000), **MUSCLE** (Edgar, 2004), **PRALINE** (Simossis and Heringa, 2005), **ProbCons** (Do *et al.*, 2005) and **MAFTT** (Katoh *et al.*, 2005). A related approach is embodied in **PRANK**, a phylogeny-aware progressive alignment program that, uniquely, distinguishes insertions from deletions, and hence gives rise to alignments with very distinctive gap distributions (Löytynoja and Goldman, 2005, 2010).

The number of alignment tools available today is overwhelming. Many, like those mentioned above, use similar algorithmic approaches. To obviate the need to choose between programs, several methods have been combined into a consensus package, called **M-Coffee** – as the name suggests, this is based on T-Coffee (Wallace *et al.*, 2006). The program computes alternative alignments using eight or more of the most accurate programs, combining them into a single result that's consistent with the original alignments. In so doing, M-Coffee can produce 'better' alignments than the best individual method almost 70 per cent of the time (Wallace *et al.*, 2006). Nevertheless, this program is only as good as the methods it uses; hence it doesn't provide a panacea for the problem of computing reliable alignments for remote homologues. Some progress towards improving the alignment of distantly related sequences has been made by including structural information, albeit tempered with the usual caveat about the scarce availability of protein 3D structures, and the disparity between sequence- and structure-based alignments (this issue is discussed further in Section 5.8.4).

The above-mentioned multiple alignment programs represent just a small sample of the many that have been developed to date. More detailed information about these and other tools, how they work, and how they differ, can be found in a variety of reviews, such as those by Gotoh (1999), Wallace *et al.* (2005), Edgar *et al.* (2006) and Notredame (2007). Although many enhancements over the years have led to programs that produce alignments with arguably better quality, and sometimes with reduced computational costs, ClustalW is probably still the most popular, owing to its ease-of-use and portability across different computer systems. In 1997, a new version was released (ClustalX), wrapped in a graphical user interface, which allowed users to visualise alignments more effectively and to select, automatically re-align, and hence refine, parts of the alignment deemed to be of lower quality (Thompson *et al.*, 1997).

The need to correct errors and improve automatically-generated alignments was recognised soon after first-generation multiple alignment tools, like Clustal (Higgins and Sharp, 1988), were made available. This resulted in a proliferation of manual alignment editors, such as **SOMAP** (Parry-Smith and Attwood, 1991), **SEAVIEW** (Galtier *et al.*, 1996), **CINEMA** (Parry-Smith *et al.*, 1998; Sinnott *et al.*, 2004; Pettifer *et al.*, 2009), **ANTHEPROT** (Deleage *et al.*, 2001), and **JalView** (Clamp *et al.*, 2004); many of these, and other editors[4], are still used today, and will continue to play an important role while heuristic algorithms produce good, but not optimal, sequence alignments.

4.3.5 Alignment-based analysis methods

At the start of this chapter, we discussed the importance of adding annotation to uncharacterised sequences in order to give them meaning and to make them reusable. In the second part, we reviewed some of the fundamental tools that are routinely used to visualise and quantify the similarity between pairs and groups of sequences, in order to expose and better understand the relationships between them. This final section brings the story full circle, as we explore some of the principal ways in which the information concealed in groups of sequences can be used for the purpose of sequence annotation. It's important to understand something of the nature of this information, because protein families are hierarchical, and the functional insights we may gain from exploring them differ at different levels of the hierarchy – an overview of this complexity is given in Box 4.5.

All of these approaches use multiple alignments as their starting points; they differ both with respect to how they use regions of similarity and difference within alignments to diagnose structural, functional and evolutionary relationships, and in terms of their diagnostic strengths and weaknesses. Figure 4.24 gives an overview of the main alignment-based methods, which can be broadly categorised according to whether they exploit single motifs (*e.g.*, consensus expressions), multiple motifs (*e.g.*, fingerprints), or complete domains (*e.g.*, profiles). These methods are described in the sections that follow.

Consensus expressions

The analysis method that is conceptually most easy to understand characterises a group of sequences, or some part of a group of sequences, in terms of their alignment consensus. Consider the (admittedly rather trivial) example in Figure 4.25, which shows a 20-residue chunk of an alignment of highly similar haemoglobin α subunits. The consensus sequence beneath the alignment notes those columns in which residues are completely conserved – the paler the background colour, the more conserved the alignment position (here, 13 positions are fully conserved).

---

[4]http://en.wikipedia.org/wiki/List_of_alignment_visualization_software

Box 4.5 Protein family hierarchies

When protein sequences are aligned, different types of relationship become apparent. Often, we can discern regions of the alignment that are shared by all of the constituent sequences, while other regions may be common to only a few of them. If we are to try to make structural, functional and evolutionary inferences from alignments, it's important to understand the level of the hierarchy to which our inferences relate, because the functional conclusions we can draw safely are related to the level of granularity at which we view the alignment. Figure 4.23 illustrates four different levels of a family hierarchy.

In Figure 4.23, motifs are coloured according to the level of the hierarchy they represent. The purple motifs are those that delineate a superfamily; orange, blue, green, pink and yellow motifs denote families; shades of those colours are subfamily-specific motifs; and grey motifs pinpoint a domain family.

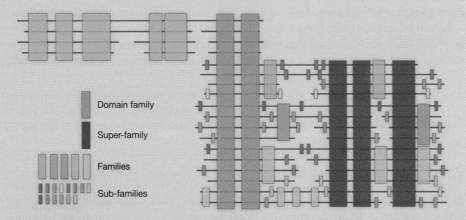

Figure 4.23

The hierarchical nature of protein families.

Understanding the hierarchy is important from both diagnostic and functional points of view. For example, if we want to be able to identify members of a superfamily, we need to focus on those regions of the alignment that are characteristic of all family members, otherwise the discriminator we develop is likely to be too selective – *i.e.*, it will be biased to a particular family, or perhaps to a few families, but will not be representative of the entire superfamily. Conversely, if we want to be able to identify members of a subfamily, then we must focus on those regions of the alignment that are unique to the specific subfamily, otherwise the discriminator will not be selective enough, but will cross-react with other closely-related subtypes. Finally, if we want to be able to identify members of a domain family, then clearly we have to focus on regions of the alignment that are characteristic of the domain, otherwise, again, the discriminator may be selectively biased towards the domains found in only a subset of sequences that actually possess the domain.

From a functional perspective, members of a superfamily may share the same generic, high-level function (*e.g.*, signal transduction), members of a family may share a specific function (*e.g.*, binding to a particular ligand), while members of a subfamily may exploit the same pathways by which the proteins effect their functions at the molecular and cellular level (*e.g.*, binding to a specific G protein and stimulating a given intracellular cascade). Members of a domain family, by contrast, will share the function of the domain (*e.g.*, facilitating protein–protein interaction), independently of the high-level functions of their host proteins.

The diagnostic approaches used by family-based resources like PROSITE, PRINTS and Pfam target different levels of the protein hierarchy, with different measures of success and different functional consequences. For InterPro, which endeavours to amalgamate these different perspectives, the challenges are legion, not least because it isn't always clear where the boundaries of a particular discriminator have been drawn (usually, with confusing consequences). Some of these issues are explored further in this chapter.

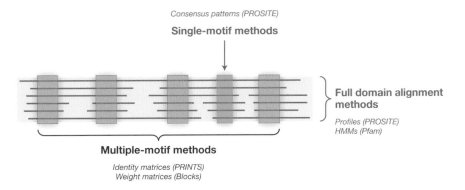

Figure 4.24

The main alignment-based methods for diagnosing members of protein families, and the databases they underpin.

HBA_MARMA	K	T	Y	F	P	H	F	D	L	S	H	G	S	A	Q	I	Q	G	H	G
HBA_SPEPA	K	T	Y	F	P	H	F	D	L	S	H	G	S	A	Q	V	Q	G	H	G
HBA_SPEBE	K	T	Y	F	P	H	F	D	L	S	H	G	S	A	Q	L	Q	G	H	G
HEA_TAMMR	K	T	Y	F	P	H	F	D	L	S	H	G	S	A	Q	V	Q	G	H	G
HBA_SPETO	K	T	Y	F	P	H	F	D	L	S	H	G	S	A	Q	V	Q	G	H	G
HBA_SPECI	K	T	Y	F	P	H	F	D	L	S	H	G	S	A	Q	V	Q	G	H	G
HBA_PANTR	K	T	Y	F	P	H	F	D	L	S	H	G	S	A	Q	V	K	G	H	G
HBA_HUMAN	K	T	Y	F	P	H	F	D	L	S	H	G	S	A	Q	V	K	G	H	G
HBA1_OTOCR	K	T	Y	F	P	Q	F	D	L	S	H	G	S	A	Q	V	K	G	H	G
HBA_MACMU	K	T	Y	F	P	H	F	D	L	S	H	G	S	A	Q	V	K	G	H	G
HBA_ANTPA	K	T	Y	F	P	H	F	D	L	S	H	G	S	A	Q	V	K	G	H	G
HBA_MYOVE	K	T	Y	F	P	H	F	D	L	S	H	G	S	A	Q	V	K	G	H	G
HBA_CYNSP	K	T	Y	F	P	H	F	D	L	A	H	G	S	P	Q	V	K	G	H	G
HBA_GORGO	K	T	Y	F	P	H	F	D	L	S	H	G	S	A	Q	V	K	G	H	G
HBA_MACNE	K	T	Y	F	P	H	F	D	L	S	H	G	S	A	Q	V	K	G	H	G
HBA_SACMY	K	T	Y	F	P	H	F	D	L	S	H	G	S	A	Q	V	K	G	H	G
HBA_ROUAE	K	T	Y	F	P	H	F	D	L	S	H	G	S	A	Q	V	K	G	H	G
HBAZ_OTOCR	K	T	Y	F	P	H	F	D	L	S	H	G	S	T	Q	V	K	G	H	G
HBA_TARBA	K	T	Y	F	P	H	F	D	L	S	H	G	S	S	Q	V	K	G	H	G
HBA_TAPGE	K	T	Y	F	P	H	F	D	L	S	H	G	S	S	Q	V	K	G	H	G
HBA_CERAT	K	T	Y	F	P	H	F	N	L	S	H	G	S	D	Q	V	K	G	H	G
HBA_MANSP	K	T	Y	F	P	H	F	N	L	S	H	G	S	D	Q	V	K	G	H	G
HBA_PAPCY	K	T	Y	F	P	H	F	D	L	S	H	G	S	D	Q	V	N	K	H	G
HBA_PAPAN	K	T	Y	F	P	H	F	D	L	S	H	G	S	D	Q	V	N	K	H	G
HBA_PONPY	K	T	Y	F	P	H	F	D	L	S	H	G	S	A	Q	V	K	D	H	G
HBA_NYCCO	K	T	Y	F	P	H	F	D	L	S	H	G	S	A	Q	V	K	A	H	G
HBA_LORTA	K	T	Y	F	P	H	F	D	L	S	H	G	S	A	Q	V	K	A	H	G
Consensus	K	T	Y	F	P		F		L		H	G	S		Q				H	G

Figure 4.25

Alignment (partial) of haemoglobin α chains, with its consensus beneath.

With this information, we can derive a diagnostic expression, which we can use to discover other sequences that share the same features. We could just use the conserved residues, and mark the non-conserved columns with a wild card, x (which permits any residue to occur), thus:

K-T-Y-F-P-x-F-x-L-x-H-G-S-x-Q-x-x-x-H-G

Such an expression would identify any sequence containing exactly these residues at the specified positions (KTYFP...). However, a sequence containing serine in place of threonine at position 2 would not be matched, even though this is a biologically conservative replacement. This is because the outcome of this type of pattern matching is binary: *i.e.*, a sequence either matches the expression exactly or it is counted as a mismatch. This approach gave rise to the PROSITE 'pattern' database, and its simplicity doubtless led to its wide appeal.

In the example above, the wild cards in the expression lend it a degree of flexibility but, in so doing, they reduce its **specificity**. For relatively long, well-conserved sequence fragments, like our α haemoglobin example, this is unlikely to cause problems. However, the shorter the sequence, and the lower the degree of conservation in the alignment, the greater the risk of identifying sequences that are similar but biologically unrelated to the expression. For example, consider the expression used to characterise members of the parathyroid hormone family (PS00335)[5], which could be found in the November 2015 release of PROSITE:

```
V-S-E-x-Q-x(2)-H-x(2)-G
```

In this 11-residue pattern, almost half the residues are degenerate (anything goes): when scanned against the November 2015 release of UniProtKB/Swiss-Prot, four of the 21 matches are found to be false. If we reduce the length of the expression by removing the first two residues, this generates a staggering 724 matches, most of which (707) are false. This demonstrates the importance of ensuring that consensus expressions are sufficiently long to be diagnostically useful. Another way to increase their specificity is to replace wild cards with biologically meaningful residue groups. This involves examining the types of amino acid residue that occupy particular alignment columns, and grouping those that share physicochemical properties. Legitimate sequence matches are then restricted to those that contain residues specified in the groups. In this way, the α haemoglobin expression above can be re-written using the following notation, as specified by the PROSITE syntax[6]:

```
K-T-Y-F-P-[HQ]-F-[DN]-L-[SA]-H-G-S-x-Q-[IVL]-[KQN]-x-H-G
```

Here, it's biologically appropriate to group HQ, DN, SA, IVL and KQN, because residues within these groups are broadly similar (to verify that this is true, compare how the PAM 250 and BLOSUM 62 matrices score their respective substitutions). For the remaining two positions in the alignment, however, the observed residues are APTSD and GKDA (columns 14 and 18 respectively). It's much less obvious that these groups are internally consistent, in terms of shared properties, so it's reasonable to use wild cards for these alignment positions, and the specificity of the expression overall will not be significantly sacrificed.

Twenty years ago, when sequence databases were smaller and less diverse than they are today, simple consensus patterns were able to perform well; this is no longer true across the board, because their diagnostic abilities are being challenged by the size and contents of current sequence databases. A PROSITE pattern whose diagnostic power has been eroded over the years is that for rhodopsin-like GPCRs (PS00237[7]), as shown

---

[5] http://prosite.expasy.org/PS00335
[6] http://prosite.expasy.org/scanprosite/scanprosite-doc.html#pattern_syntax
[7] http://prosite.expasy.org/cgi-bin/prosite/get-prosite-entry?PS00237

```
(a)   PA   [GSTALIVMC]-[GSTAPDE]-{EDPKRH}-x(2)-[LIVMNG]-x(2)-[LIVMFT]-[GSTAN]-
      PA   [LIVMFYWAS]-[DEN]-R-[FYWCH]-x(2)-[LIVM].
      NR   /RELEASE=20,22654;
      NR   /TOTAL=116(116); /POSITIVE=115(115); /UNKNOWN=0(0); /FALSE_POS=1(1);
      NR   /FALSE_NEG=0(0);

(b)   PA   [GSTALIVMFYWC]-[GSTANCPDE]-{EDPKRH}-x-{PQ}-[LIVMNQGA]-{RK}-{RK}-[LIVMFT]-
      PA   [GSTANC]-[LIVMFYWSTAC]-[DENH]-R-[FYWCSH]-{PE}-x-[LIVM].
      NR   /RELEASE=2015_11,549832;
      NR   /TOTAL=2126(2122); /POSITIVE=1976(1973); /UNKNOWN=0(0);
      NR   /FALSE_POS=150(149); /FALSE_NEG=449; /PARTIAL=32;
```

Figure 4.26

PROSITE pattern for rhodopsin-like GPCRs from release 8.0, November 1991 (a) and 20.120, November 2015 (b). The PA lines show the consensus expression; the NR lines indicate the release of the sequence database that was searched (Swiss-Prot in (a), UniProtKB/Swiss-Prot in (b)), and numerical results of the search – figures in parentheses indicate the number of proteins in which the matches were found: *e.g.*, 2,126 matches in 2,122 sequences indicates that four proteins match the pattern twice.

in Figure 4.26. This 17-residue fragment encapsulates the C-terminal end of the third TM domain of the rhodopsin-like G protein-coupled receptors (GPCRs) – this bears a conserved arginine residue, believed to be the G protein interaction site. Several features are worthy of note. Some alignment positions are degenerate: *e.g.*, positions 1 and 11 originally allowed nine different amino acid residues to occur, and in the November 2015 version allow up to 12; six wild cards were used in 1991, but only two in 2015 – the remaining four tolerate any residue except the two specified in curly brackets. Only one position is conserved.

The database searched in 1991 (Swiss-Prot) contained 22,654 sequences; that in November 2015 (UniProtKB) contained 549,832 sequences. The consequences of the ~24-fold increase in the size of the underlying sequence database are witnessed in the Numerical Results (NR). In 1991 (Figure 4.26(a)), the pattern matched 116 sequences: of these, 115 were correct, and one was false. In 2015 (Figure 4.26(b)), of 2,126 matches to the pattern, 1,976 were matched correctly; 150 were incorrect; 449 family members were missed; and there were 32 partial matches (*i.e.*, sequence fragments). These results illustrate that a match to a pattern isn't necessarily correct and a mismatch isn't necessarily false. Moreover, taken together, the figures show that, during the last 20 years, the overall error rate of this expression has degraded from <1 per cent to ~30 per cent.

Nevertheless, the current expression represents the best compromise between the ability of the pattern to diagnose true family members (**true-positive matches**) correctly and its tendency to match non-members (**false-positive matches**) by mistake, while minimising the number of true family members that it fails to match (**false negatives**). This delicate balancing has required successive revisions to the expression, to better reflect the superfamily divergence witnessed in the growing sequence database. The tension between the tendency of diagnostic tools to match sequences incorrectly and to fail to match true family members lies at the heart of efforts to optimise the diagnostic potency of sequence analysis techniques (see Box 4.6). For consensus expressions in particular, false negatives are a fundamental limitation, because patterns can be optimised to recognise known sequence variants, but can't be trained to identify unknown, or previously unseen, variants – we can only know that a pattern has failed to match a sequence if we have prior knowledge about it, or have gained additional information from other sources.

Box 4.6 Resolving true and false matches

A common task of sequence analysis is to design computational tools to help identify database sequences that are functionally or structurally related to some, usually uncharacterised, query sequence. The challenge is to determine which of the matched database sequences are truly related (so-called true-positive matches) and which are unrelated (the true-negative sequences) – see Figure 4.27(a). In any database search, at a given scoring threshold (the green bar in Figure 4.27(b)), it is likely that some unrelated sequences will be identified erroneously (false-positive matches), and some related sequences won't be identified at all (false-negative sequences).

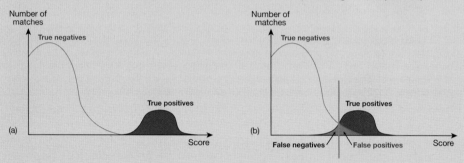

Figure 4.27

Discriminating true-positive from true-negative sequences.

    The best diagnostic performance is achieved when all (or most) true family members are captured, with few (or no) false matches or missed sequences. In practice, this means obtaining the best possible resolution between the score distributions of true-positive and true-negative sequences, reducing (or, better, eliminating) the overlap between them. This is important, because, for sequences within the region of overlap, it can be difficult or impossible to determine which are really correct and which are genuinely false. Statistical approaches are used to assign confidence levels to such matches; however, while matches above the scoring threshold may be statistically significant, they are not necessarily biologically relevant. This is a tough sequence analysis nut to crack.

    To address some of these issues, PROSITE began to complement some of its weaker patterns with diagnostically more powerful profiles (Bairoch and Bucher, 1994). We discuss profiles at the end of the chapter. Before doing so, however, we will briefly explore one other use of consensus expressions. The patterns we've described above are family-specific – they are designed to concisely capture the conserved essence of a protein family, and to use this short (usually 15–25 residue) distillate to recognise other members of the same family. But patterns can also be used to pinpoint non-family-specific features of sequences, such as those that define sugar-attachment sites, glycosylation, phosphorylation or other **PTM** sites, and so on. These features are generally much smaller (sometimes only 3–4 residues in length) – some typical examples are given in Table 4.1. The same syntax is used as before: residues within square brackets are allowed at a particular position; those in curly brackets are disallowed at that position; x denotes any residue; and numbers in parentheses indicate the number of residues of the specified type.

    The short, non-specific nature of expressions like those in Table 4.1 renders them diagnostically unreliable. In isolation, matches to them are therefore relatively meaningless and can only indicate possible locations of such sites, should such functionality

Table 4.1

Example functional sites and consensus patterns used to detect them.

Functional site	Pattern
Amidation	x-G-[RK]-[RK]
ATP/GTP-binding (P-loop)	[AG]-x(4)-G-K-[ST]
N-glycosylation	N-{P}-[ST]-{P}
N-myristoylation	G-{EDRKHPFYW}-x(2)-[STAGCN]-{P}
Casein kinase II phosphorylation	[ST]-x(2)-[DE]
Protein kinase C phosphorylation	[ST]-x-[RK]

be relevant to the proteins in which they're found (which can only be verified experimentally). Consequently, when searching PROSITE, these frequently occurring patterns should be 'switched off' (the Web pages provide an option to do this), unless there's good reason to think that the query sequence may be phosphorylated, glycosylated, or whatever.

At one time, such patterns were called 'rules', to differentiate them from the family-specific expressions; although this term now seems to have fallen out of favour, you may still find articles, books and other documents that use it. However, the term should not be confused with **ProRule**: this is a database containing additional information to try to enhance the diagnostic performance of various PROSITE profiles and patterns – *e.g.*, such as the positions of structurally or functionally critical amino acid residues, and the conditions they must fulfill in order to play their particular biological roles (Sigrist *et al.*, 2005).

Fingerprints

As we've just seen, the simplistic nature of consensus expressions has diagnostic limitations: they perform well in identifying short, conserved sequence features, but are less efficient at diagnosing members of divergent superfamilies. One of their principal drawbacks hinges on the philosophy of using single motifs. In practice, when we examine multiple sequence alignments, it is usual to find not one, but several motifs. Diagnostically, it makes sense to use many, or all, of them to build a characteristic signature or 'fingerprint' of the family – intuitively, we know that the more information there is in a fingerprint (in this case, the more motifs), the greater its potency. As diagnostic tools, however, fingerprints are more than just collections of motifs – the order in which motifs appear in alignments and the distances between them are also key contributors to the underlying analytical model. The concept of a protein fingerprint was introduced in Box 3.4 in fairly abstract terms. Here, we add flesh to the concept and explain how fingerprints are created and used in practice.

Figure 4.28 illustrates part of an alignment of haemoglobin α chains, with four motifs highlighted (the rectangular boxes). From **seed alignments** like this, ungapped motifs are excised and used to scan a version of UniProtKB from which sequence fragments have been removed. The database-scanning algorithm translates the motifs into a set of simple identity or frequency matrices – no scoring matrix is used to weight the results, as shown in Figure 4.29. This means that the resulting matrices are relatively sparse, and hence that the searches are fairly **selective**. It's partly this selectivity that lends fingerprints their ability to differentiate highly similar sequences and to resolve them into closely related families and subfamilies; the corollary of this is that the method is

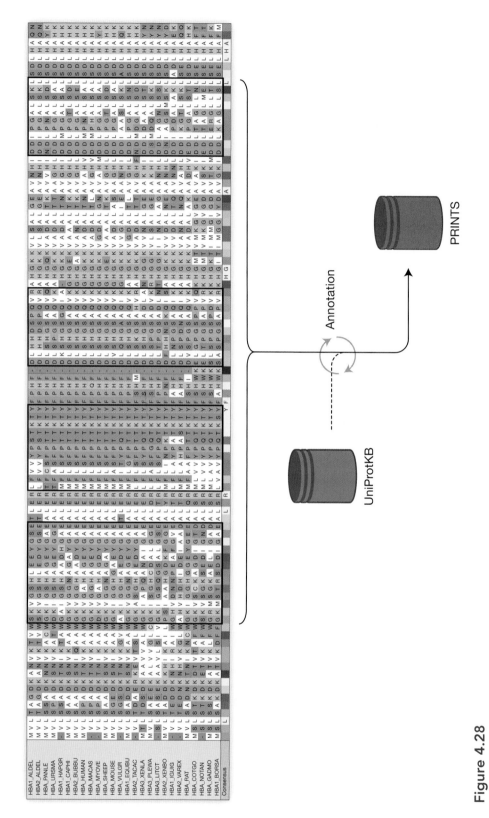

Figure 4.28

Alignment (partial) of haemoglobin α chains, showing four motifs (boxed regions) that contribute to the family fingerprint. The motifs are used to search UniProtKB iteratively (denoted by the grey arrows in the centre of the diagram), and are then annotated prior to deposition in PRINTS.

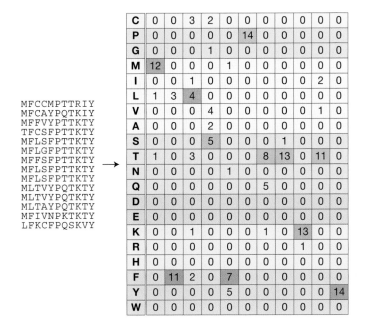

```
MFCCMPTTRIY
MFCAYPQTKIY
MFFVYPTTKTY
TFCSFPTTKTY
MFLSFPTTKTY
MFLGFPTTKTY
MFFSFPTTKTY      →
MFLSFPTTKTY
MFLSFPTTKTY
MLTVYPQTKTY
MLTVYPQTKTY
MLTAYPQTKTY
MFIVNPKTKTY
LFKCFPQSKVY
```

C	0	0	3	2	0	0	0	0	0	0	0
P	0	0	0	0	0	14	0	0	0	0	0
G	0	0	0	1	0	0	0	0	0	0	0
M	12	0	0	0	1	0	0	0	0	0	0
I	0	0	1	0	0	0	0	0	0	2	0
L	1	3	4	0	0	0	0	0	0	0	0
V	0	0	0	4	0	0	0	0	0	1	0
A	0	0	0	2	0	0	0	0	0	0	0
S	0	0	0	5	0	0	0	1	0	0	0
T	1	0	3	0	0	0	8	13	0	11	0
N	0	0	0	0	1	0	0	0	0	0	0
Q	0	0	0	0	0	0	5	0	0	0	0
D	0	0	0	0	0	0	0	0	0	0	0
E	0	0	0	0	0	0	0	0	0	0	0
K	0	0	1	0	0	0	1	0	13	0	0
R	0	0	0	0	0	0	0	0	1	0	0
H	0	0	0	0	0	0	0	0	0	0	0
F	0	11	2	0	7	0	0	0	0	0	0
Y	0	0	0	0	5	0	0	0	0	0	14
W	0	0	0	0	0	0	0	0	0	0	0

Figure 4.29

Example motif and the frequency matrix derived from it, with amino acids broadly grouped according to their physicochemical properties.

less well suited to diagnosing members of divergent superfamilies because, as we saw in Section 4.3.1, the scoring potential of sparse matrices is quite poor.

Figure 4.29 illustrates a single fingerprint motif drawn from a seed alignment of haemo-globin α chains – it sits in the same location as the second motif seen in Figure 4.28; its associated frequency matrix is shown on the right. The number of sequences in the seed alignment was 14, so the maximum score for a fully conserved alignment column is 14; the most conserved residue at each position is highlighted. Many of the substitutions can be seen to be within property groups, or within groups whose properties are known to overlap. Two positions are completely conserved, at positions 6 and 11.

Usually, following the first database scan, the search identifies additional sequences that weren't in the seed alignment. The algorithm therefore adds the new sequence information to the motifs, re-calculates the frequency matrices and searches the database again. In this way, the motifs grow iteratively, becoming more mature with each database pass. In real terms, as more family members are incorporated into the fingerprint, this process successively shifts the seed frequencies towards values that are more representative of the family as a whole – this point is illustrated in Figure 4.30. After a number of iterations, the motif has almost doubled in size, and the frequency distribution has changed accordingly. The total number of matches is 27, so this is the new maximum score for a conserved alignment column; the most conserved residues at each position remain the same as before, although their numbers have increased. Many of the observed substitutions still respect the known property groups and their overlaps. Now, however, only tyrosine in position 11 is fully conserved.

This cycle of iterative refinement continues until convergence – *i.e.*, the point at which no new family members are identified between successive scans (at this point, the fingerprint is ready to be annotated prior to deposition in the PRINTS database).

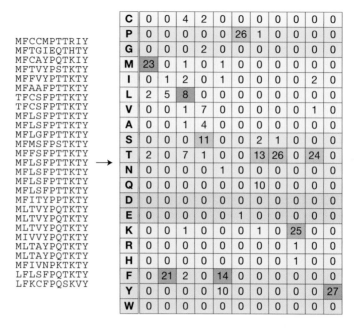

```
MFCCMPTTRIY
MFTGIEQTHTY
MFCAYPQTKIY
MFTVYPSTKTY
MFFVYPTTKTY
MFAAFPTTKTY
TFCSFPTTKTY
TFCSFPTTKTY
MFLSFPTTKTY
MFLSFPTTKTY
MFLGFPTTKTY
MFMSFPSTKTY
MFFSFPTTKTY
MFLSFPTTKTY  →
MFLSFPTTKTY
MFLSFPTTKTY
MFLSFPTTKTY
MFITYPPTKTY
MLTVYPQTKTY
MLTVYPQTKTY
MLTVYPQTKTY
MIVVYPQTKTY
MLTAYPQTKTY
MLTAYPQTKTY
MFIVNPKTKTY
LFLSFPQTKTY
LFKCFPQSKVY
```

C	0	0	4	2	0	0	0	0	0	0	0
P	0	0	0	0	0	26	1	0	0	0	0
G	0	0	0	2	0	0	0	0	0	0	0
M	23	0	1	0	1	0	0	0	0	0	0
I	0	1	2	0	1	0	0	0	0	2	0
L	2	5	8	0	0	0	0	0	0	0	0
V	0	0	1	7	0	0	0	0	0	1	0
A	0	0	1	4	0	0	0	0	0	0	0
S	0	0	0	11	0	0	2	1	0	0	0
T	2	0	7	1	0	0	13	26	0	24	0
N	0	0	0	0	1	0	0	0	0	0	0
Q	0	0	0	0	0	0	10	0	0	0	0
D	0	0	0	0	0	0	0	0	0	0	0
E	0	0	0	0	0	1	0	0	0	0	0
K	0	0	1	0	0	0	1	0	25	0	0
R	0	0	0	0	0	0	0	0	1	0	0
H	0	0	0	0	0	0	0	0	1	0	0
F	0	21	2	0	14	0	0	0	0	0	0
Y	0	0	0	0	10	0	0	0	0	0	27
W	0	0	0	0	0	0	0	0	0	0	0

Figure 4.30

Extended motif from Figure 4.29, and the frequency matrix derived from it, with amino acids broadly grouped according to their physicochemical properties.

For the haemoglobin α chain fingerprint, 242 sequences were found to match all the motifs after nine database searches: the maximum score for the conserved tyrosine at the end of the motif is thus 242, and again, the frequency distribution across the rest of the motif changes accordingly – overall, however, the matrix remains relatively sparse.

The full potency of the fingerprint method is gained not just from iterative enhancement of the motifs, but from the mutual context provided by motif neighbours. This is important because fingerprints inherently provide a biological context to motifs that are matched in the correct order and with appropriate intervals between them (see Box 4.7). This allows sequences to be characterised even when parts of a fingerprint are absent: e.g., a sequence that matches four of five motifs may still be diagnosed as a family member, if the motifs are matched in the correct order and the distances between them are consistent with those of true neighbouring motifs in known family members. This scenario is illustrated in Figure 4.33. The plot on the left-hand side shows the result of scanning the human haemoglobin α chain (P69905[8]) with the α haemoglobin fingerprint, ALPHAHAEM (PR00612[9]): strong matches are seen to each of the five fingerprint motifs. By contrast, the plot on the right-hand side shows the result of scanning an Antarctic dragonfish haemoglobin α chain (P23016[10]) with the same fingerprint: here, only four of the five motifs are matched. Although the sequence fails to make a significant match with motif 1, it's nevertheless clearly a family member, judging by the locations and order of the motifs that it does match.

---

[8] http://www.uniprot.org/uniprot/P69905
[9] http://www.bioinf.manchester.ac.uk/cgi-bin/dbbrowser/PRINTS/DoPRINTS.pl?cmd_a=Display&qua_a=/Full&fun_a=Code&qst_a=ALPHAHAEM
[10] http://www.uniprot.org/uniprot/P23016

Box 4.7 Visualising protein fingerprints

Protein fingerprints are most easily understood when viewed graphically. For a given fingerprint, a graph is created in which the query sequence is plotted along the *x*-axis, and scores with the constituent fingerprint motifs are plotted on the *y*-axis. For each motif, a window whose width corresponds to the width of the motif is slid along the query sequence, and its identity score is calculated using the associated frequency matrix derived from the motifs stored in the PRINTS database. Where a window scores above a given threshold (20 per cent identity), a coloured block is shown to mark the location of the match. If we were simply to plot such motif matches linearly, as shown in Figure 4.31, we could not know whether all the motifs had been matched, unless we used some other visual way to depict 'missing motifs'.

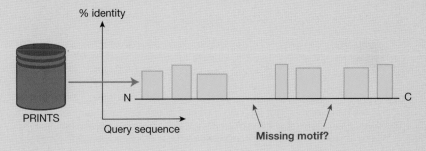

Figure 4.31
Plotting matches to consecutive motifs in a fingerprint.

An alternative is to plot the result of each motif scan separately. For true-positive matches, we then expect to see a complete set of blocks running diagonally across the plot, from N- to C-terminus. As shown in Figure 4.32, we can immediately see that the query sequence on the left matches all the fingerprint motifs, while the sequence on the right fails to match motifs 4 and 5.

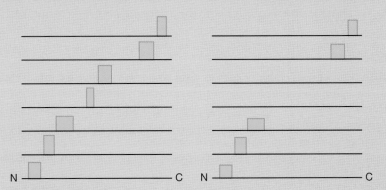

Figure 4.32
Plots showing complete (left) and partial (right) fingerprint matches.

Fingerprint graphs like this facilitate rapid visual inspection of query sequences, allowing ready identification of complete matches, and of partial matches that fail to match one or more motifs. This renders fingerprinting a more powerful diagnostic tool than methods that use only single motifs.

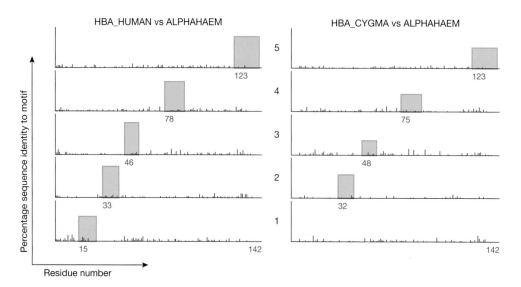

Figure 4.33

Graphical results of scanning haemoglobin α chains from human (left-hand plot) and the Antarctic dragonfish (*Cygnodraco mawsoni*) (right-hand plot) with the alpha haemoglobin fingerprint, ALPHAHAEM: the human sequence makes a complete match, the dragonfish sequence makes a partial match.

The ALPHAHAEM fingerprint identifies several such partial matches; the graphical tool, GraphScan[11], provides an elegant means of visualising the quality of the matches, in terms of the number of motifs matched, the individual motif scores, the background noise, *etc.* Figure 4.34 shows another of the partial matches.

Here, the partial match is with a haemoglobin β chain from the pale-throated three-toed sloth (P14526[12]) – only motifs 4 and 5 are matched, with relatively poor scores. However, if we scan the same sequence with the β haemoglobin fingerprint, BETAHAEM (PR00814[13]), we see a striking complete positive match. This illustrates the power of fingerprints to differentiate between highly similar sequences from the same family or superfamily (here, the globin superfamily). By contrast with the consensus expressions discussed previously, false negatives tend to be less of an absolute limitation for fingerprints, because divergent family outliers are more likely to be identified as partial matches. Ideally, we'd like to be able to capture such outliers as complete matches, but the use of frequency matrices limits the overall scoring potential of the method, making fingerprints less sensitive than profile-based approaches.

Profiles and hidden Markov models

We saw at the start of this section that the main alignment-based pattern-recognition methods make use of the sequence information encompassed either in single motifs, in multiple motifs or in complete domains. Techniques in this latter category are profiles and hidden Markov models.

---

[11] http://www.bioinf.manchester.ac.uk/cgi-bin/dbbrowser/fingerPRINTScan/GRAPHScan.cgi
[12] http://www.uniprot.org/uniprot/P14526
[13] http://www.bioinf.manchester.ac.uk/cgi-bin/dbbrowser/PRINTS/DoPRINTS.pl?cmd_a=Display&qua_a=/Full&fun_a=Code&qst_a=BETAHAEM

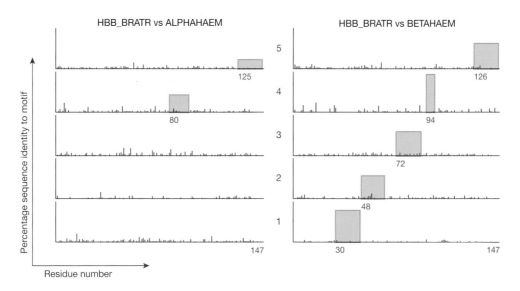

Figure 4.34

Graphical results of scanning the haemoglobin β chain from the pale-throated three-toed sloth (*Bradypus tridactylus*) with the α haemoglobin fingerprint, ALPHAHAEM (left), with which it makes a partial match and the β haemoglobin fingerprint, BETAHAEM (right), with which it makes a complete match.

Profiles were first introduced into PROSITE in 1994, in an attempt to offer diagnostic alternatives for some of its weaker patterns (Bairoch and Bucher, 1994). Profiles are essentially complex scoring tables or matrices that encapsulate not only the conserved motifs within sequence alignments, but also the regions in which insertions/deletions (**indels**) are found to accumulate. The generalised profile syntax is an adaptation from the profiles introduced by Gribskov *et al.* (1987). The profile itself encodes alignments as a series of match positions (M), such as those that occur in conserved motifs, and insert positions (I), where additional residues can be inserted (Bucher and Bairoch, 1994; Bucher *et al.*, 1996). When traversing an alignment, it is possible to move from a match position to another match position, or from a match to an insert position; or, a residue may appear to have been deleted. At each match or insert position, there is a match-extension and an insert-extension score per residue; there's also a residue-independent deletion extension score at each match position. State transitions occur between any two consecutive extension steps: *e.g.,* match-insert (MI), match-delete (MD), insert-match (IM), delete-match (DM), *etc.*

Part of the PROSITE profile for the globin superfamily is illustrated in Figure 4.35: this corresponds to the N-terminal half of the alignment shown in Figure 4.28, up to the first gap and including the first two motifs of the α haemoglobin fingerprint. Two reasonably long match regions are seen, separated by a small insert region. Within the conserved blocks, small indels aren't completely forbidden, but are strongly impeded by large deletion and insertion penalties (D=-20, I=-20), and even larger match-insert, match-delete transition penalties (MI=-210, MD=-210) defined in the DEFAULT line. These strict values are superseded by more permissive values in the gapped regions of the alignment (*e.g.,* I=-5, MI=-55, MD=-55), and in the match regions (*e.g.,* D=-6).

The parameters listed for each match position include a range of position-specific scores, including initiation and termination scores, state-transition scores, insertion/

```
...
/GENERAL_SPEC: ALPHABET='ABCDEFGHIKLMNPQRSTVWYZ'; LENGTH=154;
/DISJOINT: DEFINITION=PROTECT; N1=1; N2=154;
/NORMALIZATION: MODE=1; FUNCTION=LINEAR; R1=-0.8705306; R2=0.0209303; TEXT='NScore';
/CUT_OFF: LEVEL=0; SCORE=424; N_SCORE=8.0; MODE=1; TEXT='!';
/CUT_OFF: LEVEL=-1; SCORE=353; N_SCORE=6.5; MODE=1; TEXT='?';
/DEFAULT: D=-20; I=-20; MI=-210; MD=-210; IM=0; DM=0;
/I: I=-6;
/M: SY='A'; M=7,-7,-8,-10,-10,-8,3,-12,-4,-8,-6,-4,-6,-10,-10,-10,3,4,3,-14,-10,-10; D=-6;
/I: I=-6; MI=-59; MD=-59;
/M: SY='H'; M=1,-3,-21,0,-6,-20,0,2,-16,-10,-16,-10,-4,0,-8,-12,-2,-9,-11,-23,-13,-8; D=-6;
/M: SY='L'; M=-10,-27,-2,-30,-21,4,-28,13,-26,36,22,-27,-30,-18,-20,-26,-10,7,-24,-4,-20;
/M: SY='T'; M=2,-7,-14,-9,-9,-13,-2,-16,-13,-15,-10,-10,-1,-15,-9,-13,17,18,-6,-31,-15,-9;
/M: SY='P'; M=-5,8,-30,15,10,-28,-17,-12,-18,-6,-19,-17,-3,16,-4,-15,-4,-9,-18,-31,-20,1;
/M: SY='A'; M=10,-7,-21,-8,4,-23,1,-10,-17,-8,-11,-9,-5,-13,5,-10,4,-5,-14,-24,-17,4;
/M: SY='E'; M=-11,10,-30,17,37,-36,-19,5,-27,9,-21,-12,2,-6,36,3,0,-10,-30,-27,-15,36;
/M: SY='K'; M=-2,-9,-27,-12,-4,-18,-17,-15,-22,17,-21,-12,-8,-14,-1,13,-6,-1,-15,4,-5,-2;
/M: SY='T'; M=1,-5,-21,-15,-10,-14,-20,-14,5,-12,-5,-1,1,-14,-1,-13,2,8,0,-25,-10,-7;
/M: SY='N'; M=-2,1,-21,-6,-5,-14,-14,-8,-9,3,-3,-4,7,-19,-5,-1,-7,-6,-11,-26,-11,-5;
/M: SY='V'; M=-3,-30,-16,-33,-30,0,-33,-30,36,-23,13,13,-27,-27,-27,-23,-13,-3,44,-27,-7,-30;
/M: SY='K'; M=-10,-11,-26,-14,-6,-14,-25,-15,-5,13,-5,0,-8,-16,-3,12,-11,-2,-4,-21,-7,-6;
/M: SY='S'; M=4,5,-18,-1,2,-23,-9,-7,-19,5,-24,-13,11,-12,7,0,15,10,-15,-30,-16,4;
/M: SY='A'; M=20,-13,-14,-17,-9,-15,-11,-16,-2,-12,-3,-3,-12,-16,-4,-15,4,0,4,-24,-14,-7;
/M: SY='W'; M=-16,-33,-43,-35,-26,6,-20,-28,-18,-18,-18,-18,-33,-26,-18,-18,-29,-16,-25,118,23,-18;
/M: SY='G'; M=12,-11,-21,-12,-7,-22,21,-19,-18,-13,-16,-12,-8,-16,-13,-17,1,-10,-8,-23,-22,-10;
/M: SY='P'; M=-10,-5,-33,0,15,-31,-20,-9,-26,20,-27,-13,-6,23,12,7,-7,-10,-26,-25,-18,12;
/M: SY='V'; M=-3,-30,-11,-31,-30,12,-30,-29,26,-21,10,9,-29,-30,-31,-20,-11,-1,43,-24,-4,-30;
/I: I=-5; MI=-55; MD=-55;
/M: SY='L'; M=-2,-5,-4,-5,-4,2,-5,-4,4,-5,9,4,-5,-5,-4,-4,-5,-2,2,-4,0,-4; D=-5;
/I: I=-5; MI=-55; MD=-55;
/M: SY='K'; M=-7,3,-20,2,6,-12,-2,0,-18,7,-16,-10,6,-11,1,2,-3,-7,-17,-12,0,3; D=-5;
/I: I=-5; MI=-55; MD=-55;
/M: SY='A'; M=18,-5,-15,-7,4,-12,-7,-7,-12,-5,-12,-10,-5,-9,-2,-11,7,0,-8,-15,-4,0; D=-5;
/M: SY='H'; M=-11,19,-22,15,2,-22,-10,28,-25,-6,-24,-14,24,-16,2,-4,8,2,-23,-36,-7,0;
/M: SY='V'; M=0,-23,-22,-27,-24,-4,-12,-20,15,-20,2,4,-18,-23,-20,-22,-10,-8,16,-16,1,-24;
/M: SY='P'; M=-7,-5,-29,2,1,-26,2,-16,-20,-10,-22,-16,-8,15,-11,-16,-5,-11,-16,-29,-24,-6;
/M: SY='E'; M=-5,2,-24,3,19,-24,-6,-11,-25,5,-20,-15,0,-8,3,-2,5,7,-19,-27,-17,11;
/M: SY='H'; M=-7,-12,-22,-14,-12,-8,-21,20,-1,-15,-9,0,-4,-20,-6,-12,0,-4,-1,-24,9,-12;
/M: SY='G'; M=-3,-7,-24,-10,-14,-22,29,-17,-29,-10,-22,-15,0,-17,-13,-5,5,4,-19,-23,-21,-14;
/M: SY='T'; M=2,-7,-19,-11,-8,-19,-6,1,-13,-10,-13,-6,-4,16,0,-9,4,8,-6,-26,-10,-5;
/M: SY='E'; M=-12,16,-26,24,28,-28,-17,-4,-29,7,-21,-19,5,-7,8,6,2,1,-23,-31,-17,17;
/M: SY='F'; M=3,-25,-17,-33,-25,31,-24,-23,9,-23,6,2,-19,-23,-28,-21,-11,-6,12,-8,7,-25;
/M: SY='L'; M=-5,-26,-20,-28,-20,20,-26,-13,10,-24,28,10,-24,-27,-20,-18,-22,-9,4,-7,16,-20;
/M: SY='I'; M=-7,-17,-27,-20,-11,-13,-17,-22,11,-8,-2,3,-13,-17,-11,-13,-12,-10,8,-23,-10,-13;
/M: SY='R'; M=-15,5,-27,4,1,-20,-17,-3,-22,15,-14,-10,9,-19,3,29,-8,-8,-19,-26,-12,0;
/M: SY='L'; M=-9,-29,-17,-31,-24,16,-29,-20,20,-25,30,19,-27,-29,-23,-19,-22,-7,20,-19,1,-23;
/M: SY='F'; M=-17,-30,-20,-37,-27,60,-30,-20,6,-30,21,6,-23,-30,-34,-20,-23,-10,3,1,21,-27;
/M: SY='E'; M=2,-3,-20,-7,2,-14,-16,-12,-10,-2,-2,-5,-2,-15,-4,-5,-4,1,-9,-25,-12,-1;
/M: SY='S'; M=-2,-5,-21,-9,-10,-18,-3,-15,-8,-4,-17,-8,3,-17,-8,-8,5,-1,-3,-29,-15,-10;
/M: SY='H'; M=-7,-12,-22,-17,-13,13,-21,25,-13,-15,-7,-3,-6,-21,-11,-7,-4,-11,-10,21,-13;
/M: SY='P'; M=-11,-21,-37,-14,-4,-16,-21,-20,-17,-13,-25,-17,-20,74,-14,-20,-11,-10,-26,-25,-22,-13;
/M: SY='E'; M=5,-6,-25,-7,12,-18,-14,-12,-21,0,-16,-15,-8,-11,1,0,-2,-3,-17,-1,-11,6;
/M: SY='T'; M=14,-8,-10,-16,-14,-11,-16,-22,-3,-12,-6,-6,-8,-14,-14,-15,12,27,9,-27,-13,-14;
/M: SY='R'; M=-12,-1,-27,-4,5,-24,-19,-2,-18,17,-13,-4,3,-16,18,20,-8,-9,-19,-23,-10,11;
/M: SY='P'; M=-12,3,-31,10,19,-28,-19,-9,-26,3,-23,-19,-3,22,3,1,-2,-3,-25,-30,-20,9;
/M: SY='F'; M=-15,-14,-24,-14,-13,19,-26,-11,-3,-11,10,1,-15,-24,-17,-10,-19,-10,-5,-8,14,-14;
/M: SY='F'; M=-20,-27,-21,-36,-26,67,-29,-17,-4,-22,6,-1,-17,-29,-33,-8,-19,-10,-3,6,25,-26;
/M: SY='D'; M=-8,13,-27,23,15,-31,-11,-8,-28,-4,-29,-23,4,17,0,-11,8,-2,-24,-36,-23,6;
/M: SY='M'; M=-10,-12,-23,-15,-8,-1,-20,2,-9,2,-4,7,-8,-18,-4,0,-9,-7,-8,-20,1,-6;
/M: SY='F'; M=-15,-26,-22,-33,-27,50,-12,-20,-4,-28,9,-1,-18,-28,-34,-20,-18,-12,-4,0,15,-27;
/I: I=-4; MI=-42; MD=-42;
...
```

Figure 4.35   Part of PROSITE profile GLOBIN (PS01033)[i] (note that this is just one way of representing a profile, a handy format that allows the data to be stored and re-used).

[i]http://prosite.expasy.org/PS01033

match/deletion extension scores, and so on. The overall similarity score is then the sum of the scores assigned to all components of the alignment that can be scored, including the beginning, all extension steps, all state transitions, and the end (Bucher and Bairoch, 1994; Bucher *et al.*, 1996).

A consensus sequence runs vertically down the left-hand side of the profile (HLT-PAEKT….). It's difficult to reconcile this with the consensus that's just discernible in Figure 4.28, largely because those haemoglobin α chains constitute just a small subset of the sequences matched by this very potent profile (which includes α, β, γ, δ and ε haemoglobin chains, together with myoglobins, **neuroglobins, cytoglobins, leghaemoglobins, erythrocruorins,** and so on): 1,152 family members in total.

Conceptually, profiles are analogous to **Hidden Markov Models** (HMMs), which are the basis of the Pfam database. The main difference between them relates to how the parameters are interpreted. Most profile scores can be mapped to parameters of HMMs as defined by Krogh *et al.* (1994); the principal difference is that HMM parameters represent log probabilities. The mathematical theory behind HMMs is substantial and has been described in numerous excellent books. Here, it is sufficient to note that HMMs are probabilistic models that consist of several interconnecting states – in essence, they can be thought of as linear chains of match, delete and insert states that encode conserved domains in sequence alignments. As with profiles, match states are assigned to conserved residue positions, insert states allow insertions relative to match states, and delete states allow match positions to be skipped. Hence, building an HMM from a multiple alignment requires each position to be assigned to a match, delete or insert state, as illustrated in Figure 4.36.

The complexity of HMMs renders them very powerful tools for encapsulating the sequence information in multiple alignments; like profiles, they find their strength in modelling divergent domain families and superfamilies, diagnosing outliers with much greater sensitivity than can be achieved with consensus expressions or fingerprints. However, their sensitivity also brings a number of issues. The most important caveat is that the methods used to define the diagnostic HMMs in Pfam are not as closely supervised as the manual methods used in PROSITE and PRINTS. Family membership is determined by specific cut-off values[14]. Superfamilies and domain families with thousands of members call into question how confident we can be that all sequences

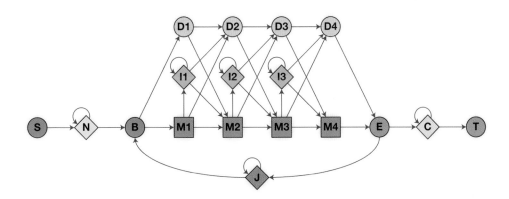

Figure 4.36

Linear hidden Markov model. M are match states; I are insert states; D are delete states; B and E define the beginning and end of the domain; J allows the domain to be repeated; N and C allow for N- and C-terminal extensions; and S and T denote the start and termination of the model respectively.

[14] http://pfam.xfam.org/help?tab=helpScoresBlock#tabview=tab5

captured above the threshold are true-positives – in Pfam, no false-positive matches or false-negative sequences are listed; all sequences gathered above the given cut-off are thereby asserted to be true. There were 17,947 matches to the Globin (PF00042[15]) HMM from the May 2015 release of Pfam, bearing witness to the size of the superfamily, and presenting a significant number of sequences to check. But this number is small compared to the largest family in the database[16] (the **ATP**-binding domain of **ABC transporters** (PF00005[17])): how confident can we be that all 1,466,247 matches to this HMM are correct? The real problem is, if false matches do sneak under the radar of the HMM cut-offs, then they will successively corrupt the model, such that future updates on later releases of the sequence database will identify further sequences with similarity to those masquerading as true matches (the same problem arises with profiles, where the effects are referred to as 'profile dilution').

So far, in discussing the various diagnostic models that have been developed to recognise members of the globin superfamily, we've noted 242 full fingerprint matches (with 51 true partial matches) to the α haemoglobin fingerprint, 1,152 matches to the globin profile, and 17,947 matches to the globin HMM. It's challenging to understand how these numbers relate to each other, and important to understand something about what they mean. Crucially, they refer to different sets of sequences in radically different versions of the underlying sequence database: the fingerprint encodes only the α haemoglobin subset of sequences in Swiss-Prot/TrEMBL in June 1999, which is the last time this fingerprint was updated; the profile relates to the wider superfamily of globins in UniProtKB/Swiss-Prot in November 2015; and the HMM relates to globins in UniProtKB in the May 2015 release of Pfam, which was based on the July 2014 release of UniProtKB. The numbers are therefore not directly comparable.

We also saw that there's no consensus expression in PROSITE for the globins, presumably because members of the superfamily are too divergent for a pattern to be diagnostically useful; for the same reason, there's no superfamily fingerprint in PRINTS, but instead there are family-specific fingerprints for haemoglobin α and β chains, for myoglobin, for erythrocruorins, plant globins, and so on, which the other methods don't provide; by contrast, the profile and HMM offer superfamily signatures, because these methods are ideally suited to encoding members of divergent superfamilies and domain families. Here, then, we get a sense of the complementarity of single-motif, multiple-motif and domain-based approaches, and the different diagnostic opportunities they afford. We can perhaps also begin to understand the importance of their integration within InterPro, so that users can both analyse sequences using a range of different, but complementary, tools and compare the results meaningfully, because InterPro synchronises all of the component diagnostic models against the same version of the same source sequence database: UniProtKB. For the globins (Globin, IPR000971[18]), this results in 11,500 matches, derived from hits to the Globin HMM and to the GLOBIN profile in the October 2015 release of InterPro (based on the October 2015 release of UniProtKB). This figure is thus more up to date than that quoted in Pfam; however, users are left to guess why there are such disparities between InterPro's 11,500 matches and the 1,152 matches reported for the PROSITE profile, and the 17,947 reported for

[15] http://pfam.xfam.org/family/PF00042#tabview=tab3
[16] http://pfam.xfam.org/family/browse?browse=top%20twenty
[17] http://pfam.xfam.org/family/PF00005
[18] http://www.ebi.ac.uk/interpro/entry/IPR000971

Pfam's HMM. InterPro's updated number of α haemoglobins is 1,789 (Haemoglobin, alpha, IPR002338[19]), derived from the ALPHAHAEM fingerprint.

A final point to note about integrating such diagnostic signatures is that Inter-Pro also includes information from structure-based databases. We saw in Box 4.3 that the 3D structures of distantly related proteins are better conserved than their sequences, and that some relationships are only evident at the structural level, unde-tectable by even the most sensitive sequence-based methods. For the globins, Inter-Pro provides information about sequences related to the globin fold, as defined by the Gene3D and **SUPERFAMILY** databases. Specifically, InterPro entry 'Globin, struc-tural domain' (IPR012292[20]) reports 20,061 proteins derived from Gene3D (Globins, G3DSA: 1.10.490.10[21]), and entry Globin-like (IPR009050[22]) reports 25,964 proteins derived from SUPERFAMILY (Globin-like superfamily, SSF46458[23]). Notice that the numbers are different from each other, suggesting that these databases have differ-ent perspectives on how to define globin-like domains. Notwithstanding this discrep-ancy, however, more importantly, the numbers are also significantly greater than those derived from the sequence-based methods, showing that structure-based approaches can penetrate far deeper into the Twilight and Midnight zones, offering much broader perspectives on the evolutionary histories of protein architectures. This uniquely com-prehensive overview of protein family relationships has made InterPro the pre-eminent resource for sequence annotation, both at the level of individual protein sequences and at the level of whole genomes and meta-genomes (Hunter *et al.*, 2012).

Another important aspect of annotation that we've not touched on in this chapter concerns how pertinent information is extracted from the scientific literature and added to database entries. This brings challenges of its own; some of these are explored in detail in Chapter 8.

4.4 Summary

This chapter discussed the importance of annotation, and introduced some of the main sequence analysis methods that are used to characterise and annotate sequence data today. Specifically, we saw that:

1 By annotating 'naked' sequences, we render them more meaningful and hence both useful and reusable;
2 For nucleotide sequences, annotation can include, amongst other things: details of gene structures, such as the locations of exons and introns, and how to join these to form contiguous coding sequences; the names of genes and their products; the name of the parent organism, and so on;
3 The need to structure annotation, to render the information machine-accessible, led to the development of the first database flat-file formats;
4 Flat-file formats (*e.g.*, such as the EMBL format) use a series of specific tags to indicate the type of information held in each database field;

---

[19] http://www.ebi.ac.uk/interpro/entry/IPR002338
[20] http://www.ebi.ac.uk/interpro/entry/IPR012292
[21] http://www.cathdb.info/cathnode/1.10.490.10
[22] http://www.ebi.ac.uk/interpro/entry/IPR009050
[23] http://supfam.cs.bris.ac.uk/SUPERFAMILY/cgi-bin/scop.cgi?ipid=SSF46458

5 Amongst the most important tags for linking information within and between databases are those of an entry's identifier (ID) and accession number (AC);

6 Information-retrieval software can exploit such tags to allow information within flat-file databases to be indexed, interconnected and interrogated;

7 To quantify the similarity between sequences, scoring or substitution matrices such as the PAM and BLOSUM series are used;

8 Dotplots are simple visual ways of assessing the similarity of pairs of sequences – plots are 'clean' when unitary scores are used, but noisy when similarity matrices are used;

9 FASTA and BLAST are pairwise comparison tools that have been optimised for database searching – the (mathematical) significance of matches they return can be assessed by means of p-values or e-values;

10 Matches that have significant p- or e-values may not be biologically relevant – more sensitive diagnostic tools may be needed to verify their biological significance;

11 There are many different tools for multiple sequence alignment – most are based on a type of progressive algorithm and use heuristics to return good (but not necessarily optimal) alignments in reasonable time-frames;

12 Multiple alignments underpin the main sequence-based methods used to diagnose protein family membership;

13 The main methods exploit single motifs (consensus expressions or patterns), multiple motifs (fingerprints) and complete domains (profiles, HMMs);

14 These methods have different diagnostic strengths and weaknesses: patterns perform well in identifying small functional sites and closely related protein families; fingerprints perform well at diagnosing protein families and subfamilies; profiles and HMMs have greatest strength in diagnosing members of protein superfamilies and domain families;

15 InterPro amalgamates a unique range of such diagnostic tools, providing the world's most comprehensive package for sequence data annotation.

4.5 References

Altschul, S.F., Gish, W., Miller, W. *et al.* (1990) Basic local alignment search tool. *Journal of Molecular Biology*, **215**(3), 403–410.

Altschul, S.F., Madden, T.L., Schäffer, A.A. *et al.* (1997) Gapped BLAST and PSI-BLAST: a new generation of protein database search programs. *Nucleic Acids Research*, **25**(17), 3389–3402.

Bairoch, A. and Bucher, P. (1994) PROSITE: recent developments. *Nucleic Acids Research*, **22**(17), 3583– 3589.

Bucher, P. and Bairoch, A. (1994) A generalized profile syntax for biomolecular sequence motifs and its function in automatic sequence interpretation. *Proceedings of the 2nd International Conference on Intelligent Systems for Molecular Biology (ISMB 1994)* 53–61.

Bucher, P., Karplus, K., Moeri, N. and Hofmann, K. (1996) A flexible motif search technique based on generalized profiles. *Journal of Computational Chemistry*, **20**(1), 3–23.

Cserzo, M., Eisenhaber, F., Eisenhaber, B. and Simon, I. (2002) On filtering false positive trans-membrane protein predictions. *Protein Engineering, Design and Selection*, **15**, 745–752.

Dayhoff, M.O., Schwartz, R.M. and Orcutt, B.C. (1978) A model of evolutionary change in proteins. In *Atlas of Protein Sequence and Structure*, Vol. 5, Suppl. 3 (ed. M.O. Dayhoff) pp. 345–352. National Biomedical Research Foundation, Washington, DC.

Deléage, G., Combet, C., Blanchet, C. and Geourjon, C. (2001) ANTHEPROT: an integrated protein sequence analysis software with client/server capabilities. *Computers in Biology and Medicine*, **31**(4), 259–267.

Do, C.B., Mahabhashyam, M.S., Brudno, M. and Batzoglou, S. (2005) ProbCons: probabilistic consistency-based multiple sequence alignment. *Genome Research*, **15**, 330–340.

Doolittle, R.F. (1986) *Of URFs and ORFs: a primer on how to analyse derived amino acid Sequences*, pp. 11–12. University Science Books, Mill Valley, CA.

Edgar, R.C. (2004) MUSCLE: a multiple sequence alignment method with reduced time and space complexity. *BMC Bioinformatics*, **5**, 113.

Edgar, R.C. and Batzoglou, S. (2006) Multiple sequence alignment. *Current Opinion in Structural Biology*, **16**, 368–373.

Etzold, T. and Argos, P. (1993) SRS – an indexing and retrieval tool for flat file data libraries. *Computer Applications in the Biosciences*, **9**(1), 49–57.

Etzold, T., Ulyanov, A. and Argos, P. (1996) SRS: information retrieval system for molecular biology data banks. *Methods in Enzymology*, **266**, 114–128.

Galtier, N., Gouy, M. and Gautier, C. (1996) SEAVIEW and PHYLO_WIN: two graphic tools for sequence alignment and molecular phylogeny. *Computer Applications in the Biosciences*, **12**(6), 543–548.

Gotoh, O. (1999) Multiple sequence alignment: algorithms and applications. *Advances in Biophysics*, **36**, 159–206.

Gribskov, M., McLachlan, A.D. and Eisenberg, D. (1987) Profile analysis: detection of distantly related proteins. *Proceedings of the National Academy of Sciences of the U.S.A.*, **84**(13), 4355–4358.

Hekkelman, M.L. and Vriend, G. (2005) MRS: a fast and compact retrieval system for biological data. *Nucleic Acids Research*, **33**(Web Server issue), W766–769.

Henikoff, S. and Henikoff, J. (1992) Amino acid substitution matrices from protein blocks. *Proceedings of the National Academy of Sciences of the U.S.A.*, **89**, 10915–10919.

Henikoff, S. and Henikoff, J. (1993) Performance evaluation of amino acid substitution matrices. *Proteins*, **17**(1), 49–61.

Higgins, D.G. and Sharp, P.M. (1988) CLUSTAL: a package for performing multiple sequence alignment on a microcomputer. *Gene*, **73**(1), 237–244.

Hogeweg, P. and Hesper, B. (1984) The alignment of sets of sequences and the construction of phyletic trees: an integrated method. *Journal of Molecular Evolution*, **20**(2), 175–186.

Hunter, S., Jones, P., Mitchell, A. *et al.* (2012) InterPro in 2011: new developments in the family and domain prediction database. *Nucleic Acids Research*, **40**(Database issue), D306–312.

Katoh, K., Kuma, K., Toh, H. and Miyata, T. (2005) MAFFT version 5: improvement in accuracy of multiple sequence alignment. *Nucleic Acids Research*, **33**, 511–518.

Krogh, A., Brown, M., Mian, I.S. *et al.* (1994) Hidden Markov models in computational biology. Applications to protein modeling. *Journal of Molecular Biology*, **235**(5), 1501–1531.

Lipman, D.J. and Pearson, W.R. (1985) Rapid and sensitive protein similarity searches. *Science*, **227**(4693), 1435–1441.

Löytynoja, A. and Goldman, N. (2005). An algorithm for progressive multiple alignment of sequences with insertions. *Proceedings of the National Academy of Sciences of the U.S.A.*, **102**(30), 10557–10562.

Löytynoja, A. and Goldman, N. (2010). webPRANK: a phylogeny-aware multiple sequence aligner with interactive alignment browser. *BMC Bioinformatics*, **11**, 579.

Maizel, J.V. Jr. and Lenk, R.P. (1981) Enhanced graphic matrix analysis of nucleic acid and protein sequences. *Proceedings of the National Academy of Sciences of the U.S.A.*, **78**(12), 7665–7669.

Needleman, S.B. and Wunsch, C.D. (1970) A general method applicable to the search for similarities in the amino acid sequence of two proteins. *Journal of Molecular Biology*, **48**(3), 443–453.

Notredame, C. (2007) Recent evolutions of multiple sequence alignment algorithms. *PLoS Computational Biology*, **3**(8), e123.

Notredame, C., Higgins, D.G., Heringa, J. (2000) T-Coffee: a novel method for fast and accurate multiple sequence alignment. *Journal of Molecular Biology*, **302**, 205–217.

Parry-Smith, D.J. and Attwood, T.K. (1991) SOMAP: a novel interactive approach to multiple protein sequences alignment. *Computer Applications in the Biosciences*, 7(2), 233–235.

Parry-Smith, D.J., Payne, A.W., Michie, A.D. and Attwood, T.K. (1998) CINEMA – a novel colour INteractive editor for multiple alignments. *Gene*, **221**(1), GC57–63.

Pearson, W.R. and Lipman, D.J. (1988) Improved tools for biological sequence comparison. *Proceedings of the National Academy of Sciences of the U.S.A.*, 85(8), 2444–2448.

Pettifer, S., Thorne, D., McDermott P. *et al.* (2009) Visualising biological data: a semantic approach to tool and database integration.*BMC Bioinformatics*, **10**, S18.

Rost, B. (1998) Marrying structure and genomics. *Structure*, 6, 259–263.

Sigrist, C.J., De Castro, E., Langendijk-Genevaux, P.S. *et al.* (2005) ProRule: a new database containing functional and structural information on PROSITE profiles. *Bioinformatics*, **21**(21), 4060–4066.

Simossis, V.A. and Heringa, J. (2005) PRALINE: a multiple sequence alignment toolbox that integrates homology-extended and secondary structure information. *Nucleic Acids Research*, 33, W289–W294.

Sinnott, J.R., Pettifer, S.R. and Attwood, T.K. (2004) Introduction to the CINEMA5 sequence alignment editor. *EMBNet News*, **10**(3).

Smith, T.F. andWaterman, M.S. (1981) Identification of common molecular subsequences. *Journal of Molecular Biology*, **147**(1), 195–197.

Thompson, J., Higgins, D. and Gibson, T. (1994) CLUSTAL W: improving the sensitivity of progressive multiple sequence alignment through sequence weighting, position-specific gap penalties and weight matrix choice. *Nucleic Acids Research*, 22, 4673–4690.

Thompson, J.D., Gibson, T.J., Plewniak, F. *et al.* (1997) The CLUSTAL_X windows interface: flexible strategies for multiple sequence alignment aided by quality analysis tools. *Nucleic Acids Research*, 25(24), 4876–4782.

Wallace, I.M., Blackshields, G. and Higgins, D.G. (2005) Multiple sequence alignments. *Current Opinion in Structural Biology*, 15, 261–6.

Wallace, I.M., O'Sullivan, O., Higgins, D.G. and Notredame, C. (2006) M-Coffee: combining multiple sequence alignment methods with T-Coffee. *Nucleic Acids Research*, 34, 1692–1699.

Wilbur, W.J. and Lipman, D.J. (1983) Rapid similarity searches of nucleic acid and protein data banks. *Proceedings of the National Academy of Sciences of the U.S.A.*, 80(3), 726–730.

4.6 Quiz

This multiple-choice quiz will test how much you've remembered about the sequence analysis and annotation methods described in this chapter.

1 Which of the following statements are correct?
   A Annotation helps computers to process sequence data.
   B Annotation can be free-text notes.
   C Annotation must be structured in order to be machine-accessible.
   D Annotation helps humans to understand sequence data.

2 A commonly used database flat-file format is:
   A FASTA
   B NBRF-PIR
   C EMBL
   D SRS

3 Which of the following statements are correct?
   A The ID tag stores the accession number of the database entry.
   B Database accession numbers never change.
   C Database identifiers never change.
   D None of the above.

4 The database tag, FT is:
   A a Function-Tag describing the functions of sequences.
   B a Free-Text tag describing the functions of sequences as free text.
   C a Formatted-Text tag describing the functions of sequences using controlled
       vocabularies.
   D the Feature-Table tag describing notable characteristics or features of sequences.

5 Which of the following statements are correct?
   A Unitary matrices are sparse.
   B Unitary matrices score all identical matches equally.
   C The diagnostic power of unitary matrices is poor.
   D Unitary matrices score all non-identical matches equally.

6 Which of the following statements are correct?
   A PAM 250 relates to short evolutionary distances.
   B PAM 250 became the default scoring matrix in many analysis and alignment
       packages because it identifies fewer false matches.
   C PAM 250 is a better scoring matrix than BLOSUM 62.
   D None of the above.

7 Which of the following statements are false?
   A A false-negative match is a match returned by a database search in error.
   B A false-negative match is a true sequence missed by a database search.
   C A false-negative match is a true sequence matched by a database search.
   D None of the above.

8 Which of the following are alignment tools?
   A FASTA
   B PRANK
   C CINEMA
   D BLAST

9 Which of the following statements are false?
   A Consensus expressions or patterns are probabilistic models.
   B HMMs are more selective than profiles.
   C Fingerprints use PAM 250 scoring.
   D Profile pattern-matching is binary.

10 Which of the following statements are correct?
    A InterPro is used for sequence annotation.
    B InterPro integrates only sequence-based diagnostic methods.
    C InterPro synchronises and updates its component diagnostic methods with
        respect to UniProtKB.
    D InterPro is diagnostically more powerful than its component methods.

4.7   Problems

1   For the sequence shown in Figure 4.1(a) (reproduced here), we generated a translation product by beginning the translation process from the first nucleotide base. Use an online translation tool[24], to discover what happens when the translation is started from the second or third base positions, and also in the three reverse frames. Are there any other legitimate peptide or protein translation products? (This search tool[25] will help with your answer.)

```
gggtgcaaccgtgaagtccgccacgaccgcgtcacgacaggagccgaccagcgacacccagaaggtgc
gaacggttgagtgccgcaacgatcacgagttttttcgtgcgcttcgagtggtaacacgcgtgcacgcat
cgacttcaccgcgggtgtttcgacgccagccggccgttgaaccagcaggcagcgggcatttacagccg
ctgtggcccaaatggtggggtgcgctattttggtatggtttggaatccgcgtgtcggctccgtgtctg
acggttcatcggttctaaattccgtcacgagcgtaccatactgattgggtcgtagagttacacacata
tcctcgttaggtactgttgcatgttggagttattgccaacagcagtggaggggggtatcgcaggcccag
atcaccggacgtccggagtggatctggctagcgctcggtacggcgctaatgggactcgggacgctcta
tttcctcgtgaaagggatgggcgtctcggacccagatgcaaagaaattctacgccatcacgacgctcg
tcccagccatcgcgttcacgatgtacctctcgatgctgctggggtatggcctcacaatggtaccgttc
ggtgggggagcagaaccccatctactgggcgcggtacgctgactggctgttcaccacgccgctgttgtt
gttagacctcgcgttgctcgttgacgcggatcagggaacgatccttgcgctcgtcggtgccgacggca
tcatgatcgggaccggcctggtcggcgcactgacgaaggtctactcgtaccgcttcgtgtggtgggcg
atcagcaccgcagcgatgctgtacatcctgtacgtgctgttcttcgggttcacctcgaaggccgaaag
catgcgccccgaggtcgcatccacgttcaaagtactgcgtaacgttaccgttgtgttgtggtccgcgt
atcccgtcgtgtggctgatcggcagcgaaggtgcgggaatcgtgccgctgaacatcgagacgctgctg
ttcatggtgcttgacgtgagcgcgaaggtcggcttcgggctcatcctcctgcgcagtcgtgcgatctt
cggcgaagccgaagcgccggagccgtccgccggcgacggcgcggccgcgaccagcgactgatcgcaca
cgcaggacagccccacaaccggcgcggctgtgttcaacgacacacgatgagtcccccactcggtcttg
tactc
```

2   Figure 4.10 shows 'the EMBL database entry for human α-globin'. What are the identifier, accession number and sequence version, and from which version of EMBL does the sequence derive?

3   How many hydrophobic domains appear in each of the profiles in Figure 4.37? What do the plots reveal about the nature of the two proteins? Explain your reasoning.

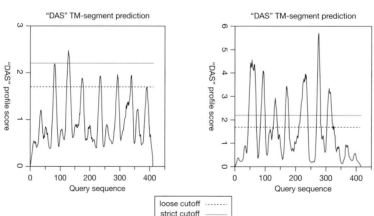

Figure 4.37

TM profiles.

---

[24] http://web.expasy.org/translate
[25] http://prosite.expasy.org/scanprosite

4 Using the range of analysis tools and databases described in this chapter, what is the likely function of the sequence below[26] and to what family might it belong. Explain your reasoning.

```
MSDERRLPGSAVGWLVCGGLSLLANAWGILSVGAKQKKWKPLEFLLCTLAATHMLNVAVP
IATYSVVQLRRQRPDFEWNEGLCKVFVSTFYTLTLATCFSVTSLSYHRMWMVCWPVNYRL
SNAKKQAVHTVMGIWMVSFILSALPAVGWHDTSERFYTHGCRFIVAEIGLGFGVCFLLLV
GGSVAMGVICTAIALFQTLAVQVGRQADRRAFTVPTIVVEDAQGKRRSSIDGSEPAKTSL
QTTGLVTTIVFIYDCLMGFPVLVVSFSSLRADASAPWMALCVLWCSVAQALLLPVFLWAC
DRYRADLKAVREKCMALMANDEESDDETSLEGGISPDLVLERSLDYGYGGDFVALDRMAK
YEISALEGGLPQLYPLRPLQEDKMQYLQVPPTRRFSHDDADVWAAVPLPAFLPRWGSGED
LAALAHLVLPAGPERRRASLLAFAEDAPPSRARRRSAESLLSLRPSALDSGPRGARDSPP
GSPRRRPGPGPRSASASLLPDAFALTAFECEPQALRRPPGPFPAAPAAPDGADPGEAPTP
PSSAQRSPGPRPSAHSHAGSLRPGLSASWGEPGGLRAAGGGGSTSSFLSSPSESSGYATL
HSDSLGSAS
```

[26] http://www.uniprot.org/uniprot/Q6NV75.fasta

Chapter 5

The gap

5.1 Overview

In this chapter, we drill down into some of the issues that have arisen as the slow dawn of bioinformatics has given way to the roller-coaster of modern biology. In particular, we examine the consequences of the piecemeal evolution of biological databases and software tools as they've been forced to respond to each new data 'crisis', and to adapt to tackle bigger, more complex problems. Specifically, we look more deeply at issues inherent in finding and reconstructing genes from genomic data, problems of sequence and database entry versioning, the challenges in interpreting sequence alignments, the difficulties in defining protein families and functions, and the thorny problem of nomenclature.

By the end of the chapter, you should have an appreciation of the importance of **metaphor** and semantics; you should also have insights into how semantic technologies are gaining traction in addressing the challenges of high-throughput biological data management and analysis.

5.2 Bioinformatics in the 21st century

In what has been coined 'the heroic age' of complete genome sequencing, bioinformatics was hailed as the harbinger of unparalleled advances in the biomedical sciences. The human genome was expected to yield abundant **drug targets**, allowing 21st-century biotechnology to expand the number and kinds of drugs available; along with bioinformatics, it was predicted to bring forth breakthrough products in the fight against **cancer**, **AIDS**, **Alzheimer's** and other killer diseases. The bioinformatics revolution was foreseen as ushering in a new era in pharmaceutical research – **personalised medicine** – where adverse reactions to drug regimes would be eliminated, and therapeutic strategies would be tailored to individual **genotypes** (Cantor, 2000). Bioinformatics would also provide tools to tackle the complexity of holistic biological research – **systems biology** – and the methodology both to make accurate predictions of protein structures from their sequences, and to design drugs based on simulation of small molecules docking to predicted protein architectures (Wallace, 2001). Overall, bioinformatics would dramatically improve our understanding in areas such as the regulation of gene

Bioinformatics challenges at the interface of biology and computer science: Mind the Gap. First Edition.
Teresa K. Attwood, Stephen R. Pettifer and David Thorne. Published 2016 © 2016 by John Wiley and Sons, Ltd.
Companion website: www.wiley.com/go/attwood/bioinformatics

expression, protein structure determination, comparative evolution and drug discovery (Roos, 2001).

This isn't the first time that predictions like this have been made. Some years before, Rose and Creamer (1994)

> knowingly crawled out to the C-terminus of a precarious limb to predict that the problem of predicting protein conformation from the amino acid sequence will be solved by the decade's end.

As they so presciently stated at the time, 'predictions have a way of coming back to twit would-be seers' – and theirs was no exception. So, why have so many of the early predictions not yet materialised? Part of the answer lies in the fact that getting computers to solve biological problems is really quite hard.

Let's just think about the predictions made for the role of bioinformatics in genomic medicine, drug discovery and design: realistically, what might such processes entail? Some of the key steps might involve, first, identifying specific genes involved in a given disease and identifying the encoded proteins. To those proteins, functions would have to be assigned and their structures predicted, modelled, or determined experimentally. Relevant biochemical pathways would have to be elucidated, and key regulatory proteins isolated as potential drug targets. Compounds that interact with the targets would then have to be identified, and structural information used to refine binding affinity and *in vitro* data and/or *in silico* prediction to optimise ADME-Tox profiles (Editorial, 2001).

The problem is, our understanding of many of these processes is incomplete, and many of the computational and conceptual tools we have to tackle them are still rather primitive. For example, gene-prediction tools aren't completely reliable, and it isn't always clear what we mean by 'gene' anyway; function-prediction tools aren't completely reliable, and it isn't always clear what we mean by 'function'; and structure-prediction methods aren't reliable (see Box 5.1), and it isn't always clear what we mean by 'structure' (Attwood, 2000). In the following sections, we examine some of these issues in more detail.

5.3 Problems with genes

In analysing genomes, it was quickly discovered that it's actually much harder to count genes than was first thought, largely owing to the difficulty of building gene models computationally. More than fifteen years ago, a critical assessment of gene-prediction software was made, based on a well-characterised region of the *Drosophila* genome. The results suggested that, while the ability of the software to identify coding nucleotides was good, all of the tools had problems assembling complete genes, and the best were incorrect more than 50 per cent of the time: overall, intron–exon structures were correctly predicted for only ~40 per cent of genes, the methods missed 5–44 per cent of genes, and wrongly identified 10–55 per cent (Reese, 2000).

At around this time, before the human genome sequence was published, a online sweepstake was opened to allow researchers to test their mettle by predicting the number of genes 'our genome' might contain: as we saw in Box 2.8, estimates varied wildly. The upper bound – ~300,000 – probably points more to a rather arrogant presumption of the much greater sophistication of humans relative to other

Box 5.1 Issues with structures and structure prediction

In the wake of the genome era, bioinformatics was predicted to provide a variety of revolutionary tools, including those able to make accurate predictions of protein structures from their linear sequences (Wallace, 2001). The grail of 3D structure prediction was spurred by **Anfinsen**'s early studies on the folding of ribonuclease, which led him to assert that, at least in the case of small globular proteins, the native structure is determined by the amino acid sequence alone – Anfinsen's Dogma[a] (Anfinsen, 1973).

This simple idea stimulated the development of a range of structure-prediction methods, including knowledge-based approaches that use information from the structure databases to build models, and computationally-intensive strategies that attempt to simulate the physical and chemical forces involved in protein folding. However, despite more than 40 years of research, the problem of predicting protein structures from their sequences remains unsolved: homology modelling and threading techniques generally produce low resolution models; no method gives accurate predictions for remote homologues. For small proteins, *ab initio* methods can generate models with segments 'resembling' the correct fold, but results deteriorate for larger proteins (*i.e.*, with more than 100 residues). It turns out that knowledge-based methods that combine information from a variety of different approaches give best results; but for modelling and fold recognition, most advances have accrued from balancing better algorithms with appropriate levels of manual intervention.

This shouldn't surprise us. Prediction methods aren't accurate to atomic resolution partly because we don't fully understand the protein folding problem – how primary structure determines tertiary structure – *i.e.*, we can't yet read the language sequences use to create their folds. Another part of the problem is that the beautiful structures we find enshrined in PDB files are static snapshots of dynamic biological objects that have been frozen in time (although, granted, solution structures do convey more of the dynamic nature of protein folds). So, when we try to predict structures from their sequences, we're attempting to capture a particular moment in the complex folding dance[b] of a given protein – not surprisingly, this is difficult to achieve.

It was hoped that the Protein Structure Initiative (PSI) would lessen our reliance on prediction by providing structures or models for proteins in all completed genomes. But, as we saw in Chapter 3, this lofty goal has been tempered by an acceptance that the structures we can determine are limited to those that respond to the available biophysical techniques (proteins that crystallise, that are small enough to be solved by NMR, *etc.*); membrane proteins remain difficult, as they're intractable to conventional crystallisation techniques, and most are too large to be solved by NMR.

A key motivating principle of the PSI was that knowledge of all these protein structures would divulge their molecular functions – but this too was over-simplistic. By January 2016, the functions of more than a third of structures determined by the PSI (1,682/4,873) remained unknown – indeed, researchers are invited to become 'functional sleuths'[c] to help functionally characterise them. It's important to put the value of protein structure in perspective. Protein folds provide scaffolds that can be decorated in different ways by different sequences to confer different functions. Knowing both the fold *and* the function allows us to piece together the mechanistic processes by which the structure effects its function at the molecular level. Structures provide one part of a complex biological puzzle. Without supporting information, knowing structure alone does not inherently tell us function – for a good example, see how Zarembinski *et al.* struggled to determine the precise biochemical function and cellular role of a hypothetical protein, despite having determined its molecular structure (Zarembinski *et al.*, 1998). Knowledge of the scaffold of an uncharacterised protein becomes a useful piece of the functional jigsaw only in relation to other molecular, cellular and functional clues.

[a]https://en.wikipedia.org/wiki/Anfinsen%27s_dogma
[b]https://www.youtube.com/watch?v=meNEUTn9Atg
[c]http://sbkb.org/functionalsleuth

organisms than it does to flaws in the software used to produce these wayward predictions. Regardless, since 2000, gene-prediction approaches have improved, particular genomic data-sets have been re-analysed, and gene counts have been subject to iterative refinement. When the draft human genome was made available, the estimated number of genes fell to 30,000–35,000; five months later, the numbers crept up again, to 65,000–75,000 (Wright *et al.*, 2001). Almost a decade later, the number of predicted human genes weighed in more modestly between 18,877 and 20,251 (Pertea and Salzberg, 2010).

Aside from the difficulty of accurately locating intron–exon boundaries, another issue is that the notion of what constitutes a gene has been rather fluid. Opinions have differed as to whether it should be considered a heritable unit relating to an observable **phenotype**, or a packet of genetic information that encodes RNA, a protein or several proteins – as we saw in Chapter 1, genes may encode multiple splice forms. Alternative splicing was first noted in the **calcitonin** gene: the six exons of the human gene harbour two isoforms that span exons 1-4, and a third that spans exons 1–3, 5 and 6, skipping exon 4 – marking its significant difference from the others, the third isoform is termed **calcitonin gene-related peptide**. These differences are shown in Figure 5.1.

The extent of alternative splicing in some genes makes this calcitonin example appear trivial: *e.g.*, the complexity of the *Drosophila* **Dscam** gene led to a prediction that it may have 38,016 splice variants (Schmucker *et al.*, 2000)! To further confound matters, genes can exist in multiple copies, and they don't have to be expressed to exist (it's little wonder, then, that a definition and a set of guidelines had to be provided for the gene-prediction sweepstake to indicate what would actually count as a gene in the contest). Needless to say, getting computers to handle such convolutions poses enormous challenges. The conceptual baggage bound up with our understanding of the simple term 'calcitonin gene' presents a virtually unnavigable minefield of uncertainty and ambiguity to a machine; deconvoluting the concept into a technical specification that would allow a computer to find and reliably construct a given calcitonin gene (or any alternatively-spliced gene) becomes a significant undertaking.

Calcitonin, *Homo sapiens* (canonical isoform) - P01258, CALC_HUMAN
```
MGFQKFSPFL ALSILVLLQA GSLHAAPFRS ALESSPADPA TLSEDEARLL LAALVQNYVQ
MKASELEQEQ EREGSSLDSP RSKRCGNLST CMLGTYTQDF NKFHTFPQTA IGVGAPGKKR
DMSSDLERDH RPHVSMPQNA N
```

Calcitonin, *Homo sapiens* (isoform 2) - P01258-2, CALC_HUMAN
```
MGFQKFSPFL ALSILVLLQA GSLHAAPFRS ALESSPADPA TLSEDEARLL LAALVQNYVQ
MKASELEQEQ EREGSSLDSP RSKRCGNLST CMLGTYTQDF NKFHTFPQTA IGVGAPGKKR
DMSSDLERDH RPHNHCPEES L
```

Calcitonin gene-related peptide 1, *Homo sapiens* - P06881, CALCA_HUMAN
```
MGFQKFSPFL ALSILVLLQA GSLHAAPFRS ALESSPADPA TLSEDEARLL LAALVQNYVQ
MKASELEQEQ EREGSRIIAQ KRACDTATCV THRLAGLLSR SGGVVKNNFV PTNVGSKAFG
RRRRDLQA
```

Figure 5.1

Splice forms of the human calcitonin gene (CALC_HUMAN, P01258[i]; CALCA_HUMAN, P06881[ii]). Sequence differences are highlighted (red) with respect to isoform 1, the canonical form.

[i] P01258 (http://www.uniprot.org/uniprot/P01258#P01258)
[ii] P06881 (http://www.uniprot.org/uniprot/P06881#P06881-1)

5.4 Problems with names

In the Preface, we mentioned that getting humans to agree on what words mean is hard; getting them to understand when they're using the same word to mean different things is harder; getting a machine to understand the difference is harder still. For machines, precision is vital; unfortunately, however, adherence to standard nomenclatures has, at best, been patchy in the life sciences.

Consider the protein, 'DNA replication licensing factor MCM4', shown in Table 5.1: this has five different protein and six different gene names in three different species. With a bit of work, humans can unravel this kind of intricacy relatively easily; but once we involve computers in the loop of comprehension, the task becomes more complicated, and the level of precision required to specify a particular protein or gene becomes much more difficult to achieve. For example, if we were to use a text-mining algorithm to scour the literature for protein 'disc proliferation abnormal', we would miss other mentions of the same protein (cell division control protein 54, CDC21 homolog, P1-CDC21); similarly, if we were to search for the gene *dpa*, we would miss mentions of *Mcm4*, *HCD21*, *CDC54*, etc. Only if the software had been programmed to use comprehensive synonym dictionaries could we be sure that we'd not missed a relevant gene or protein in another species – how else would a computer know that *Mcm4* in mouse is the 'same' gene as *dpa* in the fly? To be absolutely confident in our results, of course, we'd need access to the ultimate gene and protein synonym repository for all species, the creation and maintenance of which would be a logistical nightmare, given the dearth of nomenclature standards or, more to the point, our failure to adhere consistently to those that do exist[1].

Inconsistent use of identifiers and accession numbers throws up related issues. Let's briefly consider ovine rhodopsin. This protein first entered the PIR database with identifier (ID) OOSH, accession number (AC#) A03155. Subsequent changes to its sequence and annotation led its AC# to be changed to A93264, A90319 and then A30407. Later, the same protein also appeared in Swiss-Prot with ID OPSD$SHEEP, AC# P02700[2]. Following a database-wide modification to the format of Swiss-Prot's identifiers, this ID changed to OPSD_SHEEP; finally, the PIR and Swiss-Prot entries acceded to

Table 5.1

Synonymous gene and protein names for *Mcm4* in three different species.

Species	Gene	Synonym	Protein	Alternative Name	UniProtKB ID, AC#
Mus musculus	Mcm4	Cdc21, Mcmd4	DNA replication licensing factor MCM4	CDC21 homolog, P1-CDC21	MCM4_MOUSE, P49717[a]
Saccharomyces cerevisiae	MCM4	CDC54, HCD21	DNA replication licensing factor MCM4	Cell division control protein 54	MCM4_YEAST, P30665[b]
Drosophila melanogaster	dpa		DNA replication licensing factor MCM4	Protein disc proliferation abnormal	MCM4_DROME, Q26454[c]

[a] http://www.uniprot.org/uniprot/P49717
[b] http://www.uniprot.org/uniprot/P30665
[c] http://www.uniprot.org/uniprot/Q26454

---

[1] *Michael Ashburner once observed that "biologists would rather share a toothbrush than share a gene name", a behaviour that's led to a nomenclature blight across the field (Pearson, 2001).*
[2] http://www.uniprot.org/uniprot/P02700

Table 5.2

Database identifiers and accession numbers for ovine rhodopsin.

Database	ID	AC#
PIR	OOSH	A03155
		A93264
		A90319
		A30407
Swiss-Prot	OPSD$SHEEP	P02700
	OPSD_SHEEP	
UniParc		UPI000059C30D
		UPI0000130E18

UniProtKB, where changes and refinements continued to be made. By October 2015, UniProtKB archived *98 versions* of the entry for ovine rhodopsin, including these three different IDs and five different AC#s; the sequence itself is also stored in UniParc, with AC#s UPI000059C30D[3] and UPI0000130E18[4], the latter being the active sequence of record up to January 2016 – see Table 5.2.

Such complications pose similar problems for text-mining algorithms to those mentioned above for the gene *Mcm4*, with its multiple synonyms. Hence, for example, suppose we want to find all instances of ovine rhodopsin in the literature, and we use the current UniProtKB/Swiss-Prot identifier, OPSD_SHEEP. Unfortunately, the result wouldn't be comprehensive because it would miss all examples of OPSD$SHEEP and of OOSH. Alternatively, we could use the current UniProtKB/Swiss-Prot accession number, P02700; but again, such a search would miss any of the other possible accession numbers listed in Table 5.2. As before, to be confident of achieving comprehensive results, we'd need to use all of the identifiers or all of the accession numbers known to us. Fortunately, UniParc provides an archive of protein sequence histories, including changes to their identifiers and accession numbers. Nevertheless, reconstructing those histories accurately requires significant human effort.

5.5 Problems with sequences

Related to the nomenclature issues mentioned above are more subtle problems. For example, consider this: when we refer to the sequence of ovine rhodopsin, which sequence do we really mean? This might sound like a pedantic question. After all, we probably just mean the current, most up-to-date version; but the '*current version*' of a sequence is a historical artefact – the current version of ovine rhodopsin in 1986 was different from the 'current version' in 1991. In the mid 1980s, the former is likely to have been a version of the sequence based on the 1986 PIR entry (AC# A03155, ID OOSH) – see Figure 5.2; during the 1990s and onwards, however, the latter is likely to have been a version based on the 1991 PIR entry (AC# A30407, ID OOSH) – see Figure 5.3.

---

[3] http://www.uniprot.org/uniparc/UPI000059C30D
[4] http://www.uniprot.org/uniparc/UPI0000130E18

```
ID    OPSD$SHEEP      STANDARD;       PRT;     315 AA.
AC    P02700;
DT    21-JUL-1986  (REL. 01, CREATED)
DT    21-JUL-1986  (REL. 01, LAST SEQUENCE UPDATE)
DT    01-NOV-1990  (REL. 16, LAST ANNOTATION UPDATE)
DE    RHODOPSIN (FRAGMENTS).
OS    SHEEP (OVIS ARIES).
OC    EUKARYOTA; METAZOA; CHORDATA; VERTEBRATA; TETRAPODA; MAMMALIA;
OC    EUTHERIA; ARTIODACTYLA.
...
DR    PIR; A03155; OOSH.
DR    PROSITE; PS00237; G_PROTEIN_RECEPTOR.
DR    PROSITE; PS00238; OPSIN.
KW    PHOTORECEPTOR; RETINAL PROTEIN; TRANSMEMBRANE; GLYCOPROTEIN; VISION;
KW    PHOSPHORYLATION; LIPOPROTEIN; G-PROTEIN COUPLED RECEPTOR.
FT    DOMAIN        1       36        EXTRACELLULAR.
FT    TRANSMEM      37      61
FT    DOMAIN        62      73        CYTOPLASMIC.
FT    TRANSMEM      74      98
FT    DOMAIN        99      >111      EXTRACELLULAR.
FT    NON_CONS      111     112
FT    DOMAIN        <112    120       CYTOPLASMIC.
FT    TRANSMEM      121     144
FT    DOMAIN        145     169       EXTRACELLULAR.
FT    TRANSMEM      170     197
FT    DOMAIN        198     219       CYTOPLASMIC.
FT    TRANSMEM      220     243
FT    DOMAIN        244     251       EXTRACELLULAR.
FT    TRANSMEM      252     276
FT    DOMAIN        277     315       CYTOPLASMIC.
FT    CARBOHYD      2       2         BY HOMOLOGY.
FT    CARBOHYD      15      15        BY HOMOLOGY.
FT    BINDING       263     263       RETINAL CHROMOPHORE.
FT    BINDING       289     289       PALMITATE (BY HOMOLOGY).
FT    BINDING       290     290       PALMITATE (BY HOMOLOGY).
SQ    SEQUENCE   315 AA;   35281 MW;   558239 CN;
      MNGTEGPNFY VPFSNKTGVV RSPFEAPQYY LAEPWQFSML AAYMFLLIVL GFPINFLTLY
      VTVQHKKLRT PLNYILLNLA VADLFMVFGG FTTTLYTSLH GYFVFGPTGC NSNFRFGENH
      AIMGVAFTWV MALACAAPPL VGWSRYIPQG MQCSGALYFT LKPEINNESF VIYMFVVHFS
      IPLIVIFFCY GQLVFTVKEA AAQQQESATT QKAEKEVTRM VIIMVIAFLI CWLPYAGVAF
      YIFTHQGSDF GPIFMTIPAF FAKSSSVYNP VIYIMMNKQF RNCMLTTLCC GKNPLGDDEA
      STTVSKTETS QVAPA
//
```

Figure 5.2

Excerpt from the Swiss-Prot entry for ovine rhodopsin current during the 1980s. Compare the highlighted lines with those in Figure 5.3.

Compare the highlighted lines in Figures 5.2 and 5.3. Note that the Swiss-Prot ID and AC# are the same, but the underlying sequence is different (one has 315[5] amino acids, the other 348[6]) because the two versions were based on different PIR entries, each of which used the same identifier (OOSH) but different AC# (A03155 and A30407). This challenges our understanding of the role of primary database IDs and AC#s – if they don't chart database entry versions, what do they mean? Here, OPSD$SHEEP/P02700 points to different sequences – only if we refer to the sequence versions in UniProt's history file can we pinpoint when the sequence changed – *i.e.*, between versions 6 and 7. Clearly, with effort, we can piece together the history of 'the current sequence of ovine rhodopsin[7]'

---

[5] http://www.uniprot.org/uniprot/P02700.txt?version=6
[6] http://www.uniprot.org/uniprot/P02700.txt?version=7
[7] http://www.uniprot.org/uniprot/P02700

```
ID   OPSD$SHEEP      STANDARD;      PRT;     348 AA.
AC   P02700;
DT   21-JUL-1986  (REL. 01, CREATED)
DT   01-FEB-1991  (REL. 17, LAST SEQUENCE UPDATE)
DT   01-FEB-1991  (REL. 17, LAST ANNOTATION UPDATE)
DE   RHODOPSIN.
OS   SHEEP (OVIS ARIES).
OC   EUKARYOTA; METAZOA; CHORDATA; VERTEBRATA; TETRAPODA; MAMMALIA;
...
DR   PIR; A30407; OOSH.
DR   PROSITE; PS00237; G_PROTEIN_RECEPTOR.
DR   PROSITE; PS00238; OPSIN.
KW   PHOTORECEPTOR; RETINAL PROTEIN; TRANSMEMBRANE; GLYCOPROTEIN; VISION;
KW   PHOSPHORYLATION; LIPOPROTEIN; G-PROTEIN COUPLED RECEPTOR.
FT   DOMAIN        1      36        EXTRACELLULAR.
FT   TRANSMEM     37      61
FT   DOMAIN       62      73        CYTOPLASMIC.
FT   TRANSMEM     74      98
FT   DOMAIN       99     113        EXTRACELLULAR.
FT   TRANSMEM    114     140
FT   DOMAIN      141     152        CYTOPLASMIC.
FT   TRANSMEM    153     176
FT   DOMAIN      173     202        EXTRACELLULAR.
FT   TRANSMEM    203     230
FT   DOMAIN      231     252        CYTOPLASMIC.
FT   TRANSMEM    253     276
FT   DOMAIN      277     284        EXTRACELLULAR.
FT   TRANSMEM    285     309
FT   DOMAIN      310     348        CYTOPLASMIC.
FT   CARBOHYD      2       2        BY HOMOLOGY.
FT   CARBOHYD     15      15        BY HOMOLOGY.
FT   BINDING     296     296        RETINAL CHROMOPHORE.
FT   BINDING     322     322        PALMITATE (BY HOMOLOGY).
FT   BINDING     323     323        PALMITATE (BY HOMOLOGY).
FT   DISULFID    110     187        BY HOMOLOGY.
SQ   SEQUENCE   348 AA;   38891 MW;   680445 CN;
     MNGTEGPNFY VPFSNKTGVV RSPFEAPQYY LAEPWQFSML AAYMFLLIVL GFPINFLTLY
     VTVQHKKLRT PLNYILLNLA VADLFMVFGG FTTTLYTSLH GYFVFGPTGC NLEGFFATLG
     GEIALWSLVV LAIERYVVVC KPMSNFRFGE NHAIMGVAFT WVMALACAAP PLVGWSRYIP
     QGMQCSCGAL YFTLKPEINN ESFVIYMFVV HFSIPLIVIF FCYGQLVFTV KEAAAQQQES
     ATTQKAEKEV TRMVIIMVIA FLICWLPYAG VAFYIFTHQG SDFGPIFMTI PAFFAKSSSV
     YNPVIYIMMN KQFRNCMLTT LCCGKNPLGD DEASTTVSKT ETSQVAPA
//
```

Figure 5.3

Excerpt from the Swiss-Prot entry for ovine rhodopsin current from the 1990s. Compare the highlighted lines with those in Figure 5.2. The clue to the sequence change is in the cross-reference to the PIR.

(*i.e.*, version 98 in UniProtKB/Swiss-Prot in January 2016; of course, by the time this book is published, a different version will be 'current' – can you find out what it is?), but this is challenging to achieve computationally, because there's no machine-readable description of the provenance trail, and hence no way for a computer to 'reason' this out for itself.

Initiatives such as the MIRIAM[8] Registry (Laibe and Novère, 2007) and Identifiers .org[9] (Juty *et al.*, 2011) will be crucial in untangling such 'identity crises' in future, as they work towards establishing a standard framework for providing persistent resolvable identifiers for biomedical data.

---

[8] http://www.ebi.ac.uk/miriam
[9] http://identifiers.org

5.6 Problems with database entries

Some of the issues we've been discussing in the previous sections are bound up with the constant state of flux of database entries. Our next example is no exception. Consider the following: when we refer to the database entry for the protein designated by AC# Q23293[10], to which protein are we really referring? In 1996, this sequence entered the first release of TrEMBL with the description, 'Similarity to mammalian μ- and κ-type opioid receptors'[11]. In the 19 years that followed, there were 107 versions of this entry, amongst which the description of the protein changed more than a dozen times. By March 2012, the UniProtKB/TrEMBL description line read, 'Protein DMSR-11', and its sequence had changed in length from 320 to 355 amino acids – a summary of these changes is given in Table 5.3.

A few aspects of the history of changes shown in Table 5.3 deserve special mention. The first description of the protein was reasonably definitive, asserting that it resembled an opioid receptor (a GPCR); over the years, the force of this assertion diminished, changing first to 'hypothetical 36.4 kDa protein'[12], then to 'putative uncharacterized protein'[13], and more recently to 'protein ZC404.11'[14] and 'protein DMSR-11'[15]. But the initial claim that the sequence was like an opioid receptor remained in the database for seven years; during this time, it's possible that any subsequent analyses based on that assertion may have led to inappropriate conclusions; worse, the original description had seven years in which to propagate to related database entries. The

Table 5.3

Database descriptions and sequence length changes for Q23293.

Database release	Description	Length
TrEMBL 1	SIMILARITY TO MAMAMALIAN MU- AND KAPPA-TYPE OPIOID RECEPTORS	320
TrEMBL 19	HYPOTHETICAL 36.4 KDA PROTEIN	320
TrEMBL 21	Hypothetical 36.4 kDa protein	320
TrEMBL 25	Hypothetical protein	320
UniProtKB/TrEMBL 2.7/27.7	Hypothetical protein ZC404.11	320
UniProtKB/TrEMBL 4.3/29.3	Hypothetical protein	320
UniProtKB/TrEMBL 12.0/37.0	Putative uncharacterized protein	320
UniProtKB/TrEMBL 14.0/39.0	SubName: Full=Putative uncharacterized protein	320
UniProtKB/TrEMBL 2012_01	SubName: Full=Protein ZC404.11	355
UniProtKB/TrEMBL 2012_03	SubName: Full=Protein DMSR-11	355
UniProtKB/TrEMBL 2015_04	SubName: Full=DMSR-11	355
UniProtKB/TrEMBL 2015_05	SubName: Full= ZC404.11	355
UniProtKB/TrEMBL 2015_08	SubName: Full=DMSR-11	355
UniProtKB/TrEMBL 2015_11	SubName: Full=DroMyoSuppressin Receptor related	355

[10] http://www.uniprot.org/uniprot/Q23293
[11] http://www.uniprot.org/uniprot/Q23293.txt?version=1
[12] http://www.uniprot.org/uniprot/Q23293.txt?version=14
[13] http://www.uniprot.org/uniprot/Q23293.txt?version=45
[14] http://www.uniprot.org/uniprot/Q23293.txt?version=74
[15] http://www.uniprot.org/uniprot/Q23293.txt?version=77

good news is that the description *'protein DMSR-11'* remained relatively consistent between March 2012 and April 2015. In May, however, it reverted to 'ZC404.11'; in July it changed back again to 'DMSR=11', and in November 2015 (entry version 106) it had evolved into the slightly more informative 'DroMyoSuppressin Receptor related'; but, of course, there are no guarantees that it won't change again at some point in the future.

This gradual change in the level of confidence ascribed to the identity of the protein probably reflects the growing appreciation of the fallibility of computer-based functional annotation methods. Regardless, the 14 changes to the description line resulted in ten unique protein descriptions in a 19-year period. Hence, while a researcher accessing the database entry for sequence Q23293[16] in November 2015 would have seen 'DroMyoSuppressin Receptor related', six months before that, he or she would have seen 'Protein ZC404.11', eight years before, he or she would have seen a 'putative uncharacterized protein', 13 years before, a 'hypothetical 36.4 kDa protein', and 15 years before, a protein with 'similarity to mammalian μ- and κ-type opioid receptors'. How, then, can we be confident in the identity of protein Q23293, or indeed of any protein, at a given snapshot in time, given the volatility of database records?

Some of the other changes in Table 5.3 are rather more subtle. Between TrEMBL releases 19 and 21, the description lines changed from being all upper-case ('HYPO-THETICAL 36.4 KDA PROTEIN'[17]) to mixed upper- and lower-case ('Hypothetical 36.4 kDa protein'[18]). Such a change would have caused problems for any naïve computer script designed to recognise the title of the protein as 'the upper-case string following the DE tag' – indeed, such scripts are doomed to fail. This is exactly what happened in 1999 when the Swiss-Prot format began a transformation from exclusively upper-case to mixed case – hundreds (perhaps thousands) of bespoke Swiss-Prot search scripts around the world broke overnight following the overhaul. Related problems are likely to have occurred following the change we see between UniProtKB/TrEMBL releases 12.0/37.0 and 14.0/39.0 from the description 'Putative uncharacterized protein' to 'SubName: Full=Putative uncharacterized protein'[19] – naïve scripts designed to expect only the protein description following the DE tag would include the string, 'SubName: Full=' as part of the protein title; hence, outputs from bespoke BLAST implementations are typically now littered with such additional 'noise'.

5.6.1 Problems with database entry formats

Up to this point, we've been tracking the history of a particular biological entity through the TrEMBL database and onwards through its journey into UniProtKB. In the process, we witnessed many changes to its database entry as our knowledge about it evolved: starting life as a putative GPCR, it eventually became known as 'DroMyoSuppressin Receptor related' – the 'same' entity, but with a different name. Earlier in the chapter, we saw how the same entity can have different names in different species, and it can have different IDs and AC#s in different databases and within the same database, making it much harder to track biological entities than it should be. Related to these problems is the issue of database formats.

[16] http://www.uniprot.org/uniprot/Q23293.txt?version=106
[17] http://www.uniprot.org/uniprot/Q23293.txt?version=14
[18] http://www.uniprot.org/uniprot/Q23293.txt?version=15
[19] http://www.uniprot.org/uniprot/Q23293.txt?version=50

In Section 4.2.2, we examined the format of the EMBL data library in some detail, largely because its simplicity and ready extensibility led several other databases subsequently to adopt and adapt the format to their own ends. GenBank, however, its sister database, had grown independently more or less in parallel, and hence used a different framework. Although we saw that the databases did use a common Feature Table format to facilitate interoperation, unfortunately, this is all they share. Consider, then, the challenge of locating sequence information relating to DMSR-11.

Within the UniProtKB/TrEMBL entry version 107, we find an embedded link to 'BX284605'. Selecting the EMBL view[20] (text option), we find that BX284605 is listed both as the entry ID and the AC; by contrast, in the GenBank view, BX284605[21] is listed as the entry 'LOCUS' and 'ACCESSION'. In EMBL, its description (DE) is 'Caenorhabditis elegans chromosome V', and in GenBank its DEFINITION is 'Caenorhabditis elegans chromosome V'. In GenBank, both the SOURCE and ORGANISM are given as 'Caenorhabditis elegans', while in EMBL, this is tagged as OS, the organism species.

Similar issues are encountered when we try to trace information in the UniProtKB and NCBI Reference Sequence (RefSeq) databases: in the former, Q23293_CAEEL is the ID and Q23293[22] the AC, and in the latter, NP_504725[23] is both the LOCUS and the ACCESSION. The description (DE) in UniProtKB/TrEMBL is 'SubName: Full=DroMyoSuppressin Receptor related', and in RefSeq, the DEFINITION is 'DroMyoSuppressin Receptor related [Caenorhabditis elegans]'. At this level, the relationship between the nucleotide and protein sequence databases isn't evident. To help make this connection, the UniProtKB/TrEMBL gene-name tag (GN) makes reference to 'ORFNames=CELE_ZC404.11'. In RefSeq, this information is held in the coding sequence (CDS) component of the FEATURES table, under the tag '/locus_tag="CELE_ZC404.11"'.

Negotiating the labyrinth of ID, AC# and nomenclature differences associated with this single entity requires significant detective work; getting computers to perform the same task requires precise translation between the minutiae of each of the different database formats and nomenclature systems. But here, we've looked at just four tightly coupled primary sequence databases; automatic data retrieval from the hundreds of genomic data repositories around the world (or even within a single institute) presents many more challenges – challenges that will ultimately require database- and application-specific semantics to be fully documented if we're to be able to help computers to 'understand' and complete the tasks we need them to perform in future.

5.7   Problems with structures

By now, it shouldn't be surprising that the issues mentioned above concerning what we actually mean by 'the name', 'the sequence' or 'the database description' of a protein also hold for what we mean by 'the structure' of a given protein. So, let's consider what we mean by 'the structure of bovine rhodopsin'. If we refer to the UniProtKB/Swiss-Prot entry, OPSD_BOVIN (P02699)[24], we find 40 cross-references to rhodopsin structures in the PDB, determined by means of X-ray diffraction, solution NMR or modelling to **resolutions**

---

[20] http://www.ebi.ac.uk/ena/data/view/BX284605&display=text
[21] http://www.ncbi.nlm.nih.gov/nuccore/BX284605
[22] http://www.uniprot.org/uniprot/Q23293.txt?version=107
[23] http://www.ncbi.nlm.nih.gov/protein/392919337
[24] http://www.uniprot.org/uniprot/P02699.txt?version=160

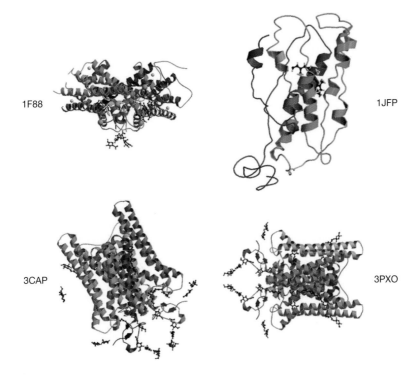

Figure 5.4

Selected structural variants of bovine rhodopsin.

Source: Protein Data Bank.

ranging from 2.20 to 4.15 Å. Closer inspection within the PDB itself reveals 45 rhodopsin structures originating from different species: 35 are forms of bovine rhodopsin. So which of the 35 bovine rhodopsin structures do we mean? The original X-ray diffraction structure at 2.8 Å resolution (1f88[25]); the solution NMR structure of dark-adapted rhodopsin (1JFP[26]); the 2.9 Å crystal structure of native opsin in its ligand-free state (3CAP[27]); the crystal structure of metarhodopsin II at 3.0 Å resolution (3PXO[28]); or any of the other 30 or so variants with or without 31 possible ligands (Figure 5.4 illustrates just a few of these variants)?

It's tempting to dismiss this as an overly convoluted example; yet, remember that the PDB contains a high level of redundancy. Consider **sperm whale** myoglobin, which was amongst the first handful of structures to enter the PDB (1MBN[29]) – so there's been lots of time for other myoglobin structures to enter the database. If we look at the UniProtKB/Swiss-Prot entry, MYG_PHYMC (P02185[30]), we see more than 200 cross-references to structures in the PDB. In the PDB itself, in December 2015, a total of 378 entries related somehow to myoglobin, associated with 83 different ligands, of which 248 derived from the sperm whale; of these, 243 originated from X-ray diffraction studies, three from **neutron diffraction** studies and one from solution NMR, with

---

[25] http://www.rcsb.org/pdb/explore/explore.do?structureId=1F88
[26] http://www.rcsb.org/pdb/explore/explore.do?structureId=1JFP
[27] http://www.rcsb.org/pdb/explore/explore.do?structureId=3CAP
[28] http://www.rcsb.org/pdb/explore/explore.do?structureId=3PXO
[29] http://www.rcsb.org/pdb/explore/explore.do?structureId=1MBN
[30] http://www.uniprot.org/uniprot/P02185.txt?version=121

resolutions varying from 1.5 to 3.0 Å. So when we talk of 'the structure of myoglobin', it's imperative to be clear which structure we really mean – if *we* don't know, a computer will find it very hard to 'guess'.

5.8   Problems with alignments

So far in this chapter, we've been looking at some of the issues relating to nomenclature, database IDs and AC#s, database entry versions, sequence and structure variants, and so on. Here, we begin to touch on slightly more abstract problems. In the field of protein sequence analysis, a common task is to compare sequences by creating an alignment of some sort. This sounds straightforward, but generating alignments involves several considerations, from the necessary input and output formats, to the underlying alignment algorithms, to the ways in which we visualise and interpret the results. Let's take a closer look.

5.8.1   Different methods, different results

Alignments can be made using different software tools, with different algorithms and different input parameters (scoring matrices, gap penalties, *etc.*); their outputs can also be visualised in many different ways. So what does 'an alignment' really mean? We can explore this question by examining the results in Figure 5.5, which shows alignments of eight receptor sequences: some are known GPCRs (OPSD, ACM1, OPRM, C5AR); others are thought to be GPCRs, or were thought to be at some time in the past (GP153, GP162, UL78, Q23293), most likely because they contain seven hydrophobic, possibly membrane-spanning regions (see Box 5.2). A FASTA-format file containing these sequences was input to three online alignment tools (ClustalW, T-Coffee, **Multalin**) using default parameters, and to a desktop application (CINEMA) that provides options to align sequences using a MUSCLE plugin. Given the size of the outputs, only part of each alignment is shown in Figure 5.5.

The first thing to note is that the alignments differ in several respects – here, we represent them more-or-less exactly as provided by the programs, in order to give as realistic an experience as possible of how hard it is to compare alignment outputs in practice, albeit using a trivially small alignment fragment. With perseverance, it's possible to see that ClustalW and Multalin show two gapped regions; MUSCLE and T-Coffee show three. Multalin inserts gaps in two sequences; the others add gaps to them all. With 10 gaps, Multalin is the most conservative gap inserter; with 58 gaps, ClustalW is the most promiscuous inserter – T-Coffee (21 gaps) and MUSCLE (28) lie somewhere in between. ClustalW has re-ordered the sequences, making it slightly more challenging to compare its output with the others; nevertheless, with a little concentration, it's clear that the alignments created by MUSCLE and T-Coffee are the most similar, both in terms of the number of gaps and where these have been inserted, while ClustalW and Multalin sit at the extremes. Bear in mind that each of the results was generated using default parameters – varying the parameters would have given rise to different alignments.

The tools used for the purpose of this illustration were chosen fairly randomly – they don't reflect an endorsement either of superiority or of inferiority of any one over any other. Many other alignment tools exist, either as desktop applications or as online

MUSCLE	Q23293	A E S L Y L A V G M A F C R Y I T L S - - S A S D S R D T W Q S P K Y A I R V A V L L C F P V F A V S S
	OPSD	I A L W S L V V - L A I E R Y V V V C - - K P M S N F R F - - G E N H A I M G V A F T W V M A L A C A A
	ACM1	A S V M N L L L - I S F D R Y F S V T - - R P L S Y R A K R - T P R R A A L M I G L A W L V S F V L W A
	OPRM	T S I F T L C T - M S V D R Y I A V C - - H P V K A L D F R - T P R N A K I I N V C N W I L S S A I G L
	C5AR	A S I L L L A T - I S A D R F L L V F - - K P I W C Q N F R - G A G L A W I A C A V A W G L A L L L T I
	GP153	A T C F S V T S - L S Y H R M W M V C - - W P V N Y R L S N - A K K Q A V H T V M G I W M V S F I L S A
	GP162	A T C F T V A S - L S Y H R M W M V R - - W P V N Y R L S N - A K K Q A L H A V M G I W M V S F I L S T
	UL78	S T A L F F L F - L I L D R L S A I S Y G R D L W H H E T R E N A G V A L Y A V A F A W V L S I V A A V
ClustalW	GP153	W M V S F I L S A L P A V G W H D T - - S E R F Y T H G C R F I V A E I G L G F G V C F L L L V G G S V A M G V I C T
	GP162	W M V S F I L S T L P S I G W H N N - - G E R Y Y A R G C Q F I V S K I G L G F G V C F S L L L L G G I V M G L V C V
	C5AR	Y A S I L L L A T I S A D R F L L V - - F K P I W C Q N F R - - - - - - - - G A G L A W I A C A V A W G L A L L L T I
	OPRM	F T S I F T L C T M S V D R Y I A V - - C H P V K A L D F R - - - - - - - - T P R N A K I I N V C N W I L S S A I G L
	OPSD	E I A L W S L V V L A I E R Y V V V - - C K P M S N F R F G - - - - - - - - - E N H A I M G V A F T W V M A L A C A
	ACM1	N A S V M N L L L I S F D R Y F S V - - T R P L S Y R A K R - - - - - - - - T P R R A A L M I G L A W L V S F V L W A
	Q23293	A E S L Y L A V G M A F C R V I T L - - S S A S D S R D T W Q - - - - - - - S P K Y A I R V A V L L C F P V F A V S S
	UL78	Y S T A L F F L F L I L D R L S A I S Y G R D L W H H E T R - - - - - - - E N A G V A L Y A V A F A W V L S I V A A V
T-Coffee	Q23293	A E S L Y L A V G M A F C R Y I T L S S A S - - D S R D T W Q S P K Y A I R V A V L L C F P V F A V S S
	OPSD	I A L - W S L V V L A I E R Y V V V C K P - - M S - - N F R F G E N H A I M G V A F T W V M A L A C A A
	ACM1	A S V - M N L L L I S F D R Y F S V T R P - - L S Y R A K - R T P R R A A L M I G L A W L V S F V L W A
	OPRM	T S I - F T L C T M S V D R Y I A V C H P - - V K A L D F R T - P R N A K I I N V C N W I L S S A I G L
	C5AR	A S I - L L L A T I S A D R F L L V F K P - - I W C Q N F - R G A G L A W I A C A V A W G L A L L L T I
	GP153	A T C - F S V T S L S Y H R M W M V C W P - - V N Y R L S - N A K K Q A V H T V M G I W M V S F I L S A
	GP162	A T C - F T V A S L S Y H R M W M V R W P - - V N Y R L S - N A K K Q A L H A V M G I W M V S F I L S T
	UL78	S T A - L F F L F L I L D R L S A I S Y G R D L W H H E T R E N A G V A L Y A V A F A W V L S I V A A V
Multalin	Q23293	A E S L Y L A V G M A F C R Y I T L S S A S D S R D T W Q S P K Y A I R V A V L L C F P V F A V S S
	OPSD	N R A R E L A A L Q G S E T P G K G G G S S S S S E R S Q P G A E G S P E T P P G R C C R C C R A P
	ACM1	R R A F T V P T I V V E D A Q G K R R S S I D G S E P A K T S L Q T T G L V T T I V F I Y D C L M G
	OPRM	R P A F E V P A I V V E D A R G K R R S S L D G S E S A K T S L Q V T N L V S A I V F L Y D S L T G
	C5AR	C L M L F V P Y Y C F R V L R G V L Q P A S A A G T G F G I M D Y V E L A T R T L L T M R L G I L P
	GP153	M I I I F F C Y G Q L V F T V K E A A A Q Q Q E S A T T Q K A E K E V T R M V I I M V I A F L I C W
	GP162	V L I I T V C Y G L M I L R L K S V R M L S G - S K E K D R N L R R I T R M V L V V V A V F I V C W
	UL78	L L T L T I C Y T F I L L R T W S R R A T R S - T K T - - - - - - - - L K V V V A V V A S F F I F W

Figure 5.5

Excerpts from alignments of eight receptors using different tools. Equivalent parts of the alignments are shown; note that ClustalW has re-ordered the sequences.

tools; and, of one thing we can be certain: each will give different outputs again from the ones shown here. This is important, because it means that, for the same set of sequences, because the patterns of conservation and gap insertion are different, the results of downstream analyses based on these outputs will also be different (for example, see Markova-Raina and Petrov, 2011).

Given that different tools give different results, how can we judge which is the 'best' alignment, the 'true' or 'right' alignment, especially when their outputs are so difficult to compare? Some stakeholders try to justify why one particular tool or algorithm is better than others, or why some types of alignment are more meaningful than others – and the opposing views sometimes collide, generating substantial amounts of heat in the process. Notable examples of such collisions include: i) the controversy that arose when, based on their alignments of DNA and protein sequences, Graur and Higgins inferred that whales are more closely related to cows than they are to **camels** or pigs (Graur and Higgins, 1994) – this finding was strongly contested by Gatesy *et al.*, who attributed the implied whale–cow relationship to an incomplete data-set, one that hadn't included **hippopotamuses** (Gatesy *et al.*, 1999); and, more recently, ii) the debate over the asserted superiority of the alignment tool, PRANK, which uses evolutionary information to place gaps, and hence allegedly prevents errors in sequence alignment and evolutionary analyses (Loytynoja and Goldman, 2008) – in a follow-up study, while Fletcher and Yang found PRANK to 'perform better' than MUSCLE, MAFTT and ClustalW, they nevertheless found that PRANK's error rates were not under control in some data-sets (Fletcher and Yang, 2010); similarly, Markova-Raina and Petrov found that PRANK performed better than T-Coffee, ClustalW, ProbCons, **AMAP** and MUSCLE, but still had a high false-positive error rate (50–55 per cent), which they considered unacceptable for most practical applications (Markova-Raina and Petrov, 2011).

Box 5.2 Membrane proteins

Cells and their organelles are shielded from their environments by **membrane**s that are impervious to most biological substances. The protection afforded by such separation, or compartmentalisation, is crucial, otherwise their internal chemical reactions would be disturbed in unpredictable ways. Yet, biological processes are dynamic and hence depend on the regulated transfer of molecules or information between cells/organelles and their surroundings. Integral membrane proteins play critical roles here as the communicators, or mediators, between the inside and outside of cells and their organelles, allowing them to react appropriately to external stimuli and translate them into specific biochemical and biophysical responses. Amongst membrane proteins are three important functional classes – transporters, receptors and enzymes – whose structural and functional differences are determined by how many times their linear sequences span the membrane.

Transporters use complex, multi-TM architectures and electrochemical gradients, or energy from chemical reactions, to usher a variety of substances in and out of cells, as illustrated schematically in Figure 5.6 (a). Enzymes (the molecules responsible for speeding up biochemical reactions) are typically not membrane bound; those that are tend to have much simpler, usually single-TM architectures, as shown in Figure 5.6 (b).

(a)   (b)

Figure 5.6

TM architectures typical of (a) transporters and (b) enzymes.

Receptors transport 'signals' rather than substrates across membranes. Many have 7TM architectures, which normally mediate cellular responses by binding a **ligand** of some kind. G protein-coupled receptors (GPCRs), which belong to the general class of 7TM receptors, have various different modes of activation, illustrated in Figure 5.7. Some bind incoming ligands within their 7TM scaffolds (a); others bind incoming ligands within extended N-terminal regions (b), which guide the ligand to interact with their external loops (c); others (the opsins) are unusual in having their ligands **covalently** bound within the TM framework. In each case, the change in conformation of the bound ligand, or the change in conformation triggered by ligand binding, causes further conformational changes to percolate through the scaffold and disrupt the bound trimeric G proteins. The dislodged G proteins dissociate, in turn scattering their message to other cellular components (d), stimulating potent intracellular signalling cascades.

Because the primary structures of membrane proteins usually snake back and forth across the membrane (hence, some receptors are referred to as 'serpentine'), tell-tale signs are left in their sequences – *i.e.*, hydrophobic regions around 20 residues long. The likelihood of a sequence being a membrane protein is therefore often predicted (inevitably, not always reliably – *e.g.*, see Wong *et al.*, 2010) by the appearance of multiple TM domains – seven in the case of GPCRs.

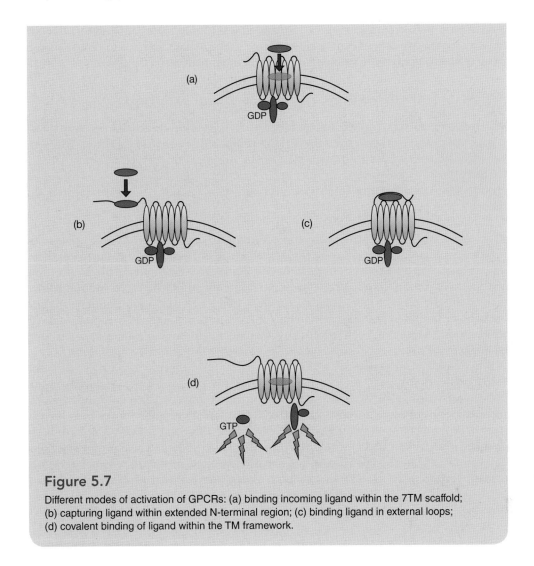

Figure 5.7

Different modes of activation of GPCRs: (a) binding incoming ligand within the 7TM scaffold; (b) capturing ligand within extended N-terminal region; (c) binding ligand in external loops; (d) covalent binding of ligand within the TM framework.

Our aim here isn't to be distracted by the minutiae of such debates; rather, we want to focus on trying to understand what different alignment outputs mean. Any attempt to interpret results like those illustrated in Figure 5.5 should be set in the context of how the underlying algorithms work, what scoring matrices were used, what gap penalties were applied, how gaps are modelled, and so on. Divorced from this context, interpretations of such outputs are, at best, likely to be misleading. The problem is that, because alignment tools are so easy to access, users seldom question their outputs. In fact, the same can be said of almost all bioinformatics tools, especially those available online, as they've been deliberately designed to provide answers as painlessly as possible – ideally, with a single click (for an entertaining vision of the future, we recommend *A history of bioinformatics (in the year 2039)*[31], in which Titus Brown refers to this

---

[31] https://www.youtube.com/watch?v=uwsjwMO-TEA

kind of push-button data analysis as 'the "easy to use" tools fiasco'). For now, in the following sections, we examine more closely some of the issues that need to be considered when trying to unravel the meaning of different alignment results.

5.8.2 What properties do my sequences share?

The task of understanding what alignments mean goes a lot deeper than just the number and placement of gaps by different algorithms. As we saw in Chapters 3 and 4, we align sequences in order to be able to compare them effectively, to detect shared characteristics, or differences, to infer their ancestry and see how they've evolved to acquire specialised functionalities and structural modifications. So, what features do the sequences in Figure 5.5 share?

Although these are more or less 'standard' outputs, it's actually virtually impossible to ascertain what the sequences have in common (it's even harder than it appears in Figure 5.5 because we've had to 'sew together' some of the aligned blocks to get them on one line). The difficulty lies in how they've been presented – mostly black letters on a white background. In practice, however, outputs like this tend to be 'consumed' by manual editing tools, which usually provide a range of visualisation options – most use colour. But different tools use different colours to depict individual amino acids or amino acid groups; as a result, comparing alignments can be challenging, as we can see from Figure 5.8.

What's clear from Figure 5.8 is the degree of confusion introduced by the different colour schemes (to make the comparison easier, the portion of the alignment is the same in each case). Here, Kyte and Zimmerman are simplistic hydrophobic/hydrophilic representations of more detailed scales (recall Chapter 2): residues in one half of each scale have been coloured white (hydrophobic), while those in the other half have been coloured green (hydrophilic). CINEMA uses the 'standard' amino acid property groups and colours illustrated in Figure 2.8. **ClustalX** is a kind of hybrid scheme, in which colours appear to be used to denote both general amino acid property groups and degree of conservation: hence, Thr and Ser are sometimes coloured green, but sometimes not; sometimes Arg and Lys are red, but sometimes not; sometimes Ala, Leu and Ile are blue, sometimes not; sometimes Asp and Glu are purple, sometimes not.

The picture is further confounded by the fact that the colours assigned to the amino acid property groups in the CINEMA and ClustalX schemes are almost complete opposites – red denotes an acid group in one, but a basic group in the other; blue denotes a basic group in one, but hydrophobic character in the other; purple denotes an aromatic group in one, but an acid group in the other; and yellow denotes the sulphur-containing amino acid Cys in one, but Pro in the other. Of course, these are just two schemes – we could have chosen others, as shown in Box 5.3. Faced with this kind of complexity, interpreting what alignments mean in terms of the characteristics they share is far from trivial.

To make any sense of the different results, a user must first understand the *metaphor* behind the colour scheme. The 'standard' scheme implemented in CINEMA was chosen to try to reduce the cognitive effort required to understand alignment outputs, by basing the colours on a scheme that many of us learned from our earliest physics and chemistry lessons: as a result, we intuitively know that red is oxygen-containing, hence acidic (negative charge); that blue is nitrogen-containing, hence basic (positive charge); that yellow is sulphur containing; and so on. In principle, therefore, we don't have to fight quite so hard to see what hides in the CINEMA alignment as we do in the ClustalX alignment.

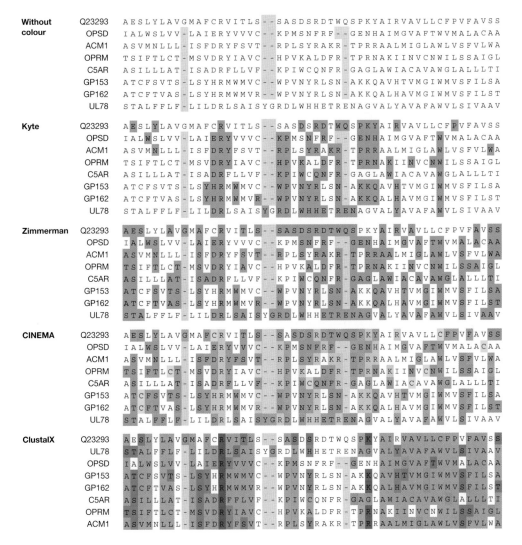

Figure 5.8

Excerpts from alignments of eight receptors using different colour schemes: Kyte and Zimmerman depict simple binary hydrophobicity scales; CINEMA and ClustalX use different amino acid property groups. Equivalent parts of the alignments are shown; note that ClustalX has re-ordered the sequences.

Let's return, then, to the question of what features are shared by the sequences shown in Figures 5.5 and 5.8. The Kyte and Zimmerman schemes should allow us to determine whether there are significant hydrophilic or hydrophobic regions (such as would allow us to infer the presence of transmembrane (TM) domains, for example). On closer inspection, we see a central hydrophilic area, flanked by more hydrophobic regions in the Kyte representation, but this isn't evident from the Zimmerman scheme. Hydrophobic regions also appear at each end of the alignment in the ClustalX scheme (the blue areas), but there is no obviously hydrophilic central region. The CINEMA scheme suggests a more hydrophobic domain at the C-terminal end of the alignment (the white/purple region), and a more hydrophilic central domain (green/blue/red region); like the Zimmerman scheme, however, it shows a mix of hydrophobic and hydrophilic residues in the N-terminal region.

Box 5.3 Colour schemes for encoding amino acid properties

Over the years, an assortment of different colour schemes has been used to denote amino acid properties, particularly for comparison purposes in multiple sequence alignments; a small sample of the available schemes is shown in Figure 5.9. The thing to notice is the lack of agreement between them: most depict cysteine (a sulphur-containing residue) as yellow, except the scheme used in ClustalX; most depict the polar neutral amino acids as green, except Taylor, which renders them in various shades of orange and magenta; most depict the basic amino acids as blue, except ClustalX, which tends to use red, the colour used by the other schemes to denote acidic properties! There is no correspondence between the colours used to denote aliphatic amino acids, nor is there much agreement on the colours used for aromatic properties, although the Taylor and ClustalX schemes use various shades of blue/green for Trp and Tyr.

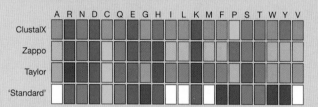

Figure 5.9

Selected colour schemes used to denote amino acid properties.

What's important here is not that one scheme is 'better' than another; the issue is the relative burden the different schemes place on users when trying to understand what the particular colour choices *mean*. This, in turn, determines how easy it is for users to switch their mental models when trying to compare, say, an alignment visualised using the ClustalX scheme versus one viewed using the Taylor scheme. The benefit of the 'standard' scheme is that it reduces cognitive burdens by using colours that are broadly intuitive, because many relate to mental models we already possess (such as positive=blue, negative=red) and map to a scheme that's already familiar to us for depicting atoms in 3D models (N=blue, O=red, S=yellow, *etc*.). Our brains are therefore freed up to concentrate on the information content of an alignment, instead of being distracted by the chore of colour-property translation.

As mentioned earlier, the sequences shown here are a mix of known and putative GPCRs – we'd therefore expect to see hydrophobic regions (TM domains) interspersed by short hydrophilic sections (loops). One of the defining characteristics of GPCRs is the presence of an acidic-Arg-aromatic triplet (sometimes termed the DRY motif) at the C-terminal end of the third TM domain – the conserved Arg is the point of interaction with G proteins. Therefore, with the standard colour scheme, we'd expect to find a Red-Blue-Purple triplet just before a hydrophilic loop region. In the CINEMA representation, this is evident in the N-terminal third of the alignment, just before the central hydrophilic domain. We see that the triplet is conserved in the known GPCRs (OPSD, ACM1, OPRM, C5AR), but not in the putative GPCRs (Q23293, GP153, GP162, UL78). This is harder to see in the ClustalX representation, as we have to up-end our mental model to look for a Purple-Red-Blueish-green triplet. Having located the relevant regions, it's then up to us to determine whether there's sufficient contextual evidence in the alignment to suggest that the putative GPCRs are likely to couple to G proteins. We will leave you to form your own conclusions!

5.8.3 How similar are my sequences?

Earlier in this section, we mentioned that one of the reasons for wishing to align sequences, aside from wanting to help detect shared structural or functional features, is to allow us to make inferences about their relatedness: *i.e.*, we often create alignments to determine how similar given sets of sequences are. So, how similar are the sequences in Figures 5.5 and 5.8? As you can see, it's really very hard to say anything meaningful. They don't look very similar, do they? But there's a twist. If we want to persuade you that the sequences are similar, we can do so by changing the colour scheme to a consensus approach. Compare Figure 5.10.

Two things are interesting about this figure. The first is how hard, relatively, the colour schemes make us work to identify the conserved residues. With the grey-scale, two positions stand out as completely conserved (R and A); those comprising predominantly V, P and W residues are also well conserved. With the PRALINE scheme, we can see that R is conserved, but it appears that the A isn't; the V and W positions are also still well conserved, but it seems that the P isn't – but overall, we have to work a bit harder to interpret what the alignment is telling us about the degree of conservation of the sequences. Part of the challenge is that (rather like the example from physics and chemistry that we saw in Section 5.8.2) we're hard-wired to equate red with hot and blue with cool/cold, so we'd usually think of a blue-to-red scale as running from

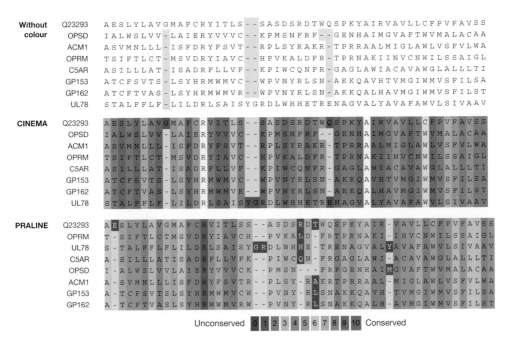

Figure 5.10

Excerpts from alignments of a group of receptors using tools with different consensus colour schemes. For comparison, the equivalent part of the alignment is shown in each case; note that PRALINE has re-ordered the sequences and the algorithm has produced a slightly different result from the MUSCLE alignment. The CINEMA colour scheme uses a kind of grey-scale (the lighter the colour, the more conserved the alignment position), while PRALINE uses a colour scheme that runs from blue to red, red being the most conserved position and blue the least conserved.

cold-to-hot – think of the colours we associate with fundamental things like fire and water/ice, and hence the colours we've appropriated to denote hot and cold taps. The PRALINE colour scale, however, runs in the opposite direction: *i.e.*, the mutational *hot*-spots (the most changeable positions) are blue, while the mutationally *cold* (least changeable) positions are red. The issue here isn't that one scheme is inherently right and another wrong; rather, it's about the mental effort required to assimilate the underlying metaphor – some metaphors are more intuitive than others (because we learned them when we were younger, because we use them in everyday life, or because our visual perception system is wired this way); their use can therefore make our lives much easier.

There is, however, a more subtle point here – it's tied up, not with the colours metaphors we use as mental short-cuts for concepts like 'hot', but with the semantics we bring to such concepts. For example, what does 'hot' mean to you? A standard dictionary definition will speak of degrees of heat or high temperature, or perhaps, say, the degree of spiciness in a curry. A dictionary of slang usage, however, will cast a much wider semantic net, capturing meanings such as 'angry', 'stolen', 'attractive', 'talented', 'well-known/popular' and, ironically, 'cool'! In the context of biological sequence and structure analysis, it could be argued that stable regions of conservation are the 'hot' thing to look for rather than mutations, because they highlight shared structural and functional features, and hence they should be red; but it could equally be argued that mutations are the 'hot' things to look for, given their potential to underpin key structural and functional changes[32], and their importance in understanding, for example, human genetics and genetic diseases like cancer[33]. Each of us carries different semantic baggage, and hence what we consider 'hot' is subjective and task dependent. Different colour schemes may therefore appear to be more or less intuitive, and hence easier or more difficult to use, depending on the interplay between our personal metaphorical and semantic mental models.

The second interesting thing about Figure 5.10 is that by changing the manner of visualisation (from colouring per residue, or residue group, to colouring per column), we can radically influence how you perceive and hence interpret the result. Relative to the alignments shown in Figures 5.5 and 5.8, the sequences in Figure 5.10 look much more similar. By colouring the alignment per column, we effectively smooth out the background noise and focus on the most conserved regions. Various software tools have been developed to exploit this fact: some use grey-scale or colour shading, like those above; others allow boxes to be drawn around regions that are the same, or similar, in order to draw our attention specifically to them. Consider the alignment shown in Figure 5.11.

This alignment, published in *The Plant Journal* in 2007, compares the protein GCR2 (Q9C929[34]) with nisin cyclase (NisC, Q48670[35]). The authors have gone to some trouble to align the sequences relative to the known crystal structure, and have marked up the α-helical and β-strand regions, and have also pinpointed the zinc ligands; the identical and similar residues are shaded black and grey. Most of us simply accept the result

---

[32] http://www.sciencedirect.com/science/article/pii/S0969212610001048
[33] http://www.hindawi.com/journals/cmmm/2014/653487/fig5
[34] http://www.uniprot.org/uniprot/Q9C929
[35] http://www.uniprot.org/uniprot/Q48670

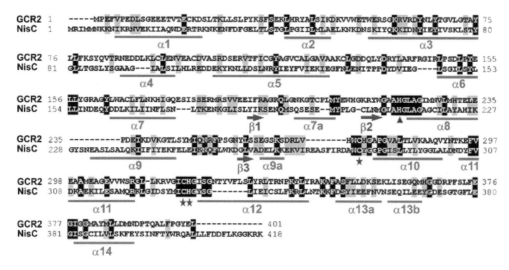

Figure 5.11

Alignment of protein GCR2, with nisin cyclase (NisC), created using ClustalW. Identical and similar amino acids are shaded black and grey, respectively. Helices and strands are marked according to the NisC crystal structure (Li *et al.*, 2006); conserved zinc ligands are noted with asterisks, the proposed catalytic base for substrate de-protonation with a triangle.

Source: Gao, Y. *et al.* (2007). Reproduced with permission from John Wiley & Sons.

as presented: our curiosity is 'satisfied' by the black and grey blocks, we have faith in the authors' explanation of the alignment annotations, and we therefore don't feel the need to investigate further. However, if we examine the alignment carefully, we discover that the N-terminal portion is misaligned. In fact, we should be able to find seven GxxG motifs, each of which lies at the N-terminus of a short helical region – three of these are missing.

It's interesting that neither the authors, the reviewers nor the journal editors spotted the mistake prior to publication, presumably owing to the suggestive power of the representation, which gave them no cause to scrutinise the result further. Had the authors used a different alignment tool (MUSCLE, T-Coffee, Multalin, for example), they'd have got different outputs; had they visualised the outputs differently, they might have spotted the errors. Unfortunately, the literature is full of alignments just like this, where the manner of presentation is sufficiently persuasive, with seductive black and grey boxes or other visual tricks, that mistakes are effectively concealed. Hence, alignments don't always mean what we think they mean, or indeed what their authors either thought they meant or intended that they should mean.

Appropriate visualisation techniques, in the right context and using the right metaphors, can help us to communicate, to understand, to discover and to learn more efficiently: 'Science needs metaphor because it provides the cognitive means to chart the unknown' (Klamer and Leonard, 1994). We ignore this at our peril, otherwise visualisations are likely to be confusing or wrong, to conceal information, or to be just pretty pictures – ultimately, to be effective, visualisations must be imbued with biological meaning.

5.8.4   How good is my alignment?

Another interesting feature of Figure 5.11 is the lack of correspondence between the sequence alignment and the structural annotations: *i.e.*, the regions of sequence aligned with helices 4, 9 and 12 contain gaps. Perhaps rather fittingly, this brings us back full circle to where we began this section, with the observation that different alignment methods produce different results. We saw examples of disparate alignments produced by tools such as ClustalW, T-Coffee, MUSCLE and so on; this gave our first hints that not all alignments are equal. Crucially, however, it isn't just a question of the alignment algorithm or parameters used – the *conceptual framework* that motivates and provides the basis for alignment is also important. If we overlook this critical point, problems are inevitable: for example, when we try to compare sequence- and structure-based alignments, we force two fundamentally different frameworks to collide.

Before we think about why this is so, we should emphasise that comparison of sequence- and structure-based alignments is common practice in bioinformatics. This is because we're often attempting to understand the evolutionary histories of proteins, and proteins are generally known to retain structural similarity even when their sequences are highly divergent. In principle, therefore, we can be a little more confident that residues from different proteins are homologous, if they can be seen to overlap in 3D (Iantorno *et al.*, 2012). Consequently, structure-based alignments have been considered the 'gold standard' (Daniels *et al.*, 2012) – it was this that provided the impetus for benchmarking studies that judge the quality of sequence-alignment algorithms, and their outputs, with reference to such alleged gold-standard, structure-based data-sets (*e.g.*, see Edgar, 2010; Iantorno *et al.*, 2012, and references therein). In fact, Edgar has suggested that publication of one of the first benchmark data-sets, BAliBase (Thompson *et al.*, 1999), 'triggered a benchmark war'[36] (Edgar, 2010). He observed that:

> Structure … has intrinsic limitations as a standard for residue alignment, [and hence that evaluation of alignment quality] is more challenging than generally realized, and skepticism is appropriate for claims that method rankings or advances can be reliably measured by current benchmarks (Edgar, 2010).

Let's now consider some of the reasons why. Normally, when we align sequences, we do so by creating a 2D grid or matrix, which we populate, row by row, with the one-letter codes that represent the constituent amino acids (or bases, if we're aligning DNA or RNA) of the sequences we wish to compare. When sequences have different lengths, we simply accommodate the discrepancy by adding gap characters, in order that equivalent residues can be brought into register – the operative word here is 'equivalent'. When we stack one residue beneath another in an alignment grid, we're making some sort of assertion about their relationship: *e.g.*, we believe that they're evolutionarily equivalent (derived from a common ancestral residue), we consider them structurally equivalent (have similar structural contexts), or we don't think they're equivalent at all, but just wish (or need) to represent them in this way for convenience. Different analysis tools, different alignment algorithms, assume different kinds of equivalence – and we need to understand this if we're to be able to interpret their outputs meaningfully. In the context of a 2D alignment grid, this means, for example, that we can align two residues

---

[36] https://robertedgar.wordpress.com/2010/05/02/multiple-protein-alignment-is-a-dead-field

that are similar in some way (size, charge, hydrophobicity, substitution probability, *etc.*), but *they may not be equivalent* within the particular alignment framework being used; equally, we can align two residues that are not similar (by whatever measure we care to choose), yet *they may be equivalent* within that framework.

Aligning proteins in 3D is necessarily different from this 2D-grid approach. Based on the geometric locations of Cα atoms in the polypeptide backbone, structural alignment programs tend to produce rigid-body transformations that align the structures in space, from which they then generate sequence alignments with apparently homologous residue equivalences (Daniels *et al.*, 2012). Clearly, the more divergent structures are, the more difficult it is to assess structural equivalence, and

> there are no unique criteria for structural similarity of individual residues; rather, structural equivalence must be defined in the context of a particular alignment protocol and requires arbitrary parameters such as distance cutoffs (Edgar, 2012).

To get a better feel for what this might mean in practice, let's return again to the example of two proteins whose sequences have different lengths. Recall that in the 2D grid, we simply inserted gap characters to bring 'equivalent' residues into register. In 3D, however, sequence-length discrepancies manifest themselves as slightly longer or shorter helices, slightly longer or shorter strands or longer or shorter loops – *there are no gaps in protein structures*. Inevitably, structure-based and sequence-based alignments are not the same thing: the former exploits a 3D-model of comparison, the latter a 2D-model in which gaps have a very specific role. When comparing proteins of different length, the gaps that sequence alignments contain tend to accumulate in regions corresponding to loops, as these have the greatest plasticity. Sometimes, however, insertions appear within alignment regions that correspond to core secondary structures, the consequences of which are profound.

Consider an insertion of a single residue within an α-helical region. In a sequence alignment, this would demand a gap to be added to the sequence of the shorter helix, in order to bring equivalent downstream residues into register with those of the longer sequence. In the structure-based alignment, however, such an insertion results in a 100° shift (in terms of angular separation – recall, α-helices contain 3.6 residues per 360° turn) of all downstream residues; thus, 'equivalent' residues in the rest of the helix become misaligned. Structurally, then, the consequences of even small insertions are significant. A more extreme example is seen in α-helix 12 in Figure 5.11, which we've highlighted in Figure 5.12.

Here, using ClustalW, seven residues have been inserted into the GCR2 sequence relative to the NisC helix. For comparison, the figure shows the same helical region aligned by MUSCLE and T-Coffee: both also insert seven residues, but distribute them in different ways (Figure 5.12 (a) to (c)). If we want to think about what this means in a 3D context, we can use a server like SWISS-MODEL[37] (Arnold *et al.*, 2006) to calculate a model for the likely structure of GCR2 based on the known structure of NisC (PDB ID 2G0D[38]). The alignment generated for the region spanning helix 12 is shown in Figure 5.12 (d) – seven residues are again inserted, but distributed in different ways from those shown in (a) to (c). The result in (d) is reproduced in (e), and shows

---

[37] http://swissmodel.expasy.org
[38] http://www.rcsb.org/pdb/explore.do?structureId=2g0d

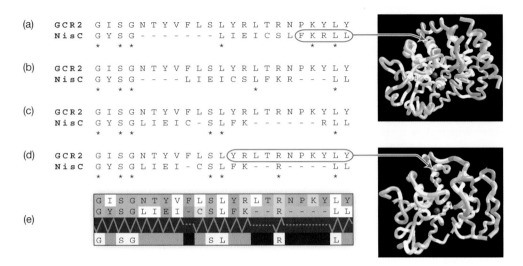

Figure 5.12

Left-hand panel: (a) Excerpt from the sequence alignment in Figure 5.11, generated using ClustalW, in which a seven-residue insertion in helix 12 is compensated by the addition of gap characters, in order to bring 'equivalent' parts of the sequences into register (asterisks denote identical residues); (b) MUSCLE, (c) T-Coffee and (d) SWISS-MODEL alignments of the same region; (e) same alignment shown in (d) represented in CINEMA ('standard' residue colours used) – the green zigzag denotes the helix, in which 'breaks' are clearly visible (the consensus is shown below). Right-hand panel: top, 3D structure of NisC (2G0D) – the view looks down the top surface of helix 12, whose C-terminal FKRLL residues are indicated by the arrow (same residue colours); bottom, 3D structure of GCR2 modelled on the NisC structure template – same view as top panel, and arrow pointing to the C-terminal YRLTRNPKYLY residues. In each case, the zinc ligand is shown as a red sphere.

how the insertion in GCR2, which is accommodated by gaps in the alignment, causes forbidden 'breaks' in the structure (the dotted lines in the green zigzag that denotes the NisC helix).

Looking at the top right-hand panel of Figure 5.12, we see a structural representation of NisC – the arrow pinpoints the location of the C-terminal FKRLL residues of helix 12. In the bottom panel, we see the modelled structure of GCR2 based on the NisC structural template – there are clearly significant differences, which is to be expected, given that the sequences share only 22 per cent identity. For helix 12, the algorithm has put six of the seven inserted residues in the helix C-terminus, causing the structure to buckle and throw out an extended loop, as highlighted by the arrow. Hence, residues YRRLY (GCR2) and FKRLL (NisC), which have been placed in equivalent positions in the sequence alignment (e) do not have equivalent structural contexts.

Structure-based alignments necessarily give different results from sequence-based alignments because the comparison frameworks are not equivalent; one is not 'better' than the other – both are valid models for the different dimensions of information they are intended to represent and for the different questions they are intended to answer. Many researchers continue to use structure-based alignments as benchmarks for the quality of sequence alignments, despite the fact that the underpinning models of comparison are designed for different purposes: sequence- and structure-based alignments are not recapitulations of the same problem, but formulations of comparison frameworks in different dimensions – one uses artificial gap characters to maximise the

correspondence of sequence landmarks on a 2D lattice, the other attempts to maximise the geometric correspondence of gross structural features (Cα backbones) in 3D space. It isn't reasonable, therefore, to expect a sequence-based alignment method to reproduce the results of a given structural alignment protocol (Edgar, 2012).

If we want to ascertain the quality of alignments, it only makes sense to do so if we compare like with like. Even then, we're likely to open a whole can of worms surrounding how we know what a good or 'correct' alignment is, because this requires us to understand what alignments mean; and, as we've seen, this is far from straightforward. We can, of course, use scoring matrices to quantify the degree of identity and/or similarity between sequences, or their overall alignment scores (Section 4.3.2); and perhaps this can tell us something useful for pairs or small numbers of closely related sequences. However, in multiple alignments, we can't always be certain to have aligned residues that are homologous, not least because alignments mask underlying evolutionary histories, and generally ignore all mutations other than substitutions, insertions and deletions (Daniels *et al.*, 2012; Edgar, 2010); and the alternative – requiring structurally equivalent residues to be aligned – is, as we've seen, also fraught with difficulty. If, when we compare and evaluate alignments, we understand in detail how they were made, the parameters and context of the comparison framework we're using, and the limitations of the alignment and comparison methods, then ascertaining the quality of an alignment becomes a little easier (although by no means trivial). Unfortunately, however, these conditions are seldom true; and if humans find it hard to appreciate the nuances, how much harder is it to teach computers to do so?

As a kind of postscript to this section, if the toy examples we've discussed here have challenged us to think more carefully about what alignments mean, consider how much more difficult the problems will be when next-generation alignment tools become mainstream – take ClustalΩ[39], for example, which has been used to generate alignments of >190,000 sequences (Sievers *et al.*, 2011). According to its authors, in benchmark tests, ClustalΩ was

> distinctly more accurate than most widely used, fast methods and comparable in accuracy to some of the intensive slow methods.

Of course, the benchmark used was a set of structure-based alignments, so the reported accuracy relates to conserved regions of sequence alignments relative to structural features, ignoring gaps, as these are not structurally conserved. Stepping aside what this really means (as we've rehearsed the arguments above), how are humans likely to cope with trying to assimilate the information hidden in, and the meaning of, alignments of hundreds of thousands of sequences, and does it even make sense[40] to try (Edgar, 2010)? And, if it does, how will computers help us to achieve this?

5.9 Problems with families

In the previous section, we dealt at some length with some of the issues bound up with creating, visualising and understanding multiple sequence alignments. This was deliberate. Alignments form the bedrock of many different sequence analysis methods: they

---

[39] http://www.ebi.ac.uk/Tools/msa/clustalo
[40] http://robertedgar.wordpress.com/2010/05/02/big-alignments-do-they-make-sense

provide springboards for understanding evolutionary relationships, they contain clues to structurally and functionally important regions and residues, they offer the means to build diagnostic signatures, and so on. It's therefore important to understand how they've been derived and how they should be interpreted, because any conclusions we draw from them, and any analyses we build on them, depend on how faithfully they reflect the 'reality' we believe them to portray.

One extremely important application of sequence alignments is in the functional characterisation of newly determined sequences emerging from the ubiquitous, highly productive world-wide genome projects. As we've seen, we align sequences in order to be able to identify shared features that might offer possible functional or structural clues, and by means of which we may group them into 'families'. As we saw in Chapter 4, different analytical methods have been devised to encode such characteristic features. These methods have different diagnostic strengths and weaknesses, and tend to give rise to different family groupings. What, then, do we mean by 'family'? The bioinformatics resource that challenges us to consider this question most carefully is probably InterPro (recall Figure 3.4). So, let's investigate a little further, using the notion of what we mean by the 'family' of GPCRs.

When we examine GPCR sequences, we see that they fall into a number of different superfamilies. Between the superfamilies, the degree of sequence similarity is negligible; what the proteins share, however, is a structural framework, formed by a bundle of seven TM domains, that allows the receptors to transduce extracellular signals into intracellular biochemical cascades (see Boxes 5.2 and 5.3). The various GPCR superfamilies include the **Glutamate receptors, Rhodopsin-like receptors, Adhesion receptors, Frizzled receptors** and **Secretin-like receptors** (sometimes referred to as GRAFS) – together, these superfamilies constitute what's known as a **clan**, members of which are believed to have a common architecture and high-level function (signal transduction), but no significant sequence similarity. A taste of some of these relationships is given in Figure 5.13.

Using their distinct diagnostic approaches, the protein family databases (PROSITE, PRINTS, Pfam, *etc.*) attempt to distinguish members of the different levels of the GPCR clan; ultimately, their results are pooled in the integrated protein family resource, InterPro – with interesting consequences.

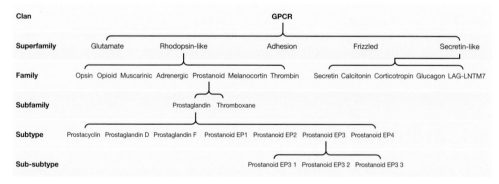

Figure 5.13

A simplified picture of some of the family relationships in the GPCR clan. The clan contains several superfamilies; within these are numerous families, some of which are shown for the rhodopsin-like and secretin-like receptors; these can be divided into subfamilies, shown here for the **prostanoid** receptors; subfamilies may be further divided into receptor subtypes, shown for the **prostaglandin** receptors; and certain subtypes of subtypes can be recognised, as here with the prostanoid EP3 receptors.

Release 50.0 of InterPro indicated that there were 107,165 members of the rhodopsin-like G protein-coupled receptor superfamily (IPR000276[41]). This figure was derived from scanning UniProtKB version 2015_02 against the relevant diagnostic signatures in PROSITE (G_PROTEIN_RECEP_F1_1, PS00237[42]), PRINTS (GPCRRHODOPSN, PR00237[43]) and Pfam (7tm_1, PF00001[44]). Drilling down to the partner databases, however, showed that the Pfam HMM matched 42,866 sequences, the PROSITE pattern had 2,123 true-positive hits, while the PRINTS signature made 2,375 full finger-print matches and ~4,000 partial matches (6,660 in total) – a combined total well short of 107,165. These results are summarised in Table 5.4.

The situation is further complicated when we see that InterPro release 50.0 contained a separate rhodopsin-like GPCR superfamily entry (IPR017452[45]) that had 114,321 members, a figure derived from the PROSITE profile G_PROTEIN_RECEPT_F1_2 (PS50262[46]). Visiting the PROSITE profile itself, however, revealed 2,433 true-positive hits, plus an equivocal 41 extra hits that might be family members. This diverse array of numbers and the similar, yet different, superfamily names appear baffling. So, why are the numbers and the names different? And what does it all mean?

Explaining some of the numerical discrepancies is relatively straightforward. Inter-Pro takes its partner database diagnostic models and scans them against whatever is the current version of UniProtKB; but there's an added twist – the way in which InterPro implements its partners' algorithms sometimes differs from the way the partners implement them. Side-stepping this issue, the most important factor is likely to be which versions of UniProtKB the various reported match-lists relate to. As mentioned above, for the examples shown in Table 5.4, the quoted InterPro results were derived from UniProtKB version 2015_02; the PROSITE pattern and profile data were derived from UniProtKB/Swiss-Prot 2015_02; the PRINTS data, were derived from much older versions of Swiss-Prot and TrEMBL (a composite of versions 55.0 and 38.0 respectively, with sequences tagged as fragments removed); the Pfam HMM was derived from Uni-ProtKB 2012_06. What is clear is that the source databases for PRINTS and PROSITE are significantly smaller than the source used by InterPro (note that PROSITE is based

Table 5.4

Rhodopsin-like GPCRs in InterPro 50.0 and its source databases.

	G protein-coupled receptor, rhodopsin-like (IPR000276) 107,165 hits	GPCR, rhodopsin-like, 7TM (IPR017452) 114,321 hits
Pfam	42,866	
PROSITE pattern	2,123	
PRINTS	6,660	
PROSITE profile		2,433 (+41)

[41] http://www.ebi.ac.uk/interpro/entry/IPR000276
[42] http://prosite.expasy.org/PS00237
[43] http://www.bioinf.manchester.ac.uk/cgi-bin/dbbrowser/PRINTS/DoPRINTS.pl?cmd_a=Display&qua_a=/Full&fun_a=Code&qst_a=GPCRRHODOPSN
[44] http://pfam.xfam.org/family/pf00001
[45] http://www.ebi.ac.uk/interpro/entry/IPR017452
[46] http://prosite.expasy.org/PS50262

Table 5.5

Families belonging to Pfam Clan GPCR_A (CL0192[i]), Family A G protein-coupled receptor-like superfamily.

7TM-7TMR_HD	7tm_1	7tm_2	7tm_3	7tm_4
7TM_GPCR_Sra	7TM_GPCR_Srab	7TM_GPCR_Srb	7TM_GPCR_Srbc	7TM_GPCR_Srd
7TM_GPCR_Srh	7TM_GPCR_Sri	7TM_GPCR_Srj	7TM_GPCR_Srsx	7TM_GPCR_Srt
7TM_GPCR_Sru	7TM_GPCR_Srv	7TM_GPCR_Srw	7TM_GPCR_Srx	7TM_GPCR_Srz
7TM_GPCR_Str	Bac_rhodopsin	Dicty_CAR	DUF1182	DUF621
Frizzled	Git3	GPR_Gpa2_C	GpcrRhopsn4	Lung_7-TM_R
Ocular_alb	Serpentine_r_xa	Sre	Srg	TAS2R
V1R				

[i]http://pfam.xfam.org/clan/CL0192

only on the Swiss-Prot component of UniProtKB), so smaller number of matches are to be expected. What we can't be sure of is the extent to which InterPro's re-implementation of the partner database models also affects the overall numbers.

The discrepancy between InterPro entries 'G protein-coupled receptor, rhodopsin-like' (IPR000276) and 'GPCR, rhodopsin-like, 7TM' (IPR017452) is perhaps more interesting: the former matched 107,165 and the latter 114,321 sequences. The question is, why database entries with such similar names matched different numbers of sequences? Part of the answer lies in the difference between the PROSITE pattern and profile designed (according to their IDs, G_PROTEIN_RECEPT_F1_1 and G_PROTEIN_RECEPT_F1_2) to identify members of the 'G-protein coupled receptors family'. As we'd expect, the profile is more permissive – or promiscuous – than the pattern. Consequently, in addition to 'mainstream' rhodopsin-like GPCRs, it also matched 41 **taste receptors**. These weren't included in the reported 2,433 true-positive set, but were annotated in the database as 'unknowns'. According to Pfam, the taste receptors belong to a separate family within the GPCR_A clan, designated TAS2R (PF05296[47]) – a list of Pfam's GPCR_A clan families is given in Table 5.5.

What may have happened, at least initially, is that because the PROSITE profile matched sequence families not matched by Pfam or PRINTS, the InterPro curators (rather than attempting to understand and resolve the discrepancy) simply made a separate entry, with a highly similar name but a different accession number. This is unfortunate. The result is especially confusing because, in InterPro release 50, IPR017452 had ~84 per cent overlap with IPR000276, the shared set of sequences constituting ~99.9 per cent of the IPR000276 set. The additional sequences matched by IPR017452 include the taste receptors matched by the profile, and, in addition, a variety of vomeronasal receptors, protoheme IX farnesyltransferases, uncharacterised proteins and other curiosities – but note that, overall, this is a much cleaner result compared to the version of this entry formerly found in InterPro release 48.0, which was confounded by bacteriorhodopsins, heat-shock proteins, sensory rhodopsins, halorhodopsins, archaerhodopsins, and even cytochrome c oxidase subunit II. If there is some debate about the relationship of vomeronasal and taste receptors to rhodopsin-like GPCRs, there is

[47]http://pfam.xfam.org/family/PF05296

certainly no doubt about bacteriorhodopsin and its close associates – they may share a 7TM scaffold, but they are not G protein-coupled receptors. Adding to the muddle, this is acknowledged in the annotation of entry IPR017452, which states that it 'spans the seven transmembrane regions of rhodopsin-like GPCRs. It also identifies some non rhodopsin-like GPCRs, including a number of taste receptors and vomeronasal receptors'. Inevitably, this has knock-on consequences, particularly for automatic annotation pipelines, which have led many thousands of UniProtKB/TrEMBL sequences to bear the rather ambiguous '*GPCR, rhodopsin-like, 7TM*' description as a result. This is a classic illustration of the dangers of blurring protein sequence/structure/function boundaries without reference to the underlying biology.

If the situation for just this one GPCR superfamily is confounding, what convolutions might we find for other members of the GPCR clan? In December 2015, a search of InterPro with keyword 'GPCR' yielded 120 matches, including 'GPCR, family 2-like' (IPR017981[48]), 'GPCR, family 2, secretin receptor' (IPR002144[49]), 'GPCR, family 2, secretin-like' (IPR000832[50]), 'GPCR, family 2, secretin-like, conserved site' (IPR017983[51]), and many other GPCR family 2 relations, but, interestingly, no overarching 'GPCR family 2' – given this small sample, we may be sure that the relationships between the sequence sets within the wider list will not be simple to ascertain.

Part of the problem is that database curators have become dependent on automatic search tools; most don't have time to analyse the results in sufficient detail to be able to relate them clearly to the underlying biology, especially when families and superfamilies are large, like the GPCRs. Hence here, for example, the clan members that Pfam lists do not all map in an intuitive way to the families listed in PRINTS and PROSITE, and discord between them has tended to result in odd disparities in InterPro, similar to the one discussed above for rhodopsin-like GPCRs. To understand the underlying evolutionary relationships for this important group of receptors, we really need to consult more detailed analyses, such as those of Nordström *et al.* (2009, 2011), to put these results on a more solid biological footing.

Before leaving this section, there is another important point to note, which relates back to our discussion in Section 5.4 about nomenclature. InterPro refers to its 'G protein-coupled receptor, rhodopsin-like' entry as a family, and to its 'GPCR, rhodopsin-like, 7TM' entry as a domain. Pfam, on the other hand, refers to its 7tm_1 entry as a family; PRINTS terms entry GPCRRHODOPSN a superfamily; and PROSITE refers to the sets of sequences identified by its G_PROTEIN_RECEPT_F1_1 and G_PROTEIN_RECEPT_F1_2 pattern and profile entries as families. Hence, for arguably the same biological concept (the group of sequences that share similarity to rhodopsin-like GPCRs), four different databases use three different terms: family, superfamily and domain. And, just to add a bit of nomenclature spice, Pfam names the *clan* that encompasses 7tm_1 and the 35 other 'families' shown in Table 5.5 as, 'Family A G protein-coupled receptor-like superfamily', thus using three different terms within the same concept. Teaching computers what we humans understand by the term 'family' is therefore considerably more difficult than we might expect.

---

[48] http://www.ebi.ac.uk/interpro/entry/IPR017981
[49] http://www.ebi.ac.uk/interpro/entry/IPR002144
[50] http://www.ebi.ac.uk/interpro/entry/IPR000832
[51] http://www.ebi.ac.uk/interpro/entry/IPR017983

5.10   Problems with functions

Faced with the rich bounty of the genome projects, our imperative is to add value to the steadily accumulating raw data by endowing them with 'meaning'. What, for example, does this or that gene product do? What is its function? Fortunately, we have at our disposal the enormous power of today's computers and a vast array of analytical tools to help answer such questions. So how do we use these technologies to infer biological functions? Function prediction is now mostly performed using automated pipelines that exploit various different types of database search: pairwise analyses are routinely performed with tools like BLAST or FASTA; family-based searches tend to be performed using resources like InterPro, PROSITE, PRINTS, Pfam, and so on; and additional information is gathered from a variety of ontological resources.

A number of issues arise when applying automated annotation approaches. As we've seen, the family-based methods differ in their diagnostic power, some (like HMMs and profiles) providing broad, high-level functional diagnoses, others (like patterns and fingerprints) providing much finer details. The databases associated with these approaches differ in terms of their family coverage; more importantly, the fundamental concept of 'family' differs between them, making comparisons difficult. InterPro was developed to try to address many of these issues, and in fact goes a long way towards achieving this; nevertheless, it also adds layers of complexity of its own.

As we saw in the previous section, one of the biggest problems is the practical definition of what constitutes a family, because the concept of 'family' encompasses sequence relationships, functional relationships, structural kinship and evolutionary ancestry. Depending on these different perspectives, the nature of defined families necessarily differs. So, for example, there's general agreement that members of a family must have a demonstrable evolutionarily relationship: this is usually established with reference to some specific sequence similarity cut-off (*e.g.*, greater than or equal to 30 per cent.). But common descent (**homology** – see Box 5.4) may be inferred in the absence of significant sequence identity: some members of the globin family, for example, share sequence identities of only ~15 per cent. So where should we draw the relatedness line? Proteins can also be grouped on the basis of structural similarities, if they have the same secondary structures in the same arrangement, with the same topological connections. But such 'fold families' may also include proteins that don't share evolutionary origins, because structural similarities can arise as a result of physicochemical preferences for certain packing arrangements and chain topologies.

Determining family membership on the basis of defined levels of sequence identity or gross structural similarity can lead to problems, because families may be highly similar in sequence but divergent in function, or similar in structure and function but divergent in sequence. An example of just how tricky both the task of functional assignment and its interpretation can be is given in Table 5.6, which charts the annotation of the **Conger eel blue-sensitive opsin** (OPSB_CONCO, O13227[52]) as it migrated from TrEMBL to Swiss-Prot to UniProtKB.

---

[52] http://www.uniprot.org/uniprot/O13227

Table 5.6

Functional annotations and database cross-references for the blue-sensitive opsin from the Conger eel, OPSB_CONCO (O13227).

Version	Date	Annotation
TrEMBL		
1	01-JUL-1997	DE OPSIN
Swiss-Prot		
3	15-JUL-1998	DE BLUE-SENSITIVE OPSIN (BLUE CONE PHOTORECEPTOR PIGMENT) CC FUNCTION: VISUAL PIGMENTS ARE THE LIGHT-ABSORBING MOLECULES THAT MEDIATE VISION. THEY CONSIST OF AN APOPROTEIN, OPSIN, COVALENTLY LINKED TO CIS-RETINAL DR PROSITE; PS00237; G_PROTEIN_RECEPTOR DR PROSITE; PS00238; OPSIN
Swiss-Prot		
4	15-DEC-1998	DE BLUE-SENSITIVE OPSIN (BLUE CONE PHOTORECEPTOR PIGMENT) CC FUNCTION: VISUAL PIGMENTS ARE THE LIGHT-ABSORBING MOLECULES THAT MEDIATE VISION. THEY CONSIST OF AN APOPROTEIN, OPSIN, COVALENTLY LINKED TO CIS-RETINAL CC TISSUE SPECIFICITY: ROD SHAPED PHOTORECEPTOR CELLS WHICH MEDIATES VISION IN DIM LIGHT DR PROSITE; PS00237; G_PROTEIN_RECEPTOR DR PROSITE; PS00238; OPSIN DR PFAM; PF00001; 7tm_1
UniProtKB/Swiss-Prot		
73	14-OCT-2015	DE RecName: Full=Blue-sensitive opsin AltName: Full=Blue cone photoreceptor pigment CC FUNCTION: Visual pigments are the light-absorbing molecules that mediate vision. They consist of an apoprotein, opsin, covalently linked to cis-retinal. CC TISSUE SPECIFICITY: Rod shaped photoreceptor cells which mediates vision in dim light DR InterPro; IPR000276; GPCR_Rhodpsn DR InterPro; IPR017452; GPCR_Rhodpsn_7TM DR InterPro; IPR001760; Opsin DR InterPro; IPR027430; Retinal_BS DR InterPro; IPR000732; Rhodopsin DR InterPro; IPR019477; Rhodopsin_N DR Pfam; PF00001; 7tm_1 DR Pfam; PF10413; Rhodopsin_N DR PRINTS; PR00237; GPCRRHODOPSN DR PRINTS; PR00238; OPSIN DR PRINTS; PR00579; RHODOPSIN DR PROSITE; PS00237; G_PROTEIN_RECEP_F1_1 DR PROSITE; PS50262; G_PROTEIN_RECEP_F1_2 DR PROSITE; PS00238; OPSIN

As shown in the table, this protein first entered TrEMBL as an opsin (sequence entry version 1) – no other annotation was provided. The following year, it became part of Swiss-Prot (entry version 3), whereupon its annotation was augmented with information informing users that this was a blue-sensitive opsin from **cone photoreceptors**, functioning as a visual pigment and mediating vision through the absorption of light. Cross-references were also provided to supporting evidence from PROSITE, whose

Box 5.4 Issues with homology

Inference of homology is fundamental to the process of sequence annotation. Unfortunately, the meaning of this term is widely misunderstood, and has consequently been widely misused, polluting the literature with statements that equate homology and similarity – they are not the same. Sequences are said to be homologous if they are related by divergence from a common ancestor. A related term, **analogy**, is used to denote the acquisition of common features (folds or functions) via convergent evolution from unrelated ancestors: *e.g.*, β-barrels occur in a variety of proteins (**trypsin**, **thrombin**, the **Tubby protein**, *etc*.), which, despite their similar architectures, share no sequence or functional similarity; similarly, the enzymes chymotrypsin and **subtilisin** share groups of catalytic residues, with near-identical spatial geometries, but no other sequence or structural similarities (*e.g.*, chymotrypsin has a barrel architecture and subtilisin a 3-layer αβα architecture).

The important point is that homology isn't a measure of similarity; it is an absolute statement that sequences have a divergent rather than a convergent evolutionary relationship – unlike similarity, it is not quantifiable. The problem is, if we use the term 'homology' when we really mean similarity (*e.g.*, 'these sequences share 5 per cent homology'), then we are pre-supposing a homologous relationship that is unproven – the sequences may share 5 per cent amino acid similarity, but they may not be homologous (the same level of sequence similarity could easily have arisen by chance).

Database search tools identify and quantify similarity between sequences; it is for users of the tools to ascertain, on the basis of other biological information, whether the similarity is significant and hence whether the sequences are likely to have derived from a common ancestor. The same arguments apply when comparing structures: structures may be similar, but their common evolutionary origin is a hypothesis that needs to be verified with additional biological evidence. When we conflate homology and analogy, things are bound to go wrong. In the past, investigations largely of globular proteins led to the broad 'common sequence/structure/function/ancestry' paradigm. A given protein fold will impose strict constraints over the packing of its hydrophobic interior; thus, two proteins sharing the same fold are likely to share the same or similar hydrophobic interactions. For hydrophobic segments of non-globular proteins, however, the dearth of available membrane protein structures has meant that the implications of similarity haven't been studied to quite the same extent; nevertheless, homology considerations have been 'silently extended' also to them (Wong *et al.*, 2010), with interesting consequences. This has particular ramifications for membrane-spanning **signal peptides** and TM helices, whose sequence similarity necessarily results from physical requirements rather than common ancestry. Matching of signal peptide/TM regions thus 'creates the illusion of matching hydrophobic cores' (Wong *et al.*, 2010), which may in turn lead to incorrect annotations – in fact, these authors discovered >1,000 problems of this sort in Pfam 23.0.

Another complicating aspect for sequence analysis is that we can recognise different types of homologue. At a simplistic level, we can define **orthologues** and **paralogues**: orthologues are homologous proteins that are considered to have arisen following speciation events – hence, they usually perform the same function in different species; paralogues, on the other hand, arise following gene-duplication events – hence, they tend to perform different but related functions within one organism. The study of orthologues allows us to chart changes in a given protein across species, while paralogues allow us to explore changes in a protein within species, as each duplicated gene follows a separate evolutionary pathway, allowing new specificities to evolve through variation and adaptation. Such complexity presents significant challenges for bioinformatics. For example, following a database search, how much functional annotation can be legitimately inherited by a query? In the past, many automatic analysis systems operated on the assumption that it was sufficient to transfer the annotation from the best-matching database hit to a given query sequence. But how can we be sure that the best match is the true orthologue and not a paralogue, whose function is probably slightly different? Unfortunately, such naïve approaches don't consider the functional plasticity engendered by divergent evolution, and have spawned numerous annotation errors in sequence databases.

patterns diagnosed the protein as a G protein-coupled receptor belonging to the opsin family. Up to this point, then, the situation seems straightforward – the protein looks like a blue-sensitive opsin.

Later that year, however, the annotation was revised again (entry version 4) with further information regarding the tissue specificity of the protein: *i.e.*, it resides in the rod-shaped photoreceptor cells, whose role is to mediate vision in dim light. What, then, is the function of the protein and what does this extra piece of annotation mean? Opsins, the visual pigments that normally mediate colour vision, are found in cone-shaped photoreceptors; rhodopsins, on the other hand, which mediate achromatic vision in dim light, are normally found in the **rod photoreceptors**. How could this protein be a cone-photoreceptor opsin, but reside in rod-shaped cells? An additional Pfam cross-reference also appears in this database record, but this doesn't help us to resolve our functional dilemma, because Pfam entry 7TM_1 makes generic diagnoses for rhodopsin-like GPCRs – it provides no family or subfamily insights.

This conflicting functional annotation persists in the 'current' UniProtKB entry, version 73. In this entry, however, we now find 14 family and domain database cross-references: one more each from PROSITE and Pfam, six from InterPro and three from PRINTS. The additional evidence from PRINTS and Pfam, and hence from InterPro, suggests that this protein has sequence characteristics that are rhodopsin-like rather than blue-sensitive opsin-like (*n.b.*, PRINTS also contains a blue-sensitive opsin fingerprint, but the sequence doesn't match it, according to the database cross-references listed here). So, is it an opsin or a rhodopsin?

Further clues can be found by referring to the paper these database entries cite – specifically, Archer and Hirano, 1996. Here, the authors explain that the Conger eel has a pure rod retina with a visual pigment maximally sensitive to blue light. So, the protein *is* found in rod photoreceptor cells (not cones, as suggested by the UniProtKB/Swiss-Prot annotation), and *does* have an amino acid sequence characteristic of rhodopsins, but this has undergone spectral fine-tuning to create a rhodopsin-like class of receptors with the functional specificity of blue-sensitive pigments.

In this fascinating scenario, we've seen how a particular protein sequence has undergone some degree of functional 'finessing' – the high-level function has remained the same (photoreception), but the specificity has altered such that the protein now appears to have the 'wrong' sequence for that function!

A rather more subtle, intriguing and perhaps more taxing situation arises when we find the same protein possessing very different functions in different contexts – this phenomenon, termed gene sharing or 'protein moonlighting'[53], is explored in more detail in Box 5.5. Unravelling how protein sequences can share high degrees of similarity yet exhibit divergent or different functions is a challenge for humans; injecting the necessary spark of intelligence to allow computers to reason with the data and make the same leap in understanding is an undertaking beyond what can be achieved at present.

---

[53] http://en.wikipedia.org/wiki/Protein_moonlighting

Box 5.5 Protein moonlighting

In 1987, members of a family of what were considered straightforward structural proteins – crystallins – were shown to exhibit taxon-specific functional differences: they were not only the main components of the vertebrate eye lens, but could also behave as enzymes (Wistow and Piatigorski, 1987). In fact, Wistow and Piatigorski showed that chicken δ2-crystallin is ~70 per cent identical to human arginosuccinate lyase – an idea of just how similar the sequences are can be gained from the alignment shown in Figure 5.14 (this includes turkey, chicken and duck δ-crystallins, and duck, chicken and goose arginosuccinate lyases – the sequences are so similar that it's extremely difficult to tell them apart). Adding to the complexity, note that UniProtKB labels ARLY2_CHICK (P05083[a]) chicken arginosuccinate lyase, but this was originally annotated as δ2-crystallin II[b]; ARLY1_CHICK (P02521[c]) is the chicken δ1-crystallin.

ARLY1_MELGA	P Q K K N P D S L E L I R S K A G R V F G R L A A I L M V L K G
ARLY1_CHICK	P Q K K N P D S L E L I R S K A G R V F G R L A A I L M V L K G
ARLY1_ANAPL	P Q K K N P D S L E L I R S K A G R V F G R L A S I L M V L K G
ARLY2_ANAPL	P Q K K N P D S L E L I R S K A G R V F G R L A S I L M V L K G
ARLY2_CHICK	P Q K K N P D S L E L I R S K A G R V F G R L A A V L M V L K G
ARLY_ANSAN	P Q K K N P D S L E L I R S K A G R V F G R L A S I L M V L K G
Consensus	P Q K K N P D S L E L I R S K A G R V F G R L A L M V L K G

Figure 5.14

Alignment of selected avian δ-crystallins and arginosuccinate lyases.

What made this observation particularly striking was not that the crystallin sequences were similar to those of enzymes, but that they were actually products of the *same genes* (Piatigorski and Wistow, 1989). Piatigorski and Wistow termed this phenomenon, whereby a gene may encode a protein that has two entirely different functions, 'gene sharing', in order to differentiate it from mechanisms such as alternative splicing, post-translational processing, gene fusion, or exon shuffling, which were known to be able to generate different functions for the same protein. Gene sharing allows a gene both to acquire and to maintain an alternative function, without the necessity for gene duplication and without loss of the original, primary function (Piatigorski and Wistow, 1989). For these proteins, then, the genes sustain sequences required to maintain both their respective catalytic functions and the intermolecular interactions necessary for lens transparency.

Following the revelation of multi-functional crystallins, many examples of proteins with two or more different functions were discovered – these were subsequently re-named 'moonlighting proteins', by analogy with people who perform multiple jobs (Jeffery, 1999; Huberts *et al.*, 2010). Studies of the growing list of such proteins have revealed that their functions can vary in response to changes in oligomeric state, cellular localisation or cell type, or even to cellular concentrations of ligands, substrates or cofactors; moreover, that a combination of these methods may be used by moonlighting proteins to swap between their functions (Jeffery, 1999).

Various consequences arise from these observations. First, they compel us to revist our understanding of concepts such as gene, gene expression and protein function – the idea that a single gene encodes a single protein with a single function is clearly over-simplistic. Second, as it's possible for a moonlighting protein to provide an unexpected link, say, between metabolic and signalling pathways, this has profound implications not just for systems biology in general but for biomedical science in particular – a moonlighting function could conceivably complicate the correlation of protein functions with clinical features of a disease, and cause a new drug, say, to target the wrong

[a] http://www.uniprot.org/uniprot/P05083
[b] http://www.uniprot.org/uniprot/P05083.txt?version=1
[c] http://www.uniprot.org/uniprot/P02521

function, with potentially toxic consequences (Jeffery, 2004). Finally, in the context of genome annotation, it shows both that a well known protein may harbour other, as yet unknown, functions, and that sequences that appear to be almost identical may not have the same function – *i.e.*, the things we think we know, we may not know, because our knowledge may be incomplete; and the things we think we've found, may not be what we think they are, because they may be moonlighting as something else! Or, as Huberts *et al.* put it,

> Moonlighting is a phenomenon that illustrates nature's ingenuity. It is a source of inspiration that should remind scientists to always keep the unexpected in mind, even on familiar ground (Huberts *et al.*, 2010).

This provides us with a salutary warning to think ever more carefully about the functions we find associated with proteins in biological databases, the vast majority of which are determined by automatic annotation transfer following similarity searches, and inheritance of GO terms; it cautions us to keep thinking critically, to remain alert to the possibility that things are not as simple as they might first appear; and it perhaps also confronts us with the question of how useful our simple computer-annotation pipelines are likely to be in future, as we discover more and more about the incredible complexity of molecular functions. Perhaps this is why the guardians of UniProt and a number of their collaborators are in the process of trying to create a new Gold Standard DataBase (GSDB), the functions of whose proteins have been determined experimentally and whose exact nucleotide/protein sequences are known for those precise experimentally determined functions; the proteins will also be accompanied by literature references so the source of this information can be traced.

5.11 Functions of domains, modules and their parent proteins

Some of the issues we've just discussed relate to the evolutionary fate of genes that acquire new functional specificities in different species. Another complication to consider is what happens when we swap a gene-family- or sequence-based perspective for a structural one, and think instead in terms of 'domain' families. The issue here is that, while domain families share sequence and structural similarity, they are likely to encompass many diverse gene families – *i.e.*, they don't shed light on their constituent functional strata, but focus the functional spotlight instead just on the domain. It's important to understand the implications of this, especially when we consider resources like InterPro. As we've seen, InterPro attempts to equate 'families' from more than a dozen resources, to provide integrated entries from which to infer the functions of newly determined sequences. However, because the merged database entries adopt different sequence- and domain-based views, rationalising their constituent families and arriving at a common understanding of their functions isn't straightforward, as was highlighted by InterPro family IPR017452.

A number of problems can arise if we don't deal effectively with the domain and/or 'modular' nature of some proteins. **Modules** are autonomous folding units that nature has used like protein building-blocks to confer a variety of functions on a parent protein, either by using multiple instances of the same module, or by using combinations of a variety of different modules to form **mosaics**. This may present difficulties for database searches, because the best hit could be a match to a single domain within a multidomain protein or to a single module within a mosaic protein. In such cases, inheriting

Table 5.7

UniProtKB/SwissProt database cross-references for human cadherin EGF LAG seven-pass G-type receptor 1 (CELR1_HUMAN, Q9NYQ6), entry version 147.

PRINTS	SMART	Pfam	PROSITE	InterPro
CADHERIN	CA	7tm_2	ASX_HYDROXYL	Cadherin
GPCRSECRETIN	EGF	Cadherin	CADHERIN_1	Cadherin-like
	EGF_Lam	DUF3497	CADHERIN_2	Cadherin_CS
	GPS	EGF	EGF_1	ConA-like_dom
	HormR	GPS	EGF_2	DUF3497
	LamG	HRM	EGF_3	EG-like_dom
		Laminin_EGF	EGF_LAM_1	EGF-like_CS
		Laminin_G_2	EGF_LAM_2	EGF-type_Asp/Asn_hydroxyl_site
			G_PROTEIN_RECEP_F2_3	EGF-laminin
			G_PROTEIN_RECEP_F2_4	GPCR_2-like
			GPS	GPCR_2_extracellular_dom
			LAM_G_DOMAIN	GPCR_2_secretin-like
				GPS
				Growth_fac_rcpt_N_dom
				Laminin_G

the annotation from the parent protein itself, and not from the domain or module, is likely to be inappropriate. Table 5.7 provides a taste of the sort of complexity that can arise, here with another GPCR example.

The human 'cadherin EGF LAG seven-pass G-type receptor 1' contains a variety of different domains/modules. Within the UniProtKB entry (CELR1_HUMAN, Q9NYQ6[54]), the structural databases, SUPERFAMILY and Gene3D, each provide annotations that point to the presence of repeated cadherin and concanavalin A-like lectin domains. As shown in Table 5.7, the protein family- and domain-based databases suggest much greater complexity. Sifting through the list of matches, we see, in addition, so-called EGF, GPS, HRM and laminin-G domains, coupled to a secretin-like GPCR 7TM scaffold. Attempting to combine these views, InterPro makes the picture seem even more convoluted, with its EG-like_dom, EGF-like CS, EGF-type_Asp/Asn_hydroxyl_site, and so on. UniProtKB annotates the function of this protein as a receptor that might have a role in cell–cell signalling during the formation of the nervous system. But this, presumably, is the function of the GPCR component; what are the functions of the protein's EGF, GPS, HRM and laminin-G domains? The outcome of database searches with the full sequence (this is a huge protein, with 3,014 amino acids) versus searches with only parts of it are likely to be very different, and drawing reliable functional conclusions from the results of either is likely to be very hard.

The problem is, nature behaves like a tinker, adapting old material to create new systems, rather than starting from scratch (Jacob, 1977). Mosaic proteins are excellent examples of this kind of tinkering, where individual modules have been recruited for

[54] http://www.uniprot.org/uniprot/Q9NYQ6.txt?version=147

context-dependent roles in a variety of different proteins (*e.g.*, EGF domains promiscuously associate with ~24,000 proteins). Such complexity presents significant challenges to computational systems, because even if we know the particular building-blocks that comprise a protein, if we don't understand their biological context, we may not be able to infer the protein's overall function. If, for example, we identify the presence of a module or domain, this tells us little about the function of the complete protein; alternatively, if we know most components of a mosaic or multi-domain protein, this doesn't allow us easily to predict whether one of the modules/domains is missing; and, as already mentioned, modules in different proteins don't always perform exactly the same function.

Appreciating the complexity of biological systems and understanding the detailed evolutionary processes that have generated biochemical, cellular and developmental innovations is essential if we are to be able to make meaningful inferences of protein function (see Box 5.6). Comparative genomics has shed light on still more complicating factors: gene functions may be redundant; non-orthologous displacement can replace genes with unrelated but functionally analogous genes; **horizontal gene transfer** can introduce genes from different phylogenetic lineages (**xenology**); and lineage-specific gene loss can eliminate ancestral genes. Genomes thus harbour many obstacles that defy an intuitive correlation between sequence, structure and function, and render reliable, fully-automatic computational assignment of protein function virtually impossible.

Box 5.6 Biological complexity

During the last few decades, advances in computational and experimental technologies have allowed us to glimpse our own genetic recipes. Yet, paradoxically, the methods we use to deduce the functions of the encoded gene products are still primarily based on quantification of sequence similarities and, often, automatic inheritance of GO terms. How has this situation arisen?

The answer lies partly in the complexity, uncertainty and unpredictability of biological systems. Biological systems are inherently hierarchical, whereby higher-order systems are built from simpler components. Each level of the hierarchy is constrained by its component parts, but new properties may also emerge, each imposing additional constraints on the system. This has important consequences: first is the necessity to analyse complex entities at all levels; second is the realisation that complex systems aren't entirely predictable. In other words, with knowledge of some level of the hierarchy, we can only make limited predictions at a higher level.

Because complex systems have emergent properties, they are more than just the sum of their parts. Rashevsky encapsulated this idea when he argued that no collection of separate descriptions or models of organisms can be assembled in such a way as to capture the organism itself (Rashevsky, 1954). Today, bioinformaticians, confronted with the complexity and uncertainty of biological systems, and the consequent difficulty of modelling them computationally, are re-discovering that biological systems aren't just 'bags of bits' that can be reassembled to create realistic pictures of the whole. So overwhelming is the degree of complexity of some processes (blood coagulation, inflammatory reactions against foreign bodies, the immunological defense mediated by the complement system, *etc.*) that,

> For the biologist, it is thus generally impossible to make a prediction, or even an inspired guess, about the nature of such molecules and their structural relations with other constituents (Jacob, 1977).

Another extremely complex system is signal transduction, mediated by the ubiquitous GPCRs. GPCRs constitute one of the largest and most diverse groups of cell-surface molecules, and provide a beautiful example of Jacob's 'molecular tinkering' – these highly successful proteins have adapted a common heptahelical framework to mediate a vast range of physiological processes, including the perception of light, taste, smell and pain, the effects of numerous therapeutic agents, the attraction of motile cells by chemotaxis, stimulation and regulation of mitosis, the entry of viruses into cells,

(continued)

and so on. Such functional diversity is achieved via interactions with a wide variety of ligands, including peptide and non-peptide **neurotransmitters**, hormones, growth factors, pheromones, **odorants** and light.

The G proteins with which GPCRs interact are constitutively-associated hetero-trimers, with α, β and γ subunits. The model for how the system works goes like this: when the receptor is charged by its ligand, a conformational change is triggered; this induces a conformational change in the α subunit, causing **GDP** to be released and replaced by **GTP**. GTP-binding induces the α subunit to dissociate from the receptor and the $\beta\gamma$ complex, allowing it to interact with intracellular molecules ('second messengers' like **cAMP** and **calcium**) that then pass the message on by altering the behaviour of a host of other cellular proteins.

Aside from G proteins, GPCRs interact with a wide range of other intracellular molecules, broadening the possible mechanisms by which they transduce environmental signals. For example, the adaptor molecule **arrestin** couples GPCRs to the activation of **Src-like kinases** and facilitates the formation of multi-molecular complexes. Other structural components (*e.g.*, polyproline-containing regions, **PDZ**, **SH2** and **SH3** domains) also facilitate coupling of GPCRs to a variety of intracellular signalling molecules. Overall, the diversity of biological responses elicited by GPCRs bears witness to the complexity of molecular mechanisms by which they transduce environmental signals.

Most of these responses don't depend on a single biochemical route, but are the integrated result of the functional activity of an intricate network of interconnected cytoplasmic signalling pathways. Individual responses depend on receptor expression levels, coupling specificity and the repertoire of signalling molecules expressed in each cellular system. The more we discover, the more labyrinthine the mechanisms of signal transduction appear to be (given this richness of interactions, additional terms – serpentine, 7TM, heptahelical receptors, *etc.* – are sometimes used to describe these proteins in order to avoid under-representing the breadth of biochemical routes they exploit).

As systems increase in complexity, it becomes increasingly difficult to predict their behaviour precisely. Perturbations to such systems have multiple responses at the level of genes, transcripts, proteins and metabolites. In studying biological systems, it is therefore inevitable that we should try to reduce them down to simpler, more manageable components (traditionally, science has advanced by seeking partial answers about phenomena that can be isolated and well characterised). Yet, in so doing, we alter the contexts in which they operate, leading to increasing levels of uncertainty in the predictions we then make about them. The situation is still more challenging when we appreciate that biological systems are also inherently dynamic – cells and their constituent molecules are in a continuous state of flux. Thus, for example, the fate of particular interactions within metabolic pathways is unlikely to be determined by the binary control of gene expression (*i.e.*, whether a gene is 'on' or 'off') but, rather, by a sequence of events that determines the level of expression relative to a particular point in time. Therefore, a consideration of both the organisation and the dynamic interactions of genes, proteins and metabolites will be necessary if we are to gain a system-level understanding of biology.

It follows that if we wish to derive insights into the ways in which genomes encode biological functions, similarity search tools alone are unlikely to be sufficient. The manifestations of biological functions are complex and ever-changing, as is the concept of what we actually mean by 'function'. Tools like BLAST, FASTA and InterPro have been the mainstays of genome annotation efforts because they reduce a complex problem to a more tractable one – that of identifying and quantifying relationships between sequences. But identifying relationships between sequences isn't the same as identifying their functions. Failure to appreciate this fundamental point has generated a number of problems for bioinformaticians and computer scientists, and demands that we adopt more rigorous approaches to the ways in which we *describe* function both for humans and for computers to interpret.

5.12   Defining and describing functions

Many of the difficulties we face are bound up with what we actually mean by 'function', and the language we use to describe it. As it should now be clear, pinning down a unique and unambiguous definition of 'function' is hard, because biological functions are multi-faceted and context dependent.

Depending on their specialisms, life scientists (geneticists, cell biologists, biochemists, biophysicists, bioinformaticians, *etc.*) tend to think of functions in different ways: some focus on the biochemical activities of proteins (*e.g.*, enzyme activity); others focus on the cellular processes in which proteins are involved, including details of catalytic mechanisms or molecular recognition; still others consider function in the genetic sense, say in relation to a particular phenotype. The problems go deeper when we consider that proteins may have multiple cellular roles; ascribing a single functional role may therefore be inappropriate – nevertheless, this often happens in databases.

To appreciate the function of a protein, we need to understand its action (its 'local function') and the role it performs in the cell (its 'integrated function'). In terms of its action, we need to know about the cellular components with which it interacts and the types of interaction involved (*e.g.*, protein–protein, protein–DNA, *etc.*). In terms of its role, we need to have an appreciation of how the cell behaves when the protein is absent. Of course, proteins don't function in isolation: they generally operate in complex systems – sometimes in several systems. Hence, function can be understood in different contexts and at different levels of granularity, from molecular and cellular, to tissue and organismal. How, then, can we synthesise this information into a meaningful one-line description that adequately encompasses the function of a protein, which is typically the approach used to annotate database entries?

Another difficulty is that, in the scientific literature and in database annotations, 'function' has been used to refer arbitrarily to biochemical activities, biological goals and cellular structures. In an attempt to introduce rigour into the field, several **ontologies** have been developed to try to better reflect biological 'reality'. These use controlled vocabularies to i) more explicitly define the relationships between gene products and their biological processes, molecular functions and cellular components (*e.g.*, the **Gene Ontology** GO); ii) encapsulate the concepts of local and integrated function (*e.g.*, **Eco-Cyc**); and iii) define relationships between enzymes, their locations in metabolic pathways, their reaction products and substrates, and the superfamilies to which they belong (*e.g.*, **KEGG**). Bio-ontologies like this are now used routinely to provide computer-accessible annotations to supplement the free-text description lines traditionally used to describe the functions of sequences in UniProtKB.

Examples of the extra granularity (and formality) this yields for three different types of protein (an enzyme, a structural protein and a carrier) are shown in Table 5.8. Note how 'fuzzy' some of UniProtKB functional descriptions are and, in some cases, how

Table 5.8

UniProtKB functional annotations and supplementary GO annotations for pepsin, haemoglobin and actin, together with annotation sources.

	Pepsin A (PEPA_PIG, P00791; version 126[i])	Source
UniProtKB Function	*Shows particularly broad specificity; although bonds involving phenylalanine and leucine are preferred, many others are also cleaved to some extent*	UniProtKB
GO: Biological Process	Digestion	UniProtKB-Keywords
GO: Cellular Component	Extracellular region	UniProtKB-Sub-cellular Location
GO: Molecular Function	Aspartic-type endopeptidase activity	UniProtKB-Keywords

(continued)

[i] http://www.uniprot.org/uniprot/P00791.txt?version=126

Table 5.8   *(continued)*

	Haemoglobin subunit alpha (HBA_HUMAN, P69905; version 136[ii])	Source
UniProtKB Function	*Involved in oxygen transport from the lung to the various peripheral tissues*	UniProtKB
GO: Biological Process	Hydrogen peroxide catabolic process	Cardiovascular Gene Ontology
	Bicarbonate transport	Reactome
	Oxidation-reduction process	Gene Ontology Consortium
	Oxygen transport	UniProtKB
	Positive regulation of cell death	Cardiovascular Gene Ontology
	Protein heterooligomerization	Cardiovascular Gene Ontology
	Response to hydrogen peroxide	Cardiovascular Gene Ontology
	Small molecule metabolic process	Reactome
GO: Cellular Component	Cytosolic small ribosomal subunit	UniProtKB
	Cytosol	Reactome
	Blood microparticle	UniProtKB
	Endocytic vesicle lumen	Reactome
	Extracellular region	Reactome
	Extracellular vesicular exosome	UniProt
	Haptoglobin-hemoglobin complex	Cardiovascular Gene Ontology
	Hemoglobin complex	UniProtKB
	Membrane	UniProtKB
GO: Molecular Function	Heme binding	InterPro
	Iron ion binding	InterPro
	Oxygen binding	InterPro
	Oxygen transporter activity	UniProtKB-Keywords
	Actin, alpha skeletal muscle (ACTS_HUMAN, P68133; version 121[iii])	Source
UniProtKB Function	*Actins are highly conserved proteins that are involved in various types of cell motility and are ubiquitously expressed in all eukaryotic cells*	UniProtKB
GO: Biological Process	Muscle filament sliding	Reactome
	Cell growth	Ensembl
	Muscle contraction	UniProtKB
	Response to extracellular stimulus	Ensembl
	Response to lithium ion	Ensembl
	Response to mechanical stiumulus	Ensembl
	Response to steroid hormone	Ensembl
	Skeletal muscle fiber development	UniProtKB
	Skeletal muscle thin filament assembly	UniProtKB
	Skeletal muscle fiber adaptation	Ensembl
GO: Cellular Component	Actin filament	UniProtKB
	Actin cytoskeleton	UniProtKB
	Blood microparticle	UniProtKB
	Cytosol	Reactome
	Extracellular space	UniProtKB
	Extracellular vesicular exosome	UniProtKB
	Sarcomere	UniProtKB
	Stress fiber	UniProtKB
	Striated muscle thin filament	UniProtKB
GO: Molecular Function	ADP binding	UniProtKB
	ATP binding	UniProtKB
	Myosin binding	UniProtKB
	Structural constituent of cytoskeleton	UniProtKB

[ii] http://www.uniprot.org/uniprot/P69905.txt?version=136
[iii] http://www.uniprot.org/uniprot/P68133.txt?version=122

little functional information they actually convey compared to the much terser, more precise GO annotations. The kind of formalism used in GO may feel a little more uncomfortable to the human reader, but is essential if we're to use computers in function-annotation tasks, and if we're to be able to more readily relate different terms or entities in different databases. Colourful, *ad hoc* nomenclature systems of the sort that appear to delight *Drosophila* geneticists, for example, are the scourge of bioinformatics (Pearson, 2001). In particular, those that include familiar, everyday words to describe biological entities (gene names like *hedgehog, hairy, ken, teashirt, cap 'n' collar, lost in space, etc.*) defy systematic analyses and quickly lose their amusement value when faced with the task of trying to train computers to discover them in scientific texts, to link them to related concepts in databases that use different nomenclature systems, and ultimately to teach computers what they mean.

5.13 Summary

This chapter introduced some of the issues that have arisen as bioinformatics tools and databases have been moulded and adapted in *ad hoc* ways to tackle new problems. Specifically, we saw that:

1 The concept of a 'gene' is fluid and may have different meanings for different scientists;

2 Genes are difficult to find and build computationally; their counts in newly sequenced genomes are therefore predictions that are subject to iterative refinement over time;

3 Genes and proteins may have different names in different species; tracking their occurrence in the literature and in databases requires the assistance of synonym dictionaries;

4 Protein sequences can change over time, and the changes are tracked by means of their database identifiers and accession numbers;

5 Occasionally, the same accession number can point to different sequences – such changes can be tracked by means of the sequence version numbers in UniProtKB, and in the protein sequence archive, UniParc;

6 The contents of database entries change over time, including the names of the entities and the data they store; this instability can make it difficult to know to what biological entity a particular database entry refers;

7 Some database entries may point to experimentally determined 3D structures; when numerous structures are linked to the same sequence, it can be difficult to know which is the most appropriate or relevant to that entry;

8 Different alignment tools give different results; understanding the results requires an understanding of the methods that gave rise to them;

9 Different alignment tools present their outputs in different ways; understanding the outputs (in terms of the properties sequences may share or how similar sequences may be) is facilitated by the use of colour schemes;

10 Different colour schemes embody different metaphors; understanding the underlying metaphor is crucial for interpreting what alignment outputs mean;

11 Structure-based alignments are fundamentally different from sequence-based alignments; they use different frames of reference and are hence not directly comparable – nevertheless, many sequence alignment tools are still benchmarked against structure-based alignments;

12   The concept of a 'family' is fluid and may have different meanings depending on whether the particular grouping has been derived from sequence- or structure-based analysis methods; merging families in resources like InterPro is therefore problematic;

13   The functions of biological molecules are evolutionarily plastic – highly similar sequences may have different functions (proteins may moonlight), different sequences may have similar functions; deriving functional annotations from sequence evidence alone may therefore sometimes be misleading;

14   The functions of multi-domain or modular proteins are multi-faceted and depend on the nature and context of their constituent domains – it's important to understand to which component of a multi-domain protein a functional annotation within a database refers;

15   The concept of 'function' has been ill-defined; ontologies have been introduced to try to formalise functional descriptions in database entries.

5.14   References

Anfinsen, C.B. (1973). Principles that govern the folding of protein chains. *Science*, **181**(4096), 223–230. doi:10.1126/science.181.4096.223

Archer, S. and Hirano, J. (1996) Absorbance spectra and molecular structure of the blue-sensitive rod visual pigment in the conger eel (*Conger conger*). *Proceedings of the Royal Society B*, **263**, 761–767.

Arnold, K., Bordoli, L., Kopp, J. and Schwede, T. (2006). The SWISS-MODEL Workspace: a web-based environment for protein structure homology modeling. *Bioinformatics*, **22**, 195–201.

Attwood, T.K. (2000) Genomics: The Babel of bioinformatics. *Science*, **290**, 471–473. doi:10.1126/science.290.5491.471

Cantor, C.R. (2000) Biotechnology in the 21st century. *Trends in Biotechnology*, **18**(1), 6–7.

Daniels, N.M., Nadimpalli, S. and Cowen, L.J. (2012) Formatt: correcting protein multiple structural alignments by incorporating sequence alignment. *BMC Bioinformatics*, **13**, 259. doi:10.1186/1471-2105-13-259

Edgar, R. (2010) Quality measures for protein alignment benchmarks. *Nucleic Acids Research*, **38**(7), 2145–2153. doi:10.1093/nar/gkp1196

Edgar, R. (2010) Big alignments – do they make sense? Available at https://robertedgar.wordpress.com/2010/05/02/big-alignments-do-they-make-sense accessed 3 December 2015.

Edgar, R. (2010) Multiple protein alignment is a dead field. Available at https://robertedgar.wordpress.com/2010/05/02/multiple-protein-alignment-is-a-dead-field accessed 3 December 2015.

Editorial. (2001) A cold dose of medicine. *Nature Biotechnology*, **19**(3), 181.

Fletcher, W. and Yang, Z. (2010) The effect of insertions, deletions and alignment errors on the branch-site test of positive selection. *Molecular Biology and Evolution*, **27**, 2257–2267.

Gao, Y., Zeng, Q., Guo, J. *et al.* (2007) Genetic characterization reveals no role for the reported ABA receptor, GCR2, in ABA control of seed germination and early seedling development in *Arabidopsis*. *The Plant Journal*, **52**, 1001–1013.

Gatesy, J., Milinkovitch, M., Waddell, V. and Stanhope, M. (1999) Stability of cladistic relationships between Cetacea and higher-level Artiodactyl taxa. *Systematic Biology*, **48**, 6–20.

Graur, D., and Higgins, D.G. (1994) Molecular evidence for the inclusion of Cetaceans within the Order Artiodactyla. *Molecular Biology and Evolution*, **11**, 357–364.

Huberts, D.H. and van der Klei, I.J. (2010) Moonlighting proteins: an intriguing mode of multi-tasking. *Biochimica Biophysica Acta*, **1803**(4), 520–525. doi:10.1016/j.bbamcr.2010.01.022

Iantorno, S., Gori, K., Goldman, N. *et al.* (2012) Who watches the watchmen? An appraisal of benchmarks for multiple sequence alignment. *arXiv*:1211.2160. doi:10.1007/978-1-62703-646-7_4

Jacob, F. (1977) Evolution and tinkering. *Science*, **196**, 1161–1166.

Klamer, A. and Leonard, T.C. (1994) So what's an economic metaphor? In *Natural Images in Economic Thought: 'Markets read in tooth and claw'* (ed. P. Mirowski), Cambridge University Press, Cambridge, New York, pp. 20–51.

Jeffery, C.J. (1999) Moonlighting proteins. *Trends in Biochemical Science*, **24**(1), 8–11.

Jeffery, C.J. (2004) Moonlighting proteins: complications and implications for proteomics research. *Drug Discovery Today: TARGETS*, **3**(2), 71–78.

Juty, N., Le Novere, N. and Laibe, C. (2011) Identifiers.org and MIRIAM Registry: community resources to provide persistent identification. *Nucleic Acids Research*, **40** (Database Issue), D580–D586.

Li, B., Yu, J. P., Brunzelle, J. S., Moll, G. N. *et al.* (2006) Structure and mechanism of the lantibiotic cyclase involved in nisin biosynthesis. *Science*, **311**, 1464–1467.

Loytynoja, A. and Goldman, N. (2008) Phylogeny-aware gap placement prevents errors in sequence alignment and evolutionary analysis. *Science*, **320**, 1632–1635.

Markova-Raina, P. and Petrov, D. (2011) High sensitivity to aligner and high rate of false positives in the estimates of positive selection in the 12 *Drosophila* genomes. *Genome Research*, **21**(6), 863–874. doi:10.1101/gr.115949.110

Nordström, K.J., Lagerström, M.C., Wallér, L.M. *et al.* (2009) The secretin GPCRs descended from the family of adhesion GPCRs. *Molecular Biology and Evolution*, **26**(1), 71–84.

Nordström, K.J., Sällman Almén, M., Edstam, M.M. *et al.* (2011) Independent HHsearch, Needleman–Wunsch-based, and motif analyses reveal the overall hierarchy for most of the G protein-coupled receptor families. *Molecular Biology and Evolution*, **28**(9), 2471–2480.

Piatigorsky, J. and Wistow, G.J. (1989) Enzyme/crystallins: gene sharing as an evolutionary strategy. *Cell*, **57**(2), 197–199.

Pearson, H. (2001) Biology's name game. *Nature*, **411**, 631–632.

Pertea, M. and Salzberg, S.L. (2010) Between a chicken and a grape: estimating the number of human genes. *Genome Biology*, **11**(5), 206. doi:10.1186/gb-2010-11-5-206

Rashevsky, N. (1954) Topology and life: in search of general mathematical principles in biology and sociology. *Bulletin of Mathematical Biophysics*, **16**, 317–348.

Reese, M.G. Hartzell, G. Harris, N.L. *et al.* (2000) Genome annotation assessment in *Drosophila melanogaster*. *Genome Research*, **10**, 483–501.

Roos, D. (2001) Bioinformatics – trying to swim in a sea of data. *Science*, **291**, 1260–1261.

Rose, G.D. and Creamer, T.P. (1994) Protein folding: predicting predicting. *Proteins*, **19**(1), 1–3.

Schmucker, D., Clemens, J.C., Shu, H. *et al.* (2000) *Drosophila Dscam* is an axon guidance receptor exhibiting extraordinary molecular diversity. *Cell*, **101**(6), 671–684.

Sievers, F., Wilm, A., Dineen, D. *et al.* (2011) Fast, scalable generation of high-quality protein multiple sequence alignments using Clustal Omega. *Molecular Systems Biology*, **7**, 539. doi:10.1038/msb.2011.75

Thompson, J.D., Plewniak, F. and Poch, O. (1999) BAliBASE: a benchmark alignment database for the evaluation of multiple alignment programs. *Bioinformatics*, **15**, 87–88.

Wallace, R.W. (2001) Bioinformatics – key to 21st century biology. BioMedNet (HMS *Beagle*), March 30, issue 99.

Wistow, G. and Piatigorsky, J. (1987) Recruitment of enzymes as lens structural proteins. *Science*, **236**(4808), 1554–1556. doi:10.1126/science.3589669

Wong, W-C., Maurer-Stroh, S. and Eisenhaber, F. (2010) More than 1,001 problems with protein domain databases: transmembrane regions, Signal peptides and the issue of sequence homology. *PLoS Computational Biology*, **6**(7), e1000867. doi:10.1371/journal.pcbi.1000867

Wright, F.A., Lemon, W.J., Zhao, W.D. *et al.* (2001) A draft annotation and overview of the human genome. *Genome Biology*, **2**(7), research0025.1-0025.18.

Zarembinski, T.I., Hung, L.W., Mueller-Dieckmann, H.J. *et al.* (1998) Structure-based assignment of the biochemical function of a hypothetical protein: a test case of structural genomics. *Proceedings of the National Academy of Sciences of the U. S. A.*, **95**(26), 15189–15193.

5.15 Quiz

This multiple-choice quiz will help you to see how much you've remembered about some of the concepts and issues described in this chapter.

1 A gene may be considered to be:
 A a heritable unit corresponding to an observable phenotype.
 B a packet of genetic information that encodes RNA.
 C a packet of genetic information that encodes a protein.
 D a packet of genetic information that encodes multiple proteins.

2 Which of the following statements about sequence database entries is true?
 A A given accession number always refers to the same sequence.
 B A given database identifier is invariant between database releases.
 C A given database description line can't change between database releases.
 D None of the above.

3 Which alignment tool uses evolutionary information to place gaps?
 A ClustalW
 B CINEMA
 C Multalin
 D PRANK

4 Which of the following amino acid colour-codings is based on a 'standard' property-based metaphor?
 A R = Red.
 B R = White.
 C R = Blue.
 D None of the above.

5 Which of these statements is true?
 A Structure-based alignments are the standard of truth for determining the accuracy of sequence alignments.
 B Structure-based alignments use a framework for comparison that differs from sequence-based alignments.
 C Structure-based alignments are more biologically meaningful than sequence-based alignments.
 D None of the above.

6 Which of the following belong to the GRAFS class of receptors?
 A Glycine (G), Arginine (R), Alanine (A), Phenylalanine (F), Serine (S)
 B Glutamine, Rhodopsin, Adhesin, Frizzled, Serine
 C Glutamate, Rhodopsin, Adhesion, Frizzled, Secretin
 D None of the above.

7 A protein family may be considered to be a group of sequences that:
   A share more than 50 per cent similarity.
   B share the same function.
   C have the same fold.
   D share less than 30 per cent identity.

8 Homology is the acquisition of common features as a result of:
   A the evolutionary process of divergence from a common ancestor.
   B gene transfer between two organisms.
   C the evolutionary process of convergence from unrelated ancestors.
   D None of the above.

9 Anfinsen's dogma states that:
   A the native structure of membrane proteins is determined by the amino acid sequence alone.
   B genetic information is transferred unidirectionally in biological systems.
   C the native structure of small globular proteins is determined by the amino acid sequence alone.
   D genetic information is transferred bidirectionally in biological systems.

10 The three ontologies within GO define:
   A biological pathway, cellular structure, molecular function.
   B biological process, cellular component, molecular function.
   C biological function, cellular component, molecular structure.
   D None of the above.

5.16 Problems

1 How many descriptive titles has carbonic anhydrase 2 (CAH2_HUMAN, P00918[55]) had during its 35-year journey from PIR, through Swiss-Prot to UniProtKB?

2 Figure 5.11 depicts an alignment of two proteins, Q9C929_ARATH (Q9C929[56]) and Q48670_9LACT (Q48670[57]), published in *The Plant Journal* in 2007. The alignment, created using ClustalW[58], contains errors that result in misalignment of three key features. What are those features?

   What does the alignment of these sequences look like when generated using MUSCLE[59], T-Coffee[60] and Multalin[61]? How well do the different algorithms perform, and which gives the 'best' result? Justify your answer fully.

---

[55] http://www.uniprot.org/uniprot/P00918.txt?version=202
[56] http://www.uniprot.org/uniprot/Q9C929
[57] http://www.uniprot.org/uniprot/Q48670
[58] http://www.ch.embnet.org/software/ClustalW.html
[59] http://www.ebi.ac.uk/Tools/msa/muscle
[60] http://www.ebi.ac.uk/Tools/msa/tcoffee
[61] http://multalin.toulouse.inra.fr/multalin/multalin.html

3 UniProtKB/TrEMBL sequence Q23239_CAEEL (Q23293[62]) is described as 'Dro-MyoSuppressin Receptor related'; when it first entered TrEMBL, however, the descriptive title of version 1[63] of the sequence was 'SIMILARITY TO MAMAMA-LIAN MU- AND KAPPA-TYPE OPIOID RECEPTORS'. Why do you think the title of the protein changed? Where does the DMSR annotation come from? What insights does this offer into how database annotations are derived?

4 Considering the same sequence (Q23293[64]) as in problem 3, use the BLAST facility in the UniProtKB entry to run a similarity search (accepting the default parameters). Select from the output (which may run over several pages) up to half-a-dozen sequences that have annotations other than 'putative' or 'hypothetical' (e.g., see if you can find examples of types of receptor). Then use an alignment tool to align your chosen sequences.

Now return to the UniProtKB BLAST facility and change the search database to UniProtKB/Swiss-Prot. Accepting default parameters, re-run the search, and select the top half-dozen sequences for alignment.

Are the alignments the same? What might account for any differences you observe? Does either set of results help to shed light on the function of the Q23293? If so, what do you think it might be, and why?

---

[62] http://www.uniprot.org/uniprot/Q23293.txt?version=107
[63] http://www.uniprot.org/uniprot/Q23293.txt?version=1
[64] http://www.uniprot.org/uniprot/Q23293

PART 2

Chapter 6

Algorithms and complexity

6.1 Overview

In this chapter, we begin to consider how computers have been brought to bear on some of the issues we've been discussing so far. We look at how to go about trying to solve problems algorithmically (from a more conceptual perspective of what computation is capable of) and, in particular, learn that computation has some surprising limits. From there, we consider how the reality of modern computer technology affects our ability to solve problems, and, conversely, how the problems faced by bioinformaticians may not be as amenable to solution as we might expect. We'll also explore some of the methods available for ameliorating such shortfalls.

We're particularly concerned here with grasping concepts, so don't be put off by the minutiae – you don't need to remember or to follow every algorithm in detail. By the end of the chapter, you should have gained an understanding of how some bioinformatics problems require highly complex solutions if they're to provide the answers we seek, and why, given the computers we have at our disposal today, some problems may not be soluble without employing sophisticated computer science techniques.

6.2 Introduction to algorithms

The process of computational problem-solving should be familiar to anybody with even a basic mathematical education. Given a real-world problem, it's the job of mathematicians (the 'computers' of these problems) to do two things: model the problem in the language of mathematics (maths), and then apply mathematical tools to that model to produce a solution. To highlight how fundamental this process is, consider the following trivial maths question:

> Sarah has nine apples. John takes two of Sarah's apples from her. How many apples does Sarah have left?

Bioinformatics challenges at the interface of biology and computer science: Mind the Gap. First Edition.
Teresa K. Attwood, Stephen R. Pettifer and David Thorne. Published 2016 © 2016 by John Wiley and Sons, Ltd.
Companion website: www.wiley.com/go/attwood/bioinformatics

Most children are taught, from a very young age, to begin by modelling this situation in the language of maths. Doing this involves transforming the real-world quantities (apples) into pure mathematical objects (numbers) – something we eventually learn to do without thinking – and then, using mathematical relationships, to describe the fruity theft. That model will then look like this:

$$9 - 2 = ?$$

From here, it's a straightforward task to apply the rules of maths to solve the model, and, in this case, discover the missing number 7. The final step, the reverse of the first, is to translate this solution back to the real-world scenario in order to discover that Sarah is now in possession of *seven apples*. As simplistic as this example clearly is, it neatly shows the two complementary tasks of problem solving: **modelling**, the process of describing a problem in abstract terms; and **processing**, the task of manipulating the model according to its rules in order to find a solution. For anything but the simplest of problems, the task of modelling is sufficiently complex to need careful thought; equally, any non-trivial problem will also require multiple processing steps in its solution. It is the description of these complex processes that we call **algorithms**.

Figure 6.1 shows a flowchart that represents the simple task of identifying the largest number in a list. Notice that, as we travel through the flowchart from the initial Start state to the final Stop state, we're remembering values (in named **variables**), and, based on those values, making decisions about which path to take. Each position in the flowchart is a possible algorithm state, but the values of variables are independent of

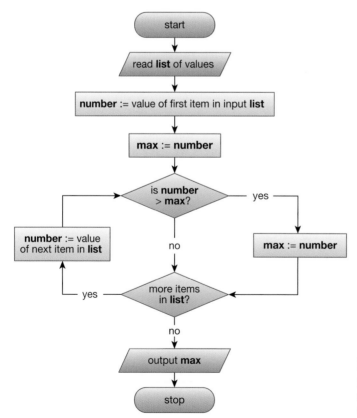

Figure 6.1

Flowchart for the task of choosing the largest of a list of numbers.

```
Let max := the first item in the input list
For each number in the rest of the input do:
   If number > max then:
     Let max := number
Return max
```

Figure 6.2

Pseudocode for the task of finding the largest in a list of numbers, where we keep track of the maximum value in the list in the variable *max*. We begin by assigning to it the first number of the input list, and then in the 'For each' loop, we iteratively compare it with every other; at each step, if the number found is greater than *max*, we update *max* to have this new maximum value. Once the loop is complete, and we've visited the entire input list, *max* is guaranteed to hold the largest of its values.

which state we're in: they retain their values between states unless explicitly modified by the algorithm. This gives rise to a potentially confusing use of the word 'state', either to mean the state of the algorithm, or the state of the variables (the data state). We will make sure the context is clear whenever we use this term.

Algorithms are often summarised by writing semi-natural-language **pseudocode**, such as that seen in Figure 6.2. There's no standard syntax in which to write pseudocode: it just needs to be simple to understand and therefore simple to follow – after all, algorithms are simply recipes for solving specific problems, and impenetrable recipes are impossible to follow.

Software engineers must choose (or design) appropriate algorithms to solve the problems they're given. Ultimately, they'll implement those algorithms in a formal programming language; however, before that step is reached, some thought is needed.

Although most algorithms are considerably more complex than the simple example introduced above, they're still mathematical constructs, and hence are described and investigated mathematically. These **formal methods** of investigation are used to quantify the efficiency of algorithms, prove or disprove their correctness, and even predict likely outcomes. Many of these formal methods are beyond the scope of this book, but an appreciation of a number of issues is important when thinking about solving problems with computers.

6.2.1 Mathematical computability

Stepping back from the question of how to go about computing solutions to problems, and looking at the theory of computation itself, various interesting points emerge. Without exception, the maths underpinning the following observations is beyond the scope of this book. We touch on the simplest aspects of **complexity theory** (a subfield of the **theory of computation**) in a more practical setting in Section 6.4, but the conclusions derived from the theories proper are important enough to state without the working-out.

Not all problems are computable

Some problems simply can't be solved, although it's rare to come across them in practice. One example of a non-computable problem is described by the **halting problem**, which states that it's impossible to decide whether some algorithms will actually complete in a finite time, or will continue running forever; for non-halting algorithms, the only way to know that they won't halt is to run them and wait and see, which will obviously take forever as well. Thankfully, this class of problem doesn't often make an appearance in practical terms, but it's a good starting point for the following observations.

Not all problems are effectively computable

Some problems, while technically computable, aren't *effectively* computable. In this case, an effectively computable problem is one that can be solved in a reasonable amount of time. This suggests that some problems, by their nature, take an unreasonable amount of time to solve.

This is sometimes a good thing – it's precisely this quality that makes current encryption techniques work: using brute force to break an encryption without the required key would take so long as to make it impractical. Taking a 128-**bit** encryption key as an example, given that there are 2^{128} different possible keys, and employing a billion computers that can test a billion keys per second each, it would still require a period of execution equivalent to a thousand times the age of the universe to test each possible key:

$$\frac{2^{128}}{10^9 \times 10^9 \times 60 \times 60 \times 24 \times 365.25} \approx \frac{3.4 \times 10^{38}}{3.1 \times 10^{25}} \approx 10^{13} \text{ years}$$

For most practical purposes, of course, taking unreasonable amounts of time to solve problems usually isn't a good thing; knowing whether the problem we're trying to solve is effectively computable is an important first step in not wasting time producing a solution that can never be employed. Often, we test our algorithms on relatively small input data – with small data, almost any algorithm will finish quickly. Knowing how an algorithm will behave when employed on realistic amounts of data is therefore crucial, especially in a world where data are being produced at an ever-increasing rate. Barely do we have the chance to appreciate the quantities of data *currently* available, before new technologies appear that threaten to bury us entirely. An example of the extreme nature of the problem can be seen in the recent work of Puckelwartz *et al.* (2014) with their **supercomputer**-powered whole-genome analysis system, which is capable of aligning and analysing 240 whole genomes in ~50 hours. Systems like this will only exacerbate the data-handling problems we're already struggling to solve.

Accuracy is sometimes negotiable

Sometimes, approximate answers are just as useful as accurate answers, or, at least, aren't entirely useless. This is a concession worth considering, because there are many situations where approximate solutions can be effectively computed, but accurate solutions can't, and a lot of time can be saved by knowingly opting for the faster, though approximate, solution.

A pertinent example is that of the **Travelling Salesman Problem** (TSP), named after the problem faced by a travelling salesman who wishes to efficiently visit each of a set of interconnected cities. If cities are thought of as nodes in a network, with the roads between them being the edges, then the TSP becomes a general graph-based problem that asks for the quickest route around a graph, such that it can be guaranteed that all nodes are visited. This problem has been known for a long time, and many complete (accurate) solutions exist; however, for very large graphs, none are effectively computable. Nevertheless, there are effectively computable solutions that use heuristics to reach *approximate* answers that can reach 99 per cent accuracy in realistic time frames[1].

---

[1] An implementation of the Lin–Kernighan heuristic approach (Helsgaun, 2000) to the TSP currently provides the best-in-show solution to the 1,904,711-node graph that represents all the cities on Earth (World TSP – http://www.math.uwaterloo.ca/tsp/world).

Sometimes, algorithm designers may need to relax their hopes of accuracy in order to solve the problem at all.

Although the TSP might seem rather abstract, it has applications in many subjects, such as integrated circuit design and, in a modified form, some multiple sequence alignment tasks. Replacing the idea of cities with possible individual residue alignments, and the idea of distance with gap penalties, solving the TSP to find the lowest overall penalty path gives us a fully gapped alignment. Therefore, it helps to be generally knowledgeable of existing algorithms to be able to recognise when they might apply to seemingly unrelated problems.

Easily verifiable ≠ effectively computable

The **P versus NP problem** is an unsolved problem in computability theory. Here, P represents the class of problems that can be effectively computed (the 'P' is short for *polynomial time*, a concept introduced later in Section 6.4.2), while NP represents the class of problems, the solutions to which can be effectively verified. To illustrate the difference between these two classes, take the following list of numbers:

$$-7, 5, 1, -4, 15, 26, -14, 66$$

Our task is to work out whether there exists a subset of these numbers that, when added together, equal zero. Feel free to take a moment to try to do just that. The larger the list, the longer it would take – exponentially longer in fact: this problem is *not* effectively computable. However, given an answer, such as the following, we can verify it very quickly indeed, simply by adding the numbers together and checking that they sum to zero:

$$5 + 1 + 15 - 7 - 14 = 0$$

This distinction is important because it can lead us to overestimate our chances of solving a problem. It can't be assumed that easily verifiable answers can be easily produced; it may be impossible to do so.

6.3 Working with computers

So far in this discussion, we've viewed both problems and their algorithmic solutions as purely mathematical entities. With the arrival of computing machines, these otherwise rather abstract and symbolic tasks become grounded in the reality of those machines' designs. Details of where and how numbers are stored, the speed of the computing machine, even the available instruction set of the machine's processor, introduce restrictions on how any algorithm can realistically be applied.

6.3.1 Discretisation of solutions

In reality, rather than floating about in our imaginations, the numbers with which algorithms work are stored in one of a number of physical substrates: depending on the machine's design, and the details of its Operating System (OS), that substrate could be the main Random Access Memory (RAM), various levels of processor cache, processor registers, or possibly on persistent storage in virtual memory systems. Equally, rather than being abstract, those numbers are actually stored in binary form as a sequence of magnetic or electric charges, with just a small number of fundamental operations available to manipulate them. Such low-level detail can have far-reaching consequences for algorithm design.

Viewed from the perspective of software engineering, even a simple task, like finding the cubed power of an integer number, is a complex multi-step process. This is partly a consequence of how memory works in modern computers, and partly because the processor's arithmetic abilities are limited to only a handful of simple (almost **axiomatic**) operations: addition, multiplication, and subtraction are provided, among others, but nothing as complex as cubic power operators. We must therefore break down the power operation into a series of multiplications. First, the binary representation of the number to be cubed must be copied from its current resting place (most likely the machine's RAM) into a register on the processor itself. Let's call this first register ax, as it acts to accumulate the result. The ax register's value must be copied into another register (bx) so that there are two copies of the original number. The ax register can now be multiplied by itself, overwriting its value with an intermediate result; then, the bx register (holding its copy of the original number) can be multiplied with this intermediate result, once again overwriting the ax register with what is now the cube of the original input. This final result is now available to be written back to main memory prior to the next calculation.

We see, then, that what can be specified and understood conceptually by the simple notation x^3, when translated into something approaching the reality of computing machines today, becomes a more complex set of tasks; and even this belies the true nature of the technology involved. Figure 6.3 shows an example of this task written in assembly language, a low-level way of specifying how a processor executes individual operations.

Of course, much effort has gone into simplifying the task of implementing algorithms, and very few occasions arise that would require knowledge of assembly language. Instead, programmers employ high-level languages that allow problems to be specified in terms that are, if not as simple as the mathematical model, much more tractable:

	Registers		Main memory	
	ax	bx	[1000h]	[1002h]
	00000000 00000000 0	00000000 00000000 0	00000000 00000111 7	00000000 00000000 0
mov ax [1000h]	00000000 00000111 7	00000000 00000000 0	00000000 00000111 7	00000000 00000000 0
mov bx ax	00000000 00000111 7	00000000 00000111 7	00000000 00000111 7	00000000 00000000 0
mul ax	00000000 00110001 49	00000000 00000111 7	00000000 00000111 7	00000000 00000000 0
mul bx	00000001 01010111 343	00000000 00000111 7	00000000 00000111 7	00000000 00000000 0
mov [1002h] ax	00000001 01010111 343	00000000 00000111 7	00000000 00000111 7	00000001 01010111 343

Figure 6.3

Assembly language formulation of the cubic power operation on the integer value 7. This example uses 16 bits to represent each number, meaning that each value is stored as two bytes. We show how the values of two different registers and two locations in main memory change during execution; those values are shown both in their binary and decimal representations. In assembly language, memory locations are denoted by hexadecimal byte addresses – hence, [1000h] represents the 4,096th byte in memory, and [1002h] the 4,098th byte.

simply writing pow(x,3) to mean raising the value of x to the power of three, frees the programmer's creative powers to concentrate on bigger problems. However, there are many cases, especially in large, complex and time-dependent applications, where an understanding of the low-level reality of computation becomes especially useful in designing and implementing efficient algorithms.

6.3.2 When computers go bad

We'll return to efficiency later in this chapter in the context of designing and implementing algorithms, but we must first concentrate on some potentially dangerous technological minefields.

Numerical overflow

As we've already mentioned, modern computers represent numbers in binary form in their memory and storage. The choice of representation depends on the platform (processor architecture) being used, and the type of number that needs to be represented. Figure 6.4 shows a variety of decimal integers represented as pure binary digits in 8-bit unsigned integer form. Note that the restriction of eight bits provides upper and lower bounds for the numbers we're able to represent: we can store anything from zero to 255.

We can see that if a number were required that fell outside the acceptable range, this representation would break down; but what exactly would happen, and how would a program cope? Well, just like on a vehicle's odometer, when the number increments beyond the available digits, the digits reset (or *overflow*) back to zero and start again. Therefore, under these circumstances, no matter how unintuitive it may be, it's possible to see that $255 + 1 = 0$. It may seem obvious that we shouldn't choose a representation that can't handle values we require; in practice, however, it can be difficult to tell what values we may need. The moment we start using functions written by someone else, what's going on 'under the hood' is no longer evident, and some functions could be introducing numerical overflow without us being aware of it. Conversely, we may make use of naïve calculations that would increase the chance of overflow, when alternative methods may be available that circumvent the risk.

Take, for example, a slightly more complicated task: finding the length of the hypotenuse of a right-angled triangle, given the lengths of the other two sides. As you probably know, we can use the Pythagorean formula $z^2 = x^2 + y^2$ to calculate the length, which, when rearranged, can be specified as:

$$z = \sqrt{x^2 + y^2}$$

	$2^7 = 128$	$2^6 = 64$	$2^5 = 32$	$2^4 = 16$	$2^3 = 8$	$2^2 = 4$	$2^1 = 2$	$2^0 = 1$
0	0	0	0	0	0	0	0	0
1	0	0	0	0	0	0	0	1
2	0	0	0	0	0	0	1	0
3	0	0	0	0	0	0	1	1
71	0	1	0	0	0	1	1	1
182	1	0	1	1	0	1	1	0
255	1	1	1	1	1	1	1	1

Figure 6.4

8-bit unsigned integer representations of various positive numbers within the range zero to 255.

Let's say that, for the values of x, y and z, we use some limited representation (similar to that introduced above) that can only represent numbers between zero and 255, and let's take the case where $x = 200$ and $y = 150$. Then,

$$z = \sqrt{200^2 + 150^2}$$
$$= \sqrt{40000 + 22500}$$
$$= \sqrt{62500}$$
$$= 250$$

From this, we can see that, even though the inputs and outputs (150, 200, 250) may be representable, intermediate steps of the calculation may produce overflow during execution. Without taking into account this sort of hidden numerical overflow, a program could produce wildly incorrect results without an obvious cause (although some programming languages raise exceptions on overflow, many don't). Fortunately, experience of such issues makes us better equipped to avoid them: as it turns out, the Pythagorean formula above can be rearranged differently, with the following simple algebra:

$$z = \sqrt{x^2 + y^2}$$
$$= \sqrt{x^2(1 + (y/x)^2)}$$
$$= |x|\sqrt{(1 + (y/x)^2)}$$

In this new formulation, as long as the input lengths are chosen such that $|x| \geq |y|$, we have mitigated the danger of overflow from squaring potentially large numbers. In fact, this particular case of overflow crops up so often that most programming languages have some sort of built-in `hypot(x,y)` function that does just this rearrangement behind the scenes. A good knowledge of the mathematical functions available to a language (not to mention of the maths itself) can help to reduce or remove such problems.

Loss of precision

One aspect of our Pythagorean example left unexplained is that the square roots of many whole numbers are not themselves whole numbers, but real (fractional) numbers; however, so far we've only mentioned integers. Integers can be precisely represented in binary, but real numbers are not so straightforward. The simplest way of demonstrating this is by asking a computer to perform some simple arithmetic with real numbers and seeing what happens. Below, we've used the interactive command prompt for the **Python** programming language to ask for the result of the trivial calculation $0.1 + 0.2$.

```
>>>0.1 + 0.2
0.30000000000000004
```

Instead of producing the seemingly obvious value of 0.3, Python returns a number very close to it: 0.30000000000000004. This represents a miniscule error, but an error nonetheless. It isn't Python's fault: we would get the same result from **Ruby**, or **JavaScript**, or any other programming language we might care to try. So how could a computer make a mistake with such a simple calculation?

	$2^{-1} = \frac{1}{2}$	$2^{-2} = \frac{1}{4}$	$2^{-3} = \frac{1}{8}$	$2^{-4} = \frac{1}{16}$	$2^{-5} = \frac{1}{32}$	$2^{-6} = \frac{1}{64}$	$2^{-7} = \frac{1}{128}$	$2^{-8} = \frac{1}{256}$
0.0	0	0	0	0	0	0	0	0
0.00390625	0	0	0	0	0	0	0	1
0.25	0	1	0	0	0	0	0	0
0.5	1	0	0	0	0	0	0	0
0.75	1	1	0	0	0	0	0	0
0.875	1	1	1	0	0	0	0	0
0.9921875	1	1	1	1	1	1	1	0
0.99609375	1	1	1	1	1	1	1	1

Figure 6.5

Binary representations of various positive fractional numbers.

The answer lies in how real numbers are stored using the 1s and 0s of our digital computer's memory. Recall the binary integer examples of Figure 6.4, where each binary digit represented a power of two (1, 2, 4, 8, *etc.*); similarly, each digit of a binary fraction represents a fractional power of two ($\frac{1}{2}, \frac{1}{4}, \frac{1}{8}, \frac{1}{16}$, *etc.*). Some examples of positive 8-bit binary fractions are shown in Figure 6.5.

The precise representation in memory is somewhat more complicated than this illustration might suggest (see Box 6.1 for an explanation of the standard **floating-point** representation); however, the most important aspect of such fractional number systems is their 'coverage'. There are a finite number of values in any finite integer range, allowing any sufficiently wide binary representation to be able to encode all its values. However, there are an infinite number of values in any non-trivial real number range, and no representation can encode an infinite number of values in a finite number of bits. This means that there are 'holes' in the real number line that a given binary fractional representation simply can't encode. In fact, necessarily, there are an infinite number of such unencodable values.

In practice, therefore, unencodable values are rounded up or down to the nearest representable number, otherwise practically no calculation would ever succeed. Unencodable numbers are so common that it's almost impossible not to come across them in everyday calculations. As we saw at the start of this section, this is true even of some of the simplest, most common decimal values: the decimal value 0.1 can't be encoded using the standard 'IEEE 754 single-precision floating-point' representation (see Box 6.1). The nearest representable number is the rather long-winded 0.100000001490116119384765625. This represents an error of more than one part in a billion. These rounding errors, introduced in the course of everyday computation with floating-point numbers, can cause problems if they're not expected. Even when they are expected, they can be difficult to manage, depending on the application: even minute errors can grow exponentially over the course of a program's lifetime, as each subsequent calculation can multiply that error while introducing yet more error itself, pushing the result further away from the ideal mathematical value.

Reclaiming precision

So what can we do to mitigate the effects of such precision loss? One answer is to increase the number of bits available to the floating-point representation. If we were to double the number of bits and use double-precision floating-point numbers, we could guarantee to encode many more numbers accurately (but still miss infinitely many). That would reduce our rounding error above to 0.000000000000000005551115123126. This is a much better result, but it is still in error. For obvious reasons, we can't simply

Box 6.1 IEEE 754 single-precision floating-point representation

How do digital computers represent real numbers in binary? Remember that a real number has two parts: a whole number part and a fractional part, separated by a radix point (known as the *decimal point* when the decimal number system is used). It is perfectly possible to set aside some bits for the whole number part, and some for the fractional part – doing this is known as *fixed-point* representation. In practice, however, this provides far too limited a range of numbers for most applications to find useful. Instead, computers today use what's called a *floating-point* representation, similar to the scientific notation of decimal numbers (*e.g.*, 1.25×10^5 instead of 125,000): a number of bits are used for the exponent (the power part), and a number of bits for the mantissa (the significant digits). This allows the radix point of the number to change, or float, depending on the value of the exponent, hence the name *floating point*.

s	exponent (e)	mantissa bits ($m_1 - m_{23}$)

| 0 | 0 | 1 | 1 | 1 | 1 | 0 | 1 | 1 | 0 | 0 | 1 | 0 | 0 | 0 | 0 | 0 | 0 | 0 | 0 | 0 | 0 | 0 | 0 | 0 | 0 | 0 | 0 | 0 | 0 | 0 | 0 |

Figure 6.6
The anatomy of the IEEE 754 single-precision floating-point number. The representation uses 32 bits: it has one sign bit, s; 8 bits for the exponent, e; and 23 bits for the mantissa, m. A negative number has a '1' for a sign bit, otherwise a '0'; the exponent is stored as an 8-bit unsigned integer; the mantissa bits encode a binary fraction.

Figure 6.6 shows the anatomy of a standard way of representing floating-point numbers, and, as an example, encodes the decimal real number 0.0703125. The first thing to notice is how much more complicated this representation is when compared to unsigned integers. There are three different blocks of bits that encode different information, one giving the sign of the number, one placing the radix point, and one representing the most significant digits. To turn those three bit ranges into their represented number, we must use the following formula:

$$r = (-1)^s \times \left(1 + \sum_{i=1}^{23} m_i 2^{-i}\right) \times 2^{e-127}$$

The sign bit, s, is either zero or one, causing the first expression $(-1)^s$ to make the number either positive or negative respectively. The 23 bits of the mantissa, $m_1 - m_{23}$, are converted into decimal by attributing to each digit the requisite fractional power of two, according to its position (just as in Figure 6.5). The eight bits of the exponent form an unsigned integer number e (just as in Figure 6.4), which has the effect of moving the radix point left or right depending on its value. Using the given formula on the bit ranges above, the floating-point number, r, can be calculated as follows:

$$r = 1 \times (1 + 0.125) \times 0.0625$$
$$= 0.0703125$$

We'll leave it with you to investigate how a value of zero is encoded using the IEEE 754 single-precision floating-point representation.

continue adding more and more bits: there will always be infinitely many numbers that we can't represent accurately. Even the decimal fraction 0.1 would need an infinite number of bits to be represented as a binary fraction, as it's infinitely recurring, as shown below (standard notation is used[2]).

$$0.1_{10} = 0.0\dot{0}01\dot{1}_2 = 0.000110011001100110011\ldots_2$$

Depending on the task to be performed, another solution at our disposal is that of rational fractions, a representation that makes use of a pair of integers – the numerator and denominator of a division. This allows relatively simple fractions to be represented exactly: 0.1 can be represented precisely as 1/10, circumventing the lack of precision in floating-point representations. Of course, given that the numerator and denominator are themselves subject to the limits described in Section 6.3.2, the set of rational numbers that can be encoded is itself limited. However, as long as the values that need to be worked on remain within the encodable range, rational types provide perfect numerical fidelity.

Unfortunately, there are occasions where we have little choice. If we want to make use of irrational numbers, like e or π, then, without the use of highly specialised mathematical languages, we can't avoid the pitfalls of floating-point representations.

6.4 Evaluating algorithms

Once we've sufficiently appreciated the limitations of computers, we can begin designing algorithms to solve our problems. Many factors play a part in how well an algorithm performs (*i.e.*, its *efficiency*) in relation both to other algorithms and to the amount of information it must process. This will be the focus of the following sections.

6.4.1 An example: a sorting algorithm

As an example on which to base our discussion, we'll examine the deceptively simple task of sorting a list of numbers; in the world of bioinformatics, this may represent anything from similarity scores in a BLAST result, to computed energy values of protein folds. In everyday life, we don't often have to sort objects, although we all do it on occasion: ordering playing cards during a card game; managing our physical music collection; or, for the computationally-minded, compiling a shopping list in the order in which products are likely to be found in-store. Regardless of how we decide what it means to be ordered, we can and do sort things quite instinctively. The problem with many instinctive tasks is that it's often harder to write instructions on how to perform them than it is just to perform them. As an exercise, try to do just that: formulate a method of sorting a list of numbers of arbitrary length.

As it happens, there are many ways to design a sorting algorithm, each with its own advantages and disadvantages. Here, we'll introduce one of the simplest sorting algorithms: *bubble sort*. The bubble sort algorithm makes progressive passes through a list, at each pass comparing consecutive pairs of numbers, swapping those that aren't locally ordered. After some number of passes, it's guaranteed to produce a correct global ordering, as each pass moves yet another item into the correct position (a little like a

---

[2] Subscripts are used to show the difference between decimal$_{10}$ and binary$_2$ numbers, while repeating digits are delineated by placing a dot above the first and last digit of the repeating range.

bubble rising to the top of a volume of water – hence the name, bubble sort). A naïve formulation of the bubble sort algorithm is described in pseudocode in Figure 6.7 – we recommend that you scrutinise its detail, as it forms the basis of subsequent discussions.

We now have a basis for introducing a few interesting qualities that can help us to produce (or choose) more efficient algorithms. First, although the idea of swapping two numbers is obviously a simple one, the reality is that it's impossible for a computer to atomically transpose two memory values. Instead, the computer must temporarily copy the value from the first item elsewhere, before overwriting that item with the second value, then finally overwriting the second item with the temporarily stored value. This triangular rotation of values is always required in such cases: even if our preferred programming language provides a handy `swap(a,b)` function, it will still be performing the same three-step process under the hood, taking longer and using up more memory than we might expect. Understanding what lies beneath a language's helpful abstractions can mean the difference between an algorithm completing in the blink of an eye, and it taking longer than our concentration span.

Second, although it may not be obvious at first, roughly half the total number of comparisons (line 11 in Figure 6.7) will *always* return false. After the first pass, the final item is necessarily the largest value in the list; therefore, on the next pass, the last two values are guaranteed to be in the correct order: there's no need to check them. In each subsequent pass, more values are bubbled into their correct order, increasing the number of pointless comparisons, each of which wastes processor time and slows the algorithm's execution.

More subtly, bubble sort exhibits *stable sorting*, such that values that are considered equal remain in the same relative order to each other once the whole list has been processed. If a different algorithm had been chosen that didn't exhibit stable sorting, and the program in which it was used implicitly relied on a stable sort, difficulties could arise from these unexpected side-effects.

Finally, one interesting feature of any algorithm is how its progression changes depending on the nature of its input. If the input to our bubble sort were already in true order, we could observe that there had been no swaps during the first pass and simply stop the execution there, the rest of the passes having become superfluous. Conversely, if the input were in reverse order to begin with, this algorithm would pathologically perform more

```
 1 function swap(integer a, integer b):
 2     let t=a
 3     a=b
 4     b=t
 5 end function
 6
 7 function sort(list l)
 8     let n=length(l)
 9     repeat n-1 times:
10         for each element in [0, 1 ... n-2] as i:
11             if l[i] > l[i+1]:
12                 swap(l[i], l[i+1])
13             end if
14         end for
15     end repeat
16 end function
```

Figure 6.7

Pseudocode for a naïve formulation of the bubble sort algorithm. Following the programming convention, zero-based indices are used to show the items in the list: `l[0]` would be the first item of the list, and `l[n-1]` the last item in a list of length `n`.

swap operations than there were values to sort. Hence, knowing the nature of the input to a particular problem can inform us about how best to solve that problem.

6.4.2 Resource scarcity: complexity of algorithms

In addition to the subtleties of bubble sort introduced above, there are more important qualities of algorithms that result directly from the reality of modern computation. The most obvious resource that a process requires is time: there will usually be a time-scale in which a task is expected to complete. This is normally informed by some interaction with a human being (who, for example, might not wish to wait ten minutes while the computer sorts a spreadsheet's values), but can also be restricted by the needs of other computer systems. Many biological databases, as discussed in Chapter 3, hold vast quantities of data, and are growing exponentially. Taking, for example, ENA's tally of >230 terabytes of data (Section 3.12), it's clear that developing timely algorithms for dealing with those data is of paramount importance.

The speed of an algorithm can be described by its **time complexity**: a measure of how well it performs, on average, in relation to the size of the data it's meant to work on. This measure is often described using the 'big O notation,' which is mathematical terminology used to describe the behaviour of a function as a particular value (in our naïve bubble sort example, the size of the input) tends to infinity. In general, it can be thought of as a measure of how the algorithm copes with scaled-up input.

Counting the number of comparisons required in a sorting algorithm is often the most useful measure of performance in time; in the algorithm shown in Figure 6.7, $n - 1$ pairs are compared with each other in $n - 1$ passes. Hence, for a list of size n, $(n - 1)(n - 1)$ comparisons are required to sort the list. The largest-order component of the expanded formula, $n^2 - 2n + 1$, is n^2; therefore, as n tends to infinity, this component dominates the behaviour of the algorithm. This gives our bubble sort formulation a time complexity of $O(n^2)$. Even if the naïvety were removed, and the algorithm implemented as efficiently as possible, thereby speeding up the general case, it would still exhibit a time complexity of $O(n^2)$. Table 6.1 shows the ordering of some common big O magnitudes, and how they measure an algorithm's behaviour (here, the time it takes to complete) as the input size changes.

For small input sizes, this tends not to be an issue: modern computers are often fast enough, and have enough RAM, to be able to mitigate the effects of inefficient algorithms. However, with increased input sizes, complexity begins to dominate, and

Table 6.1
Descriptions of how some big O magnitudes behave when the size of input changes. The list begins with the best possible behaviour, $O(1)$, and gets progressively worse.

Complexity	Description
$O(1)$	The measure remains *constant* regardless of the size of input.
$O(\log n)$	The measure behaves *logarithmically* in relation to the input size.
$O(n)$	The measure is *linearly* proportional to the input size.
$O(n \log n)$	*Linearithmic* behaviours fall between linear and polynomial.
$O(n^2)$	Doubling the input size quadruples the measure (called *polynomial*).
$2^{O(n)}$	The measure increases *exponentially* compared with input size.
$O(n!)$	The measure increases *factorially* compared with input size.

efficiency becomes more important. A time complexity of $O(n^2)$ means that each time the size of input doubles, the time taken to process it effectively quadruples: this increase could cripple a program if it were expected to process very large amounts of data.

The other important resource an algorithm must use efficiently is space: during execution, an algorithm must have free access to a certain amount of physical memory. The amount of space required again depends on how much data must be manipulated, and how that manipulation takes place. How the required amount of space changes according to the size of the input is the **space complexity** of an algorithm, and is also described using the big O notation, albeit as a measure of memory usage per item rather than of time taken. Even our naïve bubble sort actually does very well when measured this way: aside from the space required to store the original list (which doesn't feature in our calculation of memory usage), only one more value in memory is required – the temporary value used during the swap operation. If each item takes up one arbitrary memory unit, then this algorithm uses only one extra memory unit to perform its sorting function on a list of size n. This constant term gives bubble sort a best-in-show space complexity of $O(1)$: it will always use the same amount of space, regardless of input size.

From this, it's easy to see that an algorithm may be efficient in terms of one resource, but terribly inefficient in terms of another. Indeed, choosing (or designing) the right algorithm for a given task will inevitably involve a compromise between different competing resources, a compromise that must be seen in the context of the specific problem at hand and of its likely inputs.

6.4.3 Choices, choices

The example we've been using in this chapter is of a very simple problem (sorting a list of numbers) and a very simply formulated solution (the bubble sort algorithm). If this were the only available solution, our jobs would be much more straightforward. However, there are many sorting algorithms, each with their own qualities, quirks and complexities in space and time: some work in place, others rely on the fact that their input arrives item by item; some work better with already partially ordered lists, others rely on specific data. Some of the most often used sorting algorithms and their related complexities are shown in Table 6.2, which illustrates the worst-, average- and best-case time complexities, and the space complexity, all in big O notation.

Table 6.2
Complexities of some common sort algorithms.

Algorithm	Time complexity			Space complexity
	Worst-case	Average-case	Best-case	
Bubble sort[i]	$O(n^2)$	$O(n^2)$	$O(n)$	$O(1)$
Quicksort	$O(n^2)$	$O(n \log n)$	$O(n \log n)$	$O(\log n)$
Insertion sort	$O(n^2)$	$O(n^2)$	$O(n)$	$O(1)$
Merge sort	$O(n \log n)$	$O(n \log n)$	$O(n \log n)$	$O(n)$
Heapsort[ii]	$O(n \log n)$	$O(n \log n)$	$O(n \log n)$	$O(1)$

[i]This bubble sort is not the naïve formulation described in Figure 6.7, but rather, one with all the described shortcomings rectified.
[ii]Although, superficially, the heapsort algorithm would seem to be the best of all worlds, it requires a very specific data structure (a heap) on which to work that limits its applicability (see Section 6.5).

We see, then, that when choosing or designing an appropriate algorithm to solve a particular problem, we must take into account a range of criteria that may not be apparent on first inspection. Commonly, these include:

- *Size of input.* This directly affects the speed of the algorithm: the more data there are to be processed, the longer the processing takes.
- *Scalability of input.* Beyond speed is the question of scalability: if the input will always be on a small scale, time and space complexities can be practically ignored; if, however, the algorithm could receive a wide range of input sizes, then complexity becomes an issue. Thus, it's prudent to take into account the possibility of future scales of data, in order not to choose an algorithm that will eventually become a hindrance.
- *Nature of input.* Sometimes, the exact nature of the input (such as whether a list is already sorted) can detrimentally affect the efficiency of an algorithm; it may be necessary to consider whether those cases are likely to occur often.
- *Frequency.* If an algorithm must execute only once a month, its efficiency in time may not be too big an issue. However, if it must be executed every time a user moves the mouse pointer, or every time the computer screen refreshes, or whenever some other highly frequent event occurs, efficiency in time becomes one of the most important restrictions.
- *Maintainability.* Even if an algorithm is efficient in both time and space, if it's also highly complex in terms of its required data structures and required processing, it may still be advisable to choose a less efficient but less complex alternative. Complex code often leads to more bugs, meaning potentially less stability and less maintainability.

Of course, everything we've discussed so far about choosing or designing algorithms has revolved around the deceptively simple task of sorting a sequence of items. In many ways, this is one of the simplest problems faced by a software developer; yet, choosing the right algorithm can be almost an art form.

As the lens of bioinformatics is being brought to bear on increasingly complex and ever larger amounts of data, the issues described in this chapter have become increasingly important. With the quantities of data now being produced by high-throughput technologies, the complexity of algorithms has started to become an issue. We must do everything we can to mitigate these issues if we expect our software to perform well in this new world.

6.5 Data structures

Thus far, we've treated the data on which our sorting algorithm works as an abstract list of items: although some details of the underlying representation were described in Section 6.3, the precise form the list takes in memory has been artfully ignored. Predictably, choosing how a list of items is to be stored is not as simple as we might expect: there are many different forms a list could take, each with pros and cons. These forms are called **data structures**; this section deals with their qualities and how they relate to algorithm design.

6.5.1 Structural consequences

As their name suggests, data structures store information in a structured way, for a specific purpose. Some data structures do nothing more than group together values that pertain to the same conceptual entity (say, an individual's name, address and date of birth) into a self-contained *record*; others represent the relationships between data, whether those data represent simple values or more complex records – such structures are often called *containers*, a general term for what is, in fact, a large and varied category of data structures.

Table 6.3 shows a number of the most commonly used containers, categorised according to how their contained items relate to one another. Each provides a different way of relating a group of items, promising different things and behaving differently when used. For example, some data structures are explicitly unordered (such as a *set*), and hence provide no guarantees about the order of the items they hold; others exhibit stable ordering (such as a *linked list*), allowing us to be sure that, for example, removing an item from the container will not change the ordering of the remaining items. Matching these qualities to the problem at hand is then relatively simple. Comparing behaviour, on the other hand, is a much more involved task; nevertheless, a container's behaviour is essentially tied to the methods used to manipulate it, whether:

- Insertion – inserting new items;
- Access – finding existing items;
- Deletion – deleting items;
- Mutation – modifying existing items;
- Restructuring – changing the relationship between existing items.

To show how these categories of manipulation allow us to evaluate behaviour, we'll now explore the inner workings of three common container types: the *vector*, the *linked list* and the *binary tree*.

Anatomy of the vector

A *vector* is a one-dimensional container that stores its items consecutively in memory, allowing the whole list to reside in a single contiguous memory block. In practice, a vector usually pre-allocates more memory than it needs, in case the number of items needs to change throughout the lifetime of the program. Figure 6.8 shows how a simple vector of integers could appear in memory. In terms of our five measures of behaviour,

Table 6.3

A list of commonly used containers, categorised according to the kinds of relationship found between the contained items.

Category	Examples
One-dimensional	*linked list, vector, set*
Associative	*ordered map, multi-map, hash-map*
Multi-dimensional	*matrix, table*
Hierarchical	*binary tree, heap*
Graphical	*graph, digraph*

capacity size · data · unused space

Figure 6.8

How a *vector* might look internally in terms of its memory usage. Most data structures hold some housekeeping information, with the vector usually needing to know how much contiguous memory has been reserved (its capacity), and how much is currently being used (its size).

the vector does quite badly in all but one of the cases. Accessing individual items according to their position – or **random access** – is extremely fast: each item takes up an equal amount of memory (let's call that s). Therefore, given the memory address of the first item (which we'll call p_0), we can calculate the memory address of the ith item, p_i, by the simple formula below. Here, the index i is zero-based, as is usual in computer science, such that $i = 0$ for the first item, $i = 1$ for the second item, and so on:

$$p_i = p_0 + (i \times s)$$

This makes accessing individual items by their positions exhibit a constant time complexity of $O(1)$: it doesn't matter how many items the vector holds, it takes the same amount of time to find the ith item, regardless of the value of i. Of course, this only helps if we already know the position of each item we wish to find. If we wanted to search for an arbitrary item, given arbitrary criteria (not knowing its position), we would have to inspect each item in turn until we found one that fulfilled those criteria, giving instead a worse time complexity of $O(n)$.

Appending an item to a vector that has enough memory reserved is also a quick operation: knowing the previous size of the vector, we can calculate the position of the next free memory location in constant time. Likewise, removing the item at the end of the vector can amount to as little as reducing its size by one, and cleaning up the now unused space; again, this can be done in constant time. Inserting or deleting items anywhere else along the vector, however, is a relatively time-consuming operation, as existing items need to be shuffled about in memory to make room for the new item, or to reclaim room from the old item. This makes vectors unsuitable for storing frequently changing sequences of complex items, unless new items are only ever appended, and old items are only ever removed from the end (this exactly describes the behaviour of a *stack*, a first-in, last-out (FILO) data structure designed with this scenario in mind). Finally, mutating individual items is straightforward, and requires no reorganisation of the vector.

One optimisation is possible: if we know beforehand how many items a vector will need to hold during its lifetime, the right amount of memory can be reserved up-front. This is important because the allocation of new memory is a relatively expensive operation, so reducing the number of memory allocations is a worthwhile task. If a vector needs to grow larger than the space reserved for it, it must then allocate a new, larger contiguous block of memory; copy the items from the old block into the new block; de-allocate the old memory block, and return it to the operating system. Although most vector implementations choose sensible sizes for those re-allocations, if the maximum size required is known beforehand, no such expensive re-allocation and copying is needed.

Anatomy of the linked list

One solution to the expense of reorganising a *vector* is the *linked list*. By contrast with the single contiguous block of memory used by a vector to store its items, a linked list stores each item in a separately allocated block of memory called a *node*, each node storing the item in question and a pointer to the next node in the list. Then, given a pointer to the initial node in the list, we can traverse the links between nodes, investigating each item in turn, the end of the list being signified by a node with a null 'next' pointer. Reorganising a list then becomes a task not of reorganising individual items, but rather, of the links between them. Figure 6.9 shows how such a list might appear in memory, and how the removal of an item can be carried out by a sequence of pointer manipulations. Given that these pointers tend to be relatively small compared with the items a linked list might hold, the benefits are clear in terms of not having to shuffle around potentially large amounts of memory.

One disadvantage of their linked nature is that random access is not directly supported: to find the ith node in a linked list, we must begin at the initial node and traverse the links until the required node is reached. For large lists, this can become an expensive operation, as it has a time complexity of $O(n)$. Having found a particular node, however, manipulation is straightforward, with insertion, deletion and mutation performing in constant time $O(1)$. Most linked list implementations store not only a pointer to the initial node, as mentioned above, but also a pointer to the final node. This makes appending items a trivial task (we needn't traverse the entire list to find the final node); coupled with the ease of removing the first item, this provides precisely the right qualities needed for a data structure known as a *queue*: a first-in, first-out (FIFO) data structure. This leads us to another disadvantage of linked lists: each insertion and deletion causes, respectively, a memory allocation and a de-allocation. If the contents of the linked list are expected to change often, with insertions and deletions occurring frequently, this overhead might become a problem.

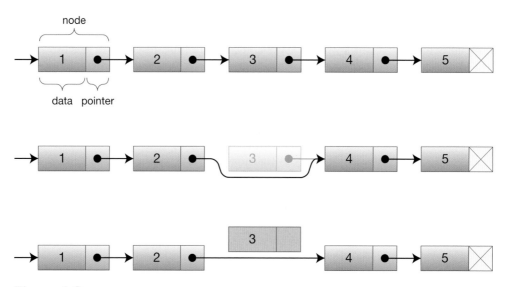

Figure 6.9

How a *linked list* might appear in memory, and how a node can be deleted with two simple pointer manipulations. Note that the nodes need not be in order in memory; in fact, they don't even have to be close by, as they are referenced directly with pointers.

One common adaptation to this data structure is to include not only a 'next' pointer in a node, but also a 'previous' pointer, so that the list can be traversed in both directions. The sort of fast bi-directional traversal provided by such *doubly linked lists* can be very useful when implementing some algorithms.

Anatomy of the binary tree

Aside from being structured quite differently in memory, the *binary tree* data structure differs from the previous examples in that it mandates an explicit ordering of its items. The ordering is explicit in two ways:

1   Items in a binary tree must be comparable, such that, given two items, A and B, we can deduce the truth of the expression $A < B$. With both the vector and linked list examples, this doesn't have to be true; the ordering of those items can be entirely arbitrary;
2   When manipulating both the items themselves and the data structure as a whole, it's up to us to maintain a correct ordering: we can't simply place an item anywhere we wish.

This may seem like a lot of work, but the result is a highly efficient and often-used data structure. Figure 6.10 shows how a binary tree is built from a hierarchy of linked nodes that, unlike the linear linked list, structures its items in a series of parent-child relationships. In a binary tree, any given node, N, will have two pointers: a *left pointer* will point to its left-hand child node, *L* and a *right pointer* will point to its right-hand child node, *R*. Most implementations also maintain a third pointer that points to its parent node, *P*. In terms of the items embedded in those nodes, according to the comparison relation introduced above (<), we find the following relationship holds true for every node in the tree:

$$L < N < R < P$$

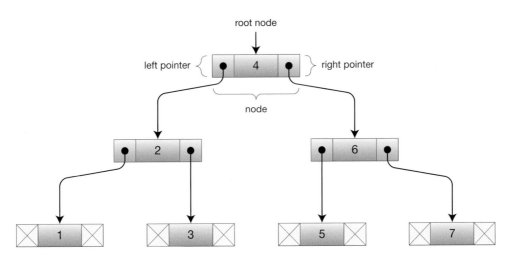

Figure 6.10

How a *binary tree* is represented in a computer's memory. Empty (or *null*) pointers signify when we reach the bottom of the tree. For the sake of simplicity, the nodes' *parent* pointers have been omitted in this diagram.

When extrapolated across the entire hierarchy, this implies that every item found within the left sub-tree of a given node, N (*i.e.*, all nodes L and below), is 'less-than' every item found within the right sub-tree (*i.e.*, all nodes R and below). This may seem overly prescriptive, but the resultant data structure exhibits one quality that makes all the work worthwhile: searching the tree for an item's existence can be performed extremely quickly.

Searching within a binary tree   Imagine a nicely balanced *binary tree* of records that has been ordered according to some unique identifier, exactly as seen in Figure 6.10. Starting at the *root node,* we can search for a given identifier by following some simple rules:

1 Let the root node be the current node under scrutiny.
2 Check whether that node's item is the item being sought. If it is, we've found the item in the tree. The search can stop.
3 Otherwise, compare that item with the item being sought.
4 If it's less than the item being sought, the item must be in the right-hand sub-tree, in which case, let the right-hand child node become the current node under scrutiny. Go to step #2 and continue the search.
5 Otherwise, the item must be in the left-hand sub-tree, in which case, let the left-hand child node become the current node under scrutiny. Go to step #2 and continue the search.
6 If there are no appropriate child nodes, the search has failed: the sought-after item does not appear in the tree.

With each subsequent node comparison, after which we choose to look at the left or right sub-trees, we're effectively halving the search space; it's this ability to 'divide and conquer' that makes binary trees so appealing for systems that need fast indices into large data-sets. For relatively stable balanced trees, doubling the size of the tree adds only a single extra comparison to perform during the search, and this gives the binary tree a logarithmic time complexity of $O(\log n)$. In practical terms, this means that searching a tree with a million items requires, at worst, twenty comparisons. Compare this with the same task but using a *linked list* of a million items: this would require, at worst, a million comparisons. For large data-sets, the difference between time complexities $O(n)$ and $O(\log n)$ can be very great indeed.

Let's take, for example, the roughly 707 million annotated sequences held in the ENA nucleotide archive in January 2016. If these entries were stored as a linked list, then searching through this list to find a particular entry would require us to visit each and every node to see if it was the one we wanted. In the worst-case scenario (the entry we want is at the end of the list, or is missing from the list entirely) we would have to perform 707 million comparison operations before we were satisfied. However, were the entries to be stored in a balanced binary tree, then each comparison would halve the search space; our worst-case scenario would drop to only 30 comparison operations (because a binary tree with a depth of 30 levels can hold ~1 billion items). This would make accessing ENA over *23 million times faster.* Quite an improvement!

Inserting an item into a binary tree   In this case, using the same approach that we'd use to find an item in the tree, we find the item that's immediately 'greater than' the item we wish to insert. It's then simply a matter of creating a new node for the inserted item, and linking into the tree appropriately, as can be seen in Figure 6.11.

Step 1

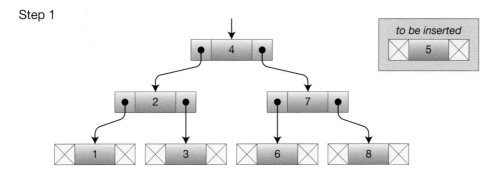

Step 2

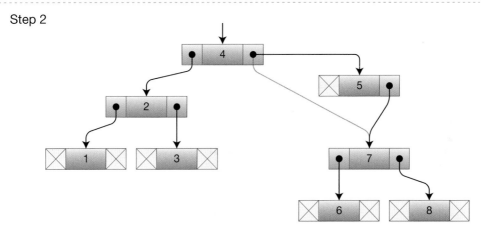

Step 3

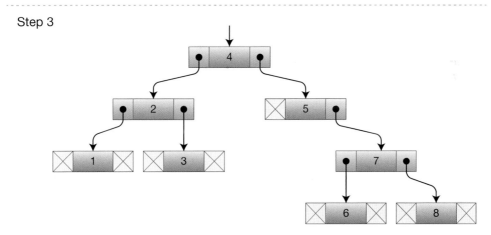

Figure 6.11

Graphical representation of insertion into in a *binary tree*. Insertions such as these tend to leave a tree unbalanced.

It's worth noting that, depending on the shape of the existing tree, the inserted item may either fill a gap already present in the structure, or supplant an existing item. Because of the initial search, insertion into a balanced binary tree exhibits a time complexity of $O(\log n)$.

Step 1

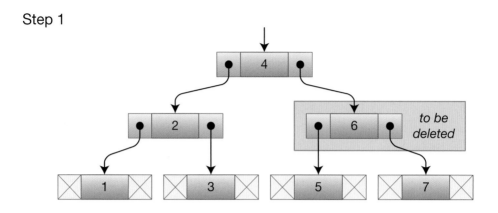

Step 2

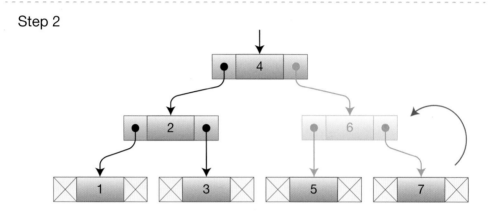

Step 3

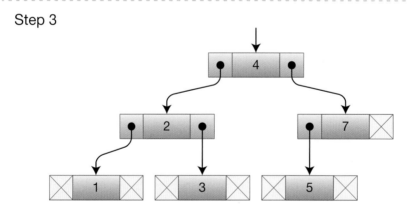

Figure 6.12

Graphical representation of deletion from a *binary tree*. Rotating sub-trees is a common occurrence when deleting items from a binary tree, and is usually more complex than this example: e.g., if the 7th node already had two children, we would then have to rotate *its* sub-tree in order not to lose nodes, and this rotation may have to propagate all the way down to the bottom of the tree.

Deleting an item from a binary tree   Step one of deleting a node is finding it, and we already know how to do that. Once we've found and removed it, we must re-link the nodes around its position in order to maintain the tree's properties. But that presents a particular problem if the deleted node has two child nodes: the deleted node's parent can only point to one node, but we now have two dangling nodes to re-link. The solution is to *rotate* the sub-trees such that they can be re-linked properly, an example of which can be seen in Figure 6.12. This is not a trivial operation, but is necessary to retain the ordering prescribed by the binary tree. The initial search and the possible rotations cause deletion from a balanced binary tree to exhibit a time complexity of $O(\log n)$.

As can be seen, maintaining the strict ordering prescribed by a binary tree requires a great deal of vigilance, and makes manipulation of the tree a relatively involved affair. The effort involved, compared to the previous simpler data structures, may seem excessive, but taking into account the performance benefits, the end really does justify the means.

The tree shown in Figure 6.10 is termed 'balanced' because there is (roughly) the same number of items on one side of the tree as the other; similarly, each sub-tree is balanced. A disadvantage of binary trees is that arbitrary insertions and deletions can break this symmetry, affecting performance along with it. If three items 'greater than' (to the right of) the root item are removed, and three items are inserted that are 'less than' (to the left of) the root item, the tree would quickly become unbalanced. This reduces the performance of searches by increasing the longest path from root to leaf; unless dealt with, this can degrade the tree's time complexity considerably. In the worst case, the tree essentially becomes a *linked list*, thereby showing a linear time complexity. It's important, then, to make sure that the tree remains well balanced, either by ensuring that the items themselves don't change, or by periodically reorganising its nodes to rebalance the tree. One approach is to use a *self-balancing* binary tree, such as a *red–black tree*, that assigns its nodes one of two values, depending on the oddness or evenness of its depth in the tree (often shown in diagrams by colouring the nodes either red or black), and uses this information to rebalance itself after every operation.

Different data structures perform differently

Take, for example, the task of finding the next item in a vector compared to the same task performed on a singly linked list. Given any item in a vector, finding the subsequent item is a simple matter of accessing the next memory location; this operation performs in constant time $O(1)$. Similarly, given a particular node in a linked list, finding the next node is a matter of following the 'next' pointer; again, this performs in constant time $O(1)$. Not particularly different so far.

Now take the converse task of finding the preceding item. For a vector, this is done in a similar way to finding the next item: we need only access the previous memory location, as we know each consecutive item is consecutive in memory. This also performs in constant time $O(1)$. However, with a singly linked list that has no 'previous' pointer, there is no direct way to find the previous node in the list. The only way of doing so is to start again at the beginning of the list, and traverse every node until we find the one whose 'next' pointer points to the current node. This performs linearly in relation to the size of the list, and so has a time complexity of $O(n)$. Table 6.4 shows a number of standard operations and their time complexities when carried out on vectors and linked lists. Given that algorithms are built from just such operations, we can see how important the choice of data structure is to algorithm performance.

Given the complexity of most things, we humans tend to think abstractly. This means that, when designing program code, we usually set aside the complex detail, and think more abstractly about the data types we use. For example, we may think of our data structure as a conceptual list, fulfilling only a handful of requirements: the items stay in the order they're inserted; we can ask for the ith item; and we can append new values. Perhaps these are the only operations that will ever be performed. However, at some point during implementation, we'll need to choose a concrete data structure to fulfil the role set out in the design.

Table 6.4

Comparison of time complexities of
various standard operations on *vectors*
and *linked lists.*

Operation	Time complexity of...	
	vector	linked list
Random access	O(1)	O(n)
Find next	O(1)	O(1)
Find previous	O(1)	O(n)
Prepend	O(n)	O(1)
Append	O(1)	O(1)
Insert	O(n)	O(n)
Delete from front	O(n)	O(1)
Delete from back	O(1)	O(n)
Delete	O(n)	O(n)

Knowledge of the underlying details then becomes important, and it's prudent to ask questions such as, is the `list` type of the Python scripting language implemented as a vector or a linked list? Without our choices being fully informed, we're unlikely to be able to implement algorithms efficiently: choosing the wrong data structure can badly affect the time complexity of an algorithm, making what might have performed logarithmically with O(log n) perform linearly as O(n) or worse.

6.5.2 Marrying form and function

As we've seen, accessing or manipulating a vector is a relatively simple task; however, when it comes to something a little more complicated, such as a binary tree, just interacting with the data structure at all requires quite specific and complicated algorithms. Conversely, we could apply a very general algorithm (such as bubble sort, introduced in Section 6.4.1) to many different data structures, but in doing so, they would probably not perform equally. If we've learned one thing so far, it's that the distinction between an algorithm and the data structure on which it works is not as clear as we might have thought.

Until now, the most complicated data structure we've discussed has been the binary tree. It's not only complicated in its structure, but also in the methods required to access and manipulate it. However, as we've just seen, if used correctly, the binary tree provides benefits that justify the effort involved, benefits that simpler data structures are unable to provide. This data structure was designed with one particular algorithm in mind. The two are different sides of the same coin: the data structure is the algorithm's footprint in memory.

There are many recurring patterns in the problems for which software is written; knowing common patterns can make software much easier to write. However, complex problems often require highly specific solutions that don't precisely match those particular patterns. In fact, if we consider only data structures that are tree-like in nature (similar to the binary tree), the landscape of possibilities is surprising; manipulating any one of these data structures requires the use of complicated algorithms designed to provide quite specific performance, often under very specific conditions:

- *2–3 tree* (algorithmically identical to the *AA tree*),
- *2–3–4 tree* (algorithmically identical to the *red–black tree*),

- *AA tree,*
- *AVL tree,*
- *B tree* (variations: *B+ tree, B*-tree*),
- *Binary tree,*
- *Fusion tree,*
- *Heap* (variations: *2–3 heap, beap, binary heap, binomial heap . . .*),
- *K–d tree,*
- *Octree,*
- *Quadtree,*
- *R-tree,*
- *Red–black tree* (variation: *LLRB tree*),
- *Scapegoat tree,*
- *Splay tree,*
- *T-tree,*
- *Top tree,*
- *Treap,*
- *Trie tree* (variation: *radix tree*).

If just one province of the landscape of useful data structures is this rich, we can see how many tried-and-tested data structures there are to choose from, and how frequently problems occur that require new and quite unique solutions to be developed. If we wish to be the designers of these new solutions, using the most efficient means available, we need to appreciate the synergy that exists between the algorithm and the data structure on which it works.

Too clever for our own good

With the help of complex data structures, and making sensible assumptions about the input, it is possible to design complex algorithms that can perform very well in both time and space. However, it could be argued that such solutions might not always be the best.

How an algorithm performs in a computer system is clearly important; however, unless it was implemented perfectly, and unless it will never be changed, somebody, at some time, will probably have to go back and modify how it works. A rule of thumb here is that the more complex an algorithm, the more fragile its implementation: fragile to unseen bugs and fragile to future changes. Sometimes, even with rigorous commenting, the only way to follow what a complex algorithm is doing is to watch the flow of control, and keep track of multiple values in memory. Sometimes, the only course of action is to throw away the algorithm and start again from scratch. In such cases, it might be sensible (depending on how important its requirements are) to choose a less complex, less efficient algorithm, if it means that it's more amenable to maintenance.

6.6 Implementing algorithms

In committing an algorithm to digital form, a new vista of interesting problems becomes apparent that can have a drastic impact on how well that algorithm performs. Some of these problems are relatively high-level, and can often be informed by the type of problem, and solution, we have before us. Choosing an appropriate programming language – if we are at liberty to do so – can lead to significant benefits in terms of efficiency and ease of development; however, in most cases, our choice is limited or even prescribed by factors beyond the algorithm.

So how can we make such decisions? Programming techniques can be categorised at every level of abstraction, on many axes, and can be either descriptive or prescriptive; such categories can be helpful when choosing the right tools for a job. These range from the high-level choice of programming *paradigm*, to what we'll call here low-level *quirks* of the language.

6.6.1 Programming paradigm

At their highest level of abstraction, programming languages can be categorised according to how multiple instructions are put together to form complex programs. **Imperative** programming languages execute commands sequentially, in a linear manner. Imperative languages are usually also **procedural** in nature, as they make use of procedures (named blocks of sequential instructions) that may be executed multiple times by being *called* from the main control flow. This parcelling up of reusable code can make writing software very efficient.

Object-oriented languages allow related groups of data to be bundled together with the logic needed to process them. *Objects* tend to manage their own state, interaction being mediated by bound procedures called *methods*, which ensure an object's internal consistency. Most imperative languages provide object-orientated facilities, although a few purely object-oriented languages do exist. This allows users to formalise and restrict the interaction between objects, preventing undesirable behaviour and allowing for better verification of a program's correctness. Object-oriented languages can therefore be said to rely on contracts between 'stateful' objects.

Functional languages organise their code around the concept of a mathematical function (similar to the procedures above), building more complex code by composing functions together and using recursive approaches to deal with sequential data. Unlike the paradigms above, functional languages are highly restrictive in terms of what can be expected when calling functions; specifically, a function that's given a particular input is always expected to return the same output, fundamentally disallowing internal state. Although this may seem a huge disadvantage, there are functional techniques that circumvent it. By contrast with the object-oriented paradigm, functional approaches can be said to centre on contracts between functions and their callers.

The final of the four language paradigms is that of **logic** programming, where code is organised around logical sentences. Logic programming is primarily used for logic-heavy tasks – such as those of artificial intelligence and natural language processing – that require very complex decisions to be made based on a wide variety of data.

These four paradigms require developers to think in fundamentally different ways when designing algorithms, and can therefore have a huge impact on the resulting formulation.

Many languages provide for more than one of these programming paradigms (see Table 6.5), either through direct support or through extensions, so it's often possible to mix different approaches in a single solution.

Paradigmatic consequences

It's important to note that, in general, the expressivities of these paradigms are equivalent, so a particular choice will not affect computability. If a problem can be solved using one of the above programming paradigms, it can be solved by any of them,

Table 6.5

How common programming languages relate to the four paradigms of algorithm design: imperative, functional, logic and object-oriented. For a given language, a paradigm may be ✓ exemplified (the language was designed around it), (✓) supported (the language provides mechanisms for its use), or effectively unsupported if left blank.

Language	Imperative	Functional	Logic	Object-oriented
C	✓			
C++	✓			✓
Haskell		✓		
Java	✓			✓
JavaScript	✓	(✓)		(✓)
Lisp		✓		
Oz			✓	
Perl	✓			(✓)
Prolog			✓	
Python	✓	(✓)		✓
Ruby	✓	(✓)		✓

although the same doesn't necessarily hold for the choice of programming language (see Section 6.6.2). Having said that, paradigmatic choice is not without consequences.

Which programming paradigm we choose – if a choice exists – depends on the nature of the problem, from which it's often evident which paradigm is the best fit: if the problem concerns interdependent stateful agents, then object-orientation is likely to be the appropriate choice; if it involves logically proving or disproving a set of mathematical rules, logic programming is likely to fit better.

By 'better', we mean a number of immediate gains: choosing the correct paradigm can drastically speed up development time, make code smaller, and allow it to run more efficiently. However, there are other good reasons for choosing one paradigm over another; for example, some methods of optimisation are more easily implemented in some paradigms than in others (see Section 6.6.4 for an example of this).

Paradigmatic reality

For better or worse, software-engineering courses tend to concentrate on imperative and object-oriented software development, and the most common languages in use follow those paradigms. Consequently, in more cases than might be ideal, imperative languages have been used to solve functional or logical problems – this can clearly be seen by the rise in availability of such things as anonymous functions (the building-blocks of the functional paradigm) in imperative languages. On the one hand, for many uses, the benefit of being able to code in multiple paradigms within the same code gives software developers greater flexibility in how they can express their solutions; on the other hand, although mixed-paradigm languages are becoming more abundant, they aren't necessarily the most efficient way of implementing those paradigms: there is potential for them to be 'Jack of all paradigms, master of none'. These issues, and the need to support or build on legacy software, will inform (or restrict) the developer's choice.

6.6.2 Choice of language

Not all programming languages are created equal. Languages, and the standard tools they make available, can differ considerably in both their expressivity and convenience.

Expressivity of programming language

In a formal mathematical sense, most programming languages have equivalent *expressivity*: *i.e.*, they're capable of solving the same set of problems, albeit potentially in different ways. There are, however, programming languages that are *less* expressive than their contemporaries: *i.e.*, there are problems that such languages can't solve, that otherwise could be solved had a different language choice been made.

At first glance, this might look like a disadvantage that would make such languages undesirable. After all, why would we want to use a language that's *less* able to solve problems? The answer lies in the Halting Problem introduced in Section 6.2.1, which stated that it's impossible to know in advance whether an algorithm will complete in a finite time, or will continue running forever. In practice, software engineers strive to avoid such situations, so the fact that they're theoretically possible seldom causes problems. However, the ability to prove mathematically that an algorithm will complete in a finite time is sometimes very important, especially in highly critical systems that require extraordinarily thorough verification (such as online financial systems like **Bitcoin**). In cases like these, standard programming languages are, at least mathematically, too unpredictable for the task. Although such cases are rare in practice, knowing they exist can inform the creation of more predictable code.

Expressivity of standard tools

By contrast with the abstract computability concerns above, there are more concrete expressivity concerns that are likely to cause problems in the field of bioinformatics. The most common comes from the use of **regular expressions** for parsing textual data. Regular expressions (see Box 6.2) allow text patterns to be matched and extracted – they're often used when parsing data from some serialised form (either a data file, or network traffic). They do, however, have limited applicability, so it's important to understand when regular expressions are *not* applicable.

Regular expressions are capable of parsing any *regular language*, a very specific term in language theory that describes mathematically the structures allowed within the language ('language' here covers not just programming languages, but any informational language). The most common example of a structure that is *not* regular is recursion: regular expressions are mathematically incapable of parsing recursive data structures. This is important, because there are many formats in use today that aren't regular languages in the formal sense, and hence regular expressions are incapable of parsing them. **XML, HTML** and **JSON**, perhaps the three most pervasive data formats in use today, are examples of *context-free languages* that can't be parsed with regular expressions. This is worth repeating solely because of the frequency with which it is attempted: *regular expressions alone cannot parse XML, HTML or JSON*. Having said that, most regular-expression libraries provide extensions that, while not sufficiently powerful to invalidate the above statement, do provide extra expressivity, and therefore allow wider application.

Most programming languages have regular-expression functionality either built in or available as a standard tool; in most cases, the particular combination of regular-expression extensions available is specific to that specific language and version. Whether those extensions provide better support for non-ASCII text, or are non-regular extensions to the language itself, knowing what's available helps us to choose an applicable language to work with. This isn't just a case of choosing a language that provides the tools we want; it is also a case of recognising that if, in future, the solution must be restated in a different language that has less expressive regular-expression support, it may be impossible to do a straight translation without substantial extra work.

For example, Ruby's regular-expression engine actually does allow recursive expressions (making them no longer regular), and so can be used to parse recursive data structures with very little code. This is perfect if we're confident that Ruby will remain in use for the foreseeable future; but if we're using Ruby solely for rapid prototyping, the

---

Box 6.2 Regular expressions

It's common in bioinformatics to search protein or nucleotide sequences for particular strings, or patterns, of amino acids or nucleotides (perhaps because those residues or bases form specific functional sites). Sometimes, we may be looking for an exact match; at other times, we may want to allow a certain degree of 'fuzziness' that captures the ambiguities and substitutions that occur in the real world of biology. Consensus sequences allow this fuzziness to be captured in a search string. For example, the consensus sequence VXRSPFZ would match the region highlighted in red in the following fragment:

MNGTEGPNFYVPFSNATG**VVRSPFE**YPQ

because the characters V, R, S, P and F in the consensus sequence explicitly match their counterparts in the fragment: X matches any residue (in this case, a V for valine), and Z matches glutamic acid or glutamine (here, the E for glutamic acid). A more complete description of the syntax commonly used to search for patterns of amino acids is given in the PROSITE user manual[a]. As well as allowing one-to-one substitutions based on IUPAC amino acid and ambiguity codes, consensus expressions can also be used to match more complex patterns. It's possible, for example, to express negation (where, for an amino acid sequence, {AM} stands for 'a single instance of any residue except alanine or methionine'), or repeats (such as A(3), which means '3 alanines in a row').

Whereas consensus expressions encode a certain amount of biological meaning in their ambiguity codes (e.g., that Z can be one of two specific amino acids), in computer science, regular expressions provide a more general-purpose framework for matching text. These are used widely throughout all aspects of bioinformatics. Although their syntax and purpose is deceptively similar, regular expressions provide a much richer language for describing general-purpose pattern matching. For example, they allow characters to be optional (by following them with a question mark '?'), repeated zero or more times (by following them with an asterisk '*'), repeated one or more times (by following them with a plus sign '+'), repeated an exact number of times (by following them with a number in braces '{3}'), grouped together with parentheses, etc. Regular expressions also have character classes that match, amongst other things, only numerical digits (\d), whitespace (\s), or word boundaries (\b); some regular-expression engines have more complicated classes, such as dash characters (\p{Pd}) and currency symbols (\p{Sc}). We can even specify an arbitrary set of characters to match such that, say, [ACT] would match either A or C or T.

The following list shows some example regular expressions, together with explanations of what they'd match:

- ENS[A-Z]*[FPTG]\d{11} means 'Match the letters ENS, followed by zero or more letters from A and Z, followed by exactly one of any of the letters F, P, T or G, followed by exactly 11 digits'. For example, this would match Ensembl gene IDs, such as ENSG00000139618 or ENSMUSG00000041147;
- PDOC\d{5} means 'Match the letters PDOC followed by exactly 5 digits', so would match PROSITE document IDs, such as PDOC00449 or PDOC50178;
- [ATGC]+ means 'Match one or more of the letters A, T, G or C', so would match any (non zero-length!) DNA nucleotide sequence consisting of the letters A, T, G or C in any order.

---

[a] http://prosite.expasy.org/scanprosite/scanprosite_doc.html

inevitable task of translating it to a different language will be made much more difficult by having to make use of less expressive tools elsewhere.

Convenience of programming language

For a variety of reasons, writing code in some languages is simply easier than in others; inevitably, however, this advantage is accompanied by disadvantages that make the choice of language less straightforward. Some examples of the competing qualities of different languages are outlined below:

- Python context blocks are denoted solely by indentation, and not by the usual braces { } found in most other languages. This reduces the size of code and can make development quicker, but introduces confusion if a combination of spaces and tabs is used for indentation.
- Perl has extended regular expressions built directly into the language, making text processing particularly efficient. However, the mechanism relies heavily on global variables and hidden side-effects, making maintenance difficult without a relatively high degree of expertise.
- Ruby has a pragmatic approach to declaring data structures that allows existing structures to be extended easily at the expense of a loss of control from the original developer.
- By contrast with the languages mentioned so far, C and C++ give very low-level, fine-tuned access to memory, allowing developers to produce exceptionally efficient code. However, as many aspects of memory management are in the developers' hands, the scope for mistakes and bugs is drastically increased.
- Java hides many of the complexities that make C and C++ difficult to learn, making programming simpler and more robust in the process. Memory management is entirely dealt with by Java, reducing the chance of mistakes, but at the same time limiting us in cases where a greater degree of control is required.

Alongside these specific comparisons, it's also worth cautioning about the ability to write compact and quick code in various languages. For example, in C, the following two excerpts of code are equivalent:

```
1  int a;
2  if (x > 3) {
3      a = 3;
4  } else if (x <=0) {
5      a = 0;
6  } else {
7      a = x;
8  }
```

```
1  int a = x >=3 ? 3 : x <=0 ? 0 : x;
```

Clearly, the second is much shorter, but it loses the clarity of the more verbose first excerpt. It's up to us to weigh the competing advantages and disadvantages of different languages when trying to choose which best serves any given problem.

6.6.3 Mechanical optimisation

It's sometimes tempting to assume that by the time the quantities of data we need to manipulate become an issue, computers will be so much larger (in terms of memory) and faster than they are today that they'll be able to cope. For many reasons, this is a false assumption, and it's worth mentioning why.

In an article published in 1965, the co-founder of Intel, Gordon E. Moore, famously formulated what has become known as Moore's Law:

> The complexity for minimum component costs has increased at a rate of roughly a factor of two per year... Certainly over the short term this rate can be expected to continue, if not to increase. Over the longer term, the rate of increase is a bit more uncertain, although there is no reason to believe it will not remain nearly constant for at least 10 years.

Moore's prediction, subsequently revised to suggest that the number of components on an integrated circuit chip would double every 18 months, has remained remarkably accurate, and also appears applicable to other measures of technology (speed, memory size, *etc.*). So, how does this affect our previous discussion of algorithm complexity? Well, in a purely mathematical way, it doesn't affect it at all. Complexity, as defined above, is a measure of how an algorithm behaves in relation to increased input size. That behaviour is independent of the speed of the computer on which it's executed: an algorithm with time complexity of $O(n^2)$ will always have time complexity of $O(n^2)$, regardless of how much faster computers may get. Yes, a faster computer will probably complete sooner, but algorithms will still degrade in the same way (logarithmic, polynomial, exponential, *etc.*) in the face of increased input size.

In absolute terms, Moore's Law suggests that, as long as our input sizes remain constant, our ability to process them will increase exponentially. Unfortunately, as discussed in Chapter 3, the amounts of data we're expected to process are also increasing exponentially, thus seemingly wiping out any benefit we may have gained from technological advances. But at least we've not taken a step back, have we? Unfortunately, maths is never so forgiving: any algorithm with a worse time complexity than $O(n)$ (or space complexity of $O(1)$) will probably degrade over time in the face of current technological advances.

To show this, let's imagine that our data acquisition is increasing at a rate equal to that of our processor speeds: *i.e.*, doubling every 18 months. An algorithm with linear time complexity, $O(n)$, will slow down in proportion to the increase in input data over time, but then be sped up by exactly the same amount owing to an identical increase in processor speed. This would leave the execution time unchanged: if, today, it can process 100GB of data in one minute, then, in 18 months' time, it will be able to process 200GB of data in one minute. However, as can be seen from Table 6.1, there are very few algorithms – even for the 'simple' task of sorting data – that could ever display a linear time complexity. The majority of such algorithms therefore have worse behaviour: $O(n \log n)$ is often the best we can hope for. In this case, these complexities actually break the symmetry between increased data and increased processing power, returning us once again to the slow march towards degradation. Box 6.3 illustrates this point with some real numbers.

Of course, all is not so bleak: there are many ways to mitigate this decline, and other technological improvements we can exploit. Some methods make use of special capabilities of processors; others use increasingly clever compiler optimisations when turning high-level algorithms into the machine code that runs on a processor. Before we look at those approaches, though, there are simpler and often more rewarding ways of increasing software efficiency.

Box 6.3 The inadequacy of Moore's Law

To see how the advantages gained by Moore's Law are lost by any time complexity class slower than $O(n)$, let's look at a specific (contrived) example. Table 6.6 shows what happens when, over time, the relative volume of data increases exponentially (a trend exhibited by most of the main bio-databases) and the relative speed of processors increases, also exponentially (according to Moore's Law).

Table 6.6

Comparison of the behaviour of linear, linearithmic and polynomial algorithms with increasing volume of data (left-hand side) and processor speed (right-hand side).

Relative volume of data	Relative time taken by algorithm (s)			Relative speed of processor (p)	Speed-adjusted relative time taken (s ÷ p)		
	$O(n)$	$O(n \log n)$	$O(n^2)$		$O(n)$	$O(n \log n)$	$O(n^2)$
1	1	1	1	1	1	1	1
2	2	2.3	4	2	1	1.15	2
4	4	5.2	16	4	1	1.30	4
8	8	11.6	64	8	1	1.45	8
16	16	25.6	256	16	1	1.60	16
32	32	56.0	1024	32	1	1.75	32
64	64	121.8	4096	64	1	1.90	64

The left half of the table shows how three classes of algorithm (linear, linearithmic and polynomial, as described in Section 6.4.2) behave when the volume of data changes; it ignores any increase in processor speed. We can see that: the time taken by the linear algorithm increases proportionally with the size of the input data; the linearithmic algorithm slowly falls behind; while the polynomial algorithm very quickly begins to take a long time to complete.

The right half of the table then introduces the fact that the speed of the processor is also increasing, alongside the volume of data. When we take this speed increase into account, we see the speed-adjusted time of each algorithm. The linear algorithm, once adjusted, takes the same amount of time to complete from now until forever (the increase in data and the increase in processor speed cancel each other out). The linearithmic algorithm's time does slowly creep up, though at a manageable pace. The polynomial algorithm, however, has started to behave badly: despite the processor's exponential speed-up, the algorithm's complexity dominates and, over time, its performance will degrade.

Writing code intelligently

The simplest way of writing more efficient code is, unsurprisingly, to understand how our chosen language works, and what we can expect from any external libraries. Three common examples should be enough to demonstrate the potential pitfalls of making incorrect assumptions, and the value of having a proper understanding of modern programming languages.

Post- versus pre-increment   Many languages provide unary operators (having only one operand) for incrementing and decrementing a value (making it one bigger or one smaller). Often, the increment operator is denoted by a double-plus symbol, a++, or by a function call, inc(a), where a represents the value to be incremented. Similarly, the decrement operator might look like a-- or dec(a). In some languages, such as C++, there are two versions of these operators: pre- and post-increment, pre- and post-decrement. The pre-increment operator, ++a, increments the specified value in

```
1 | let a=0
2 |              // a is equal to 0
3 | print a      // prints "0"
4 |              // now pre-increment a
5 | print ++a    // prints "1"
6 |              // a is now equal to "1"
7 | print a      // prints "1"
8 |              // now post-increment a
9 | print a++    // prints "1"
10 |              // a is now equal to "2"
11 | print a      // prints "2"
```

Figure 6.13

An example of pre- and post-increment, and how they differ in their evaluation.

place and returns (evaluates to) that new value. The post-increment operator, a++, first makes a copy of the value as it was before the statement, increments the variable, but returns its previous value via the copy. Figure 6.13 shows the distinction between pre- and post-increment in pseudocode. The important point here is the copying of information that occurs when using the post-increment operator: if we use this operator for the sole purpose of incrementing a value, without the need to use the return result – which is often the case when iterating over sequences – then we needlessly cause memory to be allocated and data to be copied. If the value is a simple integer number, the overhead of this may not be too great; but if that increment resides within a loop, and that loop repeats many times, this overhead can quickly become significant. Add to this the common practice of incrementing not numbers but *iterators* (fully fledged and often complex objects that keep track of a position in a data structure), and this overhead can quickly dominate a program's execution time.

Knowing and appreciating the difference between pre- and post-increment operators can prevent us from blindly using iter++ to increment an iterator, when the pre-increment version ++iter is, in most cases, the more efficient approach.

Lightening loops Loop structures, such as for or while loops, allow us to repeatedly execute some code over different data; by their very nature, they allow us to efficiently specify a large amount of processing in a relatively small amount of code. A corollary of this is that it is easy to underestimate how long a program will spend in a given loop, and what resources it may need during execution. We should therefore always be careful not to do any unnecessary processing within a loop. For example, take the hypothetical code:

```
1 | #Cycle through the list of vectors
2 | let count=vector_list.length()
3 | for index in 1..count:
4 |     # Make a rotation matrix
5 |     let m=Matrix(0, 1, 1, 0)
6 |     # Get the item of interest from the list of vectors
7 |     let v=vector_list[index]
8 |     # Multiply by the matrix and store back in the list
9 |     vector_list[index]=m*v
10 | end for
```

The thing to notice here is that the matrix that gets constructed within the loop (on line 5) is always the same matrix and it never changes during execution. We are therefore needlessly creating a new matrix every time we go around the loop, only to have it destroyed again before the next iteration. That construction and destruction is going to take time, not just because of the code it will undoubtably be executing behind the scenes, but because of the memory that needs to be allocated and deallocated each time. This problem is compounded in the case of nested loops.

Such issues can often be tackled by a simple re-ordering of code: in this case, taking the matrix construction out of the loop so that it only ever gets executed once:

```
 1 # Make a rotation matrix
 2 let m=Matrix(0, 1, 1, 0)
 3
 4 # Cycle through the list of vectors
 5 let count=vector_list.length()
 6 for index in 1..count:
 7   # Get the item of interest from the list of vectors
 8   let v=vector_list[index]
 9   # Multiply by the matrix and store back in the list
10   vector_list[index]=m*v
11 end for
```

Another good candidate for this approach is the declaration of any temporary variables used solely inside the loop. Even though their values might be different from one iteration to the next (such as v above), at least the memory they take up isn't being continually deallocated and allocated again each time around. Better yet, don't use a variable at all if it isn't needed:

```
 1 # Make a rotation matrix
 2 let m=Matrix(0, 1, 1, 0)
 3
 4 # Cycle through the list of vectors
 5 let count=vector_list.length()
 6 for index in 1..count:
 7    # Multiply the current vector by the matrix and
 8    # store it back in the list
 9    vector_list[index]=m*vector_list[index]
10 end for
```

When a function is a macro When writing software, we're likely to make use of functions written by someone else in order to provide the intended functionality. This kind of re-use is perfectly normal, accounts for a good deal of the code of any software system, and is a Good Thing. Sometimes though, not knowing the details of how such functions are written can present problems to those trying to use them (something we touched on in Section 6.3), especially if the function isn't a function at all, but a macro. Macros, short for *macroinstructions* (not to be confused with the identically named macros used in popular word-processing software), specify how certain input code can be mapped to replacement output code. The difference between using a function and using a macro is a question of when the parameters are evaluated. A function's parameters are evaluated before that function is called, while a macro is inserted into the surrounding code and the parameters are replaced before being evaluated. Simple examples can be seen in Figure 6.14 and Figure 6.15, where a function and a macro are defined, both attempting to ascertain the maximum of two values. In each case, we use them to find the maximum of the square roots of two arbitrary numbers (m and n). For the function max_f(a,b) of Figure 6.14, the parameters are evaluated before the function is called, so the two auxiliary calls to sqrt(a) and sqrt(b) are executed only once.

```
1 function max_f(integer a, integer b):
2     if a > b:
3         return a
4     else:
5         return b
6     end if
7 end function
8
9 let m=9
10 let n=25
11
12 let p=max_f(sqrt(m), sqrt(n))
```

The above function invocation is equivalent to the following code:

```
1 if 3 > 5:  // evaluates to false
2     return 3
3 else:
4     return 5 // therefore "5" is returned
5 end if
```

Figure 6.14

A simple function for calculating the larger of two numbers. The function is then called with *square root* expressions as parameters.

For the macro max_m(a,b) of Figure 6.15, however, the parameters aren't evaluated first, but rather, are substituted into the macro definition before it's executed. This leads the macro to execute one more square root than it would otherwise have needed to. In cases like these, it can be more efficient to evaluate the square roots separately, and assign their results to new variables, passing these results into the macro, rather than the unevaluated code. This would ensure that any expensive operations are executed only once.

Although the square-root function on many systems is a relatively quick operation, other more heavyweight functions would present greater overheads; and, as with the previous example of performing an operation within a loop, such overheads can grow startlingly quickly, depending on how such a macro is used. Not many languages provide macros, but knowing which do, and being aware of their existence, can save us a great deal of frustration, and present considerable speed increases.

Operator short-circuiting As a counterpoint to the previous example, where a solution is to pre-compute parameters beforehand, operator short-circuiting can be used

```
1 macro max_m(a, b):
2     if a > b:
3         return a
4     else:
5         return b
6     end if
7 end macro
8
9 let m=9
10 let n=25
11
12 let p=max_m(sqrt(m), sqrt(n))
```

The above macro invocation is equivalent to the following code:

```
1 if sqrt(m) > sqrt(n):
2     return sqrt(m)
3 else:
4     return sqrt(n)
5 end if
```

Figure 6.15

A macro for calculating the larger of two numbers. As in Figure 6.14, this macro is called using some square-rooted values, but unlike in the function example, the parameters are not evaluated, but are substituted directly, leading the *square root* operators to appear twice each in the resulting code.

to good effect to prevent evaluation of some expressions altogether. Operator short-circuiting is a method many languages employ when computing Boolean (*true* or *false*) expressions – it amounts to the premature ending of evaluation when the result is known with certainty. Such expressions can represent anything that might be true or false, and appear in one form or another in any program that has to make decisions: they might be whether a given amino acid is hydrophobic or hydrophilic, or whether the length of a protein sequence is larger than some threshold. For simplicity, in the following Boolean expressions, we replace these concepts with single-letter symbols, each of which (A–H) resolve to be either true or false:

```
A and B and C and D    (1)
E or F or G or H       (2)
```

Most programming languages would process these expressions from left to right, taking each new sub-expression (A–D, E–H) in turn and evaluating its truth. Looking at example (1), we see that first A is evaluated, followed by B, followed by C, followed finally by D. However, if A resolves to false, the values B, C and D are irrelevant, because the overall result is guaranteed to be false. In fact, as soon as we come across a sub-expression that evaluates to false, we become certain of the overall falsity of the expression. We can therefore simply skip the evaluation of those subsequent irrelevant sub-expressions. Example (2) is similar: the first sub-expression to resolve to true allows us to short-circuit the subsequent evaluations, as it can be shown that the overall expression must then evaluate to true. Of course, if the sub-expressions A–H were just Boolean variables, this doesn't gain us much; however, it's often the case that these sub-expressions are relatively heavyweight, such as a function call, or complex arithmetic. In these cases, skipping evaluation provides a powerful tool.

Of course, this behaviour happens regardless of our intent, so how can it be used intelligently to optimise some particular code? Again, we appeal to our knowledge of any given application of logic; specifically, of how heavyweight given operations are, and the likelihood of a given sub-expression being either true or false. Take example (3) below: we have four arbitrary conditions that must all be true in order for the expression to be true. In this example, we contrive two functions, oft() and sel(), that, depending on their input, return either true or false, and so can be used as conditions in logical statements:

```
oft(P,Q) and sel(R,S) and T and U  (3)
```

For this example, we shall say that oft() and sel() are both time-consuming functions of some arbitrary variables (P, Q, R and S); the other two, T and U, are simple Boolean variables. How can we rewrite this expression to make it more efficient? The simplest improvement is to re-order the sub-expressions to ensure that simpler expressions come first; this should lead to fewer evaluations of the subsequent, more heavyweight expressions. This gives us the more efficient expression seen in example (4):

```
T and U and oft(P,Q) and sel(R,S)  (4)
```

With sufficient knowledge of the application of this logic, it might be possible to reorganise this expression to further increase its efficiency. Let's assume that, in the particular domain in which this code works, T is much more likely to be true than is U;

moreover, let's assume that `oft()` is more often true than is `sel()`, which is seldom true. This subtle knowledge can lead to reorganisation of the expression (example (5)), further increasing the chance that it will short-circuit before having to do much heavy-weight processing:

U **and** T **and** sel(R,S) **and** oft(P,Q)   (5)

It's quite possible for such reorganisations to produce code that is orders of magnitude more efficient (in terms of speed of execution) than the initial naïve expression seen in example (3). However, making full use of such techniques requires knowledge of domain-specific tendencies, and would probably entail a compromise between the likelihood of a sub-expression being true and its relative complexity.

In summary, then, having a good knowledge of the details of a programming language, how its expressions are evaluated, and what side-effects various functions may present, can allow us to write more efficient code. Without such knowledge, it's perfectly possible to write software that works as intended. However, in a world of ever-increasing demands – especially in the exponentially growing field of bioinformatics – it's becoming more important that software works not only as intended, but quickly and efficiently too.

Compiler optimisation

Broadly, there are two strategies for executing the kind of high-level instructions seen throughout this chapter: *compilation* and *interpretation*. Interpreted languages – often known as *scripting* languages – are parsed and executed by a Virtual Machine (VM), a layer of software that runs between the executing program and the physical processor, translating between the high- and low-level instructions on-the-fly. Compiled languages need no such layer, as they are *compiled* into machine code (similar to that seen in Figure 6.3), meaning that they can be executed directly on the processor. There are advantages and disadvantages to both strategies: interpreted software, encumbered by its VM, often executes more slowly than compiled software, which needs no such layer; compiled software will only work on a machine that matches the processor architecture and operating system of the compiler, whereas interpreted software can run on any machine that has an appropriate VM. Choosing the right strategy for a given task is often just a case of balancing speed, portability and ease of development.

The operations we saw in Figure 6.3 (`mov`, `mul`, *etc.*) represent individual instructions that can be given to a processor, which make up the machine code into which high-level languages are compiled. The set of instructions provided by a processor is called, unsurprisingly, its *instruction set*, and there are various competing standard instruction sets in use today. Different compilers are written for different instruction sets, but, in each case, make use of the optimisations outlined above. In addition to the core instructions of the main instruction set, most processors also provide extensions that make use of highly specific circuits for performing often-used calculations.

An example extension would be one that provides SIMD functionality: *Single Instruction Multiple Data*. The instructions found in most standard instruction sets are *Single Instruction Single Data* (SISD), meaning that each individual machine-code instruction performs one operation on one item of data (this encompasses integer multiplication, as, even though two numbers are involved, only one calculation is being performed). Conversely, SIMD instructions apply one operation to a number of items of data simultaneously, thereby potentially increasing calculation speed by as many times as there are simultaneous calculations. An example of a SIMD operation is vector addition: the simultaneous pair-wise addition of two identically sized arrays of values. With SISD instructions alone, the kind of addition seen below must be performed as four separate calculations, each row addition being calculated separately and in sequence:

$$
\begin{bmatrix} 1 \\ 2 \\ 4 \\ 8 \end{bmatrix} + \begin{bmatrix} 7 \\ 3 \\ 5 \\ 1 \end{bmatrix} = \begin{bmatrix} 8 \\ 5 \\ 9 \\ 9 \end{bmatrix}
$$

Using SIMD extensions, this can be performed as one operation on the four rows simultaneously, providing a roughly four-fold increase in speed relative to the SISD formulation. Some compilers can automatically take advantage of SIMD extensions, but can only do so if code happens to fulfil certain requirements. In many circumstances, to make use of such vector arithmetic (or any of the other SIMD operations that may be available), we must explicitly write assembly language to define that part of the algorithm. This is not a trivial task, but can improve the efficiency of any code that's amenable to such optimisations.

It can be seen, then, that many strategies can be employed to better implement algorithms, from taking advantage of language quirks (or avoiding their pitfalls), to writing code that's more amenable to modern compiler optimisations, or making use of processor instruction-set extensions.

Taking a step back from these low-level concerns, let's now turn to a less subtle form of optimisation.

6.6.4 Parallelisation

Many algorithms are **sequential** by nature: they're made of a sequence of dependent steps that must be carried out, in the correct order, to solve a particular problem. Some algorithms exhibit **parallel** qualities: *i.e.*, some steps are independent of each other, and can hence be done in any order, or, crucially, *at the same time*. In our world of multi-core and multi-processor machines, server clusters and distributed systems, the ability to execute parts of an algorithm simultaneously (in parallel) allows us to dramatically reduce the time taken to solve some problems. So what kinds of algorithm can be successfully parallelised?

Divide and conquer
The bubble sort algorithm discussed in Section 6.4.1 is a good example of a sequential algorithm: each iteration around the loop assumes that all previous iterations have already completed, as it relies on the previous iteration's result in order to work properly. If the steps were done in any other order, the algorithm would fail to function properly; it can't be parallelised.

However, there's an algorithm called *Quicksort* that takes a **divide-and-conquer** approach to sorting. Each iteration of Quicksort splits the input list into two, ensuring that all the elements of the left partition (*L*) are less than all the elements of the right partition (*R*), although the partitions themselves are not internally ordered.

```
1 def quicksort(input):
2     if len(input) <= 1:
3         return input
4     f, m, b = 0, (len(input)/2), -1
5     pivot = sorted([input[f], input[b], input[m]])[1]
6     left = [l for l in input if l < pivot]
7     right = [r for r in input if r > pivot]
8     pivots = [p for p in input if p == pivot]
9     return quicksort(left) + pivots + quicksort(right)
```

Figure 6.16

A simple implementation of Quicksort. The input list is partitioned into two sub-lists, using a pivot. There are several ways to choose a pivot, such as arbitrarily choosing the first element of the input list, or the last, or a random element, or the median of the first (`f`), middle (`m`) and last (`b`) elements (the approach used in this example). Depending on the level of disorder in the input, different choices can give good or bad results, but we are aiming to try to get a relatively equal number of elements falling before and after the pivot.

Mathematically, we'd say that $(\forall l \in L)(\forall r \in R)l < r$; this can be seen in the code example in Figure 6.16. Once we have our two partitions, each partition can be sorted individually (using Quicksort), and independently, and the results concatenated to form the final ordered list.

At each iteration of Quicksort, we try to halve the number of elements we need to sort, while doubling the number of lists that require sorting, a trade-off that ultimately makes Quicksort very efficient. The combination of the $O(n)$ cost of partitioning the input lists and the $\log n$ depth of the recursive call tree (recall Section 6.4.2) gives Quicksort a time complexity of $O(n \log n)$. However, because we end up with two *independently* sortable lists after each partitioning step, these tasks can be carried out in any order, or in parallel. This means that we can take advantage of multi-threaded systems to speed up the algorithm by roughly a factor of the number of simultaneously runnable threads of execution.

Any algorithm that makes use of a divide-and-conquer approach can be similarly parallelised; knowing this can help us to choose when to take advantage of multi-core approaches and when not to. Also, with the ever-increasing number of cores becoming available in modern machines, this is a very good way of making full use of current technology.

Big data and homogenous processing

When our task is to process large amounts of data in a uniform manner, it's often possible to split the data up and process partitions of it simultaneously. Take, for example, the task of matching a short protein sequence fragment against a database such as UniProtKB. Each protein sequence from the database needs to be individually checked to see whether it contains the target fragment, but these checks are entirely independent. Whether a sequence matches bears no relation to whether any of the other sequences would match. It would therefore be theoretically possible to check all 55 million sequences from UniProtKB simultaneously. Of course, we don't tend to have access to machines that have 55 million processor cores; nevertheless, the more cores we have, the quicker we can process the database: two cores could accomplish the same task in half the time as a single core; three cores would take a third of the time, and so on. We could just keep throwing more cores at the algorithm as they became

Box 6.4 MapReduce

Anne Elk's Theory on Brontosauruses[a] states that 'All brontosauruses are thin at one end, much, much thicker in the middle, and then thin again at the far end.' Curiously, Miss Elk's theory also turns out to be true of many parallel programming tasks as well: you start out with a single piece of data to analyse that's usually in one place; you split the job of analysing the data up into little bits that can be executed independently, and spread these out over a large number of processors; and finally, you gather all the results back into a single place so that you can inspect them.

MapReduce is a paradigm for co-ordinating the processing of a parallelisable problem. It consists of two main stages. First, in the *Map* phase, a master node takes the input, splits it into smaller sub-problems, and sends these to worker nodes. Depending on the architecture of the system, a worker node may decide to split the work up further, passing these on to other nodes to create a tree-like network of workers. The worker nodes process their part of the job, and, when complete, return the results to the master node. Then, during the *Reduce* phase, the master node collects the answers to all the sub-problems and combines them in some way to form the output.

---

[a]http://en.wikipedia.org/wiki/Anne_Elk's_Theory_on_Brontosauruses

available, cutting the execution time with every new core added. Such highly parallelisable tasks are often called *embarrassingly parallel* because they're so easy to speed up in this way.

There are programming models – and associated libraries that implement them in various languages – that make this job easier, such as MapReduce, which encapsulates the two steps to the process: mapping the algorithm onto partitions of the data-set, and then reducing the output of each asynchronous task into a single result. (See Box 6.4).

Non-trivial parallelisation

Unfortunately, there are many bioinformatics problems that cannot be divided and conquered in a straightforward way, but nevertheless could benefit from parallelisation. In these cases, we cannot resort to off-the-shelf libraries, such as MapReduce, to provide this benefit; rather, we need to do the highly complex task of writing parallel programs ourselves. Unlike with MapReduce, where each sub-problem works on its own slice or copy of the data being processed, it is often the case that multiple simultaneous threads of execution must access the same data in memory, at *exactly the same time*. Making sure that two or more simultaneous memory manipulations do not conflict then becomes a task we must carry out ourselves. This usually means using locks (known as *mutexes*) to prevent such conflicting access, causing some threads to temporarily pause execution while another is updating the memory, and waking up those paused threads once the update is complete. This general dance of *synchronisation*, if not done correctly, can very easily cause parallel programs to lock forever, as two threads each wait for the other to unlock the same resource – a situation known as *deadlock*. Avoiding deadlock, while keeping even the simplest of parallel programs running smoothly, takes a great deal of knowledge and experience. If we were to expand the problem to a distributed environment, even the most experienced software engineers would have difficulty keeping up.

6.7 Summary

This chapter introduced some of the low-level issues of modelling and solving problems using computers, and showed how tasks of solving problems in biology are not necessarily straightforward, or even possible. Specifically, we saw that:

1 Some problems are not effectively computable, regardless of how clever our solution may be;
2 In some cases, the only way to get answers to our questions in a timely manner is to settle for an approximate but workable solution;
3 Just because a given solution is easy to verify, it doesn't follow that it's easily computable;
4 Algorithms take time to execute, and space to store their state; choosing the right algorithm is often a compromise between the two;
5 As algorithms are brought to bear on ever increasing volumes of data, the efficiency of those algorithms becomes more important; rarely are we lucky enough to be able to rely on Moore's Law to help us out;
6 Understanding data structures, and choosing the correct ones, is vital to making usable software;
7 Programming languages' quirks, the paradigm(s) they follow, and the tools available to the developer, are all important factors in being able to effectively choose the right technology to solve a given problem;
8 Some algorithms can, and should, be parallelised if we expect to be able to bring them to bear on the volumes of data typical of 21st-century biology.

6.8 References

Helsgaun, K. (2000). An effective implementation of the Lin–Kernighan travelling salesman heuristic. *European Journal of Operational Research*, **126**(1), 106–130. doi:10.1016/S0377-2217(99)00284-2

Puckelwartz, M. J., Pesce, L.L., Nelakuditi, V. *et al.* (2014). Supercomputing for the parallelization of whole genome analysis. *Bioinformatics*, **30**(11), 1508–1513. doi:10.1093/bioinformatics/btu071

6.9 Quiz

This multiple-choice quiz will help you to see how much you've remembered about some of the concepts and issues described in this chapter.

1 The Halting Problem:
  A causes some non-deterministic algorithms to stop unpredictably.
  B defines a set of algorithms that never run to completion.
  C states that it is impossible to tell whether certain algorithms can complete in finite time.
  D defines a set of effectively non-computable problems.

2 The bubble sort algorithm:
   A is polynomial in time, but linear in space.
   B is in time and space.
   C has logarithmic time complexity.
   D is linear in time, but polynomial in space.

3 A vector has an access time of:
   A O(n)
   B impossible to tell, it depends on the size
   C O(1)
   D O(n^2)

4 A binary tree:
   A contains two distinct data types.
   B is always balanced.
   C has linear insert time.
   D has logarithmic search time.

5 Functional programming languages:
   A allow code to be contained within function declarations.
   B allow only stateless functions.
   C guarantee that an algorithm matches its functional definition.
   D define functions by manipulating state.

6 Moore's law:
   A predicts the complexity of sorting functions.
   B states that the spatial complexity of an algorithm is more important than its time complexity.
   C predicts the period in which the number of components on a chip will double.
   D defines the maximum speed achievable by a processor.

7 The Travelling Salesman Problem:
   A is NP complete.
   B has complexity O(n^2).
   C is not effectively computable.
   D has no known solution.

8 Java is:
   A a functional object-orientated language.
   B an imperative object-orientated language.
   C a programming paradigm.
   D a way of executing programs on remote machines.

9 Operator short-circuiting:
   A causes two operators to conflict, giving imprecise results.
   B involves cancelling out operations to give more efficient code.
   C allows the calculation of an expression to terminate as soon as a certain outcome is known.
   D provides a way of reducing the space complexity of algorithms by reducing the amount of code written.

10 Embarrassingly parallel algorithms:
  A give unpredictable results when run on a large number of processors.
  B cannot be parallelised.
  C can only be written in functional languages and are, by their nature, trivial to parallelise.
  D carry out the same task on large amounts of independent data, and so are amenable to divide-and-conquer techniques.

6.10 Problems

1 Imagine a sorting algorithm that operates by repeatedly iterating through an input list of randomly ordered positive numbers. At each pass through the input list, it finds the smallest positive number, and copies it to the end of an output list, which thus accumulates results in ascending numerical order. When a number is copied from the input to the output list, it's replaced in the input list with −1 (so as not to be considered in future iterations). What is the time and space complexity of this algorithm? What pros and cons (if any) does it have compared with bubble sort?

2 You've been asked to build an application that has the fastest possible access to just the amino acid sequence associated with an accession number for the manually annotated component of UniProtKB. It's acceptable for the program to be restarted to accommodate new UniProtKB releases (i.e., a period of downtime to reload the latest version is fine). What in-memory data structure would be most efficient for mapping between an accession number and its associated sequence?

3 Some time in a not-too-distant, but completely hypothetical, future, UniProtKB moves to a model of continuous updates, and newly curated sequences appear every few seconds. What implications might this have for the data structure you use for holding these data in memory?

4 You've been asked to parse some relatively trivial XML documents, and told to use a regular-expression library in your favourite programming language to do this. What problems might you encounter? Under what circumstances would this be a sensible approach?

5 You've been asked to implement a variation of BLAST that uses a different substitution matrix. What considerations would influence your choice of programming language? Which language is best for this task?

Chapter 7

Representation and meaning

7.1 Overview

In the previous chapter, we explored – at quite a low level – how computers manipulate and process data, and saw how apparently simple choices can have significant implications for accuracy and performance. Next, we're going to move away from the 'bare metal' of the circuit board and in the direction of higher-level biological concepts that we've covered earlier in the book. In particular, we turn our attention to the storage, representation and exchange of data, and examine how design decisions at this level can have similarly profound effects on how computer systems behave today and in the future. We begin by looking at the basic idea of identifying biological entities in a robust machine-readable way; we then look at techniques for recording the properties of those entities so as to 'future proof' them; finally, we examine how to give meaning to those representations in order to help computers help us to interpret and manage them. The chapter concludes by considering how to apply these ideas in the context of building bioinformatics applications and databases using the infrastructure of the Web.

By the end of this chapter, you should have an appreciation of what makes a good or bad identifier, and how contemporary frameworks, like XML, RDF and JSON, can be used to build robust, future-proof data-representation formats. You'll understand how ontologies can be used to give meaning to data in a machine-readable and –reasonable form, and how all these ideas can be brought together to build robust bioinformatics systems that work in harmony with the Web.

7.2 Introduction

As you read this text, all kinds of amazing things are happening inside your head. The patterns of light reflecting off the ink on the page (or the pixels, if you're reading on an electronic device) are being picked up by your retina, and interpreted somewhere inside your brain as symbols that represent letters, which, in certain patterns, form words and sentences. And somehow, by recording a series of glyphs and squiggles on consecutive pages of this book, we are able to communicate complex and probably unfamiliar ideas

Bioinformatics challenges at the interface of biology and computer science: Mind the Gap. First Edition.
Teresa K. Attwood, Stephen R. Pettifer and David Thorne. Published 2016 © 2016 by John Wiley and Sons, Ltd.
Companion website: www.wiley.com/go/attwood/bioinformatics

to you. The transition from 'data' to 'knowledge' – which applies in any situation where we're interpreting facts in order to inform future decisions – can be encapsulated in the so-called 'hierarchy of meaning', shown in Figure 7.1.

The process of turning what's written on the page (or, indeed, any kind of data) into knowledge, however, relies on far more than our ability to recognise patterns: in our interactions through this book, for example, you as the reader, and we as authors, have made all manner of assumptions about how we communicate. Some of these are so obvious that they 'go without saying': we've written this book in English, so we're assuming (certainly if you've got this far!) that you can understand this language; we've used **Latin Script** to represent the letters that form the words; and we've written numbers using Western Arabic Numerals, which, for the most part, we've represented using the decimal numeral system. Until now, we didn't have to mention this, yet things have worked just fine. As authors, we're also making a swathe of other, much harder to define, assumptions about you as a reader (that you have a certain understanding of biology and computing, a certain interest in bioinformatics, and so on), all of which need to be true to some extent for this book to make sense.

As an intelligent human, you're able to make sophisticated decisions about the clarity of the communication as you read on; if we've assumed too much background knowledge and failed to explain something in enough detail for you, you can spot this and find some other source to bolster your understanding; if we've erred the other way and assumed too little, you can easily skip forward in the text to new material. Either way,

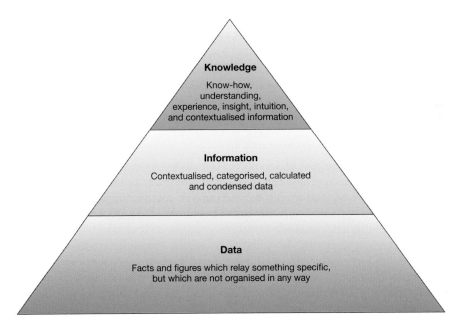

Figure 7.1

A hierarchy of meaning. Data are discrete, objective facts (rhodopsin is a protein; retinitis pigmentosa is a disease). Information gathers data together to create associations (mutations of the rhodopsin-encoding gene are implicated in retinitis pigmentosa). Knowledge applies experience and context to inform decisions, and create questions and hypotheses (can retinitis pigmentosa be cured by 'fixing' mutated rhodopsin-coding genes?)

you'll be constantly 'reading between the lines', bringing background knowledge and experience to bear on the experience of reading this – or any other – text. This process of being able to exchange ideas extends to all aspects of our lives; we humans are quite adept at dealing with the tacit and explicit assumptions, as well as the ambiguities, that arise in our day-to-day communications.

Computers, of course, have none of these faculties: once in a while, it's important to remember that, for all their apparent sophistication, the only thing that computers can really do is shuffle electrons around on a silicon chip. The fact that certain patterns of electrons can be used to represent numbers; and that numbers can be used to represent letters; and that letters can form words that, in turn, represent concepts that are meaningful to us is *merely by a series of conventions*. Some of the most basic of these conventions (*e.g.*, the presence of a certain number of electrons on a given part of the chip represents a '1' and their absence represents a '0', and, when organised in particular patterns, these binary digits represent other numbers and characters) are baked into the hardware of most computers, if not at the level of the Central Processing Unit (CPU), then certainly in the auxiliary hardware and firmware that allows a computer to start up. Other concepts, such as the idea of 'language', 'files' or 'images', appear at Operating System level; but beyond these everyday 'housekeeping concepts', a computer no more knows whether something is a protein or a gene than a light-switch knows whether it's on or off.

Take, for example, the string of characters 'valine'. Typing this into Google gives the kind of results that many readers might expect: pages describing the amino acid of that name, its chemical and biological properties, and so on. This is quite natural: Google has assumed we mean the *word* valine; but what if the intention had actually been to search with a fragment of an amino acid *sequence* using the established one-letter-code convention for the amino acid alphabet? Searching sequence databases with the same set of characters gives results with quite a different meaning: *e.g.*, nine sequences in UniProtKB/Swiss-Prot include the 'motif' V-A-L-I-N-E, as do 989 sequences in UniProtKB/TrEMBL (release 2015_02 of 4-Feb-15); this same motif can be seen in the third α-helix of the 3D structure of saccharopine dehydrogenase, chain a, from *Wolinella succinogenes* (PDB entry, 4INA[1]). Both the context and knowledge of certain conventions makes it possible for us humans to make this sort of context switch (we know what to expect from Google or from UniProtKB); for a computer, however, valine is just valine, and to 'know' the difference, it needs machine-readable context too.

In order to be able to use computers to solve biological problems, we need mechanisms for turning the tacit assumptions and implicit conventions that underpin many of the ways in which we identify, represent and exchange data, information and knowledge into explicit machine-readable forms.

In this chapter, we'll explore how the fruits of the biomedical sciences – both the raw data collected by experiment, and the data and information that result from their analysis – can be stored and retrieved again for further use, and we'll consider how computers can be used to represent and reason about 'knowledge'. We'll discuss the best technologies available to us, many found pervasively in the field already, and tease out why some approaches are better than others. In particular, we'll look at the gap between the ideal solutions available to modern computer science and the reality of what's actually in use today, and why that gulf exists. Let's begin by considering some of the key issues for effective data organisation.

---

[1] http://www.rcsb.org/pdb/explore/explore.do?structureId=4INA

7.3 Identification

The first, and perhaps most fundamental, step towards effective data management is to decide how the various objects that we want to represent are going to be *identified*. This may seem like a trivial task, but choosing the wrong identification scheme early on can cause all manner of problems (recall Section 5.4) that are particularly difficult to fix once identifiers have been released into 'the wild'. Let's look at this in a bit more detail by, first, considering qualities that *good* identifiers should have, then moving on to some practicalities of their use.

7.3.1 Namespaces

Context is everything when it comes to identification, because the same name can refer to different things in different contexts. If our context is the English language, then the word 'mist' refers to a form of aerosol; if, on the other hand, our context were the German language, it would mean a rather mild outburst, of the sort that might accompany a stubbed toe. This context, the space in which the name makes sense, is its **namespace**; we need to be able to be specific or to infer a name's namespace if we expect to know what that context is. Without context, without a namespace, names are ambiguous. When we provide a namespace with a name, we say that we are providing a **qualified identifier**.

Take, for example, the number 2907408. As with our 'valine' example earlier, without any context, we could only guess at its meaning. If we narrowed the context to databases of scientific literature, we might infer that it's an identifier from the online archive of free biomedical literature, PubMed Central[2] (in which case it would refer to an article about survival rates of childhood cancer (Frieldman *et al.*, 2010)). However, even in this more restricted domain, it could just as easily refer to a different article in a different archive (*e.g.*, one about pre-treatment of asthma found in the PubMed database (Dinh Xuan *et al.*, 1988)).

To deal with this issue, **PubMed Central**, which was formed after PubMed, did attempt to qualify their identifiers: they stipulated that their numerical identifiers should begin with the letters PMC. This means that the qualified identifier for the Dinh Xuan paper would be PMC2907408; unfortunately, however, not all establishments use the qualified form of the identifier. In most of these cases, the context is clear to us, perhaps because the surrounding natural language refers to PMC; but computers can't be expected to understand such implied context – we must specify the namespace explicitly.

7.3.2 Meaningless identifiers are a good thing

We've seen several different styles of identifier in this book so far (UniProtKB accession numbers, UniProtKB IDs, PDB IDs, PubMed IDs, and so on), and we've reflected on some of the issues that different identifier designs have created throughout the years. But what makes a good identifier?

Imagine, for a moment, that you're writing a book or a report – some reasonably lengthy document that's going to take a while to create and will need a few iterations of editing and polishing before you have a final version. You've sketched out the overall

---

[2] http://www.ncbi.nlm.nih.gov/pmc

structure (its headings, subheadings, *etc.*), and have a pretty good idea of what the figures and tables will look like. You've written the introduction, and now realise a diagram is needed to help illustrate some point or other. You create the diagram in your favourite drawing package, and save it. It's the first figure in your document, so you call it Figure1.png, and insert it into the right place in your text. A little while later, you realise this first diagram is a bit too complicated to be in the introductory material – you need a simpler overview picture early on, but the one you've just drawn will serve well later on to explain some of the detail. So you draw your simplified diagram, and are about to save it when you realise there's a problem: you already have something called Figure1.png, but now *this* is going to be the first diagram in your document. So what do you do? You could re-name the previous diagram before saving this new one – but you don't know where it's going to go in the report yet; so perhaps you call it FigureX.png for now. Or you could call your new illustration Figure0.png or NewFigure1 .png or GoesBeforeWhatWasFigure1.png; but you recognise that the more you edit the document, the more such filenames will become meaningless as they diverge from the order in which they appear in your report. A file-naming convention, designed to make it easy for humans to keep track of which file goes where, has suddenly become more of a hindrance than a help.

In this scenario, the filename is acting as an *identifier* that allows you to tell your word-processor which *thing* (here, the diagram) you want to include in a particular place in your report. The word-processor, of course, doesn't really care what you called the file – it just needs a way of figuring out uniquely which file you mean. The problem arises because your choice of filename explicitly encodes the *location* of the figure with respect to your document (at least, that was the intention). So when that location changes, there's a mismatch between what the filename appears to 'mean', and where the figure actually ends up. As the document is edited – perhaps the figures are even re-used in a different order in a different document – the 'meaning' of the filename becomes increasingly muddled. Its easy to imagine a scenario where Figure1. png is now Figure 10 in this report, and Figure 14 in some other (and a filename of OriginallyFigure1ButNowFigure10InReportAandFigure14InReportB.png is probably unhelpful).

The problem we're seeing here is caused by the fact that the chosen identifier (the filename) encodes some information about the object it identifies (the location of the illustration in the report); when the nature of the object changes, that relationship is broken. In this scenario, the meaning is being interpreted by a human, and the job of untangling the names and locations is a manual one, which, though tedious, is probably manageable. But the issue becomes pernicious when computers start interpreting the meaning of identifiers, and do so on a large scale.

For example, a putative *Arabidopsis* protein, originally deposited in TrEMBL and given identifier Q9C929[3], was eventually manually annotated and became GCR2_ARATH[4] in UniProtKB/Swiss-Prot. However, the protein had been erroneously annotated as a GPCR, and was later found to be a LanC-like protein (which makes it an enzyme, not a membrane protein); so, GCR2 (which presumably stood for the second GPCR to have been found in *Arabidopsis*, after the discovery of GCR1 – which is a GPCR!) is deeply misleading if a human or machine tries to infer meaning from the structure of the identifier.

---

[3] http://www.uniprot.org/uniprot/Q9C929.txt?version=1
[4] http://www.uniprot.org/uniprot/F4IEM5

To avoid this problem, we have to come to terms with something that, on first encounter, probably seems counter-intuitive: the best identifiers are generally those that are meaningless. Or, put another way, the only thing that an identifier should do is to identify a thing. In computer-science terminology, identifiers that are created (or, at least, are intended to be treated) like this are described as being **opaque identifiers**: we can't 'see through them' to glean any meaning or properties of the objects they identify. The problem with opaque identifiers, of course, is exactly that, on their own, they don't mean anything; therefore, to use them, some additional mechanism is required that will get us to the things to which they refer. The process of following an identifier to find the thing to which it refers is called **dereferencing** or **resolving**, an act that only makes sense within the context of a namespace: the number 356024, dereferenced via PMC, would give access to a paper by Coussens *et al.* (2004); dereferenced via the local telephone directory, it would put us in touch, by phone, with a rather good Chinese Takeaway in Glossop. The important thing to remember here is that dereferencing an identifier within a particular namespace may require using very different mechanisms: fetching a PubMed article requires a different mechanism from that of fetching a molecular structure from the PDB, or from looking up a phone number in a local directory.

Conceptually, then, to use opaque identifiers, we end up with the system presented in Figure 7.2. Given a qualified identifier (namespace + name), we can use our knowledge of the particular namespace's mechanism to dereference it to a location. It's then a matter of fetching from that location the entity in question.

Notice that, at any point, the resolution method could return a different final location, say because the structure of the PMC website changed. Equally, the resolution mechanism itself could change over time, perhaps owing to huge technical changes to the underlying database, but the identifier remains the same: only the location of the resource and/or resolution mechanism by which we find the resource changes – the resource, when we eventually get to it, should remain the same. This system allows us to decouple identification and access, thereby making our databases, and other systems, more flexible and robust to future changes. However, it does require us to carry out this dereferencing process every time it's needed, and this inevitably involves some cost in terms of time, performance and/or effort, which should be balanced against the resilience that the use of opaque identifiers gives. In our simple filename example, we could decide on a naming scheme where each image file is named by a monotonically increasing number (1.png, 2.png, *etc.*), and then maintain a separate 'lookup file' that maps from the filename to the image's latest figure number; but that would probably be a sledgehammer to crack a nut. In almost all real-world bioinformatics applications, however, the benefits of using opaque identifiers from the outset (and of providing an efficient dereferencing mechanism), almost certainly outweigh the pain of any alternative approach.

Figure 7.2

The standard model of how opaque identifiers are dereferenced.

7.3.3 Identifying things on the Web

In the early days of the Web, there were two standard and largely distinct ways of identifying things: **Uniform Resource Locators** (URLs), and **Uniform Resource Names** (URNs) . URLs were forms of identifier that carried within them the means to find resources that had specific locations on the Web; they could be used to *locate* a resource by routing instructions from a user's computer, across the Internet, to a specific server elsewhere in the world, which could then return the resource. By contrast, the less well-known URNs were a means of identifying things, often objects in the real world (like books or people) that had no real 'digital location', but to which we might want to refer in a machine-readable and unambiguous way (say, distinguishing a phone number from a database identifier). It became generally accepted that the difference between these concepts was merely that one was the combined 'name and address' of an object, while the other was just its 'name', and the two ideas were merged to form what is now known as a **Uniform Resource Identifier (URI)**.

A URI, a recommendation described by a set of standards published by the Internet Engineering Task Force (IETF), and endorsed by the WWW Consortium (W3C), is the accepted standard for identifying entities 'on the Web'. URIs unify different aspects of identifiers, allowing the encapsulation of namespaces, uniqueness and even resolution mechanisms in one standard form. A URI is really just a string of ASCII characters, prefixed with a colon-delimited *scheme* name. In other words, at the most abstract level, a URI is nothing more than:

```
scheme:someothercharacters
```

This simple convention means that a computer can inspect the characters up until the colon, determine what scheme the URI is using, and, based on that, then decide how to interpret the rest of the characters in the URI. The most common form of URI is undoubtedly the 'Web address' (which, in the early days of the Web, we'd have called a URL, but now has the more lumpen name of 'URI that uses the `http:` scheme'). This form of URI is so common that we humans have become attuned to parsing its various components quite well: most of us will instantly recognise anything in the form of 'http://blah.com/thing' as being likely to identify a Web page about 'thing' hosted by the Blah Corporation (so much so that adverts and so on often drop the 'http' prefix in the knowledge that people will take that as read). There are, however, 150 or so less familiar standard scheme names for all manner of other resources, such as fax numbers, geographical locations, email addresses, news items, virtual world coordinates, and so on (see Box 7.1).

Following the colon after the scheme name comes the rest of the URI, the exact structure of which depends on the scheme in question: in the case of a fax number, for example, it's the international dialling number for that fax machine (*e.g.*, `fax:+44145735602`); for email, on the other hand, the form is `mailto:a.person@example.com`. The specification of this structure can be found through the **Internet Assigned Names Authority** (IANA) registry, usually deposited there by some standards body like the IETF. Many URI schemes, including those used to identify Web pages or files, treat the part after the scheme name as a hierarchy that homes in on particular resources by successive narrowing of the domain. For example:

```
http://example.com/protein/gpcr/rhodopsin
```

represents the familiar HTTP 'Web address' form of a URI. After the colon come two '//' characters (which Tim Berners-Lee has since admitted are completely redundant,

Box 7.1 Scheme names

There are more than 150 'official' scheme names registered with the Internet Assigned Names Authority (IANA), and a small handful of unofficial but commonly used names. In bioinformatics, we frequently encounter:

- *http, https*: the HyperText Tranport Protocol, and its Secure equivalent, used for interactions between Web servers and clients;
- *ftp, ftps*: the File Transfer Protocol, and its Secure equivalent, still quite often used for moving large files around the Internet;
- *ssh*: Secure SHell, a protocol for allowing encrypted interactive remote login between machines;
- *git*: a distributed version-control system designed by Linus Torvalds[a], original creator of the Linux operating system kernel;
- *imap*: Internet Message Access Protocol, a protocol for e-mail retrieval and storage. Largely supersedes POP (the Post Office Protocol), although the pop: scheme remains active;
- *smtp*: the Simple Mail Transfer Protocol, used for sending email (as opposed to IMAP, which is used for retrieving it);
- *doi*: the Digital Object Identifier (see Section 7.3.6), widely used in scholarly publishing, but not an official IANA scheme at the time of writing.

Some browsers (including Safari and Chrome) allow us to type things like 'define:receptor' into their address bar; here, however, all that's happening is that the browser's treating this as a query as though we'd typed it into our search engine's input box; the colon notation is purely coincidental, and is part of the search engine's query-language syntax – it doesn't represent a scheme name.

---

[a]Git doesn't stand for anything; it's just a name (Torvalds is alledgedly quoted as saying, 'I'm an egotistical bastard, and I name all my projects after myself. First Linux, now git', though evidence for whether he really said this is scant).

but seemed like a good idea at the time), followed by the domain name of the Web server hosting the resource, followed by another '/', followed by a path the server can use to identify the specific resource it manages. Notice that this form of URI essentially carries the idea of a namespace within it, as each successive narrowing of the domain takes place within the context of the previous element of the URI (*i.e.*, 'protein' is in the namespace of 'example.com', 'gpcr' is within the namespace of 'protein', and so on). Note also that, although, in this example, we've used path components that mean something to a human (and so make the URI highly non-opaque), there's nothing in the specification of the URI standard or http: scheme that forces this to be true; the same resource could equally well have been identified by an opaque URI such as:

```
http://example.com/101127123875t7nfi1
```

which is utterly unmemorable for humans, but has all the advantages of opaque identifiers discussed previously. It's worth noting that even an unmemorable URI like this isn't actually totally opaque, because URIs such as those with the http: scheme inherently include some information about where the resource exists (in this case, at example.com); to this extent, any URI that contains a location can't be completely opaque.

The majority of these URI standards have evolved over many years, and have had the input of a wide range of people and organisations; they're therefore often considered best practice with respect to resource identification (even taking into account minor infelicities like the unnecessary '//'). If you're designing a computer system and need to identify something within it, there's probably an existing URI scheme available; if there isn't, you're either doing something technologically radical and new, or have designed your system badly.

7.3.4 Cool URIs don't change

In a 'style note'[5] from the W3C, Tim Berners-Lee describes a 'cool URI' as being 'one which does not change', pointing out that there's no good technological reason (beyond the insolvency of a domain name's owner) for a URI ever to become invalid or to dereference to the wrong thing. The note provides a list of things not to include in a URI, such as a document's authors, subject, status (draft, final, *etc.*), filename extension, and so on, all of which are very likely to change over time. Noting that 'after the creation date, putting any information in the name is asking for trouble one way or another', he implies that 'cool URIs' are essentially opaque. As tempting as it might be to create URIs that are nice for humans to read, it's best to accept that nothing that's more than a few tens of characters long is ever likely to be remembered accurately by a human anyway, and machines are just as happy with a resource called `101127123875t7nfi1` as they are with `/protein/gpcr/rhodopsin`.

7.3.5 Versioning and provenance

Almost inevitably, an entity for which we have an identifier is likely to get modified in some way, given enough time: it may be updated, or even deleted; it may be superseded by a newer entity, or related in some way to other entities. All these things are possible; each has occurred countless times through the history of every major bioinformatics database, and there's nothing special about this moment in time that prevents it from happening again; but what does this mean for our identifier?

The simplest thing that can be said is that our now-modified identifier no longer points at what it used to point at; depending on the nature of the change, this may *invalidate* the identifier. In this case, we have to make a choice: do we discard it, along with any information we may have stored about it, or do we continue to use it to represent our (now out-of-date) entity? The answer depends on both the nature of the change and the data we're identifying; either way, we run the risk of losing some crucial information. The difficulty here is the lack of context about what the change actually *means*. A database record may be deleted because a new version has been created elsewhere that replaces it, or it may have simply been retracted. In the former case, how are we to know where the new version resides? If the deleted object is no longer available, it may not be easy to discern the answer. Even if we still have access to the updated version, why was it updated? It might have been to correct minor spelling mistakes, to add literature citations, to fix a protein sequence error, or to completely reclassify a protein; somewhere on that spectrum, we could argue that the new version no longer represents the 'same thing' as the old one; but this distinction may not be obvious. The

[5] http://www.w3.org/Provider/Style/URI.html

messy evolution of data is important, and although, sometimes, an experienced human might be able to untangle the web, computers are entirely impotent without being told explicitly what that web entails.

For these reasons, versioning and provenance are important, and should be integrated into databases from their very first record. The following suggestions encapsulate what's needed and why:

- If an entity is deleted, its identifier should remain valid, and, at the very least, should resolve to something that explains *why* it was deleted, *when* and *by whom*. Ideally, the old record should remain available, but modified to include this retraction information. If it simply vanishes, we're left clueless as to the reason.
- In an ideal system, every single revision of an entity should be separately identifiable, along with the provenance that explains the history in machine-readable form. This should make it possible to traverse that history, allowing us to build up a full picture of its evolution, and allowing a computer to choose the most appropriate revision for its needs.

With this, we could then do some clever things. We could automatically update any data we've stored on our machines, if the changes made are deemed sufficiently minimal. If the changes are more substantial, we can allow users to choose, in an informed way, which records to update, and which to potentially discard. We can recommend new literature for records whose citations have changed. We can evaluate the historical path of our knowledge to gain useful insights, or to point to areas of our knowledge that need updating.

7.3.6 Case studies

Let's now explore some identifiers that appear in the field of bioinformatics.

LSIDs

Life Science Identifiers (LSIDs) are URIs that use the urn: scheme, with a namespace identifier of `lsid`. Along with a complementary resolution protocol, they were invented as a way of unifying all the disparate identifiers used by biological databases around the globe, in an attempt to tackle some of the issues we've been discussing in this chapter. The general format of an LSID, with an example from the International Plant Names Index (IPNI), is shown as:

```
urn:lsid:<AU>:<NS>:<ID>[:<VE>]
urn:lsid:ipni.org:names:1234-1
```

As we can see, LSIDs include an authority (AU), a namespace (NS) within that authority, a unique name (ID) within that namespace, and an optional version code (VE). In theory, they provide a single identifier format that could take over the job of all existing identifiers and give a single point of resolution for them all. Unfortunately, they haven't been as successful as we might hope, partly because uptake has been slow, partly through simple inertia. One criticism has been that LSID and the LSID Resolution System (LSRS), taken together, provide no demonstrable benefit over the standard infrastructure of the Web (using http: URIs as identifiers, resolving names with the Domain Name Service (DNS), *etc.*), and essentially reinvent the identification wheel. Because of this, bioinformaticians have tended to back away from using the

LSID system, although it still enjoys use in zoological and taxonomical areas of biology, where LSIDs are used to identify taxonomic names (see Page, 2006).

DOIs

In early 2000, a group of academic publishers came together to form Crossref, an independent organisation with a mission,

> to enable easy identification and use of trustworthy electronic content by promoting the cooperative development and application of a sustainable infrastructure.

As a 'trusted' DOI Registration Agency, Crossref enabled the use of a consistent, electronically resolvable identification scheme across multiple publishers from all manner of academic disciplines. Bringing together a critical mass of competing publishers to cooperate on a pre-competitive task – even one of such obvious value to all concerned – was a gargantuan task, requiring deft political manoeuvring. The process nearly foundered many times; however, the benefit of the DOI applied to scientific literature has been enormous.

A Digital Object Identifier (DOI) is a qualified identifier used to reference some digital object in a way that's independent of its location on the Web. DOIs are primarily used to identify individual scientific articles (or parts of articles) for use in citations and cross-references, but can theoretically be used to identify any locatable resource. The pure form of the DOI, and an example, is:

```
doi:<RY>.<RT>/<ID>
doi:10.1042/BJ20091474
```

Here, after the initial doi: scheme, we see three levels of identification, beginning with the number `10`, which identifies the DOI registry (RY), continuing with the number `1042`, which identifies the DOI registrant (RT), and finally the string `BJ20091474` (ID), which uniquely identifies an object within that registrant's namespace. This example refers to an article registered by the publisher Portland Press Ltd. with the Crossref registry.

Because pure DOIs aren't dereferenceable URIs, there needs to be a central registry and resolution service to map them to locations on the Web. The International DOI Foundation provides just such a service, taking a DOI and returning a URI. The above DOI is dereferenced to the following URI:

```
http://www.biochemj.org/bj/424/bj4240317.htm
```

As a convenience, Crossref provides a dereferencing service that allows DOIs for articles registered with them to be written in the form:

```
http://dx.doi.org/10.1042/BJ20091474
```

which also resolves to the same http: URI for the article in the biochemj.org domain. As mentioned previously, this URI could change at any point, say if the journal's website were reorganised, but the DOI will forever remain the same; so whereas linking to biochemj.org directly could easily break in the future, taking the small cost of the extra step of dereferencing via Crossref, or the International DOI Foundation, provides an additional level of robustness against change.

Unfortunately, versioning and provenance are absent from the DOI system, so we tend to be confronted with the same problems in literature as we are with biological databases. A single DOI often refers to the latest version of an article, with no explicit information about whether it's the initial submitted draft, a pre-print version, or the published copy. Similarly, errata and retractions aren't supported natively by DOIs. This leads to an ambiguity that can sometimes cause problems; on the whole, however, given the breadth of the domain over which it operates, the DOI system works remarkably well.

ORCIDs

Perhaps more daunting than identifying digital objects like Web pages and database records, or inanimate ones like books, is that of uniquely identifying people – not just authors of articles, but also those who've contributed to the creation of data, or who need to be credited for one reason or another for their part in a scientific result. There are many more people than there are publications; and many more institutions with international levels of politics than there are publishers; and humans have much more ambiguous names than do scientific articles (Sprouse, 2007). It's estimated that around two-thirds of the six million or so authors in MEDLINE share a family name and a first initial with at least one other author (see Box 7.2), and that, on average, any particular ambiguous name potentially refers to eight people (Torvik and Smalheiser, 2009). Numerous other authors explore this knotty problem in some depth (*e.g.*, Farkas, 2008; Torvik *et al.*, 2003; Aronson and Lang, 2010; Jonnalagadda and Topham, 2010).

Several attempts have been made to tackle this issue. In November 2010, the National Library of Medicine announced its intention to establish PubMed Author IDs; and the Dutch 'Digital Author Identifier' has been used with some success in the Netherlands. However, domain-specific or national-level identifiers have obvious limitations in an increasingly interconnected research environment. A number of large commercial publishers launched initiatives of their own, such as the 'ResearcherID' from Thompson-Reuters and Elsevier's 'Scopus Author ID', but the idea of trusting control of such an important global identifier to such large commercial enterprises has met with understandable resistance from the community, and these ID systems have seen little take-up outside their respective environments.

Perhaps the most promising move has been the creation of a not-for-profit organisation to manage the development of the Open Researcher and Contributor ID (ORCID). Formally, ORCID IDs[6] are URIs with the 'path' component being a 16-character identifier consisting of the numbers 0–9, separated in groups of four by hyphens (for example, Amos Bairoch, author of many renowned bioinformatics databases, has the ORCID ID http://orcid.org/0000-0003-2826-6444). ORCID IDs are often abbreviated by publishers to ORCID: 0000-0003-2826-6444. Technically, ORCID IDs are a subset of the International Standard Name Identifier[7] scheme used outside of academic publishing to identify contributors to television programmes, songs and newspapers, *etc*. Not only does ORCID have a pleasing acronym, but it appears to be attracting the kind of cross-publisher and international critical mass necessary to turn such an initiative into a practical reality.

InChIs

The last of our short case studies is the InChI – The IUPAC International Chemical Identifier – which is particularly interesting because it breaks most of the guidelines we've set out for designing 'good' identifiers, and yet is exceptionally useful, and is

---

[6] Although the ORCID acronym includes the term 'ID', in a self-inflicted example of RAS Syndrome (Redundant Acronym Syndrome syndrome [http://en.wikipedia.org/wiki/RAS_syndrome]), they are nevertheless referred to as ORCID IDs by their creators.

[7] http://en.wikipedia.org/wiki/International_Standard_Name_Identifier

Box 7.2 What's in a name?

It's proved difficult enough over the years to agree on identification schemes for biological objects; but at least, in these cases, we're only talking about reaching consensus on how to name things amongst a particular scientific domain. When it comes to unambiguously identifying people on a global scale, the problem is far more complex: search any social network, such as LinkedIn or Facebook, for your own name, and you'll most likely find that you're not unique.

Similar issues arise when searching bibliographic databases for our own papers, or for our colleagues' papers – how can we be sure that we've found articles from the right person? Table 7.1 shows the results of searching PubMed, at some random date in 2014, with the family names and first initials of the authors of this book and of some of our colleagues. Accordingly, S.Pettifer appears to be the least common name, and C.Taylor the most. The question is, how many of the retrieved articles were actually authored by us and our colleagues?

Table 7.1

Results of PubMed searches for a selection of author names at an arbitrary date in 2014.

Query	# of articles	Query	# of articles	Query	# of articles
S Pettifer	26	SR Pettifer	7	SR Pettifer, Manchester	6
D Thorne	104	D Thorne, Manchester	5		
T Attwood	127	TK Attwood	96	TK Attwood, Manchester	35
J Marsh	1,942	J Marsh, Manchester	21	J Marsh, Manchester, Computer Science	3
R Stevens	2,155	RD Stevens	213	RD Stevens, Manchester	8
D Robertson	2,795	DL Robertson	153	DL Robertson, Manchester	47
C Taylor	4,951	CJ Taylor	349	CJ Taylor, Manchester, Computer Science	13

Adding a second initial to any such query usually significantly reduces the problem, narrowing down the results to more reliable article sets; but there's a problem – neither Thorne nor Marsh have middle names (or, at least, neither has published with a middle name). Therefore, if we want to identify their work unambiguously, we must fine-tune the search: *e.g.*, instead of a middle initial, we can provide the name of the city/university where the authors work. This appears to work for Thorne, suggesting that he was the author of five of those original 104 articles. However, without going through the entire list by hand, we can't be sure that we've not missed something relevant in the discarded 99, perhaps articles published while he was working for a different university or in a different city.

For Marsh, we find that adding the 'Manchester' keyword hasn't worked, as the list of articles still contains the work of a mixture of 'D Marshes'. Therefore, to whittle down to the right papers, we have to further increase the specificity of the search by adding another query term – here, 'computer science'. This seems to do the trick. For Taylor, however, we've almost certainly discarded many relevant results in the 336 articles we've thrown away, and it would take much more sensitive queries to establish a more representative set of papers.

This simple exercise shows how difficult it is to unambiguously retrieve the articles of a specific author from repositories like PubMed, given that so many authors share the same name. As we've just seen, we can work around the issue using filters to narrow the search space; nevertheless, authors with common European names like 'D Jones' (9,803 PubMed articles in February 2015) and 'D Smith' (16,742 articles), or with common Asian names like 'J Wang' (59,967 articles) and 'Y Li' (56,408 articles) present much thornier problems.

increasingly commonly found in bioinformatics databases as the discipline expands into the realm of what previously would have been considered as chemistry. The InChI defines a standard way of representing small molecules, and is frequently used to identify interactions between drugs and proteins. Although InChIs are (quite reasonably) called identifiers, the term is used in a rather different sense from that which we've explored so far: rather than being a 'name' for a concept, InChIs are actually a way of explicitly encoding the atoms and bonds that form a small molecule. Every InChI starts with the string 'InChI=', followed by the version number of the InChI specification used to encode the molecule. The rest of the InChI is made up of a series of layers and sub-layers (some of which are optional) separated by '/' characters; these describe the structure and other properties of the molecule, such as its charge or stereochemistry. For example, the standard InChI for ethanol (CH_3CH_2OH) is:

```
InChI=1S/C2H6O/c1-2-3/h3H,2H2,1H3
```

and a sufficiently practised human would be able to draw the chemical structure of the molecule by interpreting the characters that form the InChI. Because InChIs contain optional components, and also because the algorithm that generates InChIs is subject to revision, they are neither opaque, unique nor immutable, so it's very dangerous to treat them as identifiers in the same was as the other case studies we've looked at. Treat them instead as a compact representation of a molecule, more akin to a FASTA file than a URI.

7.4 Representing data

The latter half of the 20th century saw most biological data being stored in human-readable 'flat' text files, each with its own bespoke syntax and format. In the absence of software and hardware required for the graphical visualisation techniques we take for granted today, data in a flat-file often simply had to be interpreted just by the act of reading the file. In consequence, many databases made use of a single data-file for each of the three stages discussed above: the file transferred to the client was an exact copy of that stored on the server, and it was this file, free of modification, that the user then read directly. Those first databases held, by comparison with what's available now, trivially small quantities of relatively simple data, and so, by and large, this one-size-fits-all approach to storing, transferring and interpreting data worked.

The fragmented evolution of bioinformatics during this time led to the introduction of many disparate data-file formats: as each new database came into being, a new format was designed to store and serve its data; and each such format compounded the problems to be faced by the next generation of bioinformaticians (there are more than 30 formats for storing protein sequence-related data alone; see Box 7.3). Most of these legacy formats fall well short of what we now consider good practice, in terms of extensibility, mutability and unambiguity, and, in part, this is because they were designed to simultaneously fulfil the three roles of storage, transfer and presentation that, in modern bioinformatics, are best considered independently.

One example of the problems caused by conflating these roles comes from the fact that most computer monitors at that time were limited to an 80-character width: as the formats were meant to be interpreted by reading them directly off the screen, this led most file formats to have similar line-width restrictions. In turn, this led to a pervasive use of tabular and columnar formats, preventing many of these formats from evolving

Box 7.3 File format proliferation

Standards are such a good thing that everyone wants to have their own! In most areas of bioin-formatics, it's common to come across a number of legacy file formats that store essentially the same types of information; having been developed by different institutions to fulfil their own specific requirements, at a time when no standards existed for such things, most are mutually incompatible, despite being put to similar uses.

The area of protein sequence analysis is a case in point, having more than 30 different file formats for handling sequence-related information: many are variations on a theme or evolutions of others; for some, their specifications are so vague that it's impossible for a computer (or indeed a human) to tell them apart. Some of these flat-file formats are shown in Table 7.2 (many more are listed in the EMBOSS analysis package (Rice, 2000)): at the top of the table, highlighted green, are a number of sequence database formats ('swiss' and 'experiment' are derivations of the EMBL entry format); in the centre of the table, highlighted pink, are several formats for storing multiple sequences, either as input to or output from sequence alignment tools; and at the bottom of the table, highlighted blue, are some formats for storing individual sequences (many of them derivatives of the original NBRF/PIR format).

Table 7.2

Selection of database flat-file formats (green), multiple sequence formats (pink) and formats used to store individual sequences (blue).

Format	Description
embl	EMBL database entry format
swiss	Swiss-Prot database entry format
rsf	RichnSequence Format from NCBI
pfam	Pfam database entry format
prints	PRINTS database entry format
genbank	GenBank database entry format
aln	Clustal alignment format
msf	Wisconsin Package GCG's Multiple Sequence Format
seqs	PIR multiple sequence format
c5a	Multiple sequence format used by CINEMA5
clustal	Multiple sequence format used by Clustal
mega	Multiple sequence format used by MEGA
phylip	Multiple sequence format used by PHYLIP
sam	Format describing alignment of sequences to a reference sequence(s)
utopia	Multiple sequence format used by Utopia
ace	Sequence format used by ACeDB
gcg	Sequence format used by the GCG package
raw	Raw representation of a sequence, without metadata
nbrf/pir	PIR sequence format
Fasta	FASTA sequence format
fastq	Format for storing sequences and their quality scores
pearson	Generic FASTA-like format to deal with bad ID lines
ncbi	NCBI evolution of the FASTA format
dbid	Modified version of the FASTA format

In the absence of any agreed standards, the consequence of this file-format proliferation means that tool developers are forced, at some considerable implementation cost, to support multiple file formats for reading and writing data in order to achieve even the most basic kind of interoperability with other systems – and all of this means that routine bioinformatics tasks are consequently much trickier and more convoluted than they should be.

```
   . . .
REMARK 290
REMARK 290 CRYSTALLOGRAPHIC SYMMETRY
REMARK 290 SYMMETRY OPERATORS FOR SPACE GROUP: P 21 21 21
REMARK 290
REMARK 290      SYMOP   SYMMETRY
REMARK 290      NNNMMM  OPERATOR
REMARK 290      1555    X,Y,Z
REMARK 290      2555    1/2-X,-Y,1/2+Z
REMARK 290      3555    -X,1/2+Y,1/2-Z
REMARK 290      4555    1/2+X,1/2-Y,-Z
REMARK 290
REMARK 290      WHERE NNN -> OPERATOR NUMBER
REMARK 290            MMM -> TRANSLATION VECTOR
REMARK 290
REMARK 290 CRYSTALLOGRAPHIC SYMMETRY TRANSFORMATIONS
REMARK 290 THE FOLLOWING TRANSFORMATIONS OPERATE ON THE ATOM/HETATM
REMARK 290 RECORDS IN THIS ENTRY TO PRODUCE CRYSTALLOGRAPHICALLY
REMARK 290 RELATED MOLECULES.
REMARK 290     SMTRY1   1  1.000000  0.000000  0.000000        0.00000
REMARK 290     SMTRY2   1  0.000000  1.000000  0.000000        0.00000
REMARK 290     SMTRY3   1  0.000000  0.000000  1.000000        0.00000
REMARK 290     SMTRY1   2 -1.000000  0.000000  0.000000       25.94976
REMARK 290     SMTRY2   2  0.000000 -1.000000  0.000000        0.00000
REMARK 290     SMTRY3   2  0.000000  0.000000  1.000000       37.59964
REMARK 290     SMTRY1   3 -1.000000  0.000000  0.000000        0.00000
REMARK 290     SMTRY2   3  0.000000  1.000000  0.000000       33.69953
REMARK 290     SMTRY3   3  0.000000  0.000000 -1.000000       37.59964
REMARK 290     SMTRY1   4  1.000000  0.000000  0.000000       25.94976
REMARK 290     SMTRY2   4  0.000000 -1.000000  0.000000       33.69953
REMARK 290     SMTRY3   4  0.000000  0.000000 -1.000000        0.00000
REMARK 290
REMARK 290 REMARK: NULL
   . . .
```

Figure 7.3

Excerpt from PDB entry 1A08[i]. The crystallographic symmetry operators for this structure are encoded in a block of pseudo-free text. All PDB records include the same textual description and tabular headings.

[i]http://www.rcsb.org/pdb/explore.do?structureId=1a08

with the data they represented – extra columns couldn't be appended without breaking the line-length limit. Inevitably, whenever such formats had to be extended, this almost always meant contravention of the original specification. A good illustration is the late inclusion of the crystallographic symmetry operators in the PDB format: the only way to include these additional data without breaking the original format was to embed them in fields originally designed for human-readable, free-text comments (because it was safe to assume that existing parsers for the format wouldn't be trying to interpret anything in the REMARK lines). An example of the resulting 'hack' is shown in Figure 7.3. As a consequence of shoe-horning the symmetry operators into the file in this way, parsers designed to access them were forced to inspect the REMARK lines (which, previously, they could safely have ignored) to try to determine whether they really contain human-readable comments or instead encode machine-readable data.

7.4.1 Design for change

While it's difficult to predict the technologies or scientific needs that may appear in the future of bioinformatics, by learning from past mistakes and making the most of contemporary techniques, we can take sensible precautions to avoid future problems.

The first thing we can anticipate is that the requirements placed on a format – any format – will inevitably evolve. Unless we've done a perfect job describing data that will never change (and we can be fairly certain that we haven't done yet), we will, at some point, have to modify the format to include new information, or remove/correct what's obsolete or wrong in some way. This means we must, at the very least, allow users to keep track of the version of the format they're using, and that means including that information *inside* the data file. It's not good enough to make that information somehow implicit from the source of the data file, as the file will probably be transferred around, and its original provenance lost; it's also no real help to include the version in the filename, as this can, and probably will, be changed many times (not to mention that files transferred across a network often don't have filenames, as it's the content that's transferred, not their metadata). The best possible place would therefore be in the data itself, preferably up-front, and in a simple way that can be guaranteed to remain unchanged in perpetuity. That way, clients designed to consume such data can check the version before having to read the rest of the file.

So, given the inevitable fact that formats will change, we must make sensible choices to try to make such changes as painless to implement as possible. Below, we list the most important considerations:

- *Version everything.* Data and data formats change over time. These changes must be captured such that we can correctly understand their meaning, and how their meaning has changed over time. Master a version-control system such as *git* or *svn*; it will feel like hard work at first, but will save you significant pain later on; and, most importantly, make sure that any formats or schemas you create, and the elements they contain, are versioned (ideally using an ontology, such as one of those described in Box 7.8).
- *Don't define arbitrary limits.* Too often, formats limit their content arbitrarily, for example, by declaring a maximum length for text strings, or enforcing upper or lower case for certain fields (recall the problems caused by the change in Swiss-Prot's format described in Section 5.6). If there are no good reasons for such limits, they shouldn't be imposed in the first place.
- *Support international character sets from day one.* At some point, someone will want to encode data from a language other than yours: whether it be the name of an author written in Japanese Kanji, or entire descriptive blocks written in a Cyrillic language, the need to support non-ASCII characters is more or less guaranteed – adding such support as an after-thought will bring nothing but pain. In practice, this means learning about how **Unicode** and its various character encodings work – this may seem daunting at first, but the effort is well worth it, and most modern programming languages provide excellent support for Unicode, either natively or through support libraries.
- *Qualify as much as possible.* If each individual datum – or some subset of data – is qualified with a name in the format, then adding new data is more straightforward: we need only choose a name for the new data and include it alongside our old data. The two-character field names of the EMBL format and its derivatives are one (admittedly limited) example of this kind of extensibility.
- *Use standards wherever you can.* Someone has probably spent a lot of time thinking about how to represent at least something closely related to what you're creating; make sure you know what formats, schemas and vocabularies already exist, and make your work compatible with these wherever possible.

Box 7.4 Machine- and human-readable formats

The terms 'machine readable' and 'human readable', when applied to electronic formats, are open to some interpretation. At some level, most digital documents can be processed and thus 'read' by a machine; similarly, most documents that aren't binary-encoded or compressed or encrypted can be, to some degree, 'read' by a human.

The real question is to what extent the format makes it *easy* rather than just *possible* for it to be interpreted and understood unambiguously by a machine or a person. The important point to keep in mind here is that humans and computers speak very different languages, and have very different requirements when it comes to reading documents. It's therefore rare, if not impossible, to have something that works well for both kinds of consumer.

In a literal sense, XML, RDF and JSON are human-readable formats, in that they are encoded as characters that a human can inspect in a text editor and, with enough experience, interpret and understand; but these formats (especially RDF) are certainly not easy to read, and they're really designed for consumption by machine. Conversely, the text-based formats for early bioinformatics data are reasonably easy for a human to read on an 80-character terminal, but as we've seen throughout the book, are actually quite fiddly to parse correctly by machine. Regardless, the days where it made sense to design a file format so that it could be read by a human are long gone, and it is better to assume that the primary consumer of data in any format will be a machine, not a person: generating a human-readable version of a machine-readable format is easy; doing the reverse is much harder.

- *Make everything easily machine readable.* It's hard to emphasise this point enough: assume that your data will eventually be consumed in huge quantities by a computer, with no human intervention, and make it *easy* not just *possible* for a machine to read it (see Box 7.4 and the following section). Also, as tempting as it may be to allow information to be encoded in a free-text 'comments' field, when that information turns out to be really interesting, you'll regret having to resort to text-mining techniques in order to extract the knowledge you've just abandoned to free-text format.

When you create data-exchange formats or schemas to be shared with others, you must expect that, at some point, other people will write code that utilises them; even if a new feature you're about to introduce is really exciting, no one will thank you if your change causes their systems to fail. In computing, a change to one part of a system that causes another to fail is known as a **breaking change**. Sometimes these are inevitable, but, with forethought, following the principles outlined in this section, and using the extensible frameworks described next, will help reduce the chances of this happening.

7.4.2 Contemporary data-representation paradigms

Reading and writing even the most trivial file format can be quite a complex task, and to understand the issues that have arisen historically, and the modern solutions to those problems, it's useful to think of the process as consisting of three distinct stages: we'll call these lexical, grammatical and semantic. The first of these – **lexical analysis** – involves converting the individual bytes or characters into what are called tokens. Imagine a file format where each line contains a list consisting of words or integer numbers separated by commas and terminated by a full stop. The lexical analysis of a line such as:

```
384, Rhodopsin, Rho.
```

would yield a sequence of tokens of the form shown in Figure 7.4.

Figure 7.4

Lexical analysis of 384, Rhodopsin, Rho.

The second stage is **grammatical analysis**, where we apply some rules governing the ordering of the tokens. In our trivial case, we want to somehow encapsulate the notion that a valid list consists of any number of 'items' separated by commas and ending with a full stop, where an 'item' is either a number or a word. For example:

```
384 Rhodopsin, Rho.
```

is lexically valid (it consists of numbers, words and commas that are all acceptable tokens), but would break our grammar rule because there's no comma between the first and second 'item'. In combination, lexical and grammatical analysis define the **syntax** of our format.

The final stage is **semantic** interpretation, where we infer meaning from a syntactically valid stream of tokens. In our hypothetical example, we might decide that the line describes a protein, that the first number defines the length of its amino acid sequence, the next word gives us its common name, and the final word its associated gene.

Even with this apparently trivial file format, there are a plethora of potential pitfalls when it comes to defining the lexical and grammatical rules, especially if we keep half an eye open for future evolution of the format. If we want to be really strict about parsing this format, we might include a rule in our code that says that each line should consist of exactly five tokens in the order described above; but then, at a later date, if we wanted to extend the format to have an additional (and perhaps optional) 'free-text comment' at the end of the line, we'd hit at least two problems:

1 we'd have to go back and modify the grammatical element of our parser to allow an extra comma and subsequent 'free text';
2 we'd have a conflict if we'd allowed commas *within* our free text, as we've chosen to use this as a token in its own right.

You'll probably have spotted that we're ignoring quite a number of details here: if we're interpreting a file character by character, how do we know that '384' forms an integer number, or 'Rhodopsin' a valid word in the first place? What about whitespaces between the different components? To answer these questions in enough detail for you to build your own parser would take several chapters, and the important message of this section is that, with modern tools for data representation, you should never have to concern yourself with doing things at this level any more; there are now far better ways to represent data than creating your own file format and parser.

The contemporary approach to storing data in files is to avoid creating new file formats altogether, and instead to use one of a handful of data-representation paradigms, such as XML (the eXtensible Markup Language), JSON (the JavaScript Object Notation) or RDF (the Resource Description Framework). It's difficult to know exactly what collective name to give these approaches: while they all address a similar goal (that of storing data), and all do so in broadly similar ways, each uses a different nomenclature: XML self-describes as a 'language', JSON as a 'notation', and RDF as a 'framework'. To avoid favouritism, we're going to stick to calling them 'paradigms'. We'll look at each of these in more detail shortly, but they all share a common principle: they provide a well-defined syntax that is devoid of any application-level semantics. This means that robust general-purpose parsers can be made available for each paradigm, allowing application programmers to focus on the important semantic components, rather than having to worry about the grubby details of things like whitespaces and international characters. As well as getting rock-solid lexical and grammatical parsing for free, the important benefit of using any of these paradigms is that they allow us to build *extensible* schemas and formats. Unlike *ad hoc* and bespoke file formats of the past – where we have no idea how some 3rd party has implemented their parser, and therefore whether a change we are introducing will break their software – with these paradigms, there's the assumption that, as long as the syntax is correct, a parser encountering application data that it doesn't understand will simply ignore it and move on to the next part of the file without crashing or even necessarily reporting an error. This means that, as long as we don't break a previous schema by changing its structure, we can easily add new information, and any client applications unaware of this new information will continue to work as though nothing had happened. The way in which these benefits are achieved will become clearer as we examine each in turn.

JSON

The simplest of these paradigms is JSON, the JavaScript Simple Object Notation, and although it contains JavaScript in its name (in which context the notation became popular), JSON parsers exist for most modern programming languages.

JSON provides a syntax for representing character strings, numbers, Boolean values, arrays (which can be thought of as lists) and dictionaries (which make associations between two values), and is extremely convenient for representing the kind of hierarchical data structures commonly found in programs and databases. Although JSON has no explicit support for namespaces, provenance or metadata, its simplicity means that building systems that generate or consume JSON can be done very quickly.

Let's take, by way of an example, a protein sequence represented in the PIR format. We've previously seen this in Figure 4.11, but it's reproduced again here for convenience:

```
>P1;HBA_HUMAN
Hemoglobin subunit alpha OS=Homo sapiens GN=HBA1 PE=1 SV=2
MVLSPADKTNVKAAWGKVGAHAGEYGAEALERMFLSFPTTKTYFPHFDLSHGSAQVKGHGKKVADALTNAVAHVDDMPN
ALSALSDLHAHKLRVDPVNFKLLSHCLLVTLAAHLPAEFTPAVHASLDKFLASVSTVLTSKYR
*
```

Recall that the PIR format starts with a '>' symbol, that 'P1' is a token telling us that the rest of the file describes a protein (rather than a nucleotide sequence or sequence fragment), and that the text following the semicolon on the first line gives an identifier for the sequence. The text on the second line is a human-readable description of the sequence, and subsequent lines contain the one-letter codes for the amino acids themselves, terminated by an asterisk.

In JSON, we can dispense with much of the syntactic fluff, as the '>', ';' and '*' characters are just there in order for parsers to be able to distinguish between the different parts of the file. All that we really care about is:

1 This is a protein
2 Its identifier is HBA_HUMAN
3 It's got a human-readable description
4 It's got an amino acid sequence

In JSON syntax, we could encode this as:

```
{
  "type" : "protein",
  "identifier" : "HBA_HUMAN",
  "description" : "Hemoglobin subunit alpha OS=Homo sapiens GN=HBA1 PE=1 SV=2",
  "sequence": "MVLSPADKTNVKAAWGKVGAHAGEYGAEALERMFLSFPTTKTYFPHFDLSHGSAQVKGH
  GKKVADALTNAVAHVDDMPNALSALSDLHAHKLRVDPVNFKLLSHCLLVTLAAHLPAEFTPAVHASLDKFLA
  SVSTVLTSKYR"
}
```

Although, in some ways, this might look more complex than the original PIR representation – and it's certainly more verbose – the huge benefit of encoding the data in this way is that there's no need to write a bespoke parser. Instead, you simply load the data into a standard JSON parser, and can then interrogate the result to retrieve the 'type', 'identifier', 'description' and 'sequence', and use them in whatever way is meaningful for your application. What we've created here is a single JSON dictionary (defined by the outer { } characters, which are part of the JSON syntax), containing four key/value pairs separated by a ':' character ('type' is a key, 'protein' is a value), one pair for each of the four properties we want to record. The keys we've chosen here are completely arbitrary, and would only mean anything to our particular application – the JSON parser doesn't care what they're called, as long as they're represented using the right syntax.

If we wanted, at a later date, to extend the number of properties we recorded for this protein, it would simply be a case of adding a new key/value pair using the JSON syntax. Tools built prior to this extension using a JSON parser would continue to work seamlessly; the JSON parser would happily read in the new key/value pairs, but the tool's internal logic would never ask for that particular bit of data, because it wasn't programmed to do so. Box 7.5 shows a more complex JSON example, illustrating the use of nested structures and arrays.

XML

The eXtensible Markup Language, XML, shares some common goals with JSON, in that it allows objects and their properties to be encoded in a well-defined language that can then be created and consumed by standard parsers. XML badges itself as a 'language', rather than as a 'notation', and is considerably richer and more complex than JSON. Based on an earlier and much more complex markup language, SGML (Goldfarb, 1990), XML presents a set of highly regular rules for how to encode hierarchically structured data.

Box 7.5 JSON: The Javascript Simple Object Notation

The JavaScript Simple Object Notation (JSON) is a serialisation format based on how values are represented in the JavaScript language. This is limited to character strings, numbers, Boolean values, arrays and dictionaries (which make associations between two values), and therefore has no built-in support for namespaces, provenance or metadata. Everything we need, we must specify ourselves, and must be explicitly encoded in the JSON itself.

At first sight, this might seem too restrictive to be of much use, but it's simple, and has excellent support in most programming languages. As a result, building systems that generate or consume JSON can be done very quickly; and, if the data are simple, without the need for complex metadata, this becomes almost trivial.

The fragment of JSON below shows a simple JSON result returned by querying **Europe PubMed Central** for UniProtKB entries that cite the article with Pubmed ID 2137202 (Dryja et al., 1990) as evidence to back up claims in the database record. Items enclosed in square brackets [ ] represent lists; those in wiggly braces { } are dictionary key/value pairs. For example, the top level of the JSON is a dictionary containing five entries with the keys dbCountList, dbCrossReferenceList, hitCount, request and version. Deeper into the JSON, dbCrossReferenceInfo contains a list, in which is one dictionary; unfortunately, in this instance, the keys used by the dictionary are rather meaningless, and the UniProtKB accession number that we might care about programmatically (P08100[a]) is simply referred to as "info1", making it very difficult for a computer to use this particular data reliably (this is an unfortunate limitation of the way Europe PMC exposes its data, and not a limitation of JSON!).

```
{
    "dbCountList": {
        "db": [
            {
                "count": 1,
                "dbName": "UNIPROTKB"
            }
        ]
    },
    "dbCrossReferenceList": {
        "dbCrossReference": [
            {
                "dbCount": 1,
                "dbCrossReferenceInfo": [
                    {
                        "info1": "P08100",
                        "info2": "Rhodopsin",
                        "info3": "Homo sapiens",
                        "info4": "UniProtKB"
                    }
                ],
                "dbName": "UNIPROTKB"
            }
        ]
    },
    "hitCount": 1,
    "request": {
        "database": "UNIPROTKB",
        "id": "2137202",
        "page": 1,
        "source": "MED"
    },
    "version": "3.0.1"
}
```

JSON retrieved from Europe PMC listing UniProtKB entries that cite Dryja et al. (1990).

[a]http://www.uniprot.org/uniprot/P08100

If we take our HBA_HUMAN example once more, a possible simple encoding in XML could look like:

```
<sequence>
    <type>protein</type>
    <identifier>HBA_HUMAN</identifier>
    <description>
        Hemoglobin subunit alpha OS=Homo sapiens GN=HBA1 PE=1 SV=2
    </description>
    <aminoacids>
        MVLSPADKTNVKAAWGKVGAHAGEYGAEALERMFLSFPTTKTYFPHFDLSHGSAQVKGHG
        KKVADALTNAVAHVDDMPNALSALSDLHAHKLRVDPVNFKLLSHCLLVTLAAHLPAEFTP
        AVHASLDKFLASVSTVLTSKYR
    <aminoacids>
</sequence>
```

In XML, data elements are surrounded by 'tags'; in our example here, <sequence> is a 'start tag', </sequence> the corresponding 'end tag', and other elements, such as 'type' and 'identifier', are nested within these. As with JSON, XML parsers don't care about what the tags are or mean, as long as they are syntactically valid and correctly nested to form a hierarchy of properties.

If that were all there were to say about XML, it wouldn't be so noteworthy. However, the simplicity of its paradigm allows us to use XML for many tasks. To begin with, we can, for a given XML application, specify precisely the expected structure of the data: *e.g.*, 'expect a *sequence* element, that holds one each of *type* and *identifier* elements'. We can use an XML Schema (XSchema) to formalise such a *schema*, somewhat recursively, in XML; this can then be used to validate that a given XML document has the expected structure, or to generate automatic parsers for arbitrary programming languages. We can specify visual styles for when we wish to view the information in an XML document in a formatted way, using the eXtensible Stylesheet Language (XSL). We can also translate between two different XML schemas (assuming the data are translatable) by using an XSL Transformation (XSLT), itself an XML document that specifies the mapping of elements and attributes needed to enact the translation. For simple queries of XML documents, we can use an XML Path (XPath), a filesystem-like path for identifying and retrieving individual elements or attributes of interest: *e.g.*, the XPath /sequence/identifier[1]/text() would match the string HBA_HUMAN in our example. For greater flexibility, we can use the XML Query language (XQuery), which allows us to reference external resources, use more advanced logic, and introduce state to the query process. To link between XML documents, allowing us to split our data and reduce redundancy, we can use the XLink protocol.

XML also provides a method for assigning namespaces to the names of elements and attributes, just by including a number of standard attributes in elements to which we wish those namespaces to apply. XML namespaces are actually just URIs; each is assigned a prefix (which can be empty, if it's the default) that can be used to qualify both element and attribute names directly.

The plethora of XML schemas available, each providing its own set of useful extensions, reduces the need to re-invent wheels. Below is a shortlist of some of the more useful XML schemas available:

- *Mathematical Markup Language (MathML)* – for logically describing mathematical equations in a way that allows both computation over the equations and conversion of them for display.

- *XML HyperText Markup Language (XHTML)* – an XML formulation of HTML4 that allows inclusion of formatted text in XML documents, which can therefore be parsed by general-purpose XML libraries. Note that HTML5, which supersedes HTML4, isn't valid XML, and is best treated as an output format, rather than being used to store data.
- *Scalable Vector Graphics (SVG)* – this XML schema allows inclusion of 2D vector graphics into documents.

Although getting started with XML is quite easy, becoming an expert, and understanding all the pros and cons to the different ways in which data can be represented in XML, takes some considerable time. There's a whirlwind tour of some of its basic properties in Box 7.6; however, to become an XML master, you'll need a proper XML tutorial.

RDF

Whereas JSON and XML represent data as a hierarchy or tree, the **Resource Description Framework (RDF)** takes a much more general approach, and considers data as an arbitrary graph describing the relationships between things. In both our JSON and XML examples, properties such as 'type' or 'id' are implicitly associated with our particular protein, because they are nested within the fragment of JSON or XML that describes it. In RDF, data structures are created by explicitly describing all the relationships between the different components. While this can end up being extremely verbose (and you'd certainly not want to try reading anything but the most trivial RDF files as a human), it does mean that RDF can be used to describe things that don't naturally form hierarchical structures.

The fragment of RDF below encodes our HBA_HUMAN example in a style known as 'Notation3':

```
@prefix p:   <http://www.example.org/sequence_details#>

<http://www.example.org/sequence#HBA_HUMAN>
    p:type "protein";
    p:description "Hemoglobin subunit alpha OS=Homo sapiens GN=HBA1 PE=1 SV=2";
    p:sequence "MVLSPADKTNVKAAWGKVGAHAGEYGAEALERMFLSFPTTKTYFPHFDLSHGSAQVKGHGKKV
               ADALTNAVAHVDDMPNALSALSDLHAHKLRVDPVNFKLLSHCLLVTLAAHLPAEFTPAVHASL
               DKFLASVSTVLTSKYR" .
```

This example makes three independent statements about the sequence HBA_HUMAN: first, that is has type 'protein'; second, that it has a particular textual description; and third, that it has a specific sequence. You can think of these as forming three 'subject–predicate–object' clauses. Any of them could be omitted, or additional ones could be added, and the RDF would still be valid. As the RDF is read by a program, the different statements are linked together to form a graph of properties and relationships. RDF graphs can be interrogated using the SPARQL query language (SPARQL, pronounced 'sparkle' is a recursive acronym for 'SPARQL Protocol and RDF Query Language). The syntax of SPARQL, and the kinds of query operation it provides, are relatively simple, as the language only has to embody the 'subject–predicate–object'; but the behaviour of a SPARQL query can be hard to predict, as the amount of work done to answer a query depends more on the topology and scale of the graph than on the structure of the query.

Box 7.6 XML: The eXtensible Markup Language

XML is a very simple framework, but one that's capable of great power and flexibility. It becomes almost trivial to include the type of information we found desirable in Section 7.4.5: if we need to keep track of the format version in use, we could simply add a `version` attribute to the root element; if we need to show provenance information, we could add attributes or elements to capture that. An example of these qualities in use in a single XML document is shown below. Line 1 tells us the version of the XML framework we're using (1.0); lines 3–6 define various namespaces, including a default namespace for all the elements that do not include an explicit prefix; lines 8–12 provide provenance information, including the current revision of this entity; line 16 defines the version of this particular file format; line 17 points us at the XSchema that defines this format. The rest of the file includes various elements from the different namespaces (session, property, *etc.*) to annotate a particular region on page 8 of a scientific journal article with a comment relating to the content.

```
 1  <?xml version='1.0' encoding='utf-8'?>
 2  <annotation
 3    xmlns:property="http://utopia.cs.manchester.ac.uk/kend/property#"
 4    xmlns:session="http://utopia.cs.manchester.ac.uk/kend/session#"
 5    xmlns:xsi="http://www.w3.org/2001/XMLSchema-instance"
 6    xmlns="http://utopia.cs.manchester.ac.uk/kend"
 7    concept="UserComment"
 8    author="http://utopia.cs.manchester.ac.uk/users/11282"
 9    created="2011-01-20T14:06:08"
10    updated="2011-01-20T14:06:08"
11    revision="1"
12    status="published"
13    id="http://utopia.cs.manchester.ac.uk/annotations/812"
14    edit="https://utopia.cs.manchester.ac.uk/kend/0.7/annotations/812"
15    context="http://utopia.cs.manchester.ac.uk/contexts/public"
16    version="0.7"
17    xsi:schemaLocation="http://utopia.cs.manchester.ac.uk/kend
18        https://utopia.cs.manchester.ac.uk/kend/0.7/xsd/annotation">
19  <properties>
20    <property:comment>
21      Does anyone know the PDB code for this figure?
22    </property:comment>
23    <session:sourceVersion>Kend/0.6</session:sourceVersion>
24  </properties>
25  <anchor>
26    <document id="http://utopia.cs.manchester.ac.uk/documents/819">
27      <extent>
28        <text>M.tb CRP/ FNR</text>
29        <bounds
30          page="8"
31          rotation="0"
32          left="260.4472"
33          top="214.578041559"
34          right="279.0992"
35          bottom="222.807653556" />
36      </extent>
37    </document>
38  </anchor>
39  </annotation>
```

An example of using XML to encode complex data.

At one level, the three paradigms outlined above achieve the same effect – they provide a way of storing data in a robust and extensible format that avoids the need to develop bespoke parsers to read and write hand-rolled formats. So while, in some sense, they are interchangeable, they are actually each best suited for slightly different purposes.

JSON is lightweight, quick to read and write, and is often sufficient for serialising the kind of data structures typically found in programs into a file format suitable for persistant storage, or for representing the results of a function call in a serialised manner. Its lightweight and comparatively terse syntax, however, tends to mean that the contents of files are not 'self describing', and don't have explicit namespaces or qualifiers, so a little extra effort is required on the part of the application programmer to make JSON files suitable for exchange between different applications. XML, by comparison, is much more verbose and has explicit support for URIs, namespaces and such; reading and writing XML files is more costly (and thus slower) than using JSON, but the results are typically better suited for exchange between applications. Finally, RDF is the most expressive and complex of these paradigms, and best suited for applications in which precision about data types and values is important, such as those that aggregate and process data from multiple sources.

7.5 Giving meaning to data

Having identifiers for things, and ways of representing data that are readily accessible to machines, gets us only so far: if we want to use machines to help us make sense of data and information, rather than merely to record it, we need to head towards what's known as 'knowledge engineering'.

To build a knowledge-base that a computer can 'understand' or reason about, we need to start with a *controlled vocabulary*: an agreed set of tokens that are allowable in a particular context, that can be associated with particular meanings, and can only change by agreement from a central authority. Having established a controlled vocabulary, we can begin to make richer associations between terms; for example, defining certain things as being synonyms or opposites of others, and eventually defining rules or axioms that govern these relationships. If the associations and rules are captured in machine-readable form, and are themselves drawn from a controlled vocabulary, then we have the basis for building an unambiguous and hence 'computable' body of knowledge.

To illustrate the process of 'bootstrapping' a machine-readable body of knowledge, imagine a scenario where a team of bioinformaticians from various international laboratories collaborates to build an integrated suite of protein sequence-analysis tools. Some of these are predictive, operating on sequences to calculate various properties (*e.g.*, predicting transmembrane domains or secondary structures); others are interactive visualisation applications, designed to allow a human to analyse and interpret the results.

Each lab has its own favourite programming language suited for a particular task, so they'll need some language-agnostic way of representing knowledge. As we've already seen, there are plenty of file formats that do a reasonable job of encapsulating amino acid sequences and associated features, so there's little point in re-inventing these, and there are also many primary databases providing sequence identifiers; but what about the names and properties of the amino acids themselves? The tools need to understand such properties in order to predict features, or to make sensible decisions about how to visualise the sequences, otherwise situations may arise where one tool believes a residue to be hydrophobic while another treats it as hydrophilic. Perhaps more importantly, all the file formats we've discussed so far simply assume that any tool consuming the data just 'knows' that 'M' is methionine; but where is this knowledge, and the fact that M has hydrophobic character, actually captured?

The reality is that, in most cases, such knowledge is buried in the logic of a particular tool/program, 'hard coded' in a series of conditional statements and bespoke data-structures. This makes it very difficult to be certain about the behaviour of tools (recall the different ways in which alignment tools such as ClustalX and CINEMA depict amino acid properties in Section 5.8.2).

It would be considerably safer and more elegant if knowledge of the various amino acid properties could be encapsulated in some machine-readable form that all the tools in the suite could agree to consume and adhere to. The first step in creating such a body of knowledge is to define the *vocabulary* in question. In this case, that's an easy task, as there are only 20 naturally occurring amino acids. Associating each of these names with a human-readable definition gives a *glossary*.

Of course there's more to amino acids than just their names – the real challenge is to capture their biochemical properties. As we've seen already, these can be neatly represented in a visual form – recall Figure 2.6. Here, the set-based metaphor of the Venn diagram provides a concise visual way of communicating to human readers a lot of information: at a glance, we can see that the amino acids fall into two sets, 'hydrophobic' and 'hydrophilic', but that these overlap (this slightly tangled love–hate relationship was explained in Section 2.2.3); we can see that these top-level sets are then sub-divided – *e.g.*, there are 'charged' residues, which are split into 'positive' and 'negative' ones, and so on. Representing this knowledge in terms of 'things' and 'relationships between things' would give us a tree-like graph, like the one shown in Figure 7.5.

This tree-like representation is certainly less visually compact than the Venn diagram, but it captures the same information, and is heading in the direction of something that's more amenable to interpretation by computer: given any property-node, for example, it's relatively straightforward to determine any sub-properties, and which amino acids have those properties; alternatively, starting with an amino acid and moving 'up' the tree, would yield a list of increasingly 'broad' properties (*e.g.*, for arginine, we'd get 'positive', 'charged' and 'hydrophilic'). It should be fairly straightforward to see how such a hierarchy could be captured in XML, JSON or RDF, and how a machine could interpret it by traversing up and down the tree-like structure.

On first inspection, it might appear that we've created a bit of a hodge-podge of different concepts and relationships here (surely 'amino acids' are not the same kind of thing as 'charged' or 'methionine'?); but thinking a little more abstractly about what we've done reveals otherwise. In trying to classify amino acids, we've ended up (perhaps not surprisingly) with a hierarchy of *classes*: we have a class of things that are amino acids, a subclass of which are those amino acids that are 'hydrophilic', a subclass of which are 'charged', and so on. It's important to remember that the things at the bottom of our diagram are also *classes of amino acid* – when we say 'valine', we are talking about the 'class of molecules that we call valine', and not a specific instance of a valine molecule. As well as having only one type of concept in our graph (*i.e.*, 'class'), we've also ended up with only one type of relationship: 'is a subclass of'.

What we've created so far is known, in knowledge-representation terms, as a **taxonomy**, a hierarchical structure that categorises concepts with shared properties in terms of classes and subclasses, with general categories at the top, and increasingly specific

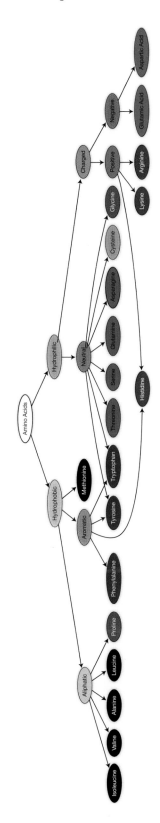

Figure 7.5

A tree-like graph of amino acids and their properties. Reading from the top down, we can 'accumulate' properties until we reach an amino acid (*e.g.*, that Aspartic Acid is an Amino Acid that is Hydrophilic, Charged and Negative).

concepts lower down the hierarchy. Many taxonomies pre-date computational knowledge representation: perhaps best known is the Linnaean hierarchy for classifying genus and species names.

Taxonomies are curiously seductive intellectual artefacts, and there's something deeply satisfying about being able to tame a tangled web of unruly ideas by arranging them in a neat hierarchy. We have to accept, however, that the desire to do so is nothing more than human predilection. Throughout history, life scientists have been concerned with documenting the minutiae of living systems in what has been described as a sort of 'biological stamp collecting' (Hunter, 2006). Although creating taxonomies may give a sense that we've mastered a particular domain, unfortunately the natural world, and all the biochemical and evolutionary forces that shape it, rarely yields concepts that fit perfectly into such tidy structures. In particular, the notion of 'hierarchy' used in taxonomies makes them fragile and difficult to adapt over time, as properties 'higher up' the tree have significant influence on how concepts are arranged lower down. Thus, mistakes in classification, or the discovery of things that don't conveniently fit anywhere in the hierarchy, can cause widespread disruption.

So let's move beyond simple taxonomies and imagine that we want to extend our system to explicitly represent the one-letter symbols used to encode amino acids in most sequence file formats. Simply including them in our tree-like diagram, as shown in Figure 7.6, would be horribly misleading, as our convention of reading 'down' the tree to give subclasses now doesn't work: it certainly doesn't make sense to say that 'P is a subclass of proline'. So we need a rather more nuanced way to capture and represent knowledge.

Until now, we've drawn our amino acid classification as a tree-like graph, albeit with the 'root' at the top and the 'leaves' at the bottom; but remember, this is merely a visual convention that was useful for illustrating the taxonomy we originally created. By adding in a different type of 'edge', representing the relationship 'has the one-letter code', we can create a basic *thesaurus* – a more general structure that allows things to be given multiple names or *labels,* and other arbitrary relationships to be explicitly recorded. Dispensing with the tree-like layout, and instead showing the relationships in a more generic graph form, gives Figure 7.7. With this kind of structure, it should be straightforward to see how additional properties or concepts could be added to the graph in a machine-readable form (consider, for example, how we might add the notion of 'essential amino acid').

Many of the so-called ontologies in the life sciences are actually examples of vocabularies, glossaries, taxonomies or thesauri. However, one further level of sophistication can be added to these to create what are often referred to as 'axiomatically rich ontologies', which are assuming increasing importance in bioinformatics (although, somewhat ironically, there's no commonly agreed name for such things). In all the structures we've examined so far, knowledge is captured by explicitly enumerating classes and relationships between them. Using an **ontology language** to write rules, and a **reasoner** to check that they are logically consistent, it's possible to add axioms to a body of knowledge, either to avoid having to list everything explicitly, or to impose logical constraints. For example, we could write axioms that assert:

1 Glutamic acid and aspartic acid are 'negative'.
2 Arginine and lysine are 'positive'.
3 Things can't be both 'positive' and 'negative'.
4 Anything that is either 'positive' or 'negative' is 'charged'.
5 All hydrophilic residues that aren't 'charged' are 'neutral'.

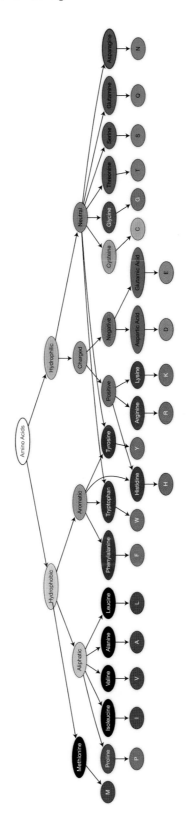

Figure 7.6

The hierarchy of amino acids and their properties, augmented with one-letter codes.

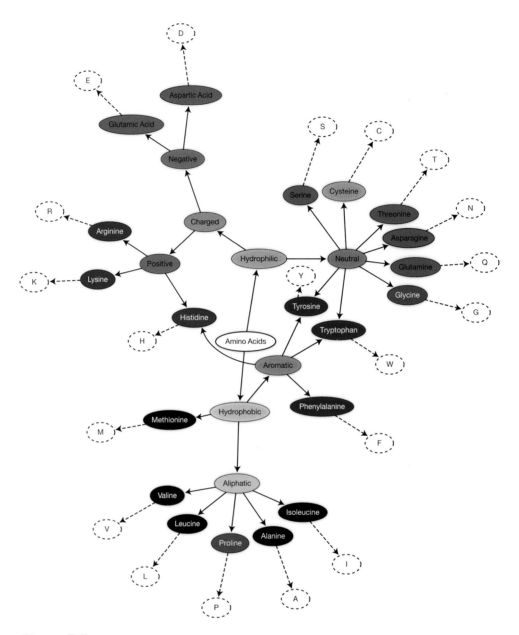

Figure 7.7
The set of relationships shown in Figure 7.6 laid-out without the 'top-down' constraint.

As well as creating richer relationships between classes, this would reduce the number of explicit relationships that we'd have to manage manually.

7.5.1 Bio ontologies in practice

Over the years, there have been many attempts to define standard formats, languages and tools for knowledge engineering, but only three – SKOS, OWL and OBO – are in widespread use in bioinformatics today. The following sections give the briefest flavours of how these are used in practice.

SKOS: the Simple Knowledge Organisation System

SKOS provides a way to create controlled vocabularies, taxonomies and thesauri. It's designed primarily for use in indexing- and searching applications: *e.g.*, a user searching a database for 'rhodopsin' may also be interested in the broader category of GPCRs – it's straightforward to capture such relationships in SKOS.

SKOS is intentionally restrictive (hence the 'simple' in its name), providing a small, controlled set of relationships between concepts. The SKOS Primer[8] describes *concepts* as:

> units of thought – ideas, meanings, or (categories of) objects and events…[that]…exist in the mind as abstract entities which are independent of the terms used to label them.

Consider the following fragment:

```
ex:Leucine rdf:type skos:Concept;
skos:prefLabel "Leucine"@en;
skos:prefLabel "Leucina"@es;
skos:prefLabel "Leucinium"@la;
skos:prefLabel "Leucin"@de;
skos:narrower ex:Aliphatic.
```

The first line defines the URI Leucine in the 'example' namespace 'ex:' as having the RDF type of a SKOS Concept. Lines 2, 3, 4 and 5 associate a 'preferred label' with the concept in English, Spanish, Latin and German, respectively (there can be, at most, one prefLabel per language). The final line defines ex:Leucine as being a 'narrower' concept than the concept ex:Aliphatic (*i.e.*, it defines Leucine as a subclass of Aliphatic). SKOS allows any number of synonymous 'alternative labels' to be defined using skos:altLabel. For example, we might add:

```
skos:altLabel "4-Methyl-L-norvaline";
skos:altLabel "H-Leu-OH";
skos:altLabel "L-Leucine";
```

to indicate that 4-Methyl-L-norvaline and H-Leu-OH are '(human) language independent' synonyms for Leucine.

OWL: the Web Ontology Language

At the other extreme of expressiveness sits the family of knowledge-representation languages known as OWL (a sort-of acronym of Web Ontology Language[9]). Whereas creating taxonomies and thesauri with SKOS is primarily about capturing concepts and simple relationships, an OWL ontology seeks to represent objects in a way that makes it possible to recognise to which category or class of objects they belong. It's this *definitional step* – how to recognise an object's class – that's the main intentional difference. OWL does this by having a logic-based language in which statements or axioms have a strict interpretation, such that a computer can draw conclusions from them. For example, we could say that in order to be considered an amino acid, a molecule must have

---

[8] http://www.w3.org/TR/skos-primer

[9] It's sometimes said that the Web Ontology Language, which would naturally have the acronym 'WOL' is known as 'OWL' because of the way in which the character Owl in Winnie The Pooh (Milne, 1926) misspelled his own name. Sadly, this isn't true: OWL was chosen as a name because it represented wisdom, sounded nice, and offered possibilities of a decent logo.

amine and carboxylic acid functional groups and specific side-chains: *i.e.*, these are *necessary* conditions. We can also say that, to recognise an object as belonging to the class of hydrophilic amino acids, it's enough that the amino acid has a hydrophilic side-chain (a *sufficient* condition). It's easy to imagine more complex combinations of conditions that will result in a tangled web of amino acid classes, each of which is defined in terms of its properties: this would be an ontology of amino acids.

Using OWL, it's possible to define classes, properties, instances and operations, such as union, intersection, complement and disjointness. By combining these, OWL allows axiomatically-rich ontologies to be created and subjected to **automated reasoning**, where a computer can check the logical consistency of an ontology (see Box 7.7). In the context of our amino acid ontology, for example, operations could be used to assert that 'charged' amino acids must be 'positive' or 'negative', but can't be both, or, indeed, anything else, thereby explicitly capturing the notion that 'positiveness' and 'negativeness' are disjoint but necessary states (in our previous representations, this idea existed only implicitly, as none

Box 7.7 Automated Reasoning and the Open and Closed World Assumptions

The Open and Closed World Assumptions in the world of formal logic and ontologies state there are two ways of looking at systems of knowledge. Under the Closed World Assumption, anything that isn't known to be true must be false; the Open World Assumption, on the other hand, says that anything that isn't known to be true is simply unknown. Take, for example, the assertion 'valine is hydrophobic'. If we ask the question 'Is valine hydrophilic?' in a Closed World context, the answer would be 'no', because we don't have either a statement saying that it is true or the means by which it can be inferred to be true; in an Open World, the answer would be 'I have no idea'.

In a Closed World, the assumption is that we have all the knowledge that there is about a certain domain captured within a system; for example, a database of all the students enrolled at a particular university can definitively answer the question, 'Is this student enrolled?': 'yes' if the person is recorded in the database as a student; 'no' for everyone else in the world. The Open World Assumption is useful when a system is trying to describe something using incomplete information. A database of sequenced genes is an example of data that should be interpreted under an Open World assumption. You can ask, 'Is there a human gene associated with diabetes', but the only sensible answers are 'yes' or 'we haven't found one yet'.

The way that logic and reasoning behave in Open and Closed Worlds differs considerably. Imagine that, starting to build an amino acid ontology from scratch, we make the assertions: 'There are two types of charge: positive and negative', 'Charged amino acids have exactly one type of charge' and then 'Arginine is a positively charged amino acid'. Next, we erroneously add the biologically incorrect assertion that 'Arginine is a negatively charged amino acid'. In a Closed World, this would be an invalid state, as we've said a charged amino acid is only allowed to have one type of charge. In an Open World, perhaps unexpectedly, this would not cause an error, but would instead result in a new inferred axiom being generated that reconciles our statements – that 'positive' and 'negative' charges must be the same thing, because we've not said explicitly otherwise! The Closed World assumption includes the idea that each named class is distinct, unless we explicitly tell our system otherwise; the Open World assumption takes the opposite view, that we don't know whether classes are the same until we're explicitly told one way or the other. Because, in most cases, ontologies describe systems that can't be – or at least aren't yet – fully encapsulated, languages like OWL and its associated reasoning tools are based on the Open World Assumption. Although it might seem daunting at first to have to explicitly define properties such as disjointness (*e.g.*, 'positive and negative are different things') for every class, ontology-building tools, like Protégé (Noy *et al.*, 2001), actually make this very easy, and also provide sanity checks, such as spotting dangling classes that haven't been linked into the ontology.

of the amino acid classes exist as subclasses of both positive and negative). Unlike SKOS, which has a restricted set of relationship types, OWL allows arbitrary associations to be created between classes, along with precisely specified rules about how those associations can be used, and should be interpreted by, machines: *e.g.*, we could capture the idea that every amino acid has a specific mass, or that there are only 20 naturally occurring amino acids.

OBO: the Open Biological and Biomedical Ontologies

The OBO ontologies play a central role in knowledge engineering in the life sciences; being rather more than just languages or formats, they require a little more context to explain them. Scientists are used to having a common language for some of the things they talk about – the prime example must be SI units. Similarly, biologists have, for a long time, had a common language for talking about the sequences of amino acid residues and nucleotide bases – the long strings of letters written in the familiar protein and nucleic acid sequence alphabets. The single-letter codes for amino acids and nucleotide bases have also allowed development of computational techniques for measuring the similarity between two or more protein or nucleic acid sequences.

In the post-genomic era, as whole genomes became readily available from many organisms, this common language also enabled measurement of similarities between genomes of different organisms. What wasn't possible was comparison of *what we knew* about those genomes – *e.g.*, what their genes did and where they did it. Each organism community had developed its own bespoke sub-language for talking about what was known about 'its' genes and gene products, creating incompatible silos of organism-specific knowledge in the process. Eventually, a group of scientists realised that a common language or vocabulary was needed to try to tackle this problem.

Thus was born the Gene Ontology, GO (The GO Consortium, 2000), a controlled vocabulary for talking about a gene product's location in a cell, its molecular function and the processes in which it participates. Now, more than 40 genome- and cross-organism databases use GO for describing the major attributes of gene products. This was largely brought about by the GO Consortium's openness – the GO was given away freely, and was open to input from anyone who used it. All of this was seen to be a Good Thing. As other scientists started making their own ontologies (Sequence Ontology (Eilbeck *et al.*, 2005), ChEBI (Hastings *et al.*, 2013), *etc.* – see Box 7.8 for other ontologies), a need was perceived to co-ordinate these activities to avoid chaos – from this grew the Open Biomedical Ontologies Consortium (Smith, 2007). One of their guiding principles was to use a common syntax (and semantics) for their ontologies. GO had its own format, which became the OBO Format; the other recommended representation was OWL. Over the years, as the need to use reasoning to facilitate the maintenance of large, complex ontologies has both grown and been realised as desirable, these have become aligned to the extent that OBO Format (OBOF) is now essentially an alternative syntax for OWL.

7.5.2 First invent the universe

One of the challenges of getting to grips with practical ontology building is deciding where to start and where to stop (or perhaps, more formally, the 'domain' and 'scope' of an ontology). What is the purpose of the ontology? What kinds of question do we want to ask of it? Who or what are the consumers and producers of its content?

Without sensible answers to these questions, there's a real danger of over- or under-engineering. Commenting on the fact that, in some extreme sense, every possible

Box 7.8 Useful ontologies and schemas

There are hundreds of life-science-related ontologies, many of which are accessible via the NCBO's BioPortal[a] system, ranging from the very broad (*e.g.*, the Gene Ontology), to the quite specific (*e.g.*, the Fungal Gross Anatomy Ontology[b]). As well as ontologies that relate specifically to the life sciences, there are a handful of schemas and ontologies that are useful to know about when building bioinformatics applications.

The *Dublin Core* (*DC*) schema is a small set of vocabulary terms that can be used to describe Web resources (video, images, Web pages, *etc.*), as well as physical resources, such as books or CDs, and objects like artworks. The vocabulary includes terms for properties such as Title, Creator, Contributor, Subject, Description, Rights.

Whereas Dublin Core mostly describes 'things', the *Friend Of A Friend* ontology (FOAF) describes people, their activities and their relationships to other people and objects.

The PROV Ontology (*PROV-O*) can be used to describe provenance information to formally record change: who did what, when and why.

CiTO, the Citation Typing Ontology, allows for the formal characterisation of citations, both factually and rhetorically, for example allowing a database record to capture the fact that it's citing an article that supports or refutes a particular property.

EDAM is an ontology of common bioinformatics operations, types of data, topics, and data formats, and is useful for classifying the behaviour of tools and services.

The *Annotation Ontology* (Ciccarese *et al.*, 2011) is a vocabulary for annotating digital objects with comments, tags, examples, errata, *etc.*

The *Vocabulary of Interlinked Datasets* (*VoID*) describes datasets expressed in RDF. VoID descriptions can be used to help machines find, interpret and integrate data. The vocabulary provides terms that cover general metadata (using Dublin Core), which protocols can be used to access the data, the schemas used to structure it, and how the data relate to other data-sets.

---

[a] http://bioportal.bioontology.org
[b] https://bioportal.bioontology.org/ontologies/FAO

concept is related in some indirect way to every other, the astronomer Carl Sagan once said, 'If you wish to make an apple pie from scratch, you must first invent the universe' (Sagan, 1980); however, while it's possible to imagine an ontology that describes aspects of quantum physics, and another that defines the relationship of short-crust pastry to various configurations of fruit-based dessert, it's hard to imagine one that *usefully* covers both. The reality is that, as much as ontology engineering is about precision and the removal of implicit assumptions, one has to draw the line somewhere.

7.6 Web services

So far in this chapter, we've explored techniques for representing and giving meaning to data, focusing on how to make both these things accessible to a computer. In this section, we'll examine how these ideas can be applied to one of the most common tasks in modern bioinformatics: creating or consuming what have become known informally as 'Web services'.

The basic idea of a Web service is straightforward enough: a program running on one machine needs to access a service provided by another program somewhere else on the

Internet. There are many circumstances where accessing a service remotely like this is preferable to creating the necessary infrastructure to do the job locally:

1 The data-sets involved may be too large to be hosted locally;
2 The data could be changing too frequently for maintenance of a local copy to be practicable;
3 The computational load could require specialist hardware;
4 The software stack required to run the service locally could be prohibitively complex to install.

At some level, accessing a Web service is conceptually the same as invoking a function or procedure: data are passed in by way of input parameters; some kind of calculation is performed; and, some time later, results are returned to the caller. Within a single program executing on a single processor, this is an almost trivially straightforward concept – but expand the notion to programs interacting in today's highly distributed, highly heterogeneous computing environment, and the story becomes much more complex. To understand what it means to interact with a Web service, it's first necessary to understand, in a bit more detail, what it means to interact with the Web; and, to do that, we need to know a little more about *distributed systems* generally, and about the architecture of the Web specifically.

It's surprisingly hard to pin down a precise definition of a distributed system; besides, being too pedantic about what is and what isn't 'distributed' tends to obscure, rather than illuminate, the points we're discussing. So rather than worrying about trying to create a rule that definitively distinguishes a distributed system from a non-distributed one, we'll try to offer some principles that will allow us to arrange systems on a kind of spectrum of 'distributedness'. The overall idea is that a distributed system consists of multiple independent, largely self-sufficient computational components that, although spatially separated, exchange information in order to co-ordinate and co-operate in achieving a common goal. The difficulty with defining a distributed system along just these lines, however, is that there are many different configurations and architectures that have some of these properties to some extent or other.

At one extreme, for example, are multi-core processors: individual CPUs that are capable of executing multiple instructions simultaneously, on which programs executing in parallel can exchange information via shared access to the system's RAM. Although it could be argued that a multi-core CPU has several 'computational components', the fact that these are highly dependent on shared hardware, and that they co-exist on a single minuscule sliver of silicon would tend to rule this out as being in any sense a *distributed* system. Stepping up in physical scale, it's increasingly common for a motherboard to support multiple CPUs (which may, in turn, be multicore devices), sharing the same RAM and other resources. Although, here, the 'spatial separation' has increased from nanometres to centimetres, it's still hard to justify this as being particularly 'distributed'. Scaling up even further, we encounter 'cluster' and 'grid' computing, where multiple computers, each with their own power supply, disk, memory, and so on, are connected by a dedicated network infrastructure in order to co-operate in solving a specific task. At this point, we're definitely in rather a grey area: after all, there's certainly a greater sense of geographic distribution here – the computers could be in different rooms or even buildings – but there's still a sense of 'centralisation' (*e.g.*, it's common for computing clusters to be 'centrally administered'). Finally, at the other extreme, we have systems like the Web – which spans the entire globe, and has no central authority or administration – where the distributed nature is more obvious.

As it's quite hard to pin down the exact nature of a distributed system by saying what it *is*, it's sometimes defined in terms of what is *isn't*, using what are commonly known as the 'Fallacies of Distributed Computing', namely that:

1 The network is reliable.
2 Latency is zero.
3 Bandwidth is infinite.
4 The network is secure.
5 Topology doesn't change.
6 There is one administrator.
7 Transport cost is zero.
8 The network is homogeneous.

According to Peter Deutsch, one of the creators of the list,

Essentially everyone, when they first build a distributed application, makes [these]... eight assumptions. All prove to be false in the long run and all cause big trouble and painful learning experiences.

An eminent computer scientist once joked that 'a distributed system is one in which the failure of a computer you didn't even know existed can render your own computer unusable' (Lamport, 1987)[10] – and for systems that make these assumptions, it's all too often the case. It's important to note that some of these fallacies can equally apply to non-distributed systems (*e.g.*, even at the level of chip design, it would be silly to assume that 'latency is zero' or that 'bandwidth is infinite'); but if we have an architecture where none of these things are true, we can be fairly certain we have a distributed system.

Another useful way to think about the 'distributedness' of an architecture is to consider it from the point of view of what happens to the overall system when a component fails, and the frequency with which that's likely to happen. In the case of a multicore processor, or a multi-CPU motherboard, these are systems that are expected to function perfectly during a fairly well-defined life span (typically some small number of years), and eventually to fail catastrophically: all the hardware components depend to some extent on all the others, so when one bit breaks, so does the whole system. In the case of a computing cluster, there's more scope for 'graceful degradation' of service, and, in a well-designed setup, the failure of one machine needn't necessarily bring the entire cluster down. Then there's the Web: when was the last time the Web – the whole Web, not just your favourite corner of it – broke for everyone?

7.6.1 The architecture of the Web

The Web is an implementation of the Client/Server architectural **design pattern**. The idea is that a particular task is conceptually partitioned into two cooperating parts: a server implements a particular function, which it performs at the request of one or more clients. The defining characteristic of the client/server model is that each component has a particular role, as either client or server, and client components can only communicate with the server component, but not directly with each other. This is in direct contrast to the **Peer To Peer (P2P)** model, where every node is considered equal, and can communicate directly with any other node.

---

[10] http://research.microsoft.com/en-us/um/people/lamport/pubs/distributed-system.txt

The advantage of the client/server model is that its behaviour is easy to understand: a client asks for something to be done by a server; the server does it; the client gets the results and moves on; and it's clear how, in principle at least, this behaviour is similar to the act of 'calling a procedure'. The different roles of the client and the server are clearly defined, as is the point where 'control' passes from one to the other and then back again.

In the case of the Web itself, the different roles of client and server are clear: a server has the task of storing and handing out content, and a client (typically a Web browser) has the task of requesting specific resources, and then rendering the results on-screen for the user. The Web's resilience to failure comes, in part, from the fact that it consists of many servers, not just a single centralised one, with each server dealing with its own content, and handing that out to clients. But, when building a distributed system on top of the Web's infrastructure using Web services, the decision about which functionality goes 'server side' and which goes 'client side' is not always as clear-cut; and even with the Web, the distinction is slightly blurry: *e.g.*, in the original vision of the Web, browsers were simple 'thin' clients, which only had functionality for displaying hyperlinked text and graphics, with servers taking on the role of dealing with any 'logic'; modern browsers, however, include far richer functionality than this, allowing playing of videos, sound, and running executable programs (like those written in JavaScript or Flash).

7.6.2 Statelessness

The real resilience of the Web comes not just from the lack of any central authority or server, but from the fact that HTTP, the protocol used to exchange content between client and server, is *stateless*. This means that a client connects to a server, requests some content (*e.g.*, a page of HTML), and then disconnects, at which point the server 'forgets' that the client ever connected. If the user then requests a second page, the process is repeated. The server doesn't wait around between requests to see if there's another one coming soon – it simply does what it's been asked to do, then moves on. This might seem crude, or stupid – after all, it's quite common to 'click around' in a website a fair amount before following a hyperlink that goes to a different place. Wouldn't it therefore be more efficient for the server to keep a connection to a client open for a while, rather than re-doing the whole job from the start each time?

Far from being a silly way of doing things, the stateless nature of interactions between clients and servers on the Web is a fundamental reason why the Web mostly remains a reliable system that appears to work most of the time. The reason is this: the decision not to remember any state between requests means that servers don't have to deal with unpredictable behaviour from clients. Imagine a scenario where, to save on the 'start-up cost' of making a connection, a Web server 'waits around' for a while to see if a particular client is going to request a second page of content. How long should the server wait? What would happen if the user of the browser got bored, and went to make a cup of tea? Should the server try to guess that, and hold the connection open a bit longer? What happens to *other* clients that are trying to connect in the meantime, but are being queued because that user is now having tea and biscuits with colleagues, or perhaps has gone to bed, leaving the connection 'open'? The stateless nature of the communication between client and server allows the system to treat each request for content as an independent, small and 'completable' transaction that's always initialised by the client; a server's response is always 'here's the stuff you asked for', or one of several variations of 'sorry, I can't give you that'. To this extent, communicating with a Web server over HTTP can be thought of as rather like trying to communicate with an estranged friend: *you* always have to start the conversation; it won't talk to you for long; it won't call you back; and as soon as the conversation's over, it forgets it ever happened.

This means that, as far as the Web's infrastructure is concerned, every request for a resource is independent of every other: whether they come from the same or different clients, the fabric of the Web doesn't preserve any kind of state between requests. While this offers great resilience, it poses interesting challenges for the construction of 'Web services', which often inherently involve a transfer of state between programs.

7.7 Action at a distance

Throughout the history of computing, different techniques have been used to allow a program running on one computer to communicate over a network with a system running on another. In the days before the Internet made genuinely distributed systems possible, Remote Procedure Calls (RPCs) were often implemented using what are known as Object Request Brokers (ORBs) – systems that 'brokered' the exchange of data between two interacting programs. Via an ORB, the two interacting components would use an Interface Definition Language (IDL) to describe the inputs and outputs of the function to be invoked, and the ORB would then automatically generate code in the appropriate programming language, which would package up the data (a process called 'marshalling'), send it across the network and unpackage (unmarshall) it at the other end, ready for the receiving program to do its work, before repeating the process to return the results. The 'stub' or 'skeleton' code generated by the ORB would take care of mundane housekeeping tasks, like dealing with network delays or outages, checking that data weren't damaged in-transit, listening for results to be returned, and making sure that the calling program could get on with doing something else while waiting for results. The result, depicted in Figure 7.8, was a system that complemented the way in which traditional programs were constructed:

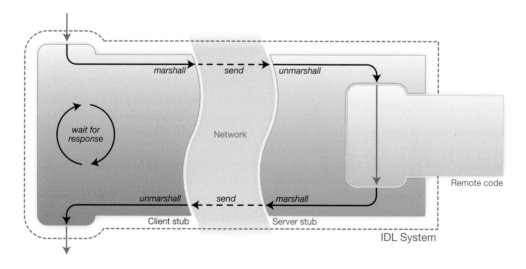

Figure 7.8

A simplified ORB-style architecture showing a 'client' and 'server' communicating by the exchange of messages.

i.e., such that a well-formed procedure call could be expected to complete reliably and within a predictable amount of time.

Most ORBs were designed in the early days of the Internet, when IP addresses were plentiful and cyber crime rare, and assumed that participating components were hosted in a safe, friendly environment on machines that could be directly accessed via static IP addresses. As the Internet became a global phenomenon, the demand for IP addresses outstripped their availability (see Box 7.9), and concerns about network security increased. Organisations were then forced to use dynamic address allocation, and to erect network firewalls to protect from hacking attempts, both of which made configuring ORBs somewhat clumsy. In consequence, the popularity of the ORB as a general-purpose RPC technique over wide-area networks diminished in favour of other architectures that could operate in this new dynamic,

---

Box 7.9 IP Addresses

Microsoft's Bill Gates reputedly said, '640k should be more than enough memory for anyone', while legend has it that IBM's Thomas J. Watson once stated, 'I think there is a world market for maybe five computers'. Although they may not have predicted quite how much information technology would change the world, it's still entertaining to imagine these pioneers of computing underestimating the impact and demand for their products so spectacularly. Somewhat disappointingly, there's no evidence that either of them ever said anything of the sort.

Where early computer scientists *did* rather badly underestimate demand, however, was in the design of the identification system used by the Internet itself: the *Internet Protocol Version 4 Address*, more commonly known as IP Address (or more accurately, IPv4 Address). IPv4 Addresses are 32-bit identifiers, usually written in human-readable form as four numbers between 0 and 255, separated by a dot (*e.g.*, 130.88.192.9). This means that there are 2^{32} possible IPv4 addresses, which gives 4,294,967,296 – or ~4.3 billion – unique addresses. Even though several large blocks of this range are reserved for special uses, the Internet's original designers, for whom, in the early 1980s, a computer was a hefty power-hungry box sitting on a desk, can perhaps be forgiven for thinking that this would be plenty of address space to share around.

What they didn't really predict was that the cost and size of computer hardware would reduce so much that we all nowadays tend to carry around several devices – smartphones, laptops, tablets – that are designed to be connected (sometimes permanently) to the Internet, and have, in our homes and offices, a growing number of other appliances – televisions, central heating controllers, security systems, and so on – that now form part of the growing 'internet of things' which also need an address. When every fridge, toaster and pencil sharpener demands an IP address, the problem will become even more acute.

Today, there are far more devices needing IP Addresses than there are IPv4 Addresses, so a number of tolerable bodges, such as CIDR (Classless Inter-Domain Routing[a]) and NAT (Network Address Translation) have been put in place while the world transitions to the next generation of IP Addressing, called IPv6[b]. IPv6 uses an 128-bit address space, which gives 2^{128} or 3.4×10^{38} possible IP addresses (or, if you like, more than 7.9×10^{28} times as many addresses as IPv4). At a rough estimate, that should be enough to allow every human that has ever lived in the history of mankind to have over 3.4×10^{27} Internet-connectable devices. Surely *that's* more than enough for anyone?

---

[a] http://en.wikipedia.org/wiki/Classless_Inter-Domain_Routing
[b] If you're wondering what happened to IPv5, the answer is an interesting one, and indicative of the complexities inherent in huge distributed systems like the Internet. The short answer is that the 'version number' forms part of the network protocol itself, and is allocated by the IANA as a kind of controlled vocabulary. The number 5 was used for what was essentially an experimental protocol, never intended for widespread use, leaving 6 as the next free value. Values from 7 to 9 have also been reserved for other purposes, so if there is a successor to IPv6 it will probably end up being called IPv10 or higher.

threat-conscious Internet. While ORBs still play an important role *within* some institutions (often those with considerable legacy hardware and software, such as the banking or finance industries), and are still used by some grid and cluster-computing systems, they're rarely seen these days as interfaces to publicly accessible bioinformatics systems on the Web.

As the popularity of the original ORBs faded, and use of the Web became more widespread, it was recognised that HTTP – originally designed primarily to deliver human-readable content via HTML to Web browsers – could be used equally well as the basis for a secure, firewall-friendly RPC paradigm more suited to the modern Internet's topology; thus was born the idea of what are now broadly referred to as 'Web services'.

The term 'Web service' was initially defined by the W3C as relating to a specific set of technologies that implemented remote procedure calling via the HTTP protocol; over time, however, the term has evolved to cover any Web-based remote procedure calling mechanism. In the following sections, we'll explore the pros and cons of three different Web service paradigms.

7.7.1 SOAP and WSDL

The original vision for Web services reflects many of the concepts used by early ORB systems. The approach revolves around two XML-based languages: WSDL (the Web Service Description Language), an IDL allowing service providers and consumers to agree on function names and parameter types; and SOAP (originally defined as the Simple Object Access Protocol, but later just as a name in its own right), a protocol for packaging up structured information to be sent over the network. In a classic example of the computer-science community's tendency to attempt to build the most generic and flexible system imaginable, the WSDL specification became extremely complex. While some of the major language toolkits sponsored by industry (namely those for Java and .net) implemented the WSDL and SOAP specifications reasonably fully and faithfully, those developed by the community for languages commonly used in bioinformatics (Perl, Python, Ruby, *etc.*) often supported only subsets of the complex specification, thus largely defeating the purposes of having a language-independent interoperability protocol in the first place. In environments where there's one administrator, an agreed programming language, reliable network and so on, the SOAP/WSDL approach can work tolerably well; however, in the more heterogeneous and anarchic world of bioinformatics services on the Internet, the downsides of its use often outweigh the benefits. Unsurprisingly, after a brief period in the limelight, the SOAP bubble burst: the use of SOAP and WSDL has dwindled in bioinformatics, and has been largely superseded by more lightweight 'Web native' approaches.

7.7.2 HTTP as an API

Whereas SOAP/WSDL-style Web services attempted to replicate traditional programmatic function or procedure calls by defining formal, rigid contracts between service suppliers and consumers, the architecture of the Web offered a rather different paradigm: it was realised that, simply with a change of mind-set, the same technologies that had been used to deliver content that *people* could read and navigate could be exploited for building programmatically accessible services, benefitting from all the security, performance and resilience of the technology that drives the human-readable Web (see Box 7.10).

Box 7.10 HTTP and RESTful Web services

HTTP is based on a controlled vocabulary of nine 'verbs' that define the actions a Web client (often a Web browser) can request to be carried out on a resource defined by a URI. Four of these verbs – HEAD, TRACE, OPTIONS, CONNECT – are essentially for 'housekeeping', allowing the client and server to optimise their communication and set up secure connections. Of the remaining four, the most common action is to GET a resource: *i.e.*, to fetch it from the server so that it can be interpreted by the client; this is what happens when we type the address of a Web page into a browser. Typically, when we fill in a form on a Web page and submit it, the browser is invoking the POST verb, which tells the server that we're handing over some data that we want it to store in relation to a particular URI. The remaining three verbs – PUT, DELETE, PATCH – allow the client to request the creation, complete removal, or the partial modification of a resource respectively, and are rarely used in regular Web-browsing sessions: generally speaking, most interactions involve GETting some content, and potentially POSTing some data back to the server (with various bits of housekeeping going on in the background). Although a typical Web-browsing session typically involves GETting a lot of content and occasionally POSTing a little back, the architecture of the Web itself is designed to allow clients to create, modify and delete resources too.

REpresentational State Transfer (commonly abbreviated to REST) is an architectural style (Fielding, 2000) designed to encapsulate how a well-behaved Web-based application should operate. The architecture is based around the idea of hierarchically arranged collections of resources, identified using URIs, which are modified strictly in line with the verbs provided by HTTP. Central to the architecture is the notion of Hypermedia As The Engine Of Application State (HATEOAS), where clients begin their interaction with a service via a fixed entry point (*e.g.*, GETting a particular URI), and then discover what further actions are available as a result of that. There are similarities between the idea of RESTful services and Linked Data, but the two paradigms are not the same (see Page *et al.*, 2011).

It's quite common for Web-based services that return data in a machine-readable format as a result of GETting a URI to be described as being RESTful; but unless the service implements all the constraints of the REST architecture, this is technically incorrect: REST is not a synonym for '*ad hoc* Web service that uses URIs to retrieve content in JSON/XML/RDF', but refers specifically to services that adhere to the HATEOAS architecture.

To understand the variations and some of the subtleties of the different Web service paradigms, it's first necessary to understand a little about HTTP itself. As the name suggests, HTTP is a protocol for transferring 'hypertext' between systems; and hypertext is simply a document (typically text, images and other media) structured in such a way as to have hyperlinks (references) to other resources. When we humans interact with the Web, our browser software uses HTTP to request a document identified by a given URL from a particular server; the server finds or creates the document (or more generally, the 'resource'), and returns it to the browser, which interprets it and renders it in a human-readable form. The 'human-readable' component of this process, however, isn't fundamental to HTTP; rather, it's just a side-effect of the fact that browsers are programmed to render documents that come back in a certain format (usually HTML) in a way that's palatable to humans. As far at HTTP is concerned, it has simply retrieved a resource and delivered it intact to the requestor.

It's a relatively small leap of the imagination to see how this set-up can be used to deliver machine-readable content by returning JSON, XML or RDF documents instead of HTML; and this, indeed, is the basis on which many '*ad hoc* Web services' are constructed. For example, pointing a Web browser at:

http://www.uniprot.org/uniprot/P02699

gives a human-readable version of the UniProtKB record with accession number P02699, whereas:

> http://www.uniprot.org/uniprot/P02699.xml

returns an XML representation of the same resource, but one that can more easily be processed programmatically. This approach of delivering machine-readable content in XML, RDF or JSON over HTTP makes it very easy to build simple tools that access online content programmatically. The simplicity of accessing a service in this way, however, comes with an important limitation: all the responsibility for interpreting the retrieved resource lies with the client. Here, this means that the client software has to understand UniProtKB's XML schema in order to be able to extract the various properties of the protein's record and do something meaningful with them. For example, inspecting the XML in a little detail, we can find a section that says:

```
<dbReference type="PDB" id="1F88">
  <property type="method" value="X-ray"/>
  <property type="resolution" value="2.80"/>
  <property type="chains" value="A/B=1-348"/>
</dbReference>
```

which we can interpret as being a link of sorts to the crystal structure of this protein, created through X-ray crystallography at a resolution of 2.80Å (and, assuming that the values for 'method', 'X-ray' and 'resolution' are available in some suitable ontology, a computer could make a similar decision); but there's nothing in the UniProtKB XML record that tells us how to then fetch the XML record for PDB's 1F88 (or even whether such a representation exists).

At this stage, the responsibility then falls back to the human programmer of the client software to read the PDB's Web service documentation to determine that – at least at the time of writing – a URL of the form:

> http://www.rcsb.org/pdb/download/downloadFile.do?fileFormat=xml&structureId=1F88

would fetch the required record, and then to explicitly write code that takes the ID value from the UniProtKB record, and 'manually' construct the appropriate PDB URL. This approach requires the human to recognise that something exists in the PDB namespace, and to manually find a suitable way of resolving the identifier within that namespace.

This somewhat fragile step, whereby the logic for linking between records has to be interpreted by humans and then explicitly written in executable code, could have been avoided by simply recording the 'outgoing' URL to the PDB record in UniProtKB's XML to create what's known as 'Linked Data'.

7.7.3 Linked Data

The term Linked Data was coined by Tim Berners-Lee in 2006 in a W3C design note relating to the idea of the 'semantic Web' – a web of data accessible to machines as well as humans. In this note, he outlined four simple rules for creating Linked Data – paraphrased slightly, these state:

1 Use URIs as names for things.
2 Use HTTP URIs so that people and machines can look up those names.

3 When someone (or thing) looks up a URI, provide useful information using standard machine-readable languages, schemas and ontologies.
4 Include links to other URIs so that more things can be discovered.

Although the idea of Linked Data is very simple in principle, its implications are profound, because it implies a different paradigm for consuming services and data on the Web. Whereas SOAP/WSDL and, to some extent, the *ad hoc* use of HTTP as an **API** (Application Programming Interface), imply a traditional programming model (call a function; make a decision; call another function), where behaviour is reliable and predictable, Linked Data allows machines to 'browse the web of data' in a similar way to which humans browse the Web. Thus, using the same stack of robust and readily available tools that allow a browser to fetch and interpret HTML for human consumption, a program can request machine-readable content, and, by inspecting the results and comparing terms with ontologies, make decisions about which further bits of data to retrieve. This 'browsing' paradigm is much more in line with the nature of the Web's underlying infrastructure.

It my appear, superficially at least, that, by looking at linked *data* on the Web, we've strayed somewhat from the original topic of Web *services*; however, in most cases, and certainly in most bioinformatics cases, the distinction between 'a service that produces some data' and 'some data that were produced by a service' is largely meaningless.

The real strength of Linked Data is that there are very few technological barriers to making content available in this style; the software for doing so is well understood and reliable, and, in most cases, only relatively minor changes would have to be made to the way in which XLM, RDF or JSON are produced. There's also very little conceptual complexity involved: the principles can be explained in four simple sentences. The barriers to moving from arbitrary 'XML over HTTP' style services are largely to do with the inertia associated with evolving any legacy system. Nevertheless, institutions such as the EBI are beginning to expose their services and data-sets as Linked Data (*e.g.*, March 2014 saw the first release of the ChEMBL database in linked RDF).

In the absence of primary database providers creating Linked Data versions of their content, various initiatives, such as bio2rdf (a community-based, open-source project) and Linked Life Data (a commercial venture from Bulgarian software company, Ontotext), have created Linked Data versions of many popular databases (Figure 7.9), enabling programs to integrate and 'browse' between what would otherwise be independent data silos.

The so-called 'Linked Data Cloud' contains RDF triples from all manner of different domains, ranging from national governmental data (*e.g.*, the UK railway stations, bus stops, airports and postcodes), geospatial information (*e.g.*, places of interest that are geographically close to one another, or information about continents and their countries, capitals and currencies), to items of more niche interest. Even in its currently emerging and somewhat anarchic state, the Linked Data Cloud provides a resource of astonishing scale and richness, which ultimately bodes well for the potential to create a Web of Knowledge readable by machines as well as humans.

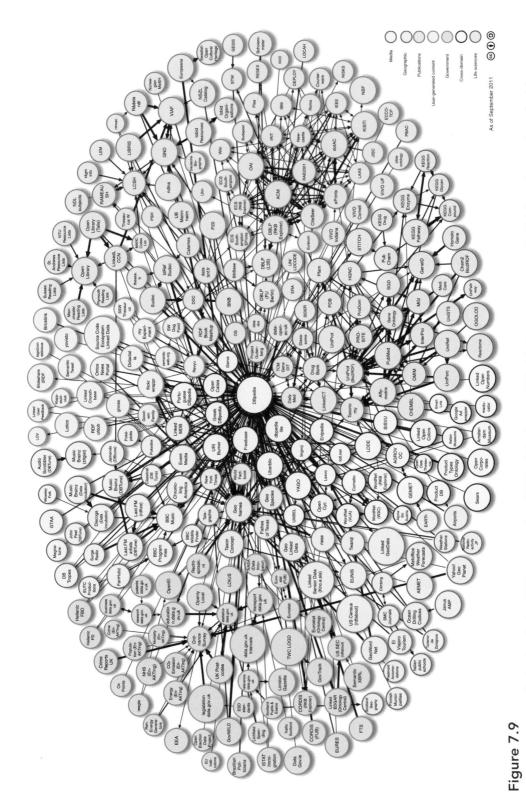

Figure 7.9

The 'linked data cloud'[i] with dbpedia (an RDF representation of Wikipedia) at its core. Linked Data resources relevant to bioinformatics are shown in pink.

Source: Linking Open Data cloud diagram 2014, by Max Schmachtenberg, Christian Bizer, Anja Jentzsch and Richard Cyganiak. Used under CC-BY-SA 3.0[ii].

[i]http://lod-cloud.net

[ii]http://creativecommons.org/licenses/by-sa/3.0

7.8 Summary

This chapter introduced some of the basic ideas involved in identifying, representing and giving meaning to bioinformatics data. Specifically we saw that:

1 Computers need explicit machine-readable context in order to be able to identify or interpret anything; although humans can deal with tacit assumptions, machines can't.
2 The best identifiers are often opaque, because these don't go stale or become inconsistent when the things they identify change.
3 Sometimes, the convenience of having non-opaque identifiers, such as for Web pages, outweighs the benefits of opacity (but this is rare).
4 Things in the real world inevitably change, so we need to build systems that take account of this.
5 But 'cool' identifiers don't change; there's no good technical reason why identifiers should ever be invalidated.
6 There's no one-size-fits-all way of representing data; it depends what we want to do with our data; but applying certain principles (like building in provenance and versioning) from day one can prevent trouble later on.
7 Paradigms such as XML, RDF and JSON can help us avoid building new file formats from scratch.
8 Knowledge can be captured in machine-readable form using technologies like SKOS, OWL and OBOF; we can then get machines to help us reason about our data and check that they make sense.
9 The various technologies that underpin the human-readable Web can also be used to create Linked Data and Web services that make data easy for machines to explore and interpret.

7.9 References

Ashburner, M., Ball, C.A., Blake, J.A. *et al.* (2000) Gene ontology: tool for the unification of biology. The Gene Ontology Consortium. *Nature Genetics*, **25**(1), 25–29.

Ciccarese, P., Ocana, M., Castro, L.J.G. *et al.* (2011) An open annotation ontology for science on Web 3.0. *Journal of Biomedical Semantics*, **2**(Suppl 2):S4.

Eilbeck K., Lewis S., Mungall C.J. *et al.* (2005) The Sequence Ontology: a tool for the unification of genome annotations. *Genome Biology*, **6**, R44.

Fielding, R.T. and Taylor, R.M. (2002) Principled design of the modern Web architecture. *ACM Transactions on Internet Technology*, **2**(2), 115–150. doi:10.1145/514183.514185

Goldfarb, D.F. (1990). *The SGML Handbook*. Oxford University Press, Oxford.

Hastings, J., de Matos, P., Dekker, A. *et al.* (2013) The ChEBI reference database and ontology for biologically relevant chemistry: enhancements for 2013. *Nucleic Acids Research*, **41**(Database Issue), 456–463.

Hunter, D.J. (2006) Genomics and proteomics in epidemiology: treasure trove or 'high-tech stamp collecting'? *Epidemiology*, **17**, 487–489. doi:http://dx.doi.org/10.1097/01.ede.0000229955.07579.f0

Milne, A.A. (1926) *Winnie the Pooh*. E. P. Dutton, London.

Noy, N.F., Sintek, M., Decker, S. *et al.* (2001) Creating Semantic Web contents with Protégé. *IEEE Intelligent Systems*, **16**(2), 60–71.

Page, K.R., De Roure, D.C. and Martinez, K. (2011) REST and Linked Data: a match made for domain driven development? In *Proceedings of the Second International Workshop on REST-ful Design* (WS-REST '11) (eds Cesare Pautasso and Erik Wilde), ACM, New York, pp. 22–25. doi:10.1145/1967428.1967435

Page, R. (2006) Taxonomic names, metadata, and the Semantic Web. *Biodiversity Informatics, North America*, 3, 1–15. Available at https://journals.ku.edu/index.php/jbi/article/view/25/12 Accessed 5 May 2016. http://dx.doi.org/10.17161/bi.v3i0.25.

Puckelwartz, M.J., Pesce, L.L., Nelakuditi, V. *et al.* (2014). Supercomputing for the parallelization of whole genome analysis. *Bioinformatics*, 30(11), 1508–1513. doi:10.1093/bioinformatics/btu071

Sagan C., (1980), *Cosmos*. Random House, New York. LCC QB44.2.S235.

Smith, B., Ashburner, M., Rosse, C. *et al.* (2007). The OBO Foundry: coordinated evolution of ontologies to support biomedical data integration. *Nature Biotechnology*, 25(11), 1251–1255, PMC 2814061. PMID 17989687. doi:10.1038/nbt1346.

7.10 Quiz

This multiple-choice quiz will help you to see how much you've remembered about some of the concepts and issues described in this chapter.

1 An opaque identifier:
   A always points to the same database entity.
   B contains concrete information about how to dereference it.
   C can't be interpreted by a human or machine without dereferencing it.
   D has no context.

2 LSIDs are:
   A a kind of URL.
   B a kind of URI.
   C both a URN and a URI.
   D the *de facto* standard used in the life sciences to identify database entries.

3 XML is:
   A a framework for encoding structured information in text files.
   B an extensible schema for describing the structure of molecular data.
   C a method of representing international characters in text files.
   D a controlled vocabulary of terms used for defining the structure of data.

4 SPARQL is used for:
   A extracting content from a graph of RDF triples.
   B parsing XML documents to extract the schema.
   C converting RDF graphs into XML hierarchies.
   D querying the content of JSON files.

5 A controlled vocabulary:
  A once created, can never be modified.
  B has a specific scientific context.
  C is an agreed set of symbols used in a particular context.
  D defines a knowledge domain in terms of axioms.

6 A thesaurus defines:
  A synonymous terms in a controlled vocabulary.
  B relationships between terms.
  C the meanings of terms in a human-readable way.
  D the meaning of terms in a machine-readable way.

7 A taxonomy defines:
  A the terms in a particular scientific domain and their meanings.
  B a hierarchical relationship between concepts.
  C axioms that determine inclusion or exclusion from an ontology.
  D a cyclic graph structure relating concepts to one another.

8 Linked Data can be represented as:
  A XML
  B RDF
  C JSON
  D any of the above.

9 Which of the following is not one of the fallacies of distributed computing?
  A The network is reliable.
  B CPU performance is unbounded.
  C Latency is zero.
  D Topology doesn't change.

10 HTTP is stateless because:
  A the protocol itself contains no concept of state.
  B the protocol prohibits the construction of stateful services.
  C none of the verbs in the protocol have any side-effects on the server.
  D in Web applications, all state is hidden by the server.

7.11 Problems

1 You're working in a lab that's assembling tens of thousands of protein sequences every day. These need to be catalogued in a database. It's been suggested that you use the MD5 hashing algorithm on the amino acid sequences to create unique identifiers for them. Investigate the MD5 cryptographic hashing algorithm. What are the pros and cons of this approach? What alternative algorithm might alleviate any problems you encounter?

2   Tim Berners-Lee has said that he regrets keeping the 'dot' notation for separating the domain name components in URIs, and has pointed out that the '//' notation is also syntactically redundant; thus the familiar form:

    http://www.uniprot.org/uniprot/P02699

could more elegantly be written:

    http:www/uniprot/org/uniprot/P02699

without any loss of information. What final inconsistency/oddity remains in this notation?

Chapter 8

Linking data and scientific literature

8.1 Overview

Throughout this book, we've been exploring the problems that arise when trying to bridge the gap between forms of representation that are meaningful to humans and those that are comprehensible to machines. In this chapter, we focus on how these issues affect how scientists, as both producers and consumers of knowledge, interact with the scientific literature and with its underlying data. In particular, we examine how records in biological databases are interwoven with prose from articles to create meaningful knowledge archives; conversely, we review how attempts are being made to weave research data more tightly back into the thread of scientific discourse.

By the end of this chapter, you should have an appreciation of the history of the scientific literature, how it's currently used to annotate and enhance biological databases, and how, in turn, it increasingly refers back to the content of those databases as authoritative data records. You'll have developed an understanding of the difficulties of linking data with the, often tacit, knowledge captured in centuries of narrative; you should also have gained insights into ways in which the literature is changing to meet the needs of the modern scientist, and the role of technology in this change. Ultimately, you should have an appreciation of some of the advances that are beginning to be made in this field, and of some of the barriers to progress.

8.2 Introduction

The story of this book concerns the gap between what bioinformaticians are trying to achieve – bringing meaning to biological data – and what computers can do to help us, today and in the future. In many ways, it's the story of the gulf between human knowledge, intuition and inspiration, on the one hand, and the unforgiving low-level pedantry of computers, on the other, and of the problems that emerge when we bring these two very different things together to support one another. We have seen, from numerous perspectives, how understanding what a computer can and can't do of its own accord

Bioinformatics challenges at the interface of biology and computer science: Mind the Gap. First Edition.
Teresa K. Attwood, Stephen R. Pettifer and David Thorne. Published 2016 © 2016 by John Wiley and Sons, Ltd.
Companion website: www.wiley.com/go/attwood/bioinformatics

is an essential part both of creating data that retain any real meaning, and of formulating algorithms that have a chance of completing in non-infinite time; and we've seen the problems that occur when these considerations are overlooked. In some senses, the story of this book boils down to a single question that we'd like humans and machines to be able to answer: 'What do my data *mean*?'

From the perspective, primarily, of biological sequences, we've looked at tools and databases, algorithms and data structures, and teased out the relationships, possibilities and pitfalls of using computers to bridge the gap between biological data and knowledge; and we've seen how the Internet and the Web, combined with tools and databases, have created a new 'knowledge infrastructure', rich both with potential and with problems. What we've only hinted at so far is the relationship between this new knowledge infrastructure and a much older one that existed – for the most part, fully-formed – long before the invention of the first computer: the scientific literature.

As we saw in Chapters 1 and 3, early sequence databases, like PIR and Swiss-Prot, were created by enthusiasts who painstakingly scoured the literature for papers containing information about newly sequenced proteins, and then manually transcribed the amino acid sequences printed therein into electronic formats. To augment and add value to the raw sequence data, they would, wherever possible, try to gather assorted facts about the protein (its subcellular location, its likely function, its post-translational modifications, and so on); by and large, they'd also record the articles they'd used as information sources. Today, biological database annotators (or biocurators, as they're also known), still spend significant amounts of time scouring the literature for facts to support, refute or somehow add extra meaning or explanation to particular database records – they do this because database entries are regarded as being more authoritative and meaningful when they can cite published experimental evidence.

The task of extricating specific facts from the literature can be quite tricky. As part of the process, biocurators have to cope with all of the vagaries of natural language, coupled with the sometimes-confounding nature of field-specific scientific jargon, and with the frequent ambiguities of biological terminology. By way of illustration, consider the following scenarios:

1 having retrieved a list of article titles and abstracts from PubMed, a biocurator interested in the function of Signal Transducer and Activator of Transcription 2 (STAT2[1]) has to decide whether the rather equivocal conclusion of the abstract, 'our data *suggest a role* for Vc in Nanog regulation networks and reveal a novel role for STAT2 in regulating Nanog expression' [our emphasis] (Wu *et al.*, 2013) is sufficiently compelling to warrant reading the full article to shed light on this putative functional role, or whether time would be better spent on more 'concrete' papers;

2 in seeking articles about opioid receptor function, a biocurator may have to decide from the opaque title, 'Exposure to chronic mild stress prevents kappa opioid-mediated *reinstatement of cocaine and nicotine place preference*' [our emphasis] (Al-Hasani *et al.*, 2013) whether this specialist paper is really likely to sequester facts pertinent to receptor function, and is hence worth taking time to read; or

3 reading a paper about Mini Chromosome Maintenance (MCM) protein complexes in *Drosophila* in search of information related to mouse DNA

---

[1] http://www.uniprot.org/uniprot/P52630

replication licensing factor, MCM4, a biocurator encountering sentences such as, 'Co-immunoprecipitation data demonstrate the existence of at least two distinct types of 600-kDa complexes, one that contains DmCDC46 and one that appears to contain both DmMCM2 and Dpa (a CDC54 homologue)' and, 'Mutants in MCM2, MCM3, MCM5/CDC46, CDC47, and CDC54 arrest with incompletely replicated DNA, indicating a role for these genes in DNA replication' may have to decide whether *dpa* in the fly is the 'same' as the murine gene *Mcm4*, and whether yeast *CDC54* protein is the same as the murine gene product – no explicit link to *Mcm4* is made either in the abstract or in the body of the paper (Su *et al.*, 1996).

So many life-science papers are published each year that it has become impossible to search and sift through the literature without the aid of computers. For the non-specialist, examples like those mentioned above would probably seem like knotty problems. Nevertheless, despite our increasing reliance on computers, and although it might not seem like it, disambiguating the seemingly unnatural natural language of scientific prose is much easier for humans than it is for machines; hence, rightly or wrongly, we still tend to place greater trust in the insight of humans over the cold logic of algorithms.

Just as the information recorded in database entries is bolstered by evidence put forward in scientific articles, all manner of published works now refer to information stored in databases. This book, for example, like many others, is replete not just with references to other scholarly works (whether they be books that exist only in printed form, or academic papers available both on paper and digitally), but also to a wealth of online resources, including biological databases. So inter-dependent are the literature and archived data that it's tempting to think that the boundaries between them are beginning to blur. However, achieving seamless links between them is a huge challenge.

Alongside all the technological innovations of recent years, the scientific literature has remained, until recently, relatively unchanged. Although machines have superficially altered the way in which scientists publish and disseminate their work, the basic act of writing, publishing and reading is fundamentally the same as it always has been. Now, however, as we search for ways to imbue raw data with greater meaning, and scientists are embracing the need to create better links between machine-processable data and the scientific literature, traditional publishing approaches are beginning to evolve in new directions.

In this final chapter, we explore this emerging landscape. We investigate the gap between the chaotic world of scientific prose and the structured world of data; and we witness the lengths to which we are compelled to go to bridge that gap, to link data with scientific knowledge such that machines and humans may extract their meaning more efficiently and incisively.

8.3 The lost steps of curators

In the early 1960s, when Dayhoff began harvesting sequences from the scientific literature, the idea of reading through dozens of articles bound in scores of journal volumes must have been daunting; but the numbers of journals were fewer back then, and manual journal scouring was the only way to keep up-to-date. Today, attempting the same task is unimaginable. Once upon a time, the traditional paper-based publishing process delivered new issues of relevant journals to local institutional libraries on a predictable schedule – typically monthly or quarterly; but, now, the Internet has removed the restrictions and bottlenecks associated with the distribution of physical journal volumes. While, at that time, it made a certain economic and practical sense to bundle articles together into issues and volumes to be transported around the world by post, the omnipresence of the Web means that the technology now exists for scholarly content to be accessible to readers as soon as it's published. When creating the *Atlas*, Dayhoff probably had a corpus of around 20 articles to scour (see Box 8.1); today, a life-science-related article appears online every 30–60 seconds! Many of these newly published papers, informally and often ambiguously, refer to the ever-growing collection of small molecules, proteins, genes and other objects of biomedical interest stored in online databases. Figure 8.1 gives a flavour of the growth of the life-science literature and of the some of today's principal biological databases with which the literature is now inextricably linked.

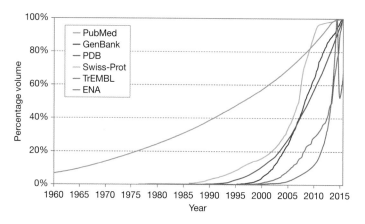

Figure 8.1

Illustration of the growth of the biomedical literature represented in PubMed and of a range of widely used biological databases (regarding the sudden massive drop in the number of entries in UniProtKB/TrEMBL, recall section 3.9).

Box 8.1 The changing literature landscape

Let's take a moment to consider how the ever-growing volumes of biomedical literature have changed the research landscape: specifically, let's think about the literature-mining task that faced Margaret Dayhoff when she was compiling her *Atlas of Protein Sequence and Structure*. We can start by using PubMed to identify articles published in, let's say, a 15-year window around the time when Dayhoff's work on the *Atlas* was in progress. Of course, PubMed wasn't available at that time, but fortunately, it now allows us to take snapshots of the literature in a number of different ways for any time period we choose to observe. So, let's consider the state of the literature from 1950 to 1965 and from 2000 to 2015, and put this in the context of articles published since 1950.

We don't know how Dayhoff searched the literature, but let's assume she had access to a range of journals and skimmed them for articles that mentioned 'protein sequence' in the title. PubMed indicates that a single article was published during the 15-year period to 1965. On closer inspection, we find that the article was about 'Computer aids to protein sequence determination', published in the *Journal of Theoretical Biology* by none other than Margaret Dayhoff (Dayhoff, 1965)! So that wasn't very fruitful; but we may guess that Dayhoff was also interested in articles describing protein 'crystal structures'. In this case, she could have found three papers published in *Nature, Science* and the *Journal of Ultrastructure Research,* respectively in 1951, 1957 and 1959. For articles mentioning 'nucleotide sequence' in the title, PubMed identifies a more encouraging set of 22 articles, published in 10 different journals (some of these are Russian and Japanese, and it's highly unlikely that Dayhoff would have had access to them – and even is she had, it's even more unlikely that she'd have been able to translate them!). Hence, overall, from 1950 to 1965, the titles of 26 articles included the terms 'protein sequence', 'protein crystal structure' and 'nucleotide sequence'. Widening the PubMed search to 'Title and abstract', that number rises to 31 (see Table 8.1). Of these, realistically, Dayhoff probably had access to around 20 articles from about half-a-dozen journals; sifting manually for those articles would have been time-consuming, but nevertheless manageable.

Table 8.1

PubMed search results with selected query terms across either the title field or both title and abstract fields for time periods 1950–1965, 2000–2015 and 1950–2015.

Article field(s)	Query terms	Period	Number of articles
Title	Protein sequence	1950–1965	1
		2000–2015	529
		1950–2015	979
	Crystal structure & protein	1950–1965	3
		2000–2015	1,174
		1950–2015	1,470
	Nucleotide sequence	1950–1965	22
		2000–2015	1,088
		1950–2015	9,288
Title/abstract	Protein sequence	1950–1965	1
		2000–2015	6,277
		1950–2015	10,629
	Crystal structure & protein	1950–1965	3
		2000–2015	14,115
		1950–2015	17,502
	Nucleotide sequence	1950–1965	27
		2000–2015	10,920
		1950–2015	35,448

Repeating these searches for articles published in the 15 years between 2000 and 2015, inevitably, gives a rather different picture. PubMed identified 2,791 articles that included those terms in the title, and, across both title and abstract fields, 31,312 articles; and looking for all articles published since 1950, 11,737 included those keywords in the title, and 63,579 across both title and abstract fields. Manual triage of article numbers on this kind of scale has clearly become impossible.

Arguably, you could say that we've inflated these numbers to make the task at hand seem more difficult than it really is – few researchers would want to scrutinise all articles published on a given topic since 1950; the last five years would seem like a more realistic target. To see how much easier the task becomes, then, let's confine the above searches to the period from 2010 to 2015. In this period, PubMed identified 815 articles that included those query terms in the title, and, across title and abstract, 8,863 – which is still far too many articles to process by hand in sensible time-scales, and orders of magnitude more than Dayhoff had to grapple with.

To understand how the literature is used to support facts or 'claims' in databases, let's consider the UniProtKB entry for human rhodopsin (P08100[2]). In 1984, Nathans and Hogness sequenced the gene, and published the result in *Proc. Natl. Acad. Sci. USA* (*PNAS*); in doing so, the nucleotide sequence was deposited in the EMBL data library. In 1988, this was translated and used as the basis of a Swiss-Prot entry[3] – Figure 8.2, top panel (A). In the process, a Swiss-Prot biocurator, with support from the original *PNAS* article and a 'more recent' 1986 paper (B), provided specific comments or 'annotations' about the *function* of the protein and its *subcellular location* (C). However, explicit links between these factual statements and their citations within the articles were not made. Recall, this was before the advent of the Web, when there was then no obvious way to do this – to verify the facts, a user would have to go back to the articles, which probably meant a trip to the library.

In the following years, the Swiss-Prot curators researched the literature in much greater detail, adding further facts to the database entry, along with citations to the most relevant papers. In 2004, the entry was assimilated into UniProtKB, where, over the years, it accumulated many more pages of such additional information: by January 2016, the UniProt entry[4] (version 182) was supported by 35 hand-selected and >100 computationally-mapped articles. Today, being Web enabled, it's possible to link some

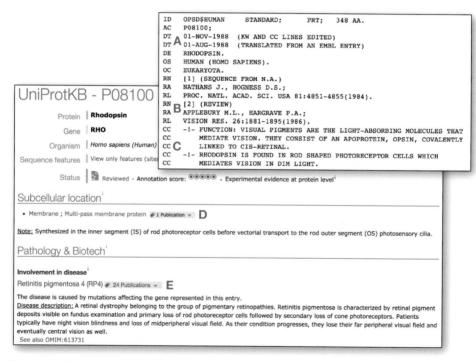

Figure 8.2

Top panel: part of the entry for human rhodopsin as it appeared in Swiss-Prot in 1988, showing annotations supported by two articles; bottom panel: part of the equivalent entry as it appeared in UniProtKB in January 2016, with specific annotations supported here by 25 articles.

---

[2] http://www.uniprot.org/uniprot/P08100
[3] http://www.uniprot.org/uniprot/P08100.txt?version=1
[4] http://www.uniprot.org/uniprot/P08100.txt?version=182

of these specific facts to their originating papers – *e.g.*, the claim that the rhodopsin protein is 'Synthesized in the inner segment (IS) of rod photoreceptor cells before vectorial transport to the rod outer segment (OS) photosensory cilia' is supported by a specific literature reference (bottom panel, D). Again, however, although this paper is cited, verifying the information would still mean reading the article to discover the context in which this claim was made.

Looking at the entry in more detail, we see that the curators have also added information about *diseases* with which rhodopsin is associated. Specifically, we learn that,

> Retinitis pigmentosa 4 (RP4). Disease description: a retinal dystrophy belonging to the group of pigmentary retinopathies. Retinitis pigmentosa is characterized by retinal pigment deposits visible on fundus examination and primary loss of rod photoreceptor cells followed by secondary loss of cone photoreceptors. Patients typically have night vision blindness and loss of midperipheral visual field. As their condition progresses, they lose their far peripheral visual field and eventually central vision as well. See also OMIM:613731[5].

These statements are supported by 24 citations (Figure 8.2, bottom panel, E). To verify the statements and understand the process that led the curator to create this particular patchwork of information, we're now faced with 24 articles from which to discover the contexts of the original claims!

To make life easier, the curators have also provided a more detailed list of known disease-associated sequence variants in a different section of the database (in the Feature Table) – Figure 8.3. Here, we learn that a variant associated with a particular type of retinitis pigmentosa (RP4), a 'proline to histidine' mutation, the 'most common variant,

Feature key	Position(s)	Length	Description	Graphical view	Feature identifier
Natural variant[i]	4 – 4	1	T → K in RP4.		VAR_004765
Natural variant[i]	15 – 15	1	N → S in RP4. 1 Publication ▾		VAR_004766
Natural variant[i]	17 – 17	1	T → M in RP4. 2 Publications ▾		VAR_004767
Natural variant[i]	23 – 23	1	P → H in RP4; most common **A** variant; leads to interaction with EDEM1 followed by degradation by the ERAD system. 4 Publications ▾ **B**		VAR_004768

Cited for: VARIANTS RP4 MET-17; HIS-23; ARG-58; SER-182 AND LEU-267.

"A dual role for EDEM1 in the processing of rod opsin."
Kosmaoglou M., Kanuga N., Aguila M., Garriga P., Cheetham M.E.
J. Cell Sci. 122:4465-4472(2009) [PubMed] [Europe PMC] [Abstract]
Cited for: CHARACTERIZATION OF VARIANT RP4 HIS-23, SUBCELLULAR LOCATION.

					VAR_004769
					VAR_004770
			lication ▾		VAR_004771
			lication ▾		VAR_004772
					VAR_004773
				| Ensembl	
			].		

Figure 8.3

Part of the Feature Table of the same entry for human rhodopsin depicted in Figure 8.2, showing details of some of its sequence variations; a variant of a particular type of retinitis pigmentosa, RP4 (A) is supported by four papers, the full literature citations for which can be accessed via a drop-down menu (B).

---

[5] http://www.omim.org/entry/613731

leads to interaction with EDEM1 followed by degradation by the ERAD system' (A). Although there are now only four supporting articles, information about what 'EDEM1' and 'ERAD' are, and other specific details of the claim can only be found by following up each citation in turn, via the pull-down menu (B), and browsing its title, abstract or indeed the full paper.

Clearly, the Swiss-Prot curators have made greater and greater use of the literature during the 27 years or so since human rhodopsin first appeared in the database; and most of that literature research has been done manually. Given the enormous growth in the number of published life-science articles, this has had several important consequences:

1 the growth of the 'gold standard' (manually annotated) component of UniProtKB (*i.e.*, Swiss-Prot) has become incredibly slow relative to the growth in the number of proteins requiring annotation in TrEMBL;
2 the proteins that can be manually annotated necessarily represent a small subset, biased towards those for which experimental data have been published;
3 the facts that are gleaned by any one curator are necessarily just a tiny fraction of the sum of the knowledge available in the entire body of life-science literature for any one protein;
4 once these 'facts' have been committed to a database, they're difficult or impossible to keep routinely up-to-date;
5 if established 'facts' are later refuted in the literature, the database entries are difficult or impossible to correct; and perhaps most serious of all,
6 the bulk of the knowledge gained even from this tiny subset of manually researched literature has probably been lost because of the dearth of computational tools available to capture this kind of information explicitly.

Overall, we're left in a paradoxical situation in which the role of database curators is to *find within* and to *distil out* pertinent information from a limited (skewed) set of published articles in order to include specific facts in databases; meanwhile, the challenge for end users, who wish to validate the 'facts' they discover in databases, is then to *re-associate* these morsels of information with their *supporting context* within the original source literature – essentially, to retrace the lost steps of the curators! Needless to say, this is an incredibly time-consuming and wasteful process.

8.4   A historical perspective on scientific literature

Against this background, and allied particularly to this last point, recent years have witnessed an increasing interest in the development of tools to extract pertinent information more efficiently from the literature, to relieve the manual burdens of database biocurators in particular and of scientists in general. The scientific literature itself is full of references to the catastrophes that await if we fail to harness the power of machines to help us to do this: *e.g.*, we read about drowning in data floods (Andrade and Sander, 1997), deluges (Hess *et al.*, 2001), surging oceans (Dubitzky, 2009), or tsunamis (Wurman, 1997), or being threatened by icebergs (Hodgson, 2001; Howe *et al.*, 2008), avalanches (Antezana *et al.*, 2009), earthquakes (Willbanks, 2007) or explosions (Diehn *et al.*, 2003) of data.

Faced with such predictions of doom, it's tempting to see the problem of harnessing the vast bodies of amassed scientific knowledge as a new phenomenon; but, throughout history, scientists have been demonstrably better at *creating* knowledge than at putting in place effective mechanisms for disseminating, accessing and exploiting it. For example, even getting access to the literature has never been straightforward. Foreshadowing current debates around matters of 'Open Access', as far back as the 'public' libraries of Ancient Rome, scholars and others were complaining that getting at the content of libraries was harder than it should be:

> ... the Palatine library was viewed as practically an extension of the house of Augustus; consequently, ... it must have been possible to control physical access to the library building when the Emperor was expected to be present. Paradoxically, then, ... access to [libraries] was, or could be, restricted" (Dix, 1994).

Perhaps surprisingly, even the phrase 'information overload' pre-dates the Internet revolution by at least a decade (Toffler, 1970); and, for example, as early as 1967, Dayhoff, bemoaning the huge challenge of keeping her *Atlas* up-to-date, observed that 'the number of known sequences is growing explosively' (letter to C. Berkley, 27 February, 1967). Although the scale and nature of the problems may have changed over time (*e.g.*, the deluge threatening to overwhelm Dayhoff amounted to a veritable dribble of 65 sequences!), it appears that there has always, for one reason or another, been more knowledge recorded by mankind than we can collectively make sense of.

Until the Internet revolution, the problem of 'knowing what we know' was primarily caused by physical restrictions – the length of time taken to manually typeset articles and ship them on paper to libraries, or the need to physically travel to a library to read their contents. Today's technology effectively removes these bottlenecks, but introduces problems of its own. Undeniably, there is a growing issue of scale here, which has been reported and analysed repeatedly in the literature (*e.g.*, Attwood *et al.*, 2009; Fraser and Dunstan, 2010). Although the details of various commentaries differ, their conclusions are consistent: we are publishing more (data and literature) than ever before; the rate of publishing is increasing; and the tools available to make sense of the growing body of knowledge continue to lag behind our needs. In some senses, the problem of dealing with the overwhelming amount of available 'stuff' doesn't matter, as long as it's easy enough to extract chunks that are relevant and useful to us; however, the goal of creating tools that are sufficiently sophisticated to be able to filter out stuff that's both relevant and of high quality remains elusive – this inability to separate out scientific wheat from chaff has been characterised as 'filter failure' (Shirky, 2008). Later in this chapter, we'll examine some potential ways forward, but only time will tell whether these are likely to mature into real 'solutions'.

Aside from scale, perhaps a more fundamental change has occurred in the way in which scientific knowledge is published. Whereas early scientific discoveries were communicated to scholarly societies as 'letters', and journals were really collections of what we might now consider quite informal exchanges between scientists, contemporary scientific articles are increasingly formalised, and reliant on data to make their case. Discoveries were once led by hypotheses, followed by experiments to create data to test the veracity of those hypotheses, but there's now a growing trend towards data-driven science – a trend neatly encapsulated in Kell and Oliver's 2004 paper, 'Here is the evidence, now what is the hypothesis?' Unlike the manually recorded data associated with early discoveries, new data-sets are often the result of automated processes or computational simulations, and are frequently orders of magnitude larger and more complex than those created in the past. The ways in which data are reported to the community are consequently changing. In early scientific articles, data were often included in the main body of the article itself. However, when data-sets grew past the point where they could reasonably be included in the main narrative without disrupting its flow, journals began publishing appendices of 'auxiliary materials'. Now, in the era of 'big data' (usually, far bigger than can be printed even on 'electronic paper'), this approach too is inadequate; increasingly, therefore, scholarly articles make reference to supplementary online resources instead, and many nuggets of scientific insight are at risk of being swamped and overlooked in the process ('big data' may have arrived, but, to quote one commentator, 'big insights have not' (Harford, 2014)).

Of course, scholarly communication, in one form or another, pre-dates computers and databases by many centuries: from 'letters' to learned societies distributed to the community on paper, through to today's online journals, the act of writing and publishing ideas and discoveries for scrutiny by the scientific community has always been a fundamental part of the scientific method. In many ways, the process of submitting an article to a scholarly journal, of having it reviewed by peers, and eventually – if it's deemed worthy – published, has changed little over the years. Perhaps not surprisingly, then, a 2012 'white paper' (the Force 11 Manifesto) opens with the following comment:

> A dispassionate observer, perhaps visiting from another planet, would surely be dumbfounded by how, in an age of multimedia, smartphones, 3D television and 24/7 social network connectivity, scholars and researchers continue to communicate their thoughts and research results primarily by means of the selective distribution of ink on paper, or at best via electronic facsimiles of the same (Allen *et al.*, 2012).

The increased reliance on publishing data to validate scientific claims has meant that articles and data-sets are becoming ever more closely intertwined (Bourne, 2005). The modern scientific article has been described as a 'story that persuades with data' (de Waard, 2014), a phrase that nicely brings together the two essential components of modern scientific writing: without the story, data are just a collection of facts with no interpretation or real meaning; without data, a story is in danger of being little more than an opinion – perhaps even a fairytale?

8.5 The gulf between human and machine comprehension

As we've seen – indeed, it's the underlying theme of the whole book – creating meaningful data on a large scale isn't trivial, but requires an understanding of the needs both of humans and of machines. The problem is, humans and machines speak very different languages: the challenge lies in finding ways to express concepts in ways that are meaningful to both. At one extreme are the subtleties and nuances of natural language, flavoured by the conventions and terminologies of specialist scientific disciplines; at the other, lie the brutal, pedantic processing capabilities of computers.

To illustrate how far apart these things are, let's consider an example that sits somewhere near the midpoint between natural language and machine code. Consider the following statement, written in a 'logic notation':

$$\forall i, j \in dom(s).i \le j \Rightarrow s[i] \le s[j]$$

To make sense of this, a human reader has first to decode the various symbols. In this toy example, this is fairly straightforward: $\forall$ means 'for all'; *dom*() refers to an abstract 'domain'; and $\Rightarrow$ means 'implies that'. The variables and 'less than or equal to' symbols carry their familiar meanings. From left to right, the statement can be read 'literally' as:

> For all pairs of things (which we will call *i* and *j*) in a domain we'll call *s*, the value of *i* being less than or equal to the value of *j* implies that the index of *i* in the domain *s* is less than or equal to the index of *j* in the domain *s*.

The statement, therefore, asserts a relationship between values, and the ordering of those values in a particular domain; but what kind of relationship? Thinking of these as being integer numbers perhaps makes the real meaning more clear: if a number is smaller than or equal to another number, then its ordering is earlier or equal to that of the other number. Or, put more simply, 'the numbers are sorted in ascending order, allowing for duplicates'. Although it's fairly easy to learn the 'vocabulary' and even 'grammar' of logic symbols, and thus to literally translate such statements, understanding the real meaning of logic notation – even for trivial examples like this one – takes considerable practice in a particular way of thinking that's alien to most scientists.

From a machine's point of view, a comparable amount of work also needs to be done to make any real use of the statement. A computer can't just execute such notation without support from automated-reasoning tools or sophisticated compilers to translate the logic into code that can be executed on a CPU – see Box 8.2. Even then, the notation says nothing about the details of the process of sorting, the time or space constraints, and so on – it simply says that the result is a sorted list of things.

This example illustrates the enormous gulf that exists between language that's comprehensible to humans and the kind of code that's executable by machines. Clearly, getting scientists to write academic articles in forms that are comprehensible to machines isn't practical. So what about the alternative: getting machines to 'understand' natural language? Perhaps not surprisingly, this too is an enormously tall order.

Let's consider an utterly trivial piece of natural language: '*the cat sat on the mat*'. A machine 'reading' this would first have to identify the objects in the sentence and, with a suitable electronic dictionary, could probably find the nouns 'cat' and 'mat'. It would then need to determine the relationship between these: in this case, 'sat' (again, using a dictionary, it could perhaps determine that 'sat' is the past tense of the verb 'to sit'). The preposition 'on' makes an association between the act of sitting, the cat and the mat; and the other words are definite articles, telling us that the sentence refers to a particular cat and a particular mat. With a few relatively straightforward rules of grammar encoded in a suitable program, we could assemble a 'parse tree' (see Figure 8.4) that represents the associations between the various parts of the sentence; and, from that, we could imagine that it would be possible for a computer to reason about the meaning

Box 8.2 Bubble sort in C

Written in C, code for bubble sort (here, operating on a small list of arbitrary numbers) that would satisfy the criterion of generating numbers 'sorted in ascending order, allowing for duplicates' looks like this:

```c
#include <stdio.h>
int main()
{
  int array[5] = {8, 5, 1, 7, 3}, n, c, d, swap;
  n = 5;
  for (c = 0 ; c < ( n - 1 ); c++)
  {
    for (d = 0 ; d < n - c - 1; d++)
    {
      if (array[d] > array[d+1]) /* For decreasing order use < */
      {
        swap = array[d];
        array[d] = array[d+1];
        array[d+1] = swap;
      }
    }
  }
  printf("Sorted list in ascending order:\n");
  for (c = 0; c < n ; c++)
  {
    printf("%d\n", array[c]);
  }
  return 0;
}
```

Compiled, and then disassembled, the first 30 or so machine-code instructions are shown below. The left column gives a memory address; the central column, the bytes in **hexadecimal** corresponding to the instruction; and the right column shows an assembly language version of the same machine-code (see Section 6.3.1). If you look carefully at rows 4 to 8, you can see the array of data containing the numbers 8, 5, 1, 7 and 3 being created in memory. After that, making sense of how the machine code corresponds to the source code in C takes considerable expertise!

```
400504:     55                    push %rbp
400505:     48 89 e5              mov %rsp,%rbp
400508:     48 83 ec 30           sub $0x30,%rsp
40050c:     c7 45 d0 08 00 00 00  movl $0x8,-0x30(%rbp)
400513:     c7 45 d4 05 00 00 00  movl $0x5,-0x2c(%rbp)
40051a:     c7 45 d8 01 00 00 00  movl $0x1,-0x28(%rbp)
400521:     c7 45 dc 07 00 00 00  movl $0x7,-0x24(%rbp)
400528:     c7 45 e0 03 00 00 00  movl $0x3,-0x20(%rbp)
40052f:     c7 45 f0 05 00 00 00  movl $0x5,-0x10(%rbp)
400536:     c7 45 f4 00 00 00 00  movl $0x0,-0xc(%rbp)
40053d:     eb 6f                 jmp 4005ae <main+0xaa>
40053f:     c7 45 f8 00 00 00 00  movl $0x0,-0x8(%rbp)
400546:     eb 4e                 jmp 400596 <main+0x92>
400548:     8b 45 f8              mov -0x8(%rbp),%eax
40054b:     48 98                 cltq
40054d:     8b 54 85 d0           mov -0x30(%rbp,%rax,4),%edx
400551:     8b 45 f8              mov -0x8(%rbp),%eax
400554:     83 c0 01              add $0x1,%eax
400557:     48 98                 cltq
400559:     8b 44 85 d0           mov -0x30(%rbp,%rax,4),%eax
40055d:     39 c2                 cmp %eax,%edx
40055f:     7e 31                 jle 400592 <main+0x8e>
400561:     8b 45 f8              mov -0x8(%rbp),%eax
400564:     48 98                 cltq
400566:     8b 44 85 d0           mov -0x30(%rbp,%rax,4),%eax
40056a:     89 45 fc              mov %eax,-0x4(%rbp)
40056d:     8b 4d f8              mov -0x8(%rbp),%ecx
400570:     8b 45 f8              mov -0x8(%rbp),%eax
400573:     83 c0 01              add $0x1,%eax
400576:     48 98                 cltq
```

and structure of the original sentence: where is 'the cat'? It's 'on the mat'. What is 'on the mat'? A cat. And so on. So far so good; but real language is seldom so simple and so unambiguous.

Let's now examine an admittedly pathological case at the other extreme. The sentence, 'Buffalo buffalo Buffalo buffalo buffalo buffalo Buffalo buffalo'[6] is a popular illustration of a seemingly intractable piece of grammatically and semantically correct English. The capitalisations offer a clue to the meaning of some instances of the word 'buffalo', Buffalo being used here as a proper noun – the name of a North American city. It's also being used in its more commonly known form as the name of a bison-like herd animal. The most obscure use is of buffalo as a verb, meaning something like 'to bully or harass'. Thus, decoded, the heavily contrived sentence reveals that 'Buffalo bison that other Buffalo bison bully, also bully Buffalo bison'. For a human reader, accurate parsing of a sentence like this requires a sound understanding of grammar, and access to a good dictionary (and, arguably, also a little geographical knowledge); for a computer to be able to deconvolute the sentence, a parse tree similar to that shown in Figure 8.4 would be required. Even slightly more realistic phrases and sentences, such as 'My hair needs cutting badly', 'Stolen dog found using Facebook' (BBC News), 'Round table talks', 'Fruit flies like a banana', and many others like them, pose entertaining problems for machine interpretation.

We might expect scientific rigour to encourage prose that's unambiguous and formally structured. However, although it may be true that extracting meaning from scientific text is perhaps more straightforward than from poetry and other forms of creative writing, the scientific literature nevertheless poses problems of its own that make its meaning hard for machines to penetrate; automating the recovery of meaning from scientific writing, *post hoc*, therefore remains challenging. Let's take a (fairly randomly selected) sentence from a scientific article:

In the present paper, we report the crystal structure of *Lp*GT, which reveals a GT-A fold, demonstrating how the enzyme interacts with UDP-glucose (Hurtado-Guerrero *et al.*, 2010).

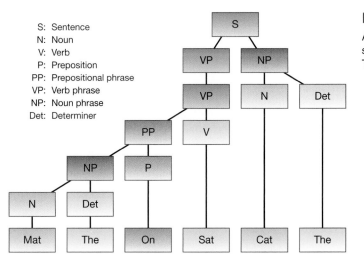

S: Sentence
N: Noun
V: Verb
P: Preposition
PP: Prepositional phrase
VP: Verb phrase
NP: Noun phrase
Det: Determiner

Figure 8.4

A parse tree for the sentence 'The Cat Sat On The Mat'.

[6] http://en.wikipedia.org/wiki/Buffalo_buffalo_Buffalo_buffalo_buffalo_buffalo_Buffalo_buffalo

For the human reader, the meaning is clear enough: the authors are describing the crystal structure of a molecule (an enzyme), which they refer to as *Lp*GT; they claim that this molecule has a GT-A fold, and that this GT-A fold is what enables the enzyme to interact with another molecule, UDP-glucose; for a computer to achieve the same level of understanding is much trickier. Leaving aside the biology for a moment, consider the grammatical structure (the syntax) of the sentence. There are three commas: these could form a list, surround a sub-clause, or just be aesthetic. Worse, they could be one of many combinations of these, with each combination yielding a different possible interpretation of the sentence (and note, we've presented here a simple sentence that we're taking to be grammatically correct, which isn't a safe assumption for innumerable sentences published in the scientific literature – if you need convincing, look at some of those highlighted in Box 8.3). Regardless, for humans, it's quite easy to spot what the authors intended; but for a machine, even this trivial syntactic construct poses problems.

Unambiguously recognising entities in scientific text turns out to be horribly difficult too. In this single sentence, for example, we have three biological concepts: '*Lp*GT', 'GT-A fold' and 'UDP-glucose'. The authors helpfully define *Lp*GT elsewhere in the paper to be '*L. pneumophila* glucosyltransferase', and in a footnote state that, 'The structural co-ordinates reported will appear in the Protein Data Bank under accession codes 2WZF[7] and 2WZG[8]', which indeed they now do. For a human, the link between the article's narrative and the electronic description of the molecules in

Box 8.3 The literature isn't always as intelligible as it should be

In recent years, the increased rate of publication of scientific articles has placed significant strain on the publishing process, so much so that, in some cases, levels of editorial processing are perhaps no longer what they used to be. In the following, randomly chosen, examples, if we mentally insert a few commas and/or hyphens, we can probably work out the intended meaning of the sentences; some, however, are trickier than others – and if we humans find them difficult to parse, consider how much more difficult it will be for a text-mining algorithm to ascertain their meaning:

* 'Treatment of primary cells from newly diagnosed CML patients in chronic phase as well as BCR-ABL+ cell lines with imatinib increased IRF-8 transcription' (Waight *et al.*, 2014);
* 'The *in vitro* blood stage antiplasmodial activity of a series of allylated chalcones based on the licochalcone A as lead molecule was investigated against chloroquine (CQ) sensitive Pf3D7 and CQ resistant PfINDO strains of *Plasmodium falciparum* using SYBR Green I assay' (Sharma *et al.*, 2014);
* 'Literatures and studies published on anti-leishmanial drug resistance, newer drug discovery for leishmanial resistance *etc.*, in PubMed, Medline and Google search and reviewed thoroughly' (Mohapatra, 2014);
* 'The sampling location in this study is a tourism hotspot with great human mobility, which may result in the tetracycline resistance gene or bacteria released from human beings into the environment (water and soil) in Tibet' (Li *et al.*, 2014);
* 'Human castration resistant prostate cancer rather prefer to decreased 5α-reductase activity' (Kosaka *et al.*, 2013).

[7] http://www.rcsb.org/pdb/explore.do?structureId=2WZF
[8] http://www.rcsb.org/pdb/explore.do?structureId=2WZG

the PDB may be clear enough; but for a machine to join together these disconnected fragments of text to make this connection is non-trivial. To muddy the waters further, it's common in some fields of biology, for example, to refer to a gene not by its given name, but by its protein product; and, as we saw in Chapter 5, the 'same' gene in different species can have different names. For a scientist in a given field, the translation from protein-name to gene-name may be implicit; for a machine, it's deeply confusing. Terminology varies dramatically from domain to domain (even the word 'domain', used here in a 'computer science' sense to mean 'scientific field', has different connotations in the life sciences, where it can be interpreted as referring to various quite different biological concepts).

If unambiguous 'entity' recognition is hard for computers, extracting scientific *assertions* at a level that's sufficiently straightforward to be processed automatically – *e.g.*, 'A does B to C' – is even harder. The difficulties here are exacerbated by what's generally regarded as publishing 'gameplay'. For example, a scientist may have concluded that A does B to C; however, making such a bold assertion in an article is risky. To be valid, the assertion must be made in context (A does B to C in the context of D and under conditions E); but the scientist, fearing that 'over claiming' (or, worse, contradicting the opinion of an anonymous reviewer) might cause the article to be rejected, may strategically soften the claim: *e.g.*, 'it is possible that, within the limits of our experimental conditions, *E*, *C* and *A* sometimes interact in a *B*-like way in a *D*-like context'. This practice is referred to as 'hedging'. In the following sentences, extracted from a range of scientific articles, we've emphasised the use of hedging phrases:

- 'Targets isolated *suggest a possible interaction* for those gene products associated with calvarial suture growth and homeostasis as well as craniosynostosis' (Cray *et al.*, 2013);
- 'Preirradiation treatment of TCE significantly reduced LPO at all the autopsy intervals in comparison to irradiated control, which testifies to our belief that one of the *possible mechanisms* of radioprotection by TCE *may be owing to* the scavenging of free radicals generated by radiation exposure and prevents the formation of endoperoxidation' (Sharma *et al.*, 2011);
- 'Expression patterns of CYC-like genes *are suggestive of participation* in the control of pseudanthium development, in a manner analogous to the distantly related Asteraceae.'(Claßen-Bockhoff *et al.*, 2013);
- 'Micromolar amounts of topical caffeine have been found to be significantly effective in inhibiting the formation of galactose cataract, strongly *suggesting its possible usefulness* against diabetic cataracts' (Varma *et al.*, 2010).

In some cases, hedged terms even appear in the article titles, as in the following examples:

- 'An evaluation of the *possible interaction* of gastric acid suppressing medication and the EGFR tyrosine kinase inhibitor erlotinib' (Hilton *et al.*, 2013);
- 'Severe hypoglycemia *due to possible interaction* between glibenclamide and sorafenib in a patient with hepatocellular carcinoma' (Holstein *et al.*, 2013);
- 'The unique pseudanthium of *Actinodium* (Myrtaceae) – morphological reinvestigation and *possible regulation by* CYCLOIDEA-like genes' (Claßen-Bockhoff *et al.*, 2013).

Possibly the most famous example of hedging to be found in the scientific literature is the oblique conclusion of Watson and Crick's paper on the structure of DNA (again with our emphasis):

> It has not escaped our notice that the specific pairing we have postulated immediately *suggests a possible copying mechanism* for the genetic material (Watson and Crick, 1953).

But note here that we, too, are hedging when we say that this is '*possibly* the most famous example of hedging': what this really means is that this is the most famous example that readily springs to our minds, but that there may be others that we don't know about – this strategy is designed to pre-empt those readers who might have been tempted to write in to correct our ignorance, had we asserted, outright, that this was *the* most famous example of scientific hedging! Such linguistic tricks have been tightly interwoven into the fabric of scientific discourse ever since researchers began committing their results to paper for others to read, and we humans have become accustomed to reading between the lines; but, for computers, it's extremely difficult to distinguish genuine scientific conditions from these mere strategies and foibles of scholarly writing.

By contrast, while reluctance to make bold statements tends to soften scientific claims, the mere act of citing previous work can lead to claim amplification. The danger here, of course, is in amplifying and propagating noisy signals. Consider this illustration from a short study conducted by de Waard *et al.* (2009):

1 Xie *et al.*'s 2009 paper, 'Drug discovery using chemical systems biology: identification of the protein–ligand binding network to explain the side effects of CETP inhibitors', contains the following fragments of text: '*Our findings suggest that*... potential side-effects of a new drug can be identified at an early stage of the development cycle and... [these can] be minimized by fine-tuning multiple off-target interactions... *The hope is that*... this can reduce both the cost of drug development and the mortality rates during clinical trials.' (Here, ellipses denote sections of text we've omitted, square brackets include text we've inserted for readability, and italic text marks hedging phrases).

2 Later in 2009, 28 articles were identified that cited Xie *et al.*, of which 19 were openly accessible for analysis. One of these (Tatonetti, 2009) included the statement, 'The recent work of Xie *et al.* [7] is another *excellent example of the use of* networks combining proteins and drugs. They investigated the reasons for serious side-effects of torcetrapib... Using this analysis, the *authors identified* possible off-targets for torcetrapib... they *showed that* torcetrapib caused more severe effects...' (again, ellipses denote omitted text, while italic text highlights amplification of the original claims).

3 By 2010, a number of articles were published that 'validated, contrasted and expanded' on Xie *at al.* (de Waard *et al.*, 2009):
   - 'Recently, Xie *et al.* [9]... identified possible off-target interactions for a set of CETP inhibitors... in contrast to the work of Xie *et al.* ... our model exploits the assumption that...'
   - 'This [Xie *et al.*'s] strategy has been applied to study... we extend this methodology...'
   - 'Another example was that of Xie *et al.* (16) in explaining the off-target effects...'

Thus, in as little as 12 months, the hedged claims in the original paper can be seen to have crept towards being established facts, increasing the likelihood that authors citing this second generation of work will accept them as 'canon'. This phenomenon of undue credence arising from repeated citation of previous articles that lack concrete scientific foundations has been referred to as 'evidence by citation' or, more whimsically, the Woozle effect[9]. Ideally, of course, authors should find and diligently read the papers they cite, following each claim back to its original source, and draw their own conclusions about the strength, context and validity of particular assertions. However, for 'hot topics' in the life sciences, the rate at which articles are published makes keeping abreast of developments exceptionally challenging; cutting corners is therefore extremely tempting. Overall, then, if unpicking the tangled threads that relate scientific claims is difficult for competent humans, how much more so will it be for machines? As Latour put it,

> ... you can transform a fact into fiction or a fiction into fact just by adding or subtracting references (Latour, 1987).

Most scientific articles, historical and contemporary, are riddled with such computational trip-hazards: *i.e.*, sentences that are readily comprehensible to literate humans but are impenetrable to machines, either because they use terms that have multiple meanings that make sense only if we understand the broader context, or because they include scientific assertions that have been smothered by cautious phraseology, or that have been inappropriately amplified – Woozled – by careless citing.

8.6 Research objects

The problems outlined in the previous section challenge the value of the scholarly literature as a mechanism for sharing and repeating scientific endeavours, especially in the 'digital age'. As we've seen, articles are riddled with uncertainty, inconsistency and 'fluff', attributes that render their latent nuggets of knowledge largely inaccessible to computers. If we're to be able to process the literature computationally, to encode the scientific data and knowledge it contains in machine-readable form, radically new approaches are required. Indeed, one commentator has noted:

> I believe that the academic paper is now obsolescent as the fundamental sharable description of a piece of research. In the future we will be sharing some other form of scholarly artefact, something which is digital and designed for reuse and to drop easily into the tooling of e-Research, and better suited to the emerging practices of data-centric researchers. These could be called Knowledge Objects or Publication Objects or whatever: I shall refer to them as Research Objects, because they capture research (De Roure, 2009).

---

[9] http://en.wikipedia.org/wiki/Woozle_effect. Unlike the Web Ontology Language (OWL), the name of which isn't related to the dyslectic tendencies of Pooh's sage friend, Owl (Section 7.5.1), the Woozle effect is named after the potentially Hostile Animal – the Woozle – whose paw-prints kept multiplying the more Pooh and Piglet tracked it around Piglet's tree-house and then around a nearby spinney of larch trees, where the Woozle appeared to become, first, two Woozles, and then three Woozles, or, possibly, two Woozles and a Wizzle.

In this statement, De Roure eloquently encapsulated a vision that had been circulating in the 'eResearch/eScience' community for some time. Specifically, he proposed that 'Research Objects' should be replayable, repeatable, reproducible, reusable, repurposable and reliable. At the time, he speculated that these were 'necessary and sufficient' properties, but later extended the idea to include six more characteristics (De Roure, 2010), categorising the resulting 12 as 'intrinsic' to Research Objects, or as relating to their 'social life' or to their use 'in practice'. He formally set out his '12 Rs' of Research Objects as shown in Figure 8.5.

It's tempting to think that many of these properties are already inherent in anything digital, and that simply moving from paper-based methods (traditional lab books, scholarly articles and journals, manuals of scientific protocols, *etc.*) will necessarily tend to satisfy several of the '12 Rs'. Superficially, there is truth in this: we have URLs to make things *referenceable*; at its heart, computer code is deterministic, so should be *repeatable* and *reproducible*; the very nature of the Web means that content is *refreshable*; we know that libraries of code are frequently shared and used in ways their authors never intended, so are *reusable* and *repurposable*; 'version control' systems for source code provide programmers with mechanisms for *recovery* and repair; and the speed at which news (or the latest video of a kitten playing a piano) spreads around the world demonstrates the Web's astonishing '*retrievability*'.

But scratch the surface of these properties, and their limitations, as direct successors to more traditional mechanisms for recording and sharing scientific knowledge, begin to show through. Bairoch's original Swiss-Prot floppy disk, for example – had it not, sadly, been mislaid years ago – could almost certainly not be read by modern PCs, most of which have not had floppy-disk drives as standard for some years (in spite of the image of a floppy disc having achieved a literally iconic status on 'save' buttons in graphical user interfaces); it is only because consistent effort has been expended over the years to keep the database safe and current that it's accessible in its current form today.

Some of the limitations of digital resources are 'social': the author of a piece of open-source software probably 'published' it as a gesture of goodwill to the community, but without either the resources or the motivation to support its use indefinitely. Others are more fundamental: hardware is constantly being improved, but inevitably degrades and eventually fails (conventional wisdom suggests that the life expectancy of hard drives is somewhere between three and five years – the blink of an eye in the context of the long-term scientific record); operating systems are constantly updated to patch security flaws, or to introduce new features, and often can't be installed on older hardware; software that relies on particular – now deprecated – features of operating systems can no longer be compiled; and, in turn, data formats that relied on now-defunct software become unreadable.

And so the conundrum: how to move beyond the relatively stable paper-based scientific literature, which has served science well for centuries, to a model of knowledge representation that is robust in the face of rapidly changing technology, and capable of supporting scientific discovery when confronted with ever-growing quantities of increasingly complex data and prose. The consequences of failing to achieve this are likely to be immediate, significant and expensive (Scott *et al.*, 2008; Begley and Ellis, 2012). Although we are a long way from a complete solution, there are many interesting developments in this area, which we'll explore in the remainder of this chapter.

1. **Repeatable** – run the experiment again. Enough information for the original researcher or others to be able to repeat the experiment, perhaps years later, in order to verify the results or validate the experimental environment. This also helps scale to the repetition of processing demanded by data intensive research.
2. **Reproducible** – enough information for an independent experiment to reproduce the results. To reproduce (or replicate) a result is for someone else to start with the description of the experiment and see if a result can be reproduced – one of the tenets of the scientific method as we know it.
3. **Reusable** – use as part of new experiments. One experiment may call upon another, and by assembling methods in this way we can conduct research, and ask research questions, at a higher level.
4. **Repurposable** – reuse the pieces in a new experiment. An experiment which is a black box is only reusable as a black box. By opening the lid we find parts, and combinations of parts, available for reuse, and the way they are assembled is a clue to how they can be reassembled.
5. **Reliable** – to trust the Research Object we must be able to verify and validate it, and derive a measure of trust of the results and methods it contains. It must also be resilient 'in the wild' and may be subject to regulatory review.
6. **Referenceable** – it is essential for a Research Object to have an identity for it to be referred to programmatically, so we can cite it and to ensure probity. Implicit in this may be versioning.
7. **Re-interpretable** – useful in and across different research communities. Papers are aimed at one target audience but we don't have that constraint with Research Objects. This relates to the notion of Boundary Objects.
8. **Respectful and Respectable** – with due attention to credit and attribution for the component parts and methods and their assembly, to the flow of intellectual property in generation of results, to data privacy, and with an effective definition of the policies for reuse.
9. **Retrievable** – if a Research Object can never be found it may as well not exist. We must be able to discover them and acquire them for use.
10. **Replayable** – a comprehensive record enables us to go back and see what happened. The ability to replay (rather than repeat) the experiment, and to focus on crucial parts, is essential for human understanding of what happened in experiments that might occur in nanoseconds or years.
11. **Refreshable** – updating a Research Object with ease when something changes, like when next year's figures come through.
12. **Recoverable** and reparable – when things go wrong we need automatic roll-back to retrace our steps, and more generally we need tools for diagnosis and repair.

Figure 8.5

De Roure's '12 Rs' of Research Objects[i].

[i]*Source:* http://www.scilogs.com/eresearch/replacing-the-paper-the-twelve-rs-of-the-e-research-record

8.7 Data publishing

Historically, scientists have tended to keep their research data securely under wraps until the time they can publish their results in a scholarly journal, cocooned within a suitably erudite narrative, so they can receive appropriate approbation from the community. The fear of being 'scooped', of seeing others make insightful leaps to exciting discoveries from one's own hard-won experimental results – and getting no credit – isn't completely unwarranted. We only have to remember Watson and Crick's controversial exploitation of Rosalind Franklin's[10] X-ray data to determine the structure of DNA (made possible when, without her knowledge, Wilkins gave Watson her key X-ray diffraction image, and Perutz gave Crick a copy of her MRC research report), and recall that Watson, Crick and Wilkins shared the Nobel Prize in 1962 for their work on nucleic acids (after Franklin's death), to understand that scooping can have significant ramifications.

---

[10] http://en.wikipedia.org/wiki/Rosalind_Franklin

Such fears have, in the past, been reinforced by several other factors: the absence of any academic incentive to publish intermediate data, the focus on 'the paper' as the only object of record that attracts kudos, and the lack of technology for publishing and formally attributing data. However, scientists' tendency to sit on important results until they've been able to place an official flag in the scientific landscape by publishing their findings in a 'reputable journal' is increasingly recognised as a serious brake on progress. Recent studies (*e.g.*, Piwowar and Vision, 2014) have shown that 'papers with publicly available datasets receive a higher number of citations than similar studies without available data'. And if – alongside the moral argument that most research is conducted with money from the public purse – such incentives aren't enough to convince scientists to part with their hard-won data, funding bodies (including the NIH, RCUK and European Commission) have begun to enforce strong rules governing the timeliness and openness of data generated by projects receiving their funding.

More recently, publishers like PLoS have put in place similar mandates that require, for any paper to be considered for publication, their referent data to be made openly and publicly available. Numerous repositories for domain-specific data-sets, such as Dryad (Vision, 2010), PANGAEA (Grobe *et al.*, 2010), DataONE (Michener, 2012), more general repositories like fig**share**, and institution-specific repositories, are being created to meet this growing need for publishing data alongside narrative; but although the assortment of carrots and sticks is growing, there remain complex political and technological challenges that will need to be addressed before it becomes commonplace to publish raw data alongside articles. While it may be fairly easy to deposit a sequence alignment, or the newly determined 3D structure of a biological molecule or chemical entity in a suitable database, it's considerably less clear how, where or by what mechanism one could make terabytes of video or other large bespoke data-sets available. Having recognised the costs of not doing it, however, the momentum behind data publishing is growing, and it seems inevitable that articles of the future will be commentaries that illuminate full data-sets, rather than mere edited highlights, as is the practice today.

8.8 Separating scientific wheat from chaff – towards semantic searches

By its nature, the scholarly record grows with time. It may be that old or 'ancient' articles are less relevant today; nevertheless, they persist as part of the scientific record. If there was ever a time when scientists could read and comprehend everything published in their field, this is certainly neither true nor possible today. If you need evidence of the truth of this, pick a biological term, and search PubMed for articles that mention it: *e.g.*, in December 2015, 'cancer' returned >3.2 million articles; 'HIV' >294,000; 'rhodopsin', >10,000; and even 'Torcetrapib', a branded drug that never even made it to market, still weighs in with >300 articles.

As scientists, we want to know the latest facts that may influence our research; and, in a world where the rate of article publication is out-pacing our ability to find and read them, we essentially want to know, 'which of the staggering number of articles should we read?' It turns out that answering this is quite hard: it presupposes that it's possible to identify all articles that are relevant in some way to a particular research domain, which, in turn, requires some level of understanding of what those articles actually contain. Not surprisingly, getting computers to do this for us is hard.

The most prominent online resource for searching the biomedical literature is the National Library of Medicine (NLM)'s PubMed. This brings together 'gold standard' citation and classification data provided by publishers via the MEDLINE database, with citations for books from the NCBI Bookshelf, articles published by NIH-funded projects, full-text Open Access publications from the PMC database (some of which may not have been indexed through MEDLINE), and 'hot-off-the-press' in-process citations for some articles that are yet to be included in the main MEDLINE index. By December 2015, PubMed included >25 million citations.

Records in PubMed – at least those indexed via MEDLINE – are classified according to MeSH (Medical Subject Headings) terms, the NLM's controlled vocabulary of biomedical concepts. PubMed's primary search interface encourages short keyword-based queries – Clarke and Wentz (2000) described these as 'telegram' searches, wherein we're challenged to consider which two or three words we'd need to transmit in order to best encapsulate a particular research question. For example, if our question were, 'How efficient is a single dose of a steroid for outpatient croup?' this might suggest the search, 'croup and outpatient'. Entered into PubMed's main search box, the MeSH vocabulary expands this short 'telegram' to a more complex query:

("croup"[MeSH Terms] OR "croup"[All Fields]) AND ("outpatients"[MeSH Terms] OR "outpatients"[All Fields] OR "outpatient"[All Fields])

This expanded query isn't hugely exciting: PubMed's processor has taken the term 'and' in our telegram to (correctly) mean the Boolean conjunctive, has opted to search all the bibliometric fields (titles, abstracts and so on) in its records, plus all the MeSH classifications associated with those records, and has included simple plurals. More interesting, perhaps, is what happens if we search for 'Torcetrapib and JTT-705'. PubMed's processor expands this as follows:

("torcetrapib"[Supplementary Concept] OR "torcetrapib"[All Fields]) AND ("dalcetrapib" [Supplementary Concept] OR "dalcetrapib"[All Fields] OR "jtt 705"[All Fields])

Notice that the research code 'JTT-705' has been cleverly expanded to include Roche's brand-name 'dalcetrapib'.

PubMed's search mechanism is a step beyond plain substring or keyword-matching, but, as we can see from its popularity, the application of some straightforward heuristics, the expansion of terms using a controlled vocabulary to include synonyms and related concepts, and some simple 'stemming' of words to deal with plurals, goes a very long way towards providing a powerful literature-search mechanism. Understanding and exploiting even a little of its behaviour can yield excellent results.

It could be argued that PubMed's approach already includes some basic 'semantic' elements, not least because it uses a controlled vocabulary to classify concepts and to expand queries. Nevertheless, the phrase 'semantic search' has come to mean something more ambitious, and is used to describe mechanisms that attempt to generate relevant results by understanding both the meaning of the contents of a searched corpus and the intent of the user. Ultimately, such systems try to give direct answers to questions like 'How efficient is a single dose of a steroid for outpatient croup?' without requiring users to consciously 'collapse' this into a simpler query, only to have that expanded again (with inevitable amplification of noise) by the chosen search engine.

General-purpose 'question-answering systems' that combine traditional key-word-based techniques with relevance-ranking, entity-recognition, data-mining, natural-language processing and ontologies are beginning to appear on the Web: Google's Knowledge Graph and Wolfram Alpha are perhaps the most prominent examples to date. It's interesting to note that pasting our full sample question directly into Google Scholar returns, as its top result (in December 2015), a link to the paper by Geelhoed *et al.*, 1996, which describes the, 'Efficacy of a small single dose of oral dexamethasone for outpatient croup'; Wolfram Alpha, on the other hand – admittedly not optimised for biomedical queries – gets into a bit of a pickle, fails to understand the query, and instead decides that the best thing would be to provide a fascinating array of facts associated with the word 'efficient'[11]. Biomedical semantic search engines, like Quertle.com and GoPubMed (Dietze and Schroeder, 2009), attempt to apply these techniques in a domain-specific way; it's probably too early to determine whether such approaches are likely to gain traction in the community. What is certain, however, is that devolving responsibility for finding relevant material to an algorithm – any algorithm – carries some risk. The more complex the algorithm, the less sure we can be that it has done the right thing: more conservative approaches might give us more manageable numbers of results, but how can we be sure that the most important thing hasn't been missed entirely?

In the move towards more 'intelligent' search engines, another – perhaps more pernicious – danger lies in wait to trap the unwary scientist. Although systems that learn the kind of results we like might seem attractive (especially to online retailers wanting to sell more products!), from a scientist's perspective, there is a fundamental difference between retrieving '*results that are relevant to my research*' and '*results that reflect the way I think*'. Whether algorithms can be created that distinguish these scenarios is an interesting question; however, as social networks and search engines attempt to record more detailed information about our browsing habits, personalised, subjective ranking of results is likely to become increasingly common, and, for scientists, dangerously subtle: searches that systematically reinforce the ideas that a scientist holds are likely to miss papers with divergent or contradictory views, potentially biasing conclusions arising from the research, and hence systematically doping the scientific record with further bias.

8.9 Semantic publication

Up to now, we've looked at some of the problems of getting machines to read natural language, and we've looked at some examples of how hard it is for humans to write pure logic. A compromise that's been proposed in various guises is to make it possible for authors to identify, and make machine-readable, only the main scientific assertions embodied in their articles. Superficially, this seems like a sensible approach, given that a fair proportion of any paper is 'padding' of one sort or another – the kind of

---

[11] As a result, we at least learn that 'efficient' is the 2,475th most common written word, was first used in English in 1398, and would score 17 in Scrabble.

'due diligence' of scientific writing, including abstracts, summaries, literature surveys, acknowledgements and so on.

The question is, how to achieve this? As we've discussed, doing so reliably in an automated, algorithmic way is extremely hard. For any given piece of text, even if we could robustly identify the various entities and processes mentioned in its sentences, and interpret its grammar with enough accuracy to collate the various assertions, how would we meaningfully work out which of these are part of the 'scientific claims' of the paper, and which are mere by-products of scholarly writing?

The alternative to identifying assertions algorithmically is to enlist the help of article authors, knowledgeable editors, or perhaps even 'the community' – so-called 'crowd-sourcing' of information has, after all, been applied successfully in many other areas. Getting authors of scientific papers to embrace the vision of producing content that's more semantically rich, however, has so far proved difficult: they seldom provide official accession numbers or unique identifiers for the entities described in their papers, or part with anything but the barest minimum of data for 'auxiliary materials'; and the prospect of them providing their core assertions in machine-friendly forms is, for the time being, almost unimaginable. Part of the resistance is cultural – in the main, authors haven't had to do this before, they don't see or understand the benefits or doing so now, and they don't have the time to learn; part also comes down to journal policy, or lack of journal policy – if journals don't uphold standards and best practices, authors are unlikely to comply. Consequently, the various recent initiatives that have tried to increase the semantic richness of articles have met with varying levels of success. Let's take a closer look.

8.9.1 Making articles 'semantic'

One of the first attempts to make articles more 'semantic' (in the sense of being 'readable' by both human and machine) was the so-called *FEBS Letters* experiment. This was a pilot project in which the journal editors worked with a small group of authors and with the curators of the MINT[12] molecular interaction database to integrate data in published articles with information stored in databases (Seringhaus and Gerstein, 2008). To make the project tractable, the initial focus was on identifying and highlighting Protein–Protein Interactions (PPIs) and Post-Translational Modifications (PTMs), as attempting to make *all* the biological information and concepts described in any given article, or set of articles, machine-readable was obviously impossible (Superti-Furga *et al.*, 2008).

At the core of the experiment was the concept of the Structured Digital Abstract (SDA) – this was a 'simple' mechanism for capturing key article facts in a machine-readable, XML-encoded summary, to render those facts accessible to text-mining tools (Seringhaus and Gerstein, 2007). The process was straightforward: participating authors were simply asked to use Excel spreadsheets to collect PPI and PTM data, together with a range of experimental, biological and technical details referred to in their articles (information about the nature of the experimental evidence, characteristics of the interacting proteins and their biological roles, the PTMs required for interaction, protein identifiers with links to MINT and UniProtKB, *etc.* (Ceol *et al.*, 2008)). An example SDA is shown in Figure 8.6. Here, beneath the conventional abstract, is the structured summary of the

---

[12] http://mint.bio.uniroma2.it/mint/Welcome.do

Abstract

Yeast two-hybrid screening was conducted using a human ovary cDNA library to search for a novel binding protein using transforming growth factor-beta stimulated clone-22 (TSC-22). The selected protein was fortilin, which has been characterized as a nuclear anti-apoptotic protein. Overexpression of fortilin in ovarian carcinoma cells reversed TSC-22-mediated apoptosis, and the inhibition of fortilin expression via small interfering RNA (siRNA) resulted in an increase in the apoptosis of ovarian carcinoma cells. Moreover, fortilin overexpression promoted the degradation of TSC-22. Thus, an interaction between fortilin and TSC-22 prevents apoptosis via the destabilization of TSC-22 in ovarian carcinoma cells.

Structured summary

   MINT-6173230, MINT-6173253:

   TSC22 (uniprotkb:Q15714) *physically interacts* (MI:0218) with *fortilin* (uniprotkb:P13693) by *co-immunoprecipitation* (MI:0019)

   MINT-6173217: *TSC22* (uniprotkb:Q15714) *binds* (MI:0407) *fortilin* (uniprotkb:P13693) by *pull-down* (MI:0096)

   MINT-6173240, MINT-6173270:

   TSC22 (uniprotkb:Q15714) *physically interacts* (MI:0218) with *fortilin* (uniprotkb:P13693) by *two-hybrid* (MI:0018)

Figure 8.6

The structured summary, with its 'normal' abstract, for a short communication in *FEBS Letters*.
Source: Lee *et al.* (2008). Reproduced with permission from Elsevier.

paper, which essentially makes three statements. The first of these asserts that Transforming growth factor-beta Stimulated Clone-22 (TSC-22) interacts physically with fortilin, as evidenced by co-immunoprecipitation studies, and described in more detail in MINT entries 6173230 and 6173253:

> MINT-6173230, MINT-6173253: TSC22[13] (uniprotkb:Q15714) physically interacts (MI:0218) with fortilin[14] (uniprotkb:P13693) by co-immunoprecipitation (MI:0019)

The stilted formulation here makes fairly uncomfortable reading, but is necessary if the facts described in the paper are to be rendered computer-accessible. Inevitably, most of the encoded relationships point to MINT entries because the pilot project was a collaboration between the Journal and the MINT curators. Nevertheless, the idea was to design the system such that it would be sufficiently robust to readily generalise to other biological relationships, and would be sufficiently easy to use to ensure wide take-up.

In spite of the alleged simplicity of the approach, the experience of processing just a handful of manuscripts wasn't encouraging (Ceol *et al.*, 2008). Only five authors agreed to participate, and although most of these needed very little help to complete their spreadsheets, one of them needed significant assistance from the MINT curators. After completion of the pilot, take-up was slow, and in the following 10 months, only 90 *FEBS Letters* papers were published with SDAs (Shotton *et al.*, 2009). Ultimately, were they to have been judged a success, SDAs would have become an integral part of MEDLINE abstracts. However, this hasn't yet happened, and most *FEBS Letters* articles still don't contain SDAs (to give an idea, of the 536 articles published in the 10 months to 2 May 2014, 78 contained SDAs).

---

[13] http://www.uniprot.org/uniprot/Q15714
[14] http://www.uniprot.org/uniprot/P13693

In a rather different project, Shotton *et al.* (2009) experimented with 'semantic enrichment' of an article by Reis *et al.* published in *PLoS Neglected Tropical Diseases (NTD)* in 2008. They chose this article because it included different data types (geospatial data, disease-incidence data, serological-assay results, *etc.*) presented in a range of formats (*e.g.*, bar charts, scatter plots, maps); perhaps more importantly, their chosen article was available in an XML format, published under a Creative Commons Licence, and hence could be modified and re-published. Their enrichments included DOIs, keyword mark-up, hyperlinks to external information resources, interactive figures, a document summary, with a tag-cloud and citation analysis, and much more. Shotton *et al.* further augmented the paper with downloadable spreadsheets containing data from the tables and figures, and 'mashups' with data from other articles and Google Maps. They also implemented a 'citation typing' ontology (Shotton, 2010) to allow collation of machine-readable metadata relating to the article and to its cited references.

A glimpse of some of this rich functionality is shown in Figure 8.7: here, from the coloured tabs at the top of the page, we've chosen to highlight entities related to 'disease' (red) and 'protein' (purple). In the Introduction to the article (right-hand panel), Reis *et al.* describe how, 'Urban epidemics of leptospirosis now occur in cities throughout the developing world during seasonal heavy rainfall and flooding', citing Ko *et al.*, 1999. To save us the trouble of having to find, possibly pay $35.95 for, download and

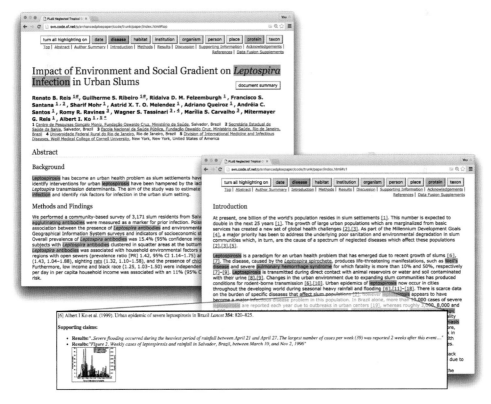

Figure 8.7

Excerpt from the *PLoS NTD* article marked up using the system developed by Shotton *et al.* (2009). Leptospirosis matter article by Reis *et al.* (2008).

Source: Shotton *et al.* (2009); Reis *et al.* (2008).

read Ko's article, 'Urban epidemic of severe leptospirosis in Brazil', Shotton *et al.* have conveniently done this for us and have provided the relevant supporting claims and figure (bottom panel, Figure 8.7):

> Severe flooding occurred during the heaviest period of rainfall between April 21 and April 27. The largest number of cases per week (39) was reported 2 weeks after this event…
>   Figure 2. Weekly cases of leptospirosis and rainfall in Salvador, Brazil, between March 10, and Nov 2, 1996.

How much easier would our lives be if all articles did this kind of background work for us! Alas, at least for now, this remains but a dream. The truth is that, while Shotton *et al.*'s study showed that significant value could be added to published articles *post hoc*, the enhancements made were both platform- and browser-dependent, and were confined to just one article. Although asserting that the approaches used here weren't 'rocket science' (they exploited standard mark-up languages, ontologies, style-sheets, programmatic interfaces, *etc.*), the authors admitted that the work was extremely manually intensive, and would necessitate significant levels of automation to bring them into mainstream publishing.

The obstacles to making the knowledge hidden in published articles machine-accessible are legion. Most of the scientific literature is therefore still 'unsemantic' (and consequently, in Shotton's view, 'antithetical to the spirit of the Web'); and it will doubtless be some time before the tools, policies and incentives necessary to encourage authors to engage with this enormously important challenge are finally put in place. Notwithstanding the difficulties, however, interest in bridging the gap between human- and machine-readable scientific content remains. The Internet and digital technologies have irrevocably changed the way in which scientists expect to use the literature, and hence publishers continue to seek innovations to satisfy the increasing demands of their authors and readers.

We've touched, here, on just a couple of the many approaches that have been used to semantically enrich scientific documents, to make the data and knowledge they contain more readily accessible and reusable. Although no universal solution has yet materialised, some common themes have emerged: most have been HTML- or XML-based, most provide hyperlinks to external websites and term definitions from relevant ontologies, and so on. However, some researchers have imagined more far-reaching possibilities, where data in online papers become 'more alive': Bourne (2005), for example, dreamed of a future in which readers could toggle between a table of numerical values and their graphical representation, allowing immediate investigation of relationships latent in the data; or between a static image of a molecule and interactive 3D views, allowing real-time exploration of molecular interactions or active sites described in the paper, without the need for specialised knowledge of esoteric data-manipulation tools – perhaps like the scenarios showcased in the 2009 article by Attwood *et al.*[15], and described in more detail in the following sections.

As it turns out, many of the article repositories, ontologies, machine-readable document standards, and so on, already exist for integrating published content with data in public databases. So why haven't publications benefitted more from the opportunities offered by such resources? Part of the reason is probably simply that the community has grown up with static manuscripts: most researchers are just used to reading articles in PDF or HTML formats (Lynch, 2007) and don't see the need to change – breaking deep-rooted cultural habits is hard. The other part of the story is that, in truth, retrospective

---

[15] http://www.biochemj.org/bj/424/0317/4240317.pdf

addition of semantics to legacy data is complex, labour-intensive and costly. The right balance therefore needs to be found between the degree of automation it's possible to inject into the process, and the degree of cultural change that can reasonably be expected within a research community that has never seen a reason to consider the relationship between data and published articles, and hence, importantly, has never had to think about the enormous task of providing the semantic context necessary to unite them.

Sharing knowledge is at the heart of scientific scholarship, and publishing articles is part of the system that helps us do this. Nevertheless, despite the 'earthquake of modern information and communication technologies', some commentators have suggested that we're still not sharing information efficiently because we lack the infrastructures that facilitate knowledge integration, and that 'we can't begin to integrate articles with databases' because 'the actors in the articles (the genes, proteins, cells and diseases) are described in hundreds of databases' (Wilbanks, 2007). Building the infrastructure highways necessary for us to be able to interconnect, openly share and integrate the knowledge held in this plethora of disparate, disconnected resources, Wilbanks warns, will be 'very, very hard'; but the rewards for doing so, and a growing, community-wide desire for progress, have stimulated some promising developments. In the next section, we showcase one such development – Utopia Documents – which led to the creation of the *Semantic Biochemical Journal*, first published by Portland Press Limited (PPL) in 2009.

8.10 Linking articles with their cognate data

Utopia Documents (Attwood *et al.*, 2009; Attwood *et al.*, 2010) is a 'smart PDF reader'. It was developed because, despite the many benefits of enriched HTML articles, like those mentioned above, most researchers still read and store papers as PDF files. PDFs have several advantages over HTML pages, not least that they're self-contained objects that can be 'owned' and stored locally – their originating websites can go offline, their original versions can be withdrawn by the journal, your subscription to that journal can expire, but you'll still have your PDF copy of the article, and can continue to access it whenever you choose. A perhaps more subtle point is that journals have, over the years, made huge investments in producing typographic formats (rendered on paper and in digital PDF) that are aesthetically pleasing and easy to read – in consequence, at least on reasonably large screens, article PDFs are simply much nicer to read than their HTML counterparts.

Utopia Documents (UD) behaves like a familiar PDF reader (Adobe Acrobat, KPDF, OS X Preview, *etc.*), but its functionality can be augmented with field-specific ontologies and plugins that use Web services 'on the fly' to retrieve and display data related to the article. With these in place, the software transforms article PDFs from static facsimiles of their printed counterparts into dynamic gateways to further knowledge, linking information (terms, phrases, objects) embedded explicitly and/or implicitly in the text to online resources, and offering seamless access to auxiliary data and interactive analysis and visualisation software – and all of this is done without compromising the integrity of the underlying PDF file.

8.10.1 What Utopia Documents does

UD inspects the typographic layout of a PDF document in order to reconstruct its 'semantic structure', separating out titles, headings, sections, subsections, and so on. Subsequently, through a combination of automatic and manual mechanisms, the software begins to augment the content seen by the human reader. It starts by treating the document as a whole – an unchanging piece of the scientific record: it constructs an article 'fingerprint', comprising its title, authors and any digital identifiers that would uniquely represent the document in online databases (DOIs, PubMed/PMC IDs, *etc.*). With these identifying features, it begins to gather up-to-date information about the overall article, calling appropriate Web services to link it to article-level or 'alternative' metrics and to biological database records that cite the article. Finally, the software invokes text- and content-mining Web services to identify objects of interest within the body of the document (*e.g.*, like gene or protein names), linking these automatically to definitions in relevant online databases.

The ability seamlessly to insert explanations of particular terms can be useful for authors and readers alike: for example, restrictive page counts can often lead to the omission of important definitions, the inclusion of which would make manuscripts more comprehensible, especially to non-experts; relative newcomers to a field may need reminding of the meaning of some of the more field-specific jargon, or may want to dig a little deeper into the subject background. UD helps here by allowing editors and authors to annotate specific terms with definitions from online resources (*e.g.*, by using SPARQL to query RDF representations of resources like Wikipedia/DbPedia, UniProtKB and PDB, which form part of the 'Linked Data Cloud' (see Sections 7.4.2 and 7.5)); moreover, it gives readers the flexibility to find definitions for themselves, and add their own annotations and comments (which are stored in RDF using the Annotation Ontology (Cicaresse, 2011)).

Aside from facilitating term look-up, UD is able to transform static tables and figures, *in situ*, into dynamic, interactive objects. To communicate the main results of an experiment, or set of experiments, as efficiently (and digestibly) as possible, figures and tables are generally just snapshots of much larger, more complex data-sets (which are nowadays often held elsewhere online): a table, for example, might focus on a pertinent fragment of a vast experimental data-set; or a figure may highlight part of a protein structure, to underscore a specific feature described in the text. The ability to link tables and figures to their associated data gives readers much richer, interactive experiences, providing opportunities both for further exploration and for much deeper investigation of a paper's principal claims.

Finally, for many readers, a given article is often a springboard to other studies cited by the authors, and diligent researchers will spend considerable amounts of time following up those citations in order to find out more. We tend to take it for granted that current articles will be available in electronic form; moreover, we also expect older articles from journal archives to be readily available – accordingly, most publishers have made substantial efforts to make their back-catalogues electronically accessible. Navigating the multitude of disparate online repositories and bibliographic tools can, however, be a complex and daunting task. UD simplifies the process of identifying related articles: where readers have legitimate access to papers, either because they're Open Access or are available by institutional licence, the software automatically links to their digital online versions.

UD thus offers a range of functionality not afforded by many PDF readers. What makes it unique is not what it does to the PDF itself (that's left unchanged, so that readers can be confident that they're seeing the 'uncontaminated' original), but how it layers and superimposes the additional tools and data on top of the PDF. We look at this in a little more detail in the next section.

8.10.2 A case study

So far, we've talked broadly about what UD allows readers and authors to do, and we've mentioned the underlying technology that permits the software to behave in this way.

In the pages that follow, we illustrate what has to happen in practice in order to bring static pages, like those you tend to view in traditional PDF readers, to life. In particular, we highlight the tension between the degree of automation it's possible to achieve with current manuscripts (most of which don't exploit data- or research-object identifiers, or otherwise adhere to data-management policies) and the level of human intervention that's actually needed to make rigorous document mark-up a practical reality.

We begin by examining part of a page from an article published in the launch issue of the *Semantic Biochemical Journal*[16] (Attwood *et al.*, 2009), which we encourage you to download and view using a traditional PDF reader. Figure 8.8 shows Figure 13 from that original article, which features a complex, composite image with screen-shots from

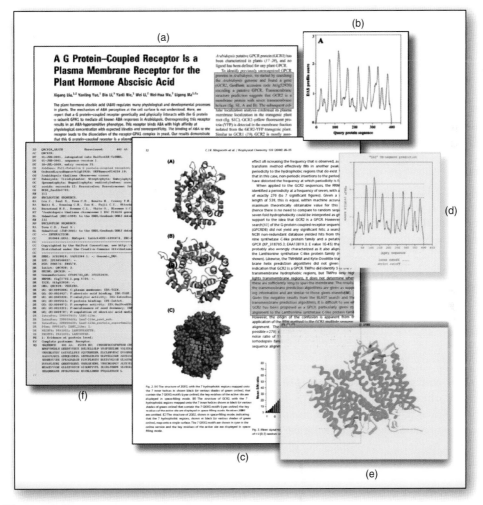

Figure 8.8

Figure 13 from the 2009 *Semantic Biochemical Journal* article by Attwood *et al.,* visualised using Adobe's Acrobat Reader.

Source: Attwood *et al.* (2009). Reproduced with permission from Portland Press Ltd.

---

[16] http://www.biochemj.org/bj/424/0317/4240317.pdf

two other journal pages (a, c), a couple of graphical outputs (b, d), a 3D molecular view (e) and an entry from an online database (f), as viewed with a traditional reader.

The background – a detective story

The story behind the composite image is fascinating. In 2007, Liu *et al.* reported the discovery of a novel plant GPCR, so-called GCR2; this was such exciting news that it warranted publication in the top journal, *Science* (Figure 8.8 (a)). The paper was short – just five pages – and hence much of the key evidence was furnished as '*Supporting Online Material*'. This 'supplementary' material included 20 pages of experimental details, 11 further figures and 18 additional bibliographic references – that's four times more information than was in the article itself. The crux of the paper was the experimental demonstration

> that a G protein–coupled receptor [GCR2] genetically and physically interacts with the G protein α subunit GPA1 to mediate all known ABA [abscissic acid] responses in *Arabidopsis*.

Their biological characterisation was 'clinched' with evidence from a bioinformatics analysis showing that 'GCR2 is a 7-transmembrane protein', illustrated by compelling hydropathy profiles in Supplementary Figure S1 A and B (Figure 8.8 (b)).

Looking back, it's very difficult to piece together what happened next. Responsibility for annotating the *Arabidopsis* genome passed from the *Arabidopsis* Genome Initiative (AGI), to The Institute for Genomic Research (TIGR) and Martinsreid Institute for Protein Sequences (MIPS), and then, after 2004, to The *Arabidopsis* Information Resource (TAIR[17]). Annotation at TAIR derives from manual scouring of the literature by the TAIR curators and from direct submissions from the research community. At first sight, what appears to have happened is that the curators annotated genomic DNA locus AT1G52920[18] and deposited the result in EMBL/GenBank/DDBJ. In June 2011, the gene product was translated and made available as an *Arabidopsis thaliana* 'G protein-coupled receptor', in UniProtKB/TrEMBL entry F4IEM5_ARATH[19].

But the history of this sequence is rather more convoluted – see Figure 8.9. Ten years earlier, in January 2000, its parent genomic sequence had been submitted to EMBL/GenBank/DDBJ by TIGR. We can trace this submission back to EMBL entry AC019018[20], whose translation product AAG52264[21] bears the annotation, 'putative G protein-coupled receptor'. What isn't apparent from this is where the annotation originated: all the entry tells us is that, 'Genes with similarity to other proteins are named after the database hits' – it doesn't tell us what version of what database was searched and/or with what version of what search tool. Nevertheless, because it first appeared in EMBL, the sequence subsequently had a separate existence in TrEMBL, where it also appeared as a 'putative G protein-coupled receptor' – entry Q9C929[22] – in June 2001.

During the 13 years following its first appearance in TrEMBL, entry Q9C929 underwent a further 67 revisions. The 35th version[23] is shown in Figure 8.8 (f) – it bears the familiar 'GPCR' annotation in its description line (red highlight, top of the panel), but

---

[17] http://www.arabidopsis.org
[18] http://www.arabidopsis.org/servlets/TairObject?type=locus&name=AT1G52920
[19] http://www.uniprot.org/uniprot/F4IEM5.txt?version=1
[20] http://www.ebi.ac.uk/cgi-bin/sva/sva.pl?index=62&view=28090129
[21] http://www.ebi.ac.uk/cgi-bin/sva/sva.pl?index=48&view=35961725
[22] http://www.uniprot.org/uniprot/Q9C929.txt?version=1
[23] http://www.uniprot.org/uniprot/Q9C929.txt?version=35

TIGR	2 JANUARY 2000	AC019018 (gene F14G24.19, translation product AAG52264)	Putative G protein-coupled receptor	G PROTEIN COUPLED RECEPTOR,G-PROTEIN COUPLED RECEPTOR 2, GCR2, GPCR	AT1GG2920	Submitted April 2011	TAIR
			EMBL/Gen	**Bank/DDBJ**			

1	01.06.2001 *TrEMBL 17.0*	Q9C929	PUTATIVE G PROTEIN-COUPLED RECEPTOR				
	2002						
6	15.12.2003 *UniProtKB/TrEMBL 1.0/25.7*	Q9C929	Putative G protein-coupled receptor				
	2004						
11	01.02.2005 *UniProtKB/TrEMBL 4.0/29.0*	Q9C929_ARATH	Putative G protein-coupled receptor				
	2006 2007 2008						
35	28.07.2009 *UniProtKB/TrEMBL 15.6/40.6*	Q9C929_ARATH	Putative G protein-coupled receptor				
	2010						
51	28.06.2011 *UniProtKB/TrEMBL 2011_07*	Q9C929_ARATH	Putative G protein-coupled receptor	G protein coupled receptor	F4IEM5_ARATH	28.06.2011 *UniProtKB/TrEMBL 2011_07*	1
	2012					2012	
68	13.11.2013 *UniProtKB/TrEMBL 2013_11*	Q9C929_ARATH	Putative G protein-coupled receptor	G-protein coupled receptor 2	F4IEM5_ARATH	13.11.2013 *UniProtKB/TrEMBL 2013_11*	23
	11.12.2013	Merged into GCR2_ARATH (F4IEM5)		**LanC-like protein GCR2** (Alt Name: G-protein coupled receptor 2)	GCR2_ARATH	11.12.2013 *UniProtKB/Swiss-Prot 2013_12*	24
				LanC-like protein GCR2 (Alt name: G-protein coupled receptor 2)	GCR2_ARATH	19.03.2014 *UniProtKB/Swiss-Prot 2014_03*	27

Figure 8.9

The provenance and evolution of the sequence referred to as GCR2.

its identifier has changed to Q9C929_ARATH. At version 68, the decision was made to merge this entry with UniProtKB/Swiss-Prot entry GCR2_ARATH. Aspects of this convoluted detective story are summarised in Figure 8.9, which traces the separate evolution of Q9C929 from the TIGR submission to EMBL/GenBank/DDBJ in 2000, and of F4IEM5 from the TAIR submission in 2011, until both were merged in December 2013.

Interestingly, the merger with GCR2_ARATH represented the 24th revision[24] of the progenitor sequence, F4IEM5_ARATH, whose annotation painted a rather different picture of the identity of GCR2. Let's dig a little further.

Shortly after publication of Liu *et al.*'s *Science* article, Illingworth *et al.* (2008) challenged their results. They demonstrated that the sequence of GCR2 is highly similar to those of lanthionine synthetase C (LanC)-like proteins, a divergent family of peptide-modifying enzymes. Strikingly, they showed that the compelling topology of the alleged transmembrane (TM) hydropathy profile was in fact the result of the seven-fold symmetry

[24] http://www.uniprot.org/uniprot/F4IEM5.txt?version=24

of the internal helical toroid of this protein's *globular* structure (the blue-green regions at the centre of the molecular structures shown in Figure 8.8 (c)). Curiously, a simple BLAST search, and/or searches of the main protein family databases, reveal the relationship with the LanC-like proteins in seconds – in consequence, note the cross-references to InterPro, Pfam and PRINTS highlighted in red towards the bottom of Figure 8.8 (f).

So why didn't Liu *et al.* perform trivial searches like this? Why were they so convinced that GCR2 was a GPCR? Part of the answer probably lies at the start of their article. They state that,

> Only one *Arabidopsis* putative GPCR protein (GCR1) has been characterized in plants…, and no ligand has been defined for any plant GPCR. To identify previously unrecognized GPCR proteins in *Arabidopsis*, we started by searching the *Arabidopsis* genome *and found a gene (GCR2, GenBank accession code At1g52920) encoding a putative GPCR* [our emphasis].

In fact, At1g52920[25] is a TAIR accession number, not a GenBank code (the clue is in the letters 'At' at the start of the code). So GCR2 had already been annotated as a possible GPCR, and Liu *et al.* saw no reason to disagree – presumably, they were motivated to find GCR2 in the wake of the earlier discovery of GCR1, and, with a ready-made candidate from TAIR, all they had to do was demonstrate G protein interaction and involvement in abscissic acid responses.

Notwithstanding the apparent failure to run a simple BLAST search, there's something a little more disturbing here. Compare the hydropathy plot (Figure 8.8 (d)) generated with the same TM prediction server, DAS (Cserzo *et al.*, 1997), that Liu *et al.* used to create their Supplementary Figure S1 A (Figure 8.8 (b)). In particular, note the omission of the significance bars in the latter, which, in the former, show that only one of the seven peaks scores above the significance threshold for TM domains, arguing strongly against this being a membrane protein. We can only speculate why Liu *et al.* removed the significance bars from the figure, and then buried the result in the middle of 20 pages of supplementary data (and it's easy to draw unflattering conclusions).

Now compare the structure of bovine rhodopsin (PDB code 1F88[26]), a *bona fide* GPCR (Figure 8.8 (e)), with the structure shown in Illingworth's paper (Figure 8.8 (c)), that of nisin cyclase (PDB code 2G0D[27]), a GCR2 homologue from the LanC-like protein family. GCR2 demonstrably lacks both sequence and structural similarity to GPCRs. Nevertheless, the relationship with GPCRs continues to be propagated through the sequence databases: as shown in Figure 8.9, in December 2013, the UniProtKB/Swiss-Prot description line was changed from 'G-protein [sic] coupled receptor 2' to the now rather confusing 'Recommended name: LanC-like protein GCR2; alternative name: G-protein [sic] coupled receptor 2; gene name: GCR2; synonyms=GPCR', as if to say that 'G protein-coupled receptor' is an acceptable alternative name – indeed, that GPCR is a *synonym* for LanC-like proteins! Worse, at the same time, in an extreme form of scientific hedging, the UniProt curators provided the following caution about the protein (which persists in version 38, in January 2016):

> was originally described as a plasma membrane G-protein [sic] coupled receptor (GPCR) that binds ABA… However, GCR2 has been controversial with respect to the reproducibility of the results… Moreover, GCR2 lacks the prototypical seven transmembrane domains of GPCRs, and is homologous to mammalian lanthionine synthetase C-like proteins.

---

[25] http://www.arabidopsis.org/servlets/TairObject?type=locus&name=AT1G52920
[26] http://www.rcsb.org/pdb/explore/explore.do?structureId=1f88
[27] http://www.pdb.org/pdb/explore/explore.do?structureId=2g0d

Thus, in spite of incontrovertible evidence, the database entry still doesn't say, categorically, that GCR2 is a globular protein, a member of the LanC-like family of peptide-modifying enzymes, unrelated in sequence and structure to 7TM GPCRs!

Animating the text

As part of the *Semantic Biochemical Journal*[28] project, the image we've just described in detail was marked up using a prototype version of UD to showcase its then emerging functionality. Figure 8.10 shows the same image viewed using UD. Notice that the image contains pale blue highlighted areas, a number of small UD icons in the far right-hand margin, and three 'play' buttons embedded within the figure itself, hints that something interactive lies beneath.

If we access this figure in the *Biochemical Journal* article (Figure 13 on page 330 of the paper) using UD, clicking on the top blue highlighted area (Figure 8.10 (a))

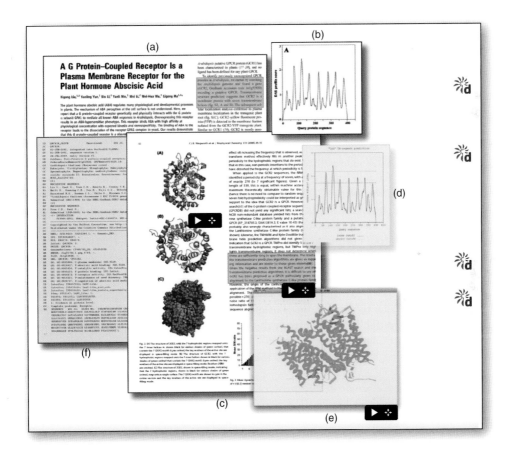

Figure 8.10

The same *Semantic Biochemical Journal* article illustrated in Figure 8.8, visualised using Utopia Documents.

Source: Attwood *et al.* (2009). Reproduced with permission from Portland Press Ltd.

---

[28]To get the most from the description that follows, we recommend that you download a PDF of the *Biochemical Journal* article and open it in Utopia Documents: www.utopiadocs.com

invokes an image pop-up of one of the hydropathy profiles presented by Liu *et al.* as part of Supplementary Figure S1 A. Clicking on the shaded hydropathy profile below (Figure 8.10 (d)) pops up a hydropathy profile for the same sequence (GCR2) using the same analysis software (DAS), allowing direct comparison of Liu *et al.*'s result with the actual result returned from the TM prediction software – from which the manipulations are immediately apparent in the published Figure.

Clicking on the 'pop out' icon on the right-hand side of the play button beneath the second blue highlighted region (Figure 8.10 (c)) pops out a new window containing a multiple sequence alignment of GCR2 with homologous members of the LanC-like enzyme family. The popped-out window is an interactive alignment editor – it's therefore possible to scroll through the alignment and hence to witness, at first hand, the similarity these sequences share and, in particular, the occurrence of several conserved GxxG motifs, which are the canonical features of these zinc-binding proteins.

Clicking on the play buttons beneath the 3D structures shown in Figures 8.10 (c) and (e) toggles each of the static images with interactive molecular views, which can be zoomed in and out – grabbing part of the molecule allows controlled rotation of the views, or can be used to set the molecules spinning. As with the alignment example above, clicking on the 'pop out' icon on the right-hand side of each of the play buttons pops the 3D molecular viewer out into a new window, which can then be re-sized and explored at will – by virtue of their greater size, the popped out views are rather easier to manipulate than the in-frame views, and hence the striking differences between the structures are easier to visualise.

Finally, clicking on the shaded panel in Figure 8.10 (f) invokes the UniProtKB/TrEMBL entry for GCR2. This allows comparison of the 'current' database entry (version 39 of F4IEM5_ARATH at the time of writing) with the version that was current at the time of publication of the 2009 *Biochemical Journal* article (version 35 of entry Q9C929_ARATH). The latter bears the GPCR annotation in its description line, while the former bears the revised description line, 'LanC-like protein GCR2', with alternative name 'G-protein [sic] coupled receptor 2' beneath it.

Making the animations work – behind the scenes

In the above scenario, we saw how UD was able to animate different components of a static, composite image: to turn pictures of globular and membrane proteins into 'live' molecules spinning in 3D, to invoke from the text an interactive alignment editor, and to allow additional images or information from a database to pop out of the page. A number of things had to happen behind the scenes to make these interactive features possible.

1. *Invoking images of hydropathy profiles.* In Figure 8.10 (a), a screen-grab of the hydropathy profile shown in Liu *et al.*'s Supplementary Figure S1 A was manually associated with a relevant portion of text (shown below), highlighted blue in the Figure 8.10 (a),

> …we started by searching the Arabidopsis genome and found a gene (GCR2, GenBank accession code At1g52929) encoding a putative GPCR. *Transmembrane structure prediction suggests that GCR2 is a membrane protein with seven transmembrane helices* (fig. S1, A and B) (our emphasis).

To achieve the same result automatically would require a text-mining tool that was sufficiently sophisticated to recognise the assertion that 'GCR2 *is a* membrane protein with seven transmembrane helices' (the sentence emphasised in the quote above), and to correctly associate it with Figure S1 (referenced in parentheses), found on page 9 of the supplementary data.

In Figure 8.10 (d), a screen-grab of a hydropathy profile created by running the DAS TM prediction tool was manually associated with its own image, highlighted blue in the figure. Ideally, to have achieved the same result automatically would have required publication of the raw output data from the DAS prediction tool, coupled with an online workflow for recreating the profile from those data.

2. *Invoking an interactive alignment editor.* In Figure 8.10 (c), the CINEMA alignment editor, pre-loaded with a range of LanC-like proteins, including GCR2 (Q9C929_ ARATH), was manually associated with a relevant chunk of text (as below), again highlighted blue in the figure,

> The results of the transmembrane prediction algorithms are given as supporting information and are similar to those given elsewhere [28]. Given the negative results from the BLAST search and the transmembrane prediction algorithms, it is difficult to see why *GCR2* has been proposed as a GPCR, particularly given its *alignment to the Lanthionine synthetase C-like protein family* (our emphasis).

To achieve this result automatically would require a text-mining tool able to recognise the entity '*GCR2*' (highlighted in the quote above), to unambiguously associate this entity with the correct version of its sequence in UniProtKB/TrEMBL or UniProtKB/Swiss-Prot (which could be any one of the 100 incarnations of Q9C929_ ARATH, F4IEM5_ARATH or GCR2_ARATH), thence to infer its relationship with the phrase '*alignment* to the Lanthionine synthetase C-like protein family' (the highlighted phrase), to correctly associate that with a pre-existing, uniquely identifiable and retrievable alignment of GCR2 and LanC-like proteins – perhaps like those from the PRINTS database (atgpcrlanc.seqs, in PIR format, or the equivalent file in a Utopia-friendly format, atgpcrlanc.utopia) – and to feed that into an alignment editor, such as CINEMA.

3. *Invoking an interactive molecular viewer.* In Figures 8.10 (c) and (e), the Ambrosia molecular viewer, pre-loaded with representative structures of a LanC-like protein and of a GPCR (respectively, *Lactococcus lactis* nisin cylase and bovine rhodopsin), was manually associated with the relevant protein structure images (the latter highlighted blue). To achieve the same result automatically would require a software tool able to identify the images depicted in Figures 8.10 (c) and (e) as the molecular structures of *Lactococcus lactis* nisin cyclase and of bovine rhodopsin, and a text-mining tool able to recognise from the caption in the original Figure 13 the phrases 'the nisin cyclase structure shown in Illingworth's paper [PDB code 2G0D (*c*)]' and 'Compare the structure of a *bona fide* GPCR [bovine rhodopsin, PDB code 1F88 (*e*)]', to make the explicit connection between the identified protein names and the PDB IDs – and then to feed those into a molecular viewer, such as Ambrosia. (Note, here, that explicit connections between nisin cyclase and PDB ID 2G0D and between GPCRs and 1F88 were not made in Illingworth's original figure caption).

4. *Invoking a database entry.* In Figure 8.10 (f), a screen-grab of the UniProtKB/ TrEMBL entry for Q9C929_ARATH, highlighted blue in the figure, was manually associated with the current entry for this sequence in UniProtKB/Swiss-Prot, GCR2_ ARATH. To achieve this result automatically would require a text-mining tool able to recognise the entity '*GCR2*' (highlighted in the quote below) found in the caption of the original Figure 13,

> Despite the obvious lack of sequence and structural similarity of *GCR2* to genuine GPCRs, and its clear affiliation with the LanC-like proteins, this error has been propagated to *the*

description line of its UniProt entry, even though the entry contains database cross-references to LanC-like proteins rather than GPCRs (f).

The software would need to infer that 'the description line of *its* UniProt entry' was a reference to GCR2 and not to GPCRs or LanC-like proteins mentioned in the same sentence, and then to unambiguously associate this entity with the correct version of its sequence in UniProtKB/TrEMBL or UniProtKB/Swiss-Prot; and finally, using a software tool able to identify the image depicted in Figure 8.10 (f) as the relevant database entry for GCR2, to associate it explicitly with its UniProtKB/TrEMBL ID or accession number.

As with Shotton's adventure into semantic publication, layering this additional narrative on top of existing traditional publications, and making precise links to definitive database entries required some non-trivial manual curation in order to make sure that the story was precisely the one we wanted to tell; neither the scholarly articles nor the databases involved yet contain enough semantic information for this to be done completely automatically. As tantalising as the vision of completely and seamlessly linking research stories with their cognate data may seem, there's still some way to go before the dream becomes a reality.

8.11 Summary

This chapter explored some of the problems that arise when trying to bridge the gaps between scientific narratives that make sense to humans and those that are comprehensible to machines, and the challenges these pose when we try to join the two together in order to add meaning to biological data. In particular, we saw that:

1 Records in biological databases cite research articles in order to lend authority to the 'facts' they include, and to provide provenance trails for their users;
2 The role of database curators is to *find within* and to *distil out* pertinent 'facts' from sub-sets of published articles in order to add them to databases;
3 To validate the 'facts' they discover in databases, users must *re-associate* database information with *supporting context* from the original source literature;
4 Most of the steps involved in manual database-annotation pipelines are currently lost;
5 Increased reliance on publishing data to validate scientific claims has meant that articles and data-sets are becoming ever more closely intertwined;
6 Initiatives are underway to weave research data more tightly back into the thread of scientific discourse;
7 Scientific articles, historical and contemporary, are riddled with computational trip-hazards;
8 Most articles use terms that have multiple meanings that make sense only if we understand the broader context;
9 Many articles include scientific assertions that have been distorted by 'hedging', or careless citing – this makes it very difficult for machines to extract their meaning;
10 Moving beyond paper-based scientific literature, to robust, digital models of knowledge representation, capable of supporting scientific discovery is challenging;
11 Sharing knowledge is at the heart of scientific scholarship, and publishing articles is part of the system that helps us do this;

12 It remains difficult to share information efficiently because we lack many of the infrastructures needed to facilitate knowledge integration;

13 Building the infrastructure highways necessary to interconnect, share and integrate the knowledge held in disparate, disconnected resources is hard;

14 The desire for progress is both community-wide and growing, and has stimulated some promising developments;

15 Utopia Documents is an application that addresses some of these challenges, but is just one of several developments in this area.

8.12 References

Al-Hasani, R., McCall, J.G. and Bruchas, M.R. (2013) Exposure to chronic mild stress prevents kappa opioid-mediated reinstatement of cocaine and nicotine place preference. *Frontiers in Pharmacology*, **4**, 96. doi:10.3389/fphar.2013.00096

Allen, B. P., Birukou, A., Blake, J.A. *et al.* (2012) The Force11 Manifesto: improving Future Research Communication and e-Scholarship. Available at http://www.force11.org/white_paper accessed 5 May 2016.

Andrade, M. and Sander, C. (1997) Bioinformatics: from genome data to biological knowledge. *Current Opinion in Biotechnology*, **8**, 675–683.

Antezana, E. Kuiper, M. and Mironov, V. (2009). Biological knowledge management: the emerging role of the Semantic Web technologies. *Briefings in Bioinformatics*, **10**(4), 392–407.

Aronson, A. R. and Lang, F. M. (2010) An overview of MetaMap: historical perspective and recent advances. *Journal of the American Medical Informatics Association*, **17**(3), 229–236.

Attwood, T. K., Kell, D., McDermott, P. *et al.* (2009) Calling International Rescue: knowledge lost in literature and data landslide! *Biochemical journal*, **424**(3), 317–333.

Attwood, T. K., Kell, D., McDermott, P. *et al.* (2010) Utopia Documents: linking scholarly literature with research data. *Bioinformatics*, **26**(18), i568–i574.

Begley, C. G. and Ellis, L. M. (2012). Drug development: raise standards for preclinical cancer research. *Nature*, **483**(7391), 531–533.

Bourne, P. (2005) Will a biological database be different from a biological journal? *PLoS Computational Biology*, **1**(3), e34.

Ceol, A., Chatr-Aryamontri, A., Licata, L. and Cesareni, G. (2008) Linking entries in protein interaction database to structured text: the FEBS Letters experiment. *FEBS Letters*, **582**(8), 1171–1177.

Clarke, J. and Wentz, R. (2000) Pragmatic approach is effective in evidence based health care. *The BMJ*, **321**(7260), 566–567.

Claßen-Bockhoff, R., Ruonala, R., Bull-Hereñu, K. *et al.* (2013) The unique pseudanthium of Actinodium (Myrtaceae) – morphological reinvestigation and possible regulation by CYCLOIDEA-like genes. *Evodevo*, **4**(1), 8. doi:10.1186/2041-9139-4-8

Cserzo, M., Wallin, E., Simon, I. *et al.* (1997) Prediction of transmembrane alpha-helices in prokaryotic membrane proteins: the Dense Alignment Surface method. *Protein Engineering*, **10**(6), 673–6.

Cray, J.J. Jr., Khaksarfard, K., Weinberg, S.M. *et al.* (2013) Effects of thyroxine exposure on osteogenesis in mouse calvarial pre-osteoblasts. *PLoS One*, **8**(7), e69067.

Dayhoff, M.O. (1965) Computer aids to protein sequence determination. *Journal of Theoretical Biology*, **8**(1), 97–112.

De Roure, D. (2009) Replacing the paper: the six Rs of the e-research record. Available at http://blog.openwetware.org/deroure/?p=56 accessed 5 May 2016.

De Roure, D. (2010) Replacing the paper: the twelve Rs of the e-research record. Available at http://www.scilogs.com/eresearch/replacing-the-paper-the-twelve-rs-of-the-e-research-record accessed 5 May 2016.

de Waard, A. (2014) Stories that persuade with data. Available at http://www.slideshare.net/anitawaard/stories-that-persuade-with-data-talk-at-cendi-meeting-january-9-2014 accessed 5 May 2016.

de Waard, A., Buckingham Shum, S., Carusi, A. *et al.* (2009) Hypotheses, Evidence and Relationships: the HypER approach for representing scientific knowledge claims. In *Proceedings of the Workshop on Semantic Web Applications in Scientific Discourse (SWASD 2009), co-located with the 8th International Semantic Web Conference* (ISWC-2009).

Diehn, M., Sherlock G., Binkley G. *et al.* (2003). SOURCE: a unified genomic resource of functional annotations, ontologies, and gene expression data. *Nucleic Acids Research*, **31**(1), 219–223.

Dietze, H. and Schroeder, M. (2009) GoWeb: a semantic search engine for the life science web. *BMC Bioinformatics*, **10** Suppl. 10, S7.

Dix, T. K. (1994) "Public Libraries" in Ancient Rome: ideology and reality. *Libraries & Culture*, **3**, 282–396.

Dubitzky, W. (2009) Editorial. *Briefings in Bioinformatics*, **10**, 343–344.

Dyer, C. (1997) Widow can be inseminated with husband's sperm. *The BMJ*, **314**(7079), 461.

Farkas, R. (2008) The strength of co-authorship in gene name disambiguation. *BMC Bioinformatics*, **9**, 69.

Fraser, A. G. and Dunstan, F. D. (2010) On the impossibility of being expert. *The BMJ*, **341**, c6815.

Geelhoed, G.C., Turner, J. and Macdonald, W.B. (1996) Efficacy of a small single dose of oral dexamethasone for outpatient croup: a double blind placebo controlled clinical trial. *The BMJ*, **313**(7050), 140–142.

Grobe, H., Sieger, R., Diepenbroek, M. and Schindler, U. (2010) PANGAEA – archive and source for data from earth system research. In *Cool Libraries in a Melting World: Proceedings of the 23rd Polar Libraries Colloquy*.

Harford, T. (2014) Big data: are we making a big mistake? *Financial Times Magazine*, March 28 2014.

Hess, K.R., Zhang, W., Baggerly, K.A. *et al.* (2001) Microarrays: handling the deluge of data and extracting reliable information. *Trends in Biotechnology*, **19**(11), 463–468.

Hilton, J.F., Tu, D., Seymour, L. *et al.* (2013) An evaluation of the possible interaction of gastric acid suppressing medication and the EGFR tyrosine kinase inhibitor erlotinib. *Lung Cancer*, **82**(1), 136–142. pii: S0169-5002(13)00265-1. doi:10.1016/j.lungcan.2013.06.008

Hodgson, J. (2001) The headache of knowledge management. *Nature Biotechnol.*, **19** Suppl, E44–46.

Holstein, A., Kovacs, P. and Beil, W. (2013) Severe hypoglycemia due to possible interaction between glibenclamide and sorafenib in a patient with hepatocellular carcinoma. *Current Drug Safety*, **8**(2), 148–152.

Howe, D., Costanzo, M., Fey, P. *et al.* (2008). Big data: the future of biocuration. *Nature*, **455**(7209), 47–50.

Illingworth, C.J.R., Parkes, K.E., Snell, C.R. *et al.* (2008) Criteria for confirming sequence periodicity identified by Fourier transform analysis: application to GCR2, a candidate plant GPCR? *Biophysical Chemistry*, **133**, 28–35.

Jonnalagadda, S.R. and Topham, P. (2010) NEMO: extraction and normalization of organization names from PubMed affiliations. *Journal of Biomedical Discovery and Collaboration*, **5**, 50–75.

Ko, A.I., Galvão Reis, M., Ribeiro Dourado, C.M. *et al.* (1999) Urban epidemic of severe leptospirosis in Brazil. *Lancet*, **354**, 820–825.

Kosaka, T., Miyajima, A., Nagata, H. *et al.* (2013) Human castration resistant prostate cancer rather prefer to decreased 5α-reductase activity. *Scientific Reports*, **3**, 1268. doi:10.1038/srep01268

Latour, B. (1987). *Science in Action: how to follow scientists and engineers through society*. Cambridge, MA: Harvard University Press. p. 33.

Lee, J.H., Rho, S.B., Park, S.-Y. and Chun, T. (2008) Interaction between fortilin and transforming growth factor-beta stimulated clone-22 (TSC-22) prevents apoptosis via the destabilization of TSC-22. *FEBS Letters*, **582**(8), 1210–1218.

Li, P., Wu, D., Liu, K., Suolang, S., He, T., Liu, X., Wu, C., Wang, Y. and Lin, D. (2014) Investigation of Antimicrobial Resistance in *Escherichia coli* and Enterococci Isolated from Tibetan Pigs. *PLoS One*, **9**(4), e95623. doi:10.1371/journal.pone.0095623

Lin, S., Wang, J., Ye, Z. *et al.* (2008). CDK5 activator p35 downregulates E-cadherin precursor independently of CDK5. *FEBS Letters*, **582**(8), 1197–1202.

Liu, X.G., Yue, Y.L., Li, B. *et al.* (2007) A G protein-coupled receptor is a plasma membrane receptor for the plant hormone abscisic acid. *Science*, **315**, 1712–1716.

Lynch, C. (2007) The shape of the scientific article in developing cyberinfrastructure. *CTWatch-Quarterly* August **2007**, 5–10.

Milne, A.A. (1926) *Winnie The Pooh*, Chapter III, In Which Pooh and Piglet Go Hunting and Nearly Catch a Woozle, pp. 32–41. E.P. Dutton, London.

Mohapatra, S. (2014) Drug resistance in leishmaniasis: newer developments. *Tropical Parasitology*, **4**(1), 4–9.

Michener, W.K., Allard, S., Budden, A. *et al.* (2012) Participatory design of DataONE – Enabling cyberinfrastructure for the biological and environmental sciences. *Ecological Informatics*, **11**, 5–15, ISSN 1574-9541. doi:http://dx.doi.org/10.1016/j.ecoinf.2011.08.007

Reis, R.B., Ribeiro, G.S., Felzemburgh, R.D. *et al.* (2008) Impact of environment and social gradient on Leptospira infection in urban slums. *PLoS Neglected Tropical Diseases*, **2**, e228.

Scott, S., Kranz, J.E., Cole. *et al.* (2008) Design, power, and interpretation of studies in the standard murine model of als. *Amyotrophic Lateral Sclerosis*, **9**(1), 4–15.

Seringhaus, M. and Gerstein, M. (2008) Manually structured digital abstracts: a scaffold for automatic text mining. *FEBS Letters*, **582**(8), 1170.

Seringhaus, M.R. and Gerstein, M.B. (2007). Publishing perishing? Towards tomorrow's information architecture. *BMC Bioinformatics*, **8**, 17.

Sharma, N., Mohanakrishnan, D., Sharma, U.K. *et al.* (2014) Design, economical synthesis and antiplasmodial evaluation of vanillin derived allylated chalcones and their marked synergism with artemisinin against chloroquine resistant strains of *Plasmodium falciparum*. *European Journal of Medicinal Chemistry*, **79C**, 350–68. doi:10.1016/j.ejmech.2014.03.079

Sharma, P., Parmar, J., Sharma, P. *et al.* (2011) Radiation-induced testicular injury and its amelioration by *Tinospora cordifolia* (an Indian medicinal plant) extract. *Evidence-Based Complementary and Alternative Medicine*, 643847. doi:10.1155/2011/643847

Shirky, C. (2008) It's not information overload. It's filter failure. Available at https://www.youtube.com/watch?v=LabqeJEOQyI accessed 5 May 2016.

Shotton, D. (2010) CiTO, the Citation Typing Ontology. *Journal of Biomedical Semantics*, **1** Suppl 1, S6.

Shotton, D., Portwin, K., Klyne, G. and Miles, A. (2009) Adventures in semantic publishing: Exemplar semantic enhancements of a research article. *PLoS Computational Biology*, **5**(4), e1000361.

Sprouse, G. D. (2007) Editorial: which Wei Wang? *Physical Review A*.

Su T. T., Feger G. O'Farrell P. H. (1996) Drosophila MCM protein complexes. *Molecular Biology of the Cell*, **7**(2), 319–329.

Superti-Furga, G., Weiland F. and Cesareni, G. (2008) Finally: The digital, democratic age of scientific abstracts. *FEBS Letters*, **582**(8), 1169.

Toffler, A. (1970). *Future Shock*. Random House, New York.

Torvik, V. I. and Smalheiser, N. R. (2009) Author Name Disambiguation in MEDLINE. *ACM Transactions on Knowledge Discovery from Data*, **3**(3), 11.

Torvik, V. I., Weeber, M., Swanson, D.R. and Smalheiser, N.R. (2003) A probabilistic similarity metric for Medline records: a model for author name disambiguation. *AMIA Annual Symposium Proceedings*, p. 1033.

Varma, S.D., Kovtun, S. and Hegde, K. (2010) Effectiveness of topical caffeine in cataract prevention: studies with galactose cataract. *Molecular Vision*, **16**, 2626–2633.

Vision, T.J. (2010). Open data and the social contract of scientific publishing. *Bioscience*, **60**(5), 330–331.

Waight, J.D., Banik, D., Griffiths, E.A. *et al.* (2014) Regulation of the IRF-8 tumor suppressor gene by the STAT5 transcription factor in chronic myeloid leukemia. *Journal of Biological Chemistry*, **289**(22), 15642–15652. doi:10.1074/jbc.M113.544320

Wan, S., Paris, C. and Dale R. (2010) Supporting browsing-specific information needs: introducing the citation-sensitive in-browser summariser. *Web Semantics: science, services and agents on the world wide web*, **8**(2–3), 196–202.

Watson, J. D. and Crick, F. H. (1953) Molecular structure of nucleic acids: a structure for deoxyribose nucleic acid. *Nature*, **171**(4356), 737–738.

Wilbanks, J. (2007) Cyberinfrastructure for knowledge sharing. *CTWatchQuarterly*, 58–66.

Wurman, R.S. (1997) *Information Architects*. Graphis Publications, New York. http://www.amazon.com/Information-Architects-Richard-Saul-Wurman/dp/1888001380

8.13 Quiz

The following multiple-choice quiz will help you to check how much you've remembered of the issues explored in this chapter concerning the challenges involved when we try to couple the scientific literature more tightly with its underlying research data.

1 Which of the following statements is true?
   A The rapid expansion of the scientific literature means that most gene and protein sequences can now be annotated by biocurators.
   B The rapid expansion of the scientific literature makes it much easier for biocurators to find pertinent information to augment database entries.
   C The structure of scientific articles makes it straightforward for claims in the literature to be captured in databases by biocurators.
   D None of the above.

2 A 'structured digital abstract':
   A contains materials, methods and a brief conclusion in that specific order.
   B makes it easier for a human to determine the content of an article without having to read it.
   C attempts to capture the main content of an article in machine-readable form.
   D describes the structure of a scientific article in machine-readable form.

3 In scholarly communication, 'hedging' means:
   A predicting the impact of biological findings.
   B 'softening' claims made in scientific articles.
   C defining the context of a scientific article.
   D causing an article to be retracted because of invalid findings.

4 MeSH is:
   A a method of identifying papers that cite one another.
   B a database of life-science citations.
   C an algorithm for calculating the impact of a scientific article.
   D a controlled vocabulary of biological concepts.

5 A Research Object is:
  A a unit of research, as defined by a peer-reviewed publication.
  B a collection of data and metadata intended to encapsulate a scientific idea.
  C a machine-readable representation of a scholarly article.
  D an object-orientated concept relating to life-science research.

6 Digital resources are sensitive to:
  A changes in operating systems.
  B changes in hardware.
  C social factors.
  D all of the above.

7 PubMed central is:
  A a version of PubMed, replicated at national levels.
  B a collection of Open Access full-text articles.
  C a central repository from which PubMed takes its metadata.
  D a repository for unpublished material.

8 A DOI can be used to identify:
  A a scholarly article.
  B the auxilliary material related to an article.
  C data-sets that form the basis of a published article.
  D all of the above.

9 'Claim amplification' is the tendency for:
  A authors to exaggerate claims in order for results to be published.
  B readers of a scientific article to believe it is saying something more concrete than the authors intended.
  C ambiguous claims to become concrete facts by repeated citation.
  D readers to believe that anything published in a 'high impact journal' is necessarily true.

10 As article search engines become more 'semantic', they will tend to return:
  A more results.
  B fewer, more targeted results.
  C a similar number of results to keyword-based techniques.
  D impossible to tell.

8.14 Problems

1 How many times has the paper 'Defrosting the Digital Library: Bibliographic Tools for the Next Generation Web' (Hull *et al.*, 2008) been cited? How many times has it been read? How reliable is the answer you've generated?

2   In the UniProtKB entry for human rhodopsin (OPSD_HUMAN)[29], we learned that the 'most common variant, leads to interaction with EDEM1 followed by degradation by the ERAD system'. Four citations support this claim: Dryja *et al.*, 1990[30]; Dryja *et al.*, 1990[31]; Sheffield *et al.*, 1991[32]; and Kosmaoglou *et al.*, 2009[33].

    What is EDEM1 and what is the ERAD system? What are the dual roles of EDEM1 in rod opsin processing? How is this relevant to retinitis pigmentosa (RP) and other protein misfolding diseases? What is the most common mutation that causes RP in North America and how might it interact with EDEM1. How much of this information is captured in the UniProtKB entry? How could the UniProt curators have made it easier for you to access this information?

3   Examine the following 'Open Access' papers. Which of these could you text mine?

- Roberts *et al.* (2014) Biochemical and genetic characterization of *Trypanosoma cruzi* N-myristoyltransferase, *Biochemical Journal*, **459**, 323–32. doi:10.1042/BJ20131033
- Chen *et al.* (2014) Small Interfering RNA Targeting Nerve Growth Factor Alleviates Allergic Airway Hyperresponsiveness. *Molecular Therapy Nucleic Acids*, **3**, e158. doi:10.1038/mtna.2014.11
- Pettifer *et al.* (2011) Ceci n'est pas un hamburger: modelling and representing the scholarly article, *Learned Publishing*[34], **24**(3), 207–20. doi: 10.1087/20110309

4   Explore the CITO ontology, which can be used to describe why something cites something else. In what way are we citing CITO in this question?

5   Take the first exemplar sentence from Box 8.3 and re-write it so that its meaning is clearer. Identify the biologically-relevant nouns and verbs; then try to construct a parse tree (along the lines of Figure 8.4) that describes the associations between the different parts of the sentence so that a computer could interpret the relationships between the parts. How sure can you be about what the text means? What does this imply for a computer's attempt at interpretation? How could the authors/editors/publishers have made this more readily digestible by a computer?

---

[29] http://www.uniprot.org/uniprot/P08100
[30] http://www.ncbi.nlm.nih.gov/pubmed/2215617
[31] http://www.ncbi.nlm.nih.gov/pubmed/2137202
[32] http://www.ncbi.nlm.nih.gov/pubmed/1897520
[33] http://www.ncbi.nlm.nih.gov/pubmed/19934218
[34] http://www.ingentaconnect.com/content/alpsp/lp

Afterword

In this book, we've meandered through the history of bioinformatics and computer science, exploring some of the discoveries and challenges that have arisen in the overlaps between these disciplines, and stopping to peer into some of the gaps that exist where they haven't quite made a perfect seal. Our starting point was an archipelago of individual research islands. For the first leg of the journey, the challenges facing the early pioneers involved cajoling the primitive computing technology of the day to store, retrieve and display the data that were being generated from isolated, largely manual experiments. Where we've ended up is a much more interconnected world in which vast amounts of data can be generated automatically, where the *relationships* between biological topics are becoming as significant as the topics themselves, and where the challenges involve using today's technology – itself a highly interconnected computing infrastructure – not simply to store and retrieve data, but to understand and extract insight and meaning from the information we've collected.

There's no doubt that we've come a long way. However, as impressive as the vista of our achievements may seem, it's also clear that this isn't the end of the journey; and it's perhaps sobering to realise that many of the challenges alluded to early in this book remain unsolved. In a 2014 'technical perspective' written thirty years or so after the first computational tools for gene and protein sequence analysis were created, James Hutchins lamented,

> Although various programs are available for exploring the properties of sets of genes, transcripts, and proteins, the ideal software tool, in my opinion, has yet to be created (Hutchins, 2014).

He went on to list five desirable properties of such Utopian software. Essentially, he said that this tool should be:

1 *free*: publicly available, cross-platform, and open source, with database architecture and algorithms described in peer-reviewed publications;
2 *comprehensive*: able to draw on a wide variety of leading database resources;
3 *updated*: regularly, coordinated with releases of major sequence and functional databases;
4 *smart*: capable of automatically recognising ID types and thus determining the relevant species and relationships between genes and their products; and
5 *flexible*: allowing a choice of analytical methods and output formats.

Bioinformatics challenges at the interface of biology and computer science: Mind the Gap. First Edition.
Teresa K. Attwood, Stephen R. Pettifer and David Thorne. Published 2016 © 2016 by John Wiley and Sons, Ltd.
Companion website: www.wiley.com/go/attwood/bioinformatics

While the desirability of the items on Hutchins' wish-list could easily be dismissed as 'obvious', the list nevertheless provides a useful vehicle both for exploring the obstacles that lie ahead on the next stage of the bioinformatics journey, and for understanding why we are where we are today, and why so many things are still not solved.

Taking these one by one, from a technical point of view, there are interesting challenges associated with each of the five properties: even with today's programming languages and support frameworks, creating cross-platform software is much harder than building something that works for just one operating system; being able to access a 'comprehensive' array of third-party databases inevitably involves dealing with multiple database paradigms and schemas; benefitting from database 'updates' necessarily involves some evolution of the consuming tool, either to cope with the growing scale of data, or to adapt to new features; 'smart' behaviour likely implies the use of complex semantic technologies; and 'flexibility' requires adherence to rigorous software-engineering principles, in order to create tools that are extensible and amenable to re-factoring to accommodate new requirements. All of these things are technically harder to achieve than hacking a hard-coded, one-off solution to a particular problem; but all are necessary to create tools that can be shared, and that have any possibility of living beyond the lifetime of a given project. The interesting point here is that these things are, from a technical perspective, just that: hard to achieve, but, as we've seen throughout this book, not impossible.

The reason why we're not awash with tools and systems that embody Hutchins' Utopian property wish-list is much more complex and subtle, and, we suggest, more to do with money, politics and, ultimately, human nature than with any specific technical hurdle. Lurking behind each of the five points is the implication that tools have to be maintained and updated in order to be useful, whether to keep them in sync with other tools, or simply to prevent them from rotting, as underlying operating systems evolve, and programming languages and frameworks rise and fall in popularity. Achieving this requires effort: not just initial effort to use the best software-engineering paradigms and the latest interoperability standards, but sustained, long-term effort – and that requires sustained, long-term funding.

The paradox is perhaps most conspicuously revealed in Hutchins' first point: funding-bodies and users want software and data to be 'free' – i.e., free in the sense of not having to pay for them, as well as free from any restrictions in the way in which they might be used. At the same time, however, software and data need to be maintained and supported indefinitely, so that they remain reliable resources on which to base current and future research. In 2005, Scott McNealy, then-CEO of Sun Microsystems, quipped that 'Open Source software is free like a puppy is free' – in other words, while you may be able to acquire your little canine companion without parting with any money, if you don't feed it and pay the vet's bills, you can't expect it to live very long. And, just like puppies, figuring out how to make software and databases free *and* sustainable presents almost insurmountable challenges, resulting in many of them being abandoned by the wayside.

To give some idea, in 2004, Bairoch and his colleagues reflected on just how tough it really is to develop and maintain high-quality databases for the life sciences (Bairoch *et al.*, 2004). They enumerated six observations that they deemed essential for would-be database developers to understand:

- Your task will be much more complex and far bigger that you ever thought it could be.
- If your database is successful and useful to the user community, then you will have to dedicate all your efforts to develop it for a much longer period of time than you would have thought possible.

- You will always wonder why life scientists abhor complying with nomenclature guidelines or standardisation efforts that would simplify your and their life.
- You will have to continually fight to obtain a minimal amount of funding.
- As with any service efforts, you will be told far more what you do wrong rather than what you do right.
- But when you will see how useful your efforts are to your users, all the above drawbacks will lose their importance!

The picture they paint is clear: maintaining public resources is a huge, complex, interminable task, a task that is much harder than it should be, both because the community avoids using standards, and because funding-bodies don't provide enough money to sustain the work!

One way to address the financial issues is to try to raise funds commercially, for example by 'dual licensing': *i.e.*, providing a resource for free to academic users but charging for its use (or for associated consultancy services) in, say, a commercial environment. This is difficult for a number of reasons. First, defining what is meant by 'commercial use' can be hard: it might be clear that a privately owned pharmaceutical company is 'commercial', but the status of many universities and research institutions is often more complex. Second, policing such agreements is challenging, and often relies on commercial organisations 'doing the right thing' and offering a fee in return for using a resource that they could easily obtain for free (this 'trust-based' model worked well for a number of years for Swiss-Prot, for example, but later had to change, when NIH funding obtained to support it stipulated that the database must be made freely available). Third, and perhaps most importantly, it relies on the resource's creators (often academics) being sufficiently entrepreneurial to make such an approach work. There's no doubt that such models can thrive. Founded in 1993, Red Hat Inc., for example, has built a successful business providing commercial support for its free, open-source Linux distribution; but the number of companies that have achieved stability off the back of free software is small compared to those with more traditional business models. In any case, it could be argued that scientists are perhaps best left to focus on science, rather than attempting to innovate in the dog-eat-dog world of big business.

A seductive alternative view is that any sufficiently important and open resource will be maintained by the community. As attractive as this may seem philosophically, the reality seems to suggest that this is also hard to achieve, and works only rarely and then only for the most prominent projects. The foundations that maintain Wikipedia and Firefox, for example, two of the world's most prominent free resources, are in a constant state of fund-raising from users to keep these important projects alive; very few smaller projects have the resources necessary to be able to manage such major fund-raising initiatives.

On balance, it's hard to avoid the conclusion that the only reliable way to keep cyberinfrastructure in good shape is for it to be funded alongside the core research that relies so fundamentally upon it. The difficulty here is that most funding agencies have mission statements that speak about improving the world through fundamental *research*: the mission of the NIH, for example, is

> to seek fundamental knowledge about the nature and behavior of living systems and the application of that knowledge to enhance health, lengthen life, and reduce illness and disability.

The problem, perhaps, is that funding infrastructure to 'keep the lights on' is a much harder and infinitely less sexy idea to sell to those who hold the purse strings than is 'curing cancer', even if one of these is a prerequisite for the other. As a result, much of the world's essential scientific cyberinfrastructure is built on the back of projects whose primary goal is fundamental research, and any infrastructures they create tend to be sort of incidental. This issue was neatly summarised by Geoffrey Bilder, when he said that

> If we funded and ran physical infrastructure the way we fund and run scholarly cyberinfra-structure, the lights and water would go out every grant cycle and we would have architects bidding to build Quonsets [sic] huts with the promise that the client would get railways and electrical grids as side-effects (Bilder, 2014).

As part of his scathing indictment of our attitude towards building and maintaining a viable scholarly cyberinfrastructure, Bilder asked bluntly, 'Why are we so crap at infrastructure? How can we get it right for a change?' The sobering truth is that we almost certainly know the answers to these questions; however, the harsh reality is that implementing the solutions requires dramatic changes in the mind-sets of academics and funding-bodies alike, and the recognition that modern scientific breakthroughs are not just the result of standing on the shoulders of giants, but also of creating the kind of cyberinfrastructural bedrock upon which our scientific titans need to root their achievements. Fortunately, it seems that this shift is beginning to take place, albeit rather slowly.

An early sign of change was the establishment, in 2002, of the European Strategy Forum on Research Infrastructures (ESFRI). ESFRI's mission is

> to support a coherent and strategy-led approach to policy-making on research infrastruc-tures in Europe, and to facilitate multilateral initiatives leading to the better use and devel-opment of research infrastructures, at EU and international level.

Deadly dull though this may sound, such high-level strategising is essential if research infrastructures are to be fully recognised for, and supported in, the vital role they play in underpinning the advancement of knowledge and technology. An ESFRI project of relevance to bioinformatics is ELIXIR, a multi-million euro, pan-European, inter-govern-mental initiative that's being created to help manage and safeguard, in the long-term, the vast quantities of biological data now being generated by publicly funded European life-science research projects. Following a five-year preparatory phase, ELIXIR went into its implementation phase at the beginning of 2013. Although still in its infancy, the project is not short of ambition, as a brief visit to its website shows:

> For the first time, ELIXIR is creating an infrastructure – a kind of highway system – that integrates research data from all corners of Europe and ensures a seamless service provision that is easily accessible to all. In this way, open access to these rapidly expanding and critical datasets will facilitate discoveries that benefit humankind.
>
> Science and technology change very quickly, and exploiting these advances can be a chal-lenge. ELIXIR partners are building an intelligent, responsive and sustainable system that will deliver the fruits of these advances to the scientists upon whom so many hopes are pinned, and whose curiosity is the very cornerstone of progress.

It remains to be seen just how (and how well) this bold vision will be put into practice. What's certain is that the realisation of ELIXIR will require continued deft political manoeuvring, a vast amount of patience, and huge financial investments: to give an idea, the UK's BBSRC made a £75 million seed-corn investment in ELIXIR in 2011, followed by a smaller, short-term grant to establish ELIXIR-UK in 2013, and more funding was sought from the EU to be able to extend ELIXIR-UK's pilot work beyond 2015. And this is just the UK. By December 2015, 12 countries had signed ELIXIR's Consortium Agreement, three held the status of 'provisional member', and four the status of 'observer'; and, as further countries weigh up the opportunities and complexities of joining versus the price of not doing so, the cost implications (financial and political) become enormous. To give a flavour of the political and financial complexities, during the summer of 2015, the ELIXIR Consortium's bid – EXCELERATE – for further EU funds to 'accelerate the implementation of the ELIXIR platforms and integrate ELIXIR bioinformatics resources into a coherent infrastructure' was successful. EXCELERATE was awarded a budget of almost €20 million, and included more than 40 partners from ELIXIR nodes in 17 member- and observer countries.

Just as ELIXIR started to unfold in Europe, a trans-US-NIH initiative – Big Data to Knowledge (BD2K) – was launched in 2012 (Margolis et al., 2014). BD2K aims to recognise and enable biomedical research as a 'digital research enterprise', in part by supporting a 'data ecosystem' able to accelerate knowledge discovery. Discussions of possible synergies between ELIXIR and BD2K are in their infancy, and it will be interesting to see what harmonious vision may emerge, if any, for sustaining a data ecosystem on a global scale. Regardless, the good news is that at least some of the bodies that fund scientific research have begun to appreciate the imperative of establishing and supporting robust, sustainable cyberinfrastructures, if science and society are to benefit from the data and tools needed to advance knowledge and understanding, and ultimately realise the vision of facilitating 'discoveries that benefit humankind'.

Conclusion

As Olga Kennard so presciently said, more than 18 years ago, 'Free access to validated and enhanced data worldwide is a beautiful dream. The reality, however, is more complex' (Kennard, 1997). As we've seen, that complexity arises from a sticky political, financial, technical and cultural gloop that tends to limit our progress, even when the solutions appear 'obvious'. Politicians need to be educated, funders and scientists need to change the way they think, the cultural barriers that divide scientific disciplines need to be torn down, and together we must start to bridge the gaps that exist between them. In the world of bioinformatics, as we've seen in this book, part of the process of bridging those cultural gaps involves recognising, on the one hand, that it's hard to teach number-crunching computers to do biology, and, on the other, that it's hard to teach users (be they life scientists or bioinformaticians) how to use and interact appropriately with computers – on both sides of the divide, there's still a lot to learn.

On the upside, we live in the most computer-savvy, digitally-aware, Internet-enabled society the planet has ever witnessed; and, fortunately, you are part of the next generation of bioinformaticians and computer scientists who, by understanding the challenges we've talked about here, will be better equipped to take the pioneering steps necessary to transform the 'beautiful dream' into an exciting new reality. En route, we counsel you to 'mind the gap', and wish you luck in what will doubtless be a challenging and inspirational journey!

References

Bairoch, A., Boeckmann, B., Ferro, S. and Gasteiger, E. (2004) Swiss-Prot: juggling between evolution and stability. *Briefings in Bioinformatics*, 5(1), 39–55.

Bilder, G. (2014) Mind the Gap: cholarly cyberinfrastructure is the third rail of funders, institutions and researchers. Why do we keep getting it wrong and what can we do about it? The Shaking it up: how to thrive in – and change – the research ecosystem workshop, Microsoft New England R&D Center, 24 November 2014. Available at www.digital-science.com/events/shaking-it-up-how-to-thrive-in-and-change-the-research-ecosystem accessed 5 May 2016.

Hutchins, J.R.A. (2014) What's that gene (or protein)? Online resources for exploring functions of genes, transcripts and proteins. *Molecular Biology of the Cell*, 25, 1187–1201.

Kennard, O. (1997) From private data to public knowledge. In The Impact of Electronic Publishing on the Academic Community, an International Workshop organised by the Academia Europaea and the Wenner-Gren Foundation, Wenner-Gren Center, Stockholm, 16–20 April, 1997 (ed. Ian Butterworth), Portland Press Ltd, London.

Margolis, R., Derr, L., Dunn, M. *et al.* (2014) The National Institutes of Health's Big Data to Knowledge (BD2K) initiative: capitalizing on biomedical big data. *Journal of the American Medical Informatics Association*, 21(6), 957–958. Available at http://www.ncbi.nlm.nih.gov/pmc/articles/PMC4215061 accessed 5 May 2016.

Glossary

A

A – the IUPAC single-letter abbreviation for alanine, or the purine base 6-aminopurine.

Å – the symbol for Ångstrom.

α-carbon (alpha-carbon) – the central tetrahedral carbon atom of an amino acid.

α-helix (alpha-helix) – a basic element of protein secondary structure with helical conformation, characterised by 3.6 amino acid residues per turn of helix.

α-solenoid (alpha-solenoid) – a protein fold characterised by α-helices arranged in an elongated curved structure.

abacus – a counting tool, with beads/tokens sliding on wires/pillars within a frame of some sort.

ABC transporter – a membrane protein that contains an ATP-Binding Cassette (ABC), a domain found in members of a large superfamily of proteins that exploit ATP hydrolysis to transport a variety of small molecules across membranes.

absorption – the process by which one substance is taken up by another, such as a drug taken up into the bloodstream in the context of pharmacology.

Absorption, Distribution, Metabolism, Excretion and Toxicity (ADME-Tox) – see ADME-Tox.

ACeDB – a *C. elegans* DataBase, originally created to manipulate, display and annotate genomic data.

accession number – a unique (and theoretically invariant) computer-readable number or code given to identify a particular entry in a particular database.

acoustic modem – a modem that's mechanically connected to a telephone via its handset (see also **modem**).

Acquired ImmunoDeficiency Syndrome (AIDS) – see AIDS.

actin – a multi-functional, globular protein found in almost all eukaryotic cells, it is a major component of the cytoskeleton, with roles in muscle contraction, cell motility and cell division.

adenine – 6-aminopurine, a purine derivative that's one of the five principal bases of nucleic acids.

adenosine 5′-triphosphate (ATP) – see ATP.

adhesion receptor – name given to a class of membrane proteins, some of which have roles in cell adhesion, that belongs to the GPCR clan.

ADME-Tox – in pharmacology, the principal factors that influence the performance and pharmacological activity of a drug: Absorption, Distribution, Metabolism, Excretion and Toxicity.

Bioinformatics challenges at the interface of biology and computer science: Mind the Gap. First Edition.
Teresa K. Attwood, Stephen R. Pettifer and David Thorne. Published 2016 © 2016 by John Wiley and Sons, Ltd.
Companion website: www.wiley.com/go/attwood/bioinformatics

aesthetics – a branch of philosophy that deals with the nature of beauty, art, culture and taste.

AIDS – Acquired ImmunoDeficiency Syndrome, a collection of symptoms caused by infection with the Human Immunodeficiency Virus (HIV), which results in disease of the immune system.

Ala – the IUPAC three-letter abbreviation for alanine.

alanine – 2-amino-propanoic acid, a chiral amino acid found in proteins.

algorithm – a series of steps that form a recipe for solving a particular problem.

aliphatic – refers to organic compounds in which the carbon atoms form usually acyclic, open (non-aromatic) chains.

alpha-carbon – see α-carbon.

alpha-helix – see α-helix.

alpha-solenoid – see α-solenoid.

Alzheimer's disease – a heterogeneous syndrome involving progressive cognitive failure, and the most common cause of senile dementia.

AMAP – a multiple sequence alignment tool based on sequence annealing.

Amazon – the world's largest online retailer, especially of books and electronic goods, and provider of cloud-computing services.

American Standard Code for Information Interchange (ASCII) – see ASCII.

amine – a weakly basic organic compound that contains an amino or substituted amino group.

amino group – the chemical group $-NH_2$ in an organic molecule.

amino acid – an organic acid with one or more amino substituents, particularly α-amino derivatives of aliphatic carboxylic acids; 20 common, naturally occurring amino acids provide the fundamental building-blocks of proteins.

amino-terminus – see N terminus.

amphipathic – refers to organic compounds that possess both hydrophobic (non-polar) and hydrophilic (polar) constituents.

analogy – the acquisition of common features (folds or functions) via convergent evolution from unrelated ancestors.

analytical engine – a mechanical general-purpose computer designed by Charles Babbage.

Anfinsen, Christian – biochemist who earned the Nobel Prize in Chemistry in 1972 for his work on ribonuclease, particularly for showing that the protein could be denatured and refolded without losing its enzymatic activity, suggesting that all the information for protein folding was encoded in the primary structure.

Ångstrom (Å) – a unit of length equal to 10^{-10} metres; used to express the sizes of atoms and molecules, and the wavelengths of electromagnetic radiation.

annotation – a note, or more usually, a set of descriptive notes, attached to text, and especially to data, to provide additional information to readers or users.

ANTHEPROT – a tool for interactive editing of multiple sequence alignments.

anticodon – a nucleotide triplet in tRNA that is complementary to a specific codon in mRNA.

API – Application Programming Interface, the specification of a software component in terms of its functions and their inputs and outputs.

Application Programming Interface (API) – see API.

Apweiler, Rolf – at the time of writing, joint associate director of the EBI.

Arabidopsis thaliana – a small flowering plant (common wall cress), used as a plant model organism owing to its relatively small genome and rapid generation time.

archaea – single-celled microorganisms that constitute one of three primary kingdoms of life.

architecture – in the CATH database, the structural similarity between proteins where there is no evidence for homology.

Arg – the IUPAC three-letter abbreviation for argininine.

arginine – 2-amino-5-guanidinopentanoic acid, an amino acid found in proteins.

aromatic – refers to organic compounds in which the atoms form one or more planar unsaturated rings (benzene is the simplest compound of this type); the term was coined to differentiate fragrant compounds from aliphatic compounds.

ARPANET – Advanced Research Project Agency NETwork, the world's first packet-switching network and one of the progenitors of the Internet.

arrestin – a member of a family of proteins that play an important role in regulating signal transduction processes.

ASCII – American Standard Code for Information Interchange, a scheme to encode characters (originally from the English alphabet, plus numerical digits and some punctuation) into 7 bits.

Asn – the IUPAC three-letter abbreviation for asparagine.

Asp – the IUPAC three-letter abbreviation for aspartic acid.

asparagine – 2-amino-3-carbamoylpropanoic acid, a chiral amino acid found in proteins.

aspartic acid – 2-aminobutanedioc acid, a chiral amino acid found in proteins.

asymmetric unit – the smallest unit into which a crystal can be divided, the unit usually solved in crystallographic structure determinations of small molecules or proteins.

Atlas of Protein Sequence and Structure – a published collection of amino acid sequences, organised by gene families, first compiled and edited by Margaret Dayhoff in 1965; the *Atlas* was the progenitor of the PIR-Protein Sequence Database (PIR-PSD).

ATP – adenosine 5′-triphosphate, a coenzyme and enzyme regulator used by cells to transport chemical energy, which may be used in active transport, converted to mechanical, electrical or light energy, or released as heat.

automated reasoning – the process of using a reasoner to check the validity of an ontology or to infer new axioms.

axiom – a fundamental rule or statement about a system that cannot be derived from other such rules or statements.

B

β-barrel (beta-barrel) – a protein supersecondary fold, or super-fold, formed from a β-sheet that twists and coils to form a closed structure via hydrogen-bonding between its first and last strands.

β-hairpin – see β-turn.

β-propeller (beta-propeller) – a protein fold characterised by a series of blade-shaped β-sheets arranged toroidally about a central axis; each component sheet tends to have four anti-parallel β-strands twisted in such a way that the first and fourth lie almost perpendicular to each other.

β-sheet (beta-sheet) – a basic element of protein secondary structure with an extended and usually twisted conformation, formed by hydrogen bonding between backbone atoms in adjacent β-strands (strands may run in parallel or anti-parallel directions).

β-strand (beta-strand) – region of a polypeptide chain with an almost fully extended conformation.

β-turn (beta-turn) – a basic element of protein secondary structure resembling a hair-pin, wherein the polypeptide chain turns back on itself, reversing its direction; often, there is a hydrogen bond between the first and fourth residues of the turn.

Babbage, Charles – mathematician and engineer who conceived the idea of a program-mable computer, and is credited with inventing the first mechanical computer.

Babylonians – inhabitants of Babylonia, an ancient nation in central-southern Mesopo-tamia (now Iraq); the Babylonians are known for developing methods to predict the motions of the planets, and for using a sexagesimal system of mathematics (a modi-fied form of which provides the basis of our systems for measuring angles, geographic coordinates and time).

back-up device – a physical storage device for copying data, particularly to allow res-toration of a computer system (data, files, *etc.*) following a crash, data loss or data corruption.

bacteria – ubiquitous group of prokaryotic microorganisms that exist as single cells or in clusters of single cells; bacteria constitute one of three primary kingdoms of life.

bacteriophage – a bacterial virus whose genetic material (circular or linear single- or double-stranded RNA or DNA) is enclosed in a protein capsid.

bacteriophage ϕX174 – a small, single-stranded DNA bacteriophage, the first DNA-based genome to be sequenced, by Fred Sanger in 1977.

bacteriophage λ – a small, double-stranded DNA bacteriophage that infects *Escherichia coli*.

bacteriorhodopsin – a light-driven proton pump used by archaea to traffick protons across membranes, creating chemical energy from the resulting proton gradient.

Bairoch, Amos – a biochemist and pioneer of bioinformatics, responsible for developing and co-developing a range of databases, including Swiss-Prot, PROSITE, ENZYME, SeqAnalRef, TrEMBL, InterPro, UniProt and neXtProt, and analysis software, PC/Gene; he was the co-founder of the Swiss Institute of Bioinformatics, and of the Gen-eBio and GeneProt companies.

base – see nucleotide base.

base-calling – see Phred base-calling.

base pair (bp) – pairing between bases in opposing strands of nucleotide sequences (DNA or RNA): adenine pairs with thymine in DNA, or with uracil in RNA; and guanine pairs with cytosine.

Basic Local Alignment Search Tool – see BLAST.

basic pancreatic trypsin inhibitor – an inhibitor that reduces the availability of biologi-cally active trypsin in a range of animals, including humans; it also weakly inhibits chymotrypsin.

baud – in telecommunications and electronics, the number of symbols or pulses per second.

BBN – Bolt, Baranek and Newman, a high-technology company, best known for its work contributing to the development of ARPANET and the Internet.

BCE – Before the Common Era, a neutral designation for the calendar era traditionally identified as Before Christ (BC).

beta-barrel – see β-barrel.

beta-hairpin – see β-hairpin.

beta-propeller – see β-propeller.

beta-sheet – see β-sheet.

beta-strand – see β-strand.

beta-turn – see β-turn.

binary notation – a positional base-2 system that represents numerical values using the symbols 0 and 1; this system is used in the logic gates of the digital electronic circuitry of computers.

biochemical pathway – a series of biochemical reactions that occur in a biological system, usually catalysed by enzymes and requiring a variety of cofactors to function correctly.

bioinformatics – a computer-based discipline at the interface of biology and computer science that deals, amongst other things, with the processes of capturing, storing, representing, retrieving, analysing and simulating biological data and systems.

biological sequence – a linearly ordered set of monomeric units, covalently linked to create a polymer: examples include the sequence of amino acids in the primary structures of proteins, linked via planar peptide bonds; and the sequence of nucleotides in DNA, linked via a sugar-phosphate backbone.

biophysics – a discipline at the interface of biology and physics that uses a range of physical techniques and methods to analyse biological systems at all levels of organisation, from the molecular scale to whole organisms and ecosystems.

biotechnology – a discipline at the interface of biology and engineering that exploits living systems (organisms, tissues, cells, *etc.*) in the development of products and services (*e.g.*, in the production of food and medicines).

biotic system – a system within a living organism.

bit – the binary unit of storage that can have one of two values: 1 or 0.

Bitcoin – an entirely virtual online currency used by various companies and institutes around the world; Bitcoin relies on high security and a global consensus to ensure the validity of its currency.

BLAST – Basic Local Alignment Search Tool, a computer program for searching nucleotide or protein sequence databases with a query sequence, developed at the NCBI in 1990.

block – an ungapped local alignment derived from a conserved region of a protein sequence alignment – equivalent to a motif – consisting of sequence segments that are clustered according to their degree of shared similarity and used to build a diagnostic family signature.

Blocks database – a database of blocks created at the Fred Hutchinson Cancer Research Centre; maintenance of the database ceased in 2007.

BLOSUM matrix – BLOcks SUbstitution Matrix, a matrix used to score protein sequence alignments, with scores for all possible amino acid substitutions based on empirical substitution frequencies observed in blocks from the Blocks database; different BLOSUM matrices relate to data-sets with different levels of sequence identity (*e.g.*, BLOSUM 62, BLOSUM 80).

blue-sensitive opsin – an opsin with characteristic short-wavelength sensitivity, responsible for perception in the blue region of the electromagnetic spectrum.

BNL – see Brookhaven National Laboratory.

Bolt, Beranek and Newman – see BBN.

bovine – of, or relating to, an ox.

bp – symbol for base pair.

Brahmagupta – a mathematician, the first to use zero as a number and to provide rules for computing with zero and negative numbers.

Brookhaven National Laboratory – a national laboratory in the USA, based in New York, originally owned by the Atomic Energy Commission and now owned by the US Department of Energy, and the former home of the Protein Data Bank (PDB).

byte – a collection of eight contiguous bits that can encode 256 different values; bytes are normally the smallest portion of storage addressable by software.

C

C – the IUPAC single-letter abbreviation for cysteine, or the pyrimidine base cytosine.

Cα – symbol for the amino acid alpha-carbon.

cadherin – a member of a family of calcium-dependent cell-adhesion proteins that functions to bind cells together within tissues.

calcium – a chemical element essential for living organisms, being involved in cell signalling, and in bone and shell mineralisation.

C-alpha – alternative name for the amino acid alpha-carbon.

cAMP – cyclic adenosine monophosphate, important for intracellular signal transduction in many different organisms.

cancer – a malignant growth of new tissue (a neoplasm) resulting from unregulated cell proliferation and growth; cancers may be classed as carcinomas (derived from epithelial tissue) or sarcomas (derived from connective tissue).

Caenorhabditis elegans – a nematode (roundworm) that inhabits temperate soil environments, used extensively as a model organism for understanding genetic control of physiology and development; *C. elegans* was the first multicellular organism whose genome was completely sequenced.

calcitonin – or thyrocalcitonin, a linear polypeptide hormone secreted by the thyroid glands, which acts in concert with parathyroid hormone to regulate calcium concentration in the blood.

calcitonin gene-related peptide – a member of the calcitonin family of polypeptides created by alternative splicing of the calcitonin gene, found in parts of the nervous system and other organs, which (amongst other functions) may play a role in pain reception.

Cambridge Crystallographic Data Centre – CCDC.

Cambridge Structural Database – see CSD.

camel – an even-toed ungulate of the genus *Camelus*, noted for its characteristic fatty deposits, or humps, on its back.

carbohydrate – or sugar, or saccharide, an organic compound based on the general formula $C_x(H_2O)_y$ (although exceptions exist, including deoxyribose); carbohydrates exist as monosaccharides, disaccharides, oligosaccharides and polysaccharides.

carboxyl – an acidic chemical group in an organic molecule consisting of carbonyl and hydroxyl constituents, usually expressed as $-COOH$ or $-CO_2H$.

carboxylic acid – an organic acid that contains one or more carboxyl groups.

carboxypeptidase – a protease enzyme that cleaves, or hydrolyses, the peptide bond of an amino acid residue at its C-terminal end.

carboxy-terminus – see C terminus.

catalyst – a substance that increases the rate of a chemical reaction but remains unchanged at the end of the reaction.

Catch-22 – a situation or dilemma made inextricable by mutually contradictory rules or constraints.

CATH – Class, Architecture, Topology, Homology, a protein-fold classification database maintained at University College London, UK, a largely automated classification of protein domains.

CCDC – Cambridge Crystallographic Data Centre, a not-for-profit crystallographic organisation based in the UK whose primary role is to compile and maintain the Cambridge Structural Database (CSD) of small molecule crystal structures.

cDNA – complementary DNA, a DNA molecule whose base sequence is complementary to that of an mRNA molecule, created by the dual action of the enzymes reverse transcriptase and DNA polymerase.

CDS – or coding DNA Sequence, the portion of an mRNA molecule that is translatable into a polypeptide; or, in DNA, an exon.

cell – the fundamental structural and functional unit, or building-block, of living organisms; it comprises a membrane-bounded mass of organelle-containing cytoplasm.

central dogma of molecular biology – the concept that genetic information is transferred unidirectionally in biological systems; more specifically, that the sequential information encoded in biopolymers cannot be transferred back from protein to protein, or from protein to nucleic acid.

Central Processing Unit – see CPU.

checksum – a fixed-size datum that's computed from a block of digital data for the purpose of detecting errors introduced during its transmission or storage.

chimera – an object or entity created by fusing parts of other different objects or entities.

chromatogram – the visual output of a chromatographic procedure (e.g., in the form of a graph or trace).

chromatography – a preparative or analytical technique for separating the components of a mixture.

chromosome – the organisational unit of a cell's hereditary material, comprising DNA and protein.

chymotrypsin – a digestive enzyme, a serine protease, formed by the action of trypsin on chymotrypsinogen; found in the pancreatic juice of vertebrates, it acts in the small intestine to hydrolyse protein.

CINEMA – Colour INteractive Editor for Multiple Alignments, a versatile tool for interactive editing of multiple sequence alignments.

clan – a group of protein families for which there are indications of evolutionary relationship (usually at the structural level), but between which there is no statistically significant sequence similarity.

class – in the CATH database, the overall secondary-structure content of a protein fold; there are four broad classes: mainly-α, mainly-β, α-β (alternating α/β or $\alpha+\beta$ elements), and low secondary-structure content.

Class, Architecture, Topology, Homology (CATH) – see CATH.

client – a component in a distributed system that interacts with a server to request functionality.

client/server – a distributed system where computation is split between a client component, which initiates requests, and a server component that satisfies them.

clinical trial – test designed to establish safety or efficacy data for drugs, therapies, devices, etc., for example involving healthy volunteers or patients with specific health conditions.

clover-leaf structure – a visual representation of the secondary structure of RNA, normally showing four base-paired helical regions connected by three loops, one of which contains the anti-codon.

ClustalΩ – a member of the Clustal family of multiple sequence alignment tools that has been scaled up to allow hundreds of thousands of sequences to be aligned within a few hours.

ClustalW – a multiple sequence alignment tool based on a progressive alignment algorithm.

ClustalX – a version of the Clustal multiple sequence alignment tool with a graphical user interface.

cluster computing – multiple computers, usually connected by a fast dedicated network, often used to solve difficult compute-intensive problems; see supercomputer.

Coding DNA Sequence – see CDS.

coding region – see CDS.

coding sequence – see CDS.

codon – a group of three nucleotide bases in an mRNA molecule that either specificies a particular amino acid, or signals the beginning or end of the message, according to its specific base sequence.

common descent (ancestry) – the concept of evolution from a common ancestor.

complexity theory – in computation, the study of the difficulty (complexity) of problems and how they behave in terms of the time they take and the space they require.

composite database – a database that amalagates different primary sources, often using criteria that determine the priority of inclusion of the sources and the level of redundancy retained in the process.

computability theory – the sub-branch of the theory of computation that deals with what is, or is not, possible to compute in terms of the models of computation that govern modern computer systems.

computational biology – a compuotional discipline that involves the development of mathematical models and application of theoretical simulations to the study of biological, behavioural and social systems.

computer graphics – visual representations created by computers, or the representation and manipulation of image data using computer hardware and software.

computer science – a scientific and mathematical discipline that deals with the theory of computation, and the design of computing machines and processes.

computer tape – punched tape (now obsolete) or magnetic tape on which data are stored digitally.

ConA – concanavalin-A, a plant carbohydrate-binding protein that stimulates cell division.

concanavalin-A (ConA) – see ConA.

conceptual translation – the computational process of translating a sequence of nucleotides into a sequence of amino acids via the genetic code.

cone photoreceptor – a light-sensitive, cone-shaped structure of the vertebrate retina that contains photopigments (opsins) responsible for vision in daylight.

Conger eel – a member of the family Congridae, a (usually) long fish; one species is an important food fish, often served on sushi.

consensus expression – or pattern, or sequence pattern, a consensus pattern of amino acid residues or nucleotide bases derived from a conserved region of a sequence alignment, used as a characteristic signature of that region or of the family of sequences that contain it; beyond a mere consensus sequence, the expression uses a formal notation to define which residues/bases or groups of residues/bases may or may not occur in different parts of the region or motif.

consensus sequence – a shorthand representation of the most frequently occurring amino acid residues or nucleotide bases found at each position in a multiple sequence alignment.

controlled vocabulary – a specific set of terms available for use within a particular context.

convergent evolution – the process of acquiring similar traits via different evolutionary pathways.

covalent – a chemical bond in which pairs of electrons are shared between two atoms.

CPU – Central Processing Unit, the hardware within a computer that's responsible for carrying out the instructions of computer programs.

Crick, Francis – molecular biologist and biophysicist known for proposing the double-helix structure of DNA, for which he was jointly awarded the Nobel Prize in Physiology or Medicine with James Watson and Maurice Wilkins in 1962; also known for introducing the concept of the 'central dogma of molecular biology'.

CSD – Cambridge Structural Database, an archive of small molecule crystal structures compiled and maintained by the Cambridge Crystallographic Data Centre (CCDC).

C terminus – or carboxy-terminus, the end of peptide or polypeptide that terminates with a free carboxyl group (-COOH).

cyanide – a chemical compound that contains a cyano group (*i.e.*, a carbon atom triple-bonded to a nitrogen atom).

cyclic adenosine monophosphate (cAMP) – see cAMP.

Cys – the IUPAC three-letter abbreviation for cysteine.

cysteine – 2-amino-3-mercaptopropanoic acid, an amino acid found in proteins.

cytochrome b_5 – an electron-transport haemoprotein found in animals, plants, fungi and certain types of photosynthesising bacteria.

cytoglobin – a member of the globin family widely expressed in all tissues, most notably exploited by marine mammals; it's believed to transfer oxygen from arterial blood to the brain, and to play a protective role under conditions of hypoxia.

cytoplasm – the gel-like substance contained within a cell's plasma membrane, excluding the nucleus but including other organelles and sub-structures.

cytosine – 4-amino-2-hydroxypyrimidine, a pyrimidine derivative that's one of the five principal bases of nucleic acids.

cytoskeleton – the cell's proteinaceous skeleton, or scaffolding, which plays a variety of roles, such as in the maintenance of cellular shape, cellular and organellar movement, cell division and so on.

D

D – the IUPAC single-letter abbreviation for aspartic acid.

data – the plural of datum, a set of numbers, characters, or values of qualitative or quantitative variables that typically result from measurements and can be visualised graphically or using images.

database – a collection of data organised in a data structure within a management system, which is typically relational or object-oriented; often used to describe data collected in aggregated flat-file records.

data integrity – the concept of data trustworthiness, in terms of completeness, currency, accuracy, consistency and validity.

data representation – the way in which data are stored in a particular computer system or database.

data security – the protection of data, usually in a database, from unauthorised actions of users.

data structure – how related data are structured in computer memory; data structures usually come with a set of algorithms for accessing, modifying and deleting the data they hold.

data warehouse – a large data store, usually constructed by bringing together and harmonising data from a variety of different sources.

Dayhoff, Margaret – a physical chemist and pioneer of bioinformatics, known for developing the first substitution matrices (the PAM matrices) and the amino acid single-letter code; she also initiated the first collection of protein sequences, which she published in the *Atlas of Protein Sequence and Structure*, which ultimately led to the creation of the PIR Protein Sequence Database (PIR-PSD).

DDBJ – the DNA Data Bank of Japan, based at the National Institute of Genetics and member of the INSDC.

DEC VT100 terminal – a video terminal, first manufactured by Digital Equipment Corporation (DEC) in 1978, whose graphical attributes led it to become the standard for terminal emulation.

deoxyribose – 2-deoxy-erythro-pentose, a monosaccharide derived from the sugar ribose by loss of an oxygen atom.

deoxyribonucleic acid (DNA) – see DNA.

dereferencing – the act of locating an object referred to by an identifier.

design pattern – a documented solution to a common design problem that is independent of any specific implementation, but which outlines the pros and cons of different approaches.

diagnostic signature – a mathematical abstraction of a conserved region or motif, or set of motifs derived from a multiple sequence alignment (*e.g.*, such as a consensus expression, fingerprint, profile or HMM), used to search either an individual query sequence or a full database for the occurrence of that same, or similar, region or motif(s).

difference engine – an automatic, mechanical calculator designed to tabulate polynomial functions.

diffraction pattern – the pattern of intensities that occurs at a receiver following the interference of waves of electromagnetic radiation (*e.g.*, light, X-rays, electrons) that have interacted with, or passed through, diffracting objects, such as closely-spaced slits or atoms in a crystal.

Digital Object Identifier (DOI) – see DOI.

dihedral angle – or torsion angle, the angle between two planes.

dinucleotide – a molecule consisting of two mononucleotides in 3′, 5′-phophodiester linkage.

dipeptide – a molecule consisting of two amino acids joined by a single peptide bond.

disease – an abnormality or impairment of the normal functioning of an organism, or of any of its parts, other than through physical injury; disease may arise as a result of environmental factors (bacteria, viruses, toxic agents, *etc.*), through inherent defects in the organism (genetic disease), or as a result of a combination of these factors.

distributed system – a set of multiple, autonomous computers that collaborate via a computer network to achieve a common goal.

distribution – in the context of pharmacology, the process by which a substance is carried to its effector site, and to other sites, such as a drug carried by the bloodstream from a muscle to other muscles and organs.

disulphide bond – a covalent bond between two sulphur atoms; in biomolecules, they are usually formed by the oxidation of sulphydryl groups in neighbouring cysteine residues.

divergent evolution – the process of acquiring similar traits via an evolutionary pathway deriving from a common ancestor.

divide and conquer – the splitting up of a large problem into successively smaller, and therefore more tractable, problems.

DNA – deoxyribonucleic acid, one of two main types of nucleic acid, consisting of a long, unbranched polymer of deoxyribonucleotides, each of whose 3′-phosphate groups is joined in 3′,5′-phosphodiester linkage to the 5′-hydroxyl group of the next; may be single-stranded or double-stranded.

DNA Data Bank of Japan – see DDBJ.

docking – the binding of a molecule to its recognition site on another molecular (often macromolecular) structure; often used to predict the specific binding orientation of drugs to their target proteins.

DOI – Digital Object Identifier, an international standard primarily used for identifying digital documents.

domain – a compact, distinct region within a protein structure that may or may not be formed from contiguous regions of an amino acid sequence: domains may be discrete entities, joined only by a flexible linking region of the chain; they may have extensive interfaces, sharing many close contacts; and they may exchange chains with domain neighbours. Domains often have discrete structural and functional roles – those formed from contiguous regions of sequence may be excised from sequence alignments and used to form diagnostic signatures for members of the same domain family.

Doolittle, Russell – a biochemist known for his work on the structure and evolution of proteins; he co-developed the hydropathy index and was part of the team that determined the structure of fibrinogen.

dotplot – a simple visual method for comparing pairs of sequences in which the sequences occupy the vertical and horizontal axes of a scoring matrix, whose cells are shaded when residues match; regions of the sequences that are identical or similar (depending on the underlying scoring matrix) appear as diagonal lines across the plot, the length and number of which provide an estimate of the degree of sequence similarity.

Down Syndrome Cell Adhesion Molecule (*Dscam*) – see *Dscam*.

draft sequence – in the context of genome sequencing, a DNA sequence, parts of which may be missing, in the wrong order, or oriented incorrectly; draft sequences are hence less accurate than finished sequences.

Drosophila melanogaster – fruit fly, used as a subject in genetic experiments, and a popular model organism by virtue of being easy to breed in laboratory conditions, by having rapid generations, by being easy to mutate, and having various visible congenital traits.

drug – a naturally occurring or synthetic substance that, when absorbed in the body of an organism, alters normal bodily function; used in the prevention and treatment of disease.

drug discovery – the process by which candidate pharmaceuticals are identified, often using high-throughput methods, such as combinatorial chemistry and high-throughput screening.

drug target – usually, a protein or a gene that interacts with, and whose function is modulated by, a particular chemical compound.

Dscam – Down Syndrome Cell Adhesion Molecule, a gene belonging to the 'Down syndrome critical region' of chromosome 21 in humans; in *Drosophila*, the gene has thousands of alternatively spliced forms.

duodenum – the first section of the small intestine of mammals, reptiles and birds.

dynamic programming – a programming approach that breaks down large, often intractable, problems into smaller, solvable sub-problems, combining the results to give an overall solution.

E

E – the IUPAC single-letter abbreviation for glutamic acid.

EBI – the European Bioinformatics Institute, the European Molecular Biology Laboratory (EMBL) hub primarily dedicated to the provision of bioinformatics services to the European community, based on the Wellcome Trust Genome Campus, Hinxton, UK.

EBV – see Epstein–Barr Virus.

EC – see European Commission.

EcoCyc – a database for the bacterium *Escherichia coli* K-12 MG1655, including hand-crafted, literature-based annotation of its entire genome, its transcriptional regulation, transporters and metabolic pathways.

E. coli – a Gram-negative gut bacterium, and probably the most widely used model organism in cell and molecular biology, biochemistry and molecular genetics.

EC system – the Enzyme Commission systematic numerical classification scheme for the chemical reactions catalysed by enzymes; EC numbers are four-tiered codes that designate the enzyme class (hydrolase, lyase, ligase, *etc.*) and progressively more detail about the nature of the reaction and substrate (*e.g.*, 3.5.1.2 are hydrolases, specifically glutaminases).

Edman degradation – a peptide sequencing method in which amino acids are sequentially removed from the N terminus.

efficacy – the ability to produce an effect; used especially in the context of the effectiveness of a drug to reproduce its intended result under ideal conditions.

EGF – Epidermal Growth Factor, a low-molecular weight, polypeptide hormone that stimulates cell growth.

electromagnetic radiation – energy transmitted and absorbed by charged particles that is progagated through space in the form of a wave.

electromagnetic spectrum – the complete frequency range of electromagnetic radiation.

electronic engineering – a discipline within the field of engineering that's concerned with the design of electronic circuits, devices and systems, using applications, principles and algorithms from fields such as solid-state physics, computer engineering, signal processing, robotics, *etc*.

electron microscopy – an imaging technique that exploits high-energy electrons to produce up to 10^7–fold sample magnifications; electron microscopes achieve greater resolution than optical microscopes because the wavelength of electrons is significantly shorter than that of visible light.

electrostatic force – the force of interaction between two point charges; positive forces are repulsive and negative forces attractive (like charges repel, unlike charges attract).

ELIXIR – a large inter-governmental project that aims to produce a pan-European research infrastructure for biological information.

email – or e-mail, or electronic mail, the method for exchanging digital messages across the Internet or other computer network (its history goes back to the early ARPANET).

EMBL – the European Molecular Biology Laboratory, an inter-governmental research organisation funded by its member states; EMBL operates across five different sites, the main one being based at Heidelberg, with four outstations, at Genoble, Hamburg, Monterotondo and Hinxton.

EMBL-Bank – or EMBL data library, the European collection of nucleotide sequences maintained at the EBI, first released in June 1982; it is a member of the INSDC.

EMBL data library (EMBL-Bank) – see EMBL-Bank.

endoplasmic reticulum – an organelle found in eukaryotic cells; different types of endoplasmic reticulum (rough, smooth, sarcoplasmic) have roles in the synthesis of proteins, lipids and membranes, metabolism of carbohydrates, regulation of calcium concentration and detoxification of drugs and poisons.

empirical – relates to the acquisition or verification of information through the act of observation or experimentation rather than theory.

ENA – the European Nucleotide Archive, which collates all publicly available nucleotide sequence data; it has three main sections: the Sequence Read Archive (SRA), the Trace Archive and EMBL-Bank.

Ensembl – a software system that creates and maintains automatic annotation for selected eukaryotic genomes, originally launched in 1999 in preparation for completion of the human genome project; it is a joint project between the EBI and Wellcome Trust Sanger Institute in the UK.

enzyme – a biological catalyst; enzymes increase the rate of biochemical reactions by acting on specific substrates – the reactions they catalyse are classified by the EC system (*e.g.*, according to whether they are hydrolases, lyases, ligases, *etc.*).

Enzyme Classification (EC) system – see EC system.

EPD – Eukaryotic Promoter Database, an annotated, non-redundant database of eukaryotic RNA polymerase II promoters, for which the transcription start sites have been experimentally verified.

Epidermal Growth Factor (EGF) – see EGF.

Epstein–Barr Virus (EBV) – or human herpesvirus 4 (HHV-4), is one of the most prevalent viruses in humans, transferred orally via saliva; it is associated with infectious mononucleosis and certain forms of cancer, and may be related to increased risk of developing some autoimmune diseases.

erythrocruorin – a large member of the globin family that functions as an oxygen carrier in invertebrates.

***Escherichia coli* (*E. coli*)** – see *E. coli*.

EST – Expressed Sequence Tag, a fragment of a cDNA sequence that results from one-shot sequencing of a cloned mRNA; ESTs, which are short (500-800 nucleotides) and of relatively poor quality, represent portions of expressed genes and are hence used in gene identification.

EU – see European Union.

eukaryote – an organism whose cells are bounded by membranes and contain a complex mass of organelle-rich cytoplasm; most notably, eukaryotic cells contain a nucleus, or nuclear envelope, housing the genetic material.

Eukaryotic Promoter Database (EPD) – see EPD.

European Bioinformatics Institute (EBI) – see EBI.

European Commission (EC) – the executive body of the European Union, responsible for its daily running, for proposing legislation, implementing decisions and upholding its treaties.

European Molecular Biology Laboratory (EMBL) – see EMBL.

European Nucleotide Archive (ENA) – see ENA.

European Union (EU) – an economic and political confederation of primarily European member states; the EU developed a free market for goods and services by adopting standardised laws in all its member states, most of whom are also partners in the Shengen Agreement, which allows free movement of people (without passport control).

Europe PubMed Central – a mirror of the NIH PubMed Central archive hosted by the European Bioinformatics Institute, and augmented with extra features, such as text-mining results.

e-value – in a database search, the number of matches with scores greater than or equal to that of the retrieved match that are expected to occur by chance in a database of the same size and composition, using the same scoring system (the closer the value to 0, the more significant the score).

evolution – the process by which inherited changes to the characteristics of biological systems across successive generations give rise to diversity, from the molecular level (DNA, protein), all the way up to individual organisms, populations and species.

excretion – the process by which waste materials and products of metabolism are eliminated from a body or organism, such as the removal of drugs and their products by the kidney, liver and lungs.

exon – an expressed region of a nucleotide sequence; in eukaryotic DNA, exons are the fragments of sequence between non-coding introns that are ligated to form the mRNA coding sequence – hence, the term can refer to the DNA sequence or to its RNA transcript, depending on context.

Expressed Sequence Tag (EST) – see EST.

eXtensible Markup Language (XML) – see XML.

F

F – the IUPAC single-letter abbreviation for phenylalanine.

false negative – in the context of database searching, a true member of a data-set that fails to meet the conditions of a search for that data-set.

false-positive match – in the context of database searching, a result that matches the conditions of search for a particular data-set, but is not a true member of the data-set.

FASTA – a computer program for searching nucleotide or protein sequence databases with a query sequence, developed in 1988; or, a sequence format suitable for database searches or use with sequence analysis tools.

feature table – a table within a biomolecular sequence database that lists known or predicted structural or functional characteristics of the sequence (*e.g.*, locations of exons, the CDS, the translation product, *etc.*, in nucleotide sequences, or the locations of transmembrane domains, metal-binding sites, PTMs, *etc.*, in protein sequences); members of the INSDC share a common feature-table definition, in order to facilitate data exchange.

fingerprint – in the context of protein family classification, a group of ungapped, aligned sequence motifs that, taken together, form a characteristic signature of a particular protein family; the concept of a fingerprint embodies not only the number of motifs it contains, but the order in which the motifs appear and the distances between them.

flat-file – a plain-text data-file that does not have structured inter-relationships; a convenient format for data interchange or analysis, flat-files may be created as output from relational databases, in a form suitable for upload to other databases or for access by analysis or query software.

floating-point – a representation that stores the significant digits of a real number separate from the position of its radix point (the point that separates the integer part from the fractional part of the number).

FlyBase – a database of *Drosophila* genes and genomes, the primary repository for genetic and molecular data for the Drosophilidae.

fMet – the IUPAC abbreviation for *N*-formyl-L-methionine.

fMet initiator – a derivative of methionine to which a formyl group has been added to the amino group, used specifically to initiate protein synthesis.

fold-classification database – a database that classifies protein tertiary structures (including domains within tertiary folds) according to architectural, topological and/or evolutionary similarities.

formal methods – mathematical methods of investigating and verifying the behaviour of both software and hardware.

formyl group – the acyl group (H-CO-) derived from formic acid.

Forsythe, George – mathematician credited with helping to establish computer science as an academic discipline in its own right.

Franklin, Rosalind – a biophysicist best known for her work on the X-ray diffraction of DNA, which led directly to Watson and Crick's opportunistic proposal for DNA's double-helical structure; also known for her pioneering work on the tobacco mosaic and polio viruses.

frizzled receptor – name given to a class of membrane proteins that function as receptors in the Wnt signalling pathway, and belong to the GPCR clan.

functional programming – the programming paradigm that treats computation as a series of side-effect-less function definitions, the execution of which provides the solution to the problem at hand.

G

G – the IUPAC single-letter abbreviation for glycine, or the purine base guanine.

gap – a 'space' character introduced into amino acid or nucleotide sequences during the process of alignment in order to bring equivalent parts of the sequences into register.

GC-content – or guanine-cytosine content, the proportion of bases in a nucleotide sequence that are guanine and cytosine; regions of high GC-content relative to the background GC-content for the entire genome are often taken as markers for the presence of genes.

GDB – Human Genome Database, a community-annotated database of human genetic and genomic data, including high-quality mapping data; GDB is now no longer maintained.

GDP – guanosine diphosphate, a metabolic precursor of guanosine triphosphate.

3Gen – third-generation sequencing, single-molecule sequencing, which includes techniques that observe single molecules of DNA polymerase sythesising single molecules of DNA; those that thread single molecules of DNA through nanopores, their base sequences being read as they pass through; and those that image single DNA molecules directly using advanced microscopy techniques – 3Gen technologies aim to reduce the cost of sequencing a single human genome to $1,000.

GenBank – the nucleotide sequence database maintained at the NCBI, first released in December 1982; it is a member of the INSDC.

gene – a molecular unit of heredity, broadly corresponding to a unit of DNA or RNA that encodes a polypeptide or functional RNA.

gene expression – the process by which information carried within a gene is used to synthesise a functional gene product (protein or RNA).

gene family – see protein family.

Gene Ontology (GO) – see GO.

Gene3D – a collection of pre-calculated structural assignments for protein sequences whose structures have not been determined, based on structural domains defined in the CATH database.

genetic code – the rules that relate the four DNA or RNA bases to the 20 amino acids; there are 64 possible three-base (triplet) sequences, or codons, each defining one amino acid – the code is degenerate because up to six codons may define one amino acid.

genome – the entirety of an organism's genetic information, encoded in the form either of DNA or of RNA.

Genome Reference Consortium (GRC) – see GRC.

genome sequencing – the laboratory process of determining the complete DNA sequence of an organism's genome.

genome sequencing project – a project that aims to sequence the complete genome of a particular organism, in order to discover and annotate its genes and other genome-encoded features (*e.g.*, the Human Genome Project).

Genome Survey Sequence (GSS) – see GSS.

genomics – the systematic sequencing, characterisation and study of complete genomes.

genotype – the complete genetic constitution of a cell, organism or individual.

Gln – the IUPAC three-letter abbreviation for glutamine.

Glu – the IUPAC three-letter abbreviation for glutamic acid.

glutamate – a salt of glutamic acid.

glutamate receptor – or metabotropic glutamate receptor, name given to a superfamily of membrane proteins that bind glutamate and mediate its effects in neurotransmission, and belong to the GPCR clan.

glutamic acid – 2-aminopentanedioic acid, a chiral amino acid found in proteins.

glutamine – 2-amino-4-carbamoylbutanoic acid, a chiral amino acid found in proteins.

Gly – the IUPAC three-letter abbreviation for glycine.

glycine – aminoethanoic acid, a chiral amino acid found in proteins.

glycoprotein – a protein containing covalently-bound oligosaccharide chains.

GO – Gene Ontology, a major initiative to standardise the description of gene and gene-product attributes that uses a set of controlled vocabularies to describe the cellular components, molecular functions and biological processes with which they're associated.

Goad, Walter – a theoretical physicist known for being one of the founders of GenBank.

Google – a multinational corporation that provides Internet-based products and services, especially Internet-search and cloud-computing facilities.

GPCR – G Protein-Coupled Receptor, member of a diverse, ubiquitous family of membrane-bound, cell-surface proteins that has adapted a characteristic heptahelical (or 7TM) architecture to mediate a range of physiological processes; GPCRs transduce extracellar signals into intracellular responses by binding a variety of ligands, and stimulating diverse intracellular pathways by coupling to different G proteins; they are of significant medical importance, being the targets of the majority of prescription drugs.

GPCR Proteolytic Site (GPS) – see GPS.

G protein – guanine nucleotide-binding protein, a protein that functions as a molecular switch.

G protein-coupled receptor (GPCR) – see GPCR.

GPS – GPCR Proteolytic Site, a cysteine-rich domain found in a number of GPCRs believed to be a site for proteolytic cleavage.

Golgi apparatus – Golgi complex or Golgi body, a membranous cytoplasmic organelle found in eukaryotic cells that consists of flattened vesicles arranged in a regular stack;

the organelle packages proteins inside the cell for transport to different locations, and is particularly important in processing proteins for secretion.

Gosling, Raymond – a physicist known for his work on the X-ray diffraction of DNA, under the supervision of Rosalind Franklin, which led directly to Watson and Crick's opportunistic proposal for DNA's double-helical structure.

Grand Canyon – an impressive canyon formed by the Colorado River in the US state of Arizona, it is considered one of the seven natural wonders of the world.

gramicidin S – a cyclic decapeptide antibiotic effective against some Gram-positive and Gram-negative bacteria, and certain fungi, it is produced by the Gram-positive bacterium *Bacillus brevis*.

grammatical analysis – identifying the meaning of a sequence of tokens in the context of parsing a program.

Gram-negative bacteria – those that do not retain dye in their cell walls following Gram staining.

Gram-positive bacteria – those that retain dye in their cell walls following Gram staining.

Gram staining – a staining procedure used to differentiate bacterial species based on whether they retain dye in their cell walls, essentially detecting cell walls that contain polymers of sugars and amino acids.

graph database – a database that uses graph structures, with nodes and edges to represent and store data.

GRC – the Genome Reference Consortium, a group of institutes that collaborate to improve the representation of reference genomes.

Greek key – a protein fold characterised by four adjacent anti-parallel β-strands and their linking hairpin loops, named after a pattern commonly found in Greek ornamental artwork.

GSS – Genome Survey Sequence, a fragment of genomic sequence that results from a genome-sequencing experiment.

GTP – guanosine triphosphate, an essential component of signal transduction (especially involving G proteins), where it is converted to guanosine diphosphate.

guanine – 2-amino-6-hydroxypurine, a purine derivative that's one of the five principal bases of nucleic acids.

guanine-cytosine content – see GC content.

guanosine diphosphate (GDP) – see GDP.

guanosine triphosphate (GTP) – see GTP.

guide tree – in phylogenetics, a tree structure or dendrogram derived from a distance matrix that helps to determine the order in which sequences should be aligned.

H

H – the IUPAC single-letter abbreviation for histidine.

haemoglobin – iron-containing, oxygen-carrying protein of the blood of vertebrates and some invertebrates, a close homologue of myoglobin.

Haemophilus influenzae – a small Gram-negative, non-motile, rod-shaped bacterium, it was the first free-living organism to have its genome fully sequenced, in 1995; originally thought to be the cause of influenza (until the viral aetiology of flu was understood), *H. influenzae* can cause serious invasive diseases.

hairpin loop – a loop formed when a linear molecular structure folds back on itself and is stabilised by intrachain hydrogen bonds; hairpin loops, or stem loops, can be found in DNA, but are more commonly found in tRNA molecules, where regions of the same strand base-pair to form a double helix (the stem), connected by an unpaired loop.

Halobacterium salinarium – an extreme halophilic marine archaeon, a Gram-negative obligate anaerobe; formerly referred to as *Halobacterium halobium*.

halting problem – the problem of being unable to know whether or not some algorithms will ever complete.

HAMAP – a system that identifies and semi-automatically annotates proteins that are part of well-conserved families or subfamilies; HAMAP is based on manually formulated family rules – it is applied to bacterial, archaeal and plastid-encoded proteins.

helix – a spiral structure in which successive turns are made at a constant angle to the base and constant distance from the axis.

heuristic – a pragmatic problem-solving technique that, while not promising an optimal outcome, provides a 'close enough' solution.

hexadecimal – the base-16 number system used to represent values stored in binary form, with a single hexadecimal digit representing four bits of information; the tenth to the fifteenth digits are represented with the letters A-F: *e.g.*, the decimal number 13810, which in binary would be 100010102, would, in hexadecimal, be 8A16.

HGP – Human Genome Project, the ambitious, 13-year-long, international project, coordinated by the US Department of Energy and the NIH, whose mission was to completely map and sequence the three billion bases of the human genome; the project began formally in 1990, the first draft of the genome was released in 2000, and the project was completed in 2003.

Hidden Markov Model (HMM) – see HMM.

hieroglyphics – the formal writing and numerical system of the ancient Egyptians.

high-throughput biology – scaled-up biological laboratory techniques made possible by increased levels of automation, usually using robots.

high-throughput DNA sequencing – a parallelised, automated process, capable of producing thousands or millions of sequences with sufficient rapidity to allow complete genomes to be sequenced.

hippopotamus – a large, semi-aquatic, herbivorous mammal from sub-Saharan Africa belonging to the family Hippopotamidae.

His – the IUPAC three-letter abbreviation for histidine.

histidine – 2-amino-3-(1 *H*-imidazol-4-yl)propanoic acid, a chiral amino acid found in proteins.

HMM – Hidden Markov Model, a probabilistic model consisting of a number of inter-connecting states denoting the match, delete or insert status at each position in a sequence alignment (a match state denotes a conserved alignment column; an insert state denotes an insertion relative to a match state; a delete state denotes a deleted match position); in the context of protein sequence analysis, HMMs are used in a similar way to profiles to encode full domain alignments.

Hogeweg, Paulien – a theoretical biologist, and progenitor of the term bioinformatics as the study of informatic processes in biotic systems.

homeostasis – the physiological and biochemical maintenance of a relatively constant internal environment in the bodies of animals.

homology – relatedness that arises as a result of the evolutionary process of divergence from a common ancestor; homology isn't a synonym for similarity – it is an inference about evolutionary relationship; in the CATH database, the homology level is equivalent to the superfamily level of SCOP.

Homo sapiens – the species to which humans belong; mammals of the primate order, believed to have originated in Africa, humans are the only living members of the genus *Homo*.

horizontal gene transfer – the transfer of genes from one organism to another (*i.e.*, gene transfer other than from parent to offspring).

hormone – a substance formed in small amounts in a specialised organ or group of cells and carried to another organ or group of cells on which it has a specific regulatory action.

Hormone Receptor Domain (HRM) – see HRM.

house of cards – a precarious structure formed from playing cards; the term is used as a metaphor for instability.

HRM – Hormone Receptor Domain, an extracellular domain found in a number of receptors that may play a role in ligand-binding.

HTTP – HyperText Transfer Protocol, the protocol used by Web-based systems to request and exchange resources.

human–computer interaction – literally, the interaction between people (users) and computers; a discipline of computer science at the interface with behavioural sciences and design, it involves the study, planning and design of both visual and non-visual human-machine interaction.

Human Genome Database (GDB) – see GDB.

Human Genome Project (HGP) – see HGP.

hydrogen bond – a weak electrostatic interaction between a hydrogen atom attached to an electronegative atom (like oxygen or nitrogen) and another electronegative atom.

hydropathy – a measure of the affinity for water of an amino acid, encompassing the hydrophobicity and hydrophilicity of its side-chain.

hydropathy profile – a graphical representation of the hydropathic nature of a protein; within a hydropathy profile, hydrophobicity values are calculated in a sliding window and plotted for each residue position of a protein sequence – resulting graphs show characteristic peaks and troughs, corresponding to the most hydrophobic and hydrophilic regions of the sequence; they are most commonly used to detect the presence of TM domains.

hydrophilic – having a strong affinity for water.

hydrophobic – having a low affinity for water.

hydrophobic effect – in the context of protein folding, it is the major driving force for folding, resulting from the tendency of non-polar, hydrophobic residues to aggregate to exclude water from the protein core.

hydroxyl – the covalently linked –OH group in a chemical compound.

HyperText Transfer Protocol (HTTP) – see HTTP.

hypoxia – a pathological condition in which the body, or part of the body, is deprived of oxygen.

I

I – the IUPAC single-letter abbreviation for isoleucine.

IANA – Internet Assigned Names Authority, the nonprofit organisation that oversees global IP address allocation.

ID – a unique code given to identify a particular entry in a particular database; typically, ID codes are designed to be more human-readable than are their corresponding accession numbers.

IDL – Interface Definition Language, a means of formally specifying the interface to a software component.

identifier (ID) – see ID.

identity matrix – see unitary matrix.

IEEE – Institute of Electrical and Electronics Engineers, a world-wide professional association that, amongst other things, acts as a standards body for many technology-related industries.

Ile – IUPAC three-letter abbreviation for isoleucine.

imidazole – an aromatic, heterocyclic compound (1,3-diaza-2,3-cyclopentadiene), weakly basic in solution; its ring-structure is the basis of histidine.

imperative programming – the programming paradigm that consists of a series of commands to be executed in sequence; this is the most common paradigm in use today.

InChI – IUPAC International Chemical Identifier, an encoding of the structure of small molecules into a compact text string.

indel – a term denoting an insertion/deletion ambiguity where, given an observed difference in length between two sequences, say, from different species, it is impossible to infer whether one species lost part of the sequence or the other species gained the extra part – *i.e.*, the phylogenetic direction of the sequence change cannot be inferred.

informatics – a broad discipline that encompasses aspects of computer science, information science, information technology, mathematics, social sciences, and so on, informatics concerns the application of computational and statistical techniques to the management, analysis and retrieval of information.

information technology – a field of engineering that concerns the design, development and management of computer-based information systems.

INSDC – the International Nucleotide Sequence Database Collaboration, an international consortium of the major nucleotide sequence-data providers; its mission is to synchronise the provision of nucleotide sequence data and to present them according to common guidelines that regulate the content and syntax of entries (including a common feature table definition); its partners are EMBL-Bank, GenBank and DDBJ.

in silico – literally, 'in silicon'; performed by means of a computer.

Institute of Electrical and Electronics Engineers (IEEE) – see IEEE.

insulin – a pancreatic hormone, responsible for the regulation of carbohydrate and fat metabolism; it mediates the uptake of glucose from the blood and its storage as glycogen in liver, muscle and fat-tissue cells; it also inhibits the release of glucagon, and hence the use of fat as an energy source.

integer – or natural number, a rational whole number.

Interface Definition Language (IDL) – see IDL.

International Nucleotide Sequence Database Collaboration (INSDC) – see INSDC.

Internet – the global network of computer networks that connects government, academic and business institutions; it grew out of local networks like ARPANET, and evolved into a world-wide network of networks when the Internet Protocol Suite (TCP/IP) was standardised in 1982.

Internet Assigned Names Authority (IANA) – see IANA.

InterPro – an integrated database of predictive protein signatures used to automatically annotate proteins and genomes – it classifies sequences at superfamily, family and sub-family levels, and predicts functional and structural domains, repeats and functional sites, using GO to formalise its functional annotations; InterPro's founding partners were PROSITE, PRINTS, Pfam and ProDom, but it now contains around a dozen partners.

intron – a region of a nucleotide sequence that is not expressed; in eukaryotic DNA, introns are the fragments of sequence between the coding exons that must be excised for the mRNA coding sequence to be generated.

invertebrate – an animal that lacks both a vertebral column (backbone) and an internal skeleton.

in vitro – literally, 'in glass'; colloquially, performed in a test-tube.

ischaemia – a restriction in blood supply to tissues, the resulting shortage of oxygen and glucose leading to tissue damage and dysfunction.

isoelectric point (pI) – see pI.

isoleucine – $(2R^*,3R^*)$-2-amino-3-methylpentanoic acid, a chiral amino acid found in proteins.

IUPAC International Chemical Identifier (InChI) – see InChI.

J

Jacquard Loom – a mechanical loom controlled by punched cards in which each row of punched holes corresponds to one row of a design that comprises multiple rows; the loom was invented by Joseph Jacquard in 1801.

JalView – a Java-based tool for editing multiple sequence alignments.

JANET – Joint Academic NETwork, a government-funded computer network in the UK dedicated to education and research; it connects all the UK's further- and higher-education organisations and the Research Councils.

Japan International Protein Information Database (JIPID) – see JIPID.

Java – an object-oriented programming language designed according to the WORA principle (Write Once, Run Anywhere); the language shares some of its syntax with C++, and is especially popular for client-server Web applications.

JavaScript – an interpreted programming language, used heavily on the Web.

JavaScript Object Notation (JSON) – see JSON.

Jelly roll – or Swiss roll, a protein fold characterised by four pairs of anti-parallel β-sheets that are wrapped in such a way as to form a barrel arrangement.

jigsaw puzzle – a tiling puzzle comprising small, inter-locking pieces each of which carries part of a picture; the pieces must be assembled to create the final complete picture.

JIPID – Japan International Protein Information Database, the protein sequence database of Japan, once part of PIR-International.

Joint Academic NETwork – see JANET.

JSON – JavaScript Object Notation, a format that uses human-readable text to store or transmit data, most-commonly using nested associative arrays.

K

K – the IUPAC single-letter abbreviation for lysine.

KEGG – a database for understanding high-level functions and utilities of biological systems (including the cell, the organism and the ecosystem) from molecular-level information (notably from large-scale molecular data-sets generated by genome sequencing and other high-throughput experimental technologies).

Kendrew, John – a biochemist and crystallographer known for his pioneering work in determining the first structures of proteins (notably, haemoglobin) at atomic resolution, for which he was jointly awarded the Nobel Prize in Chemistry with Max Perutz in 1962.

Kennard, Olga – a crystallographer and Director of the Cambridge Crystallographic Data Centre from 1965 to 1997.

kinetics – in chemistry, the study of chemical reaction rates.

knowledgebase – an information repository or database specially designed for knowledge management, providing the means for information to be collated, organised, shared and searched; knowledgebases may be designed to be human-readable or machine-readable.

knowledge representation – an aspect of artificial intelligence research concerning the means by which knowledge can best be represented in the form of symbols in order to be able to infer new knowledge.

k-tuple – in the FASTA algorithm, a short, k-letter word that determines the window size, and hence speed and sensitivity of a database search.

Kyoto Encyclopedia of Genes and Genomes (KEGG) – see KEGG.

L

L – the IUPAC single-letter abbreviation for leucine.

lactate dehydrogenase – an enzyme present in a variety of organisms that catalyses the interconversion of lactate and pyruvate; lactate dehydrogenases may be cytochrome c-dependent or NAD(P)-dependent.

LAG – Laminin-G, a globular domain found in different numbers in a variety of laminins, glycoproteins that are integral components of the structural scaffolding in almost all tissues, with diverse roles including cell migration and adhesion.

lambda function – an unnamed function that can be passed around a system and executed on arbitrary data by arbitrary components.

Laminin-G (LAG) – laminin globular (G) domain, see LAG.

LANL – Los Alamos National Laboratory, a laboratory in the USA that undertakes classified work on the design of nuclear weapons; LANL was home to the Los Alamos Sequence Database, which evolved into GenBank in 1982 – it remained the home of GenBank until the database was moved to the NCBI in 1992.

laptop – a personal computer designed for mobile use by combining most of the components of a desktop computer into a single, portable unit.

Large Hadron Collider – the largest, most powerful particle accelerator in the world, comprising a 17-mile ring of superconducting magnets, with additional accelerating units to boost the energy of the particles en route.

Latin Script – also known as Roman Script, the alphabet used in most Western and Central European languages.

lectin – a sugar-binding protein that mediates biological recognition phenomena involving cells and proteins.

leghaemoglobin – a member of the globin family that carries nitrogen or oxygen, found in the root nodules of leguminous plants.

Leu – the IUPAC three-letter abbreviation for leucine.

leucine – 2-amino-4-methylpentanoic acid, a chiral amino acid found in proteins.

lexical analysis – the process of converting a sequence of characters (typically readable by humans) into discrete tokens that have computational meaning. Often the first step in parsing the source code of programs.

Life Science Identifier (LSID) – see LSID.

ligand – usually, a small molecule that forms a complex with a biomolecule in order to effect a particular biological function.

Linked Data – a method of publishing data and resources using HTTP, RDF and URIs.

lipid bilayer – a flat, lamellar structure comprising two layers of amphipathic lipid molecules; the polar components of the lipids are oriented towards the polar environment and the non-polar tails towards the centre of the bilayer; the lipid bilayer forms a continuous protective barrier around cells and organelles.

logarithm – the power to which a fixed value (the base) must be raised in order to produce a specified number; formally, if $a = b^n$, then n is the logarithm of a to base b, or $n = log_b(a)$.

logic – in mathematics and computer science, the study of modes of reasoning and the forms that arguments may take.

logic programming – the paradigm of programming that concentrates solely on the logic of what processing is possible on the given model.

Los Alamos National Laboratory (LANL) – see LANL.

LSID – Life Science Identifier, a mechanism, based on URNs, for identifying life-science resources on the Web. Now largely deprecated in favour of HTTP URIs.

Lys – the IUPAC three-letter abbreviation for lysine.

lysine – 2,6-diaminohexanoic acid, a chiral amino acid found in proteins.

M

M – the IUPAC single-letter abbreviation for methionine.

machine learning – a discipline of computer science concerned with the design and development of algorithms that permit adaptive behaviours in response to empirical data; central to the approach is being able to automatically learn to recognise complex patterns, and to guide decision-making based on the data.

MAFFT – a multiple sequence alignment tool based on a progressive alignment algorithm.

mammoth – a species of the extinct genus *Mammuthus*, and member of the family Elephantidae, covered in long hair and possessing substantial, curved tusks.

mass spectrometry – an analytical technique for measuring the mass-to-charge ratio of charged particles, used to determine the relative masses and abundances of components of a beam of ionised molecules, or molecular fragments, produced from a sample in a high vacuum, which ultimately allows the chemical structures of such samples (*e.g.*, peptides or other chemical compounds) to be deduced.

mathematics – a discipline that evolved from counting, calculation and measurement into the use of abstraction and logical reasoning in the formal study of quantity, space, structure, motion, *etc.*

M-Coffee – an implementation of T-coffee that allows the outputs of a range of multiple sequence alignment programs (*e.g.*, MUSCLE, ClustalW, MAFFT, ProbCons) to be combined.

Medical Research Council (MRC) – see MRC.

MEDLINE – an online bibliographic database that includes information on articles published in academic journals spanning the biomedical and life sciences (including medicine, dentistry, pharmacy, biology, molecular evolution, *etc.*); the resource is searchable via the PubMed system.

membrane – see lipid bilayer.

memory – in computing, the physical devices used to store programs or data temporarily or permanently for use in a computer or other digital electronic device.

messenger RNA (mRNA) – see mRNA.

metabolic pathway – a series of interconnected biochemical reactions occurring in a biological system, usually catalysed by enzymes and requiring a variety of cofactors to function correctly.

metabolism – in the context of pharmacology, the chemical reactions and physical processes undergone by a substance, or class of substances, in a living organism; new substances created by the process of metabolism are termed metabolites.

metabolomics – the systematic, quantitative analysis of the chemical fingerprints of cellular processes – the study of a cell's small-molecule metabolite profile.

metaphor – a linguistic device, or figure of speech, that describes an entity in an expressive or symbolic way by equating it with some other, unrelated entity (*e.g.*, the dawn of bioinformatics, the roller-coaster of modern biology).

methaemoglobin – an oxidation product of haemoglobin, with the haemo group iron in the Fe3+ (ferric) state rather than the Fe2+ (ferrous) state of normal haemoglobin.

Met – the IUPAC three-letter abbreviation for methionine.

Methanocaldococcus jannachii – a thermophilic methanogenic archeaon of the class Methanococci, the first archaeon to have its genome fully sequenced.

methionine – 2-amino-4-(methylthio)butanoic acid, a chiral amino acid found in proteins.

Midnight Zone – a region of sequence identity (~0–15 per cent) in which sequence comparison techniques fail to detect structural relationships; it denotes the theoretical limit to sequence analysis techniques.

MIPS – Munich Information Center for Protein Sequences, a research centre hosted by the Institute for Bioinformatics, Neuherberg, Germany, focusing on genome-oriented bioinformatics; MIPS maintained the MIPS protein sequence database, once part of PIR-International.

mirror – an exact copy of a data-set or server, made at a different location, that facilitates access to the data-set or server by spreading traffic load.

mitochondria – semi-autonomous, self-replicating, membrane-bound organelles found in the cytoplasm of most eukaryotic cells; they are considered to be the 'power plants' of cells – the mitochondrion is the site of tissue respiration.

mobile device – in computing, usually a small hand-held device that has an operating system capable of running various types of application software.

modem – a device that both modulates an analogue signal to encode digital information and demodulates such a signal to decode the transmitted information.

modelling – the task of describing a real-world situation in an abstract formal way; this is often done using pure mathematics, graphical diagrams or programming languages.

model organism – an organism used to investigate particular biological phenomena with the expectation that insights gained will be applicable to or provide insights into the functions and behaviours of other organisms.

module – an autonomous folding unit believed to have arisen largely as a result of genetic shuffling mechanisms (*e.g.*, kringle domain, WW domain, apple domain); modules are contiguous in sequence and are often used as building-blocks to confer a variety of complex functions on a parent protein, either via multiple combinations of the same module, or by combinations of different modules to form protein mosaics.

molecular clock – the principle that the relationship between evolutionary time and distance (as determined from the study of molecular sequences) is approximately linear; it is used to estimate when events such as speciation occurred.

molecular evolution – a scientific discipline, embracing molecular biology, evolutionary biology and population genetics, that concerns the study of the processes by which changes in DNA or RNA accumulate over time giving rise to changes in genes, gene products, gene expression, gene frequency, and so on.

mosaic – a protein that comprises many different functional and structural domains or modules.

Mosaic – one of the first Web browsers, with a user-friendly graphical user interface – the first to display images inline with text – and credited with making the Web accessible to the general public.

motif – in the context of protein sequence alignment, a contiguous, conserved region, typically 10–20 amino acids in length, often denoting a key structural or functional feature of its parent proteins; motifs are sometimes exploited by pattern-recognition techniques to diagnose related sequences that share the same features.

MRC – Medical Research Council, the government agency responsible for coordinating and funding medical research in the UK.

mRNA – messenger RNA, a class of RNA molecules that carry the genetic information within DNA genes to ribosomes, the sites of protein synthesis, where the chemical message embodied in the mRNA bases is translated into a polypeptide by means of tRNA.

Muirhead, Hilary – a crystallographer, known for her work with Max Perutz in determining the 3D structure of haemoglobin; she was jointly responsible for establishing protein crystallography at Bristol, where she went on to determine the structures of a range of glycolytic enzymes.

Multalin – a multiple sequence alignment tool based on dynamic programming and clustering.

multiple alignment – the process of, or the product of the process of, comparing (aligning) more than two linear sequences of proteins or nucleic acids, usually necessitating the insertion of gaps to bring equivalent parts of the sequences into vertical register; alignments help to illustrate which features of the sequences are most highly conserved, often exploiting scoring functions to help quantify their degree of relatedness.

MUltiple Sequence Comparison by Log-Expectation (MUSCLE) – see MUSCLE.

Munich Information Center for Protein Sequences (MIPS) – see MIPS.

MUSCLE – MUltiple Sequence Comparison by Log-Expectation, a multiple sequence alignment tool based on a progressive alignment algorithm.

mutation – the process by which genetic material undergoes a detectable, heritable structural change, such as a base change in a DNA sequence, a structural change in a chromosome, addition or loss of a chromosome, and so on.

Mycoplasma genitalium – a small parasitic bacterium that lives on the ciliated epithelial cells of primate respiratory and genital tracts; it has the smallest known genome that can constitute a cell, and is hence used as a model minimal organism.

myoglobin – iron-containing, oxygen-carrying protein of the muscle tissue of vertebrates, a close homologue of haemoglobin.

N

N – the IUPAC single-letter abbreviation for asparagine.

namespace – a mechanism for grouping identifiers or names within a system to avoid clashes with the same name elsewhere, and to qualify the source of an identifier to make it unambiguous.

Napier's Bones – an abacus for calculating products and quotients of numbers, created by John Napier.

National Biomedical Research Foundation (NBRF) – see NBRF.

National Center for Biotechnology Information (NCBI) – see NCBI.

National Institute of Standards and Technology (NIST) – see NIST.

National Institutes of Health (NIH) – see NIH.

National Science Foundation (NSF) – see NSF.

NBRF – National Biomedical Research Foundation, a not-for-profit organisation whose mission was to stimulate biomedical researchers to use computers by establishing a pioneering research and development program in new areas of computer applications; the NBRF was home to the PIR-PSD.

NCBI – National Center for Biotechnology Information, part of the United States National Library of Medicine, a branch of the National Institutes of Health, located in Bethesda, Maryland, USA, and home of GenBank.

Neanderthal – or Neanderthal man, an extinct species or subspecies of the genus *Homo*, a close relative of modern humans, inhabiting Europe and parts of central and western Asia.

neuroglobin – an oxygen-binding member of the vertebrate globin family found in the brain that may play a neuroprotective role against ischaemia.

neurotransmitter – a chemical substance that transmits signals from a nerve cell to a target cell.

neutron diffraction – a complementary technique to X-ray diffraction that exploits the scattering properties of neutrons to determine the atomic structure of a material.

Next-Generation Sequencing (NGS) – see NGS.

NGS – Next-Generation Sequencing, a low-cost, high-throughput sequencing technology that parallelises the sequencing process, producing thousands or millions of sequences simultaneously (examples are 454 pyrosequencing, Illumina (Solexa) sequencing and SOLiD sequencing).

NIH – National Institutes of Health, the primary agency of the United States government responsible for biomedical and health-related research.

NIST – National Institute of Standards and Technology, a measurement standards laboratory, a non-regulatory agency of the United States Department of Commerce.

NMR – Nuclear Magnetic Resonance, a physical phenomenon in which the magnetic nuclei of atoms placed in a magnetic field of high intensity absorb and re-emit electronmagnetic radiation, having transitioned from a low- to a high-energy state; the energy is at a resonance frequency that depends on the strength of the magnetic field and the magnetic properties of the atoms – NMR is the basis both of spectroscopic techniques that provide insights into the atomic arrangements in crystals and non-crystalline materials, and of magnetic resonance imaging.

Nobel Prize – annual international awards that recognise cultural or scientific advances, bestowed by the will of Swedish chemist, Alfred Nobel, in 1895.

non-identical database – a database in which identical records have been removed to create a smaller, more efficient resource for database searching; as highly similar sequences are not removed, they are larger than non-redundant databases – NRDB, the default database for NCBI's BLAST services, is a non-identical database.

non-redundant database – a database in which identical and highly similar records have been removed to create a smaller, more efficient resource for database searching; redundancy criteria may vary in strictness depending on the degree of redundancy to be retained in the resource; NRDB, the source database for NCBI's BLAST services, the so-called Non-Redundant Database is, strictly speaking, a non-identical database.

non-template strand – or coding strand, in the context of DNA transcription, the strand of DNA that's not used as a template for RNA synthesis; it has the same base sequence as the RNA transcript (but with thymine, not uracil); its complement is the template or non-coding strand.

NRDB (non-redundant database) – see non-redundant database.

NSF – National Science Foundation, a United States government agency responsible for supporting research and education in non-medical fields of science and engineering.

N terminus – or amino-terminus, the end of a peptide or polypeptide chain that terminates with a free amino group ($-NH_2$).

Nuclear Magnetic Resonance (NMR) – see NMR.

nuclear membrane – or nuclear envelope, the membrane that envelops the nucleus of a eukaryotic cell, acting as a physical barrier to separate the contents of the nucleus from the cytoplasm; it is a double membrane comprising two lipid bilayers.

nucleic acid – a single- or double-stranded polynucleotide, which may comprise either deoxyribonucleic acids (DNA) or ribonucleic acids (RNA); concerned with the storage, transmission and transfer of genetic information, nucleic acids are essential for all known forms of life.

nucleotide base – or base, the nitrogen-containing component of nucleotides, the fundamental building-blocks of nucleic acids; the principal bases are adenine, cytosine, thymine, guanine and uracil, whose base-pairing results in characteristic nucleic acid structures – adenine and guanine are double-ring structures termed purines, while thymine, cytosine and uracil are single-ring structures termed pyrimidines.

nucleus – a membrane-bound organelle of eukaryotic cells that contains most of the genetic material, organised as linear DNA molecules complexed with proteins in the form of chromosomes; it is the site of DNA replication and RNA synthesis in the cell.

O

object-oriented database – a database-management system in which data are represented as objects, as used in object-oriented programming.

object-oriented programming – the programming paradigm that encapsulates into self-contained objects both the program's data and the methods of that data's management.

Object Request Broker (ORB) – see ORB.

OBO – Open Biomedical Ontologies, an initiative to create interoperable controlled vocabularies in the biomedical domain, or the organisation associated with this (more formally, the OBO Foundry).

odorant – a chemical compound that has a smell or odour.

oligonucleotide – a short molecular sequence that contains a small number of nucleotide units (typically, up to around 50), often chemically synthesised in a sequence-specific manner; oligonucleotides are widely used as primers, allowing DNA polymerases to extend the sequence and replicate the complementary strand.

oligosaccharide – a saccharide polymer containing a small number of simple sugars, or monosaccharides (typically, 2-10); oligosaccharides have a variety of functions, including roles in cell-cell recognition.

ontology – in computer science, the formal representation of knowledge as a set of concepts within a field of knowledge, or domain, and the relationships between those concepts; ontologies may be used to describe a domain and to reason about the entities within the domain.

ontology language – a computer language used to specify an ontology.

opaque identifiers – identifiers that reveal nothing about the object they identify without being dereferenced. Such identifiers often appear to be random characters.

Open Biomedical Ontologies (OBO) – see OBO.

Open Reading Frame (ORF) – see ORF.

Open Researcher and Contributor Identifier (ORCID) – see ORCID.

Operating System (OS) – see OS.

opsin – or photopsin, a light-sensitive visual pigment found in the cone-shaped photo-receptor cells of the retinas of most vertebrates; members of the GPCR superfamily, with characteristic 7TM architectures, opsins mediate colour vision in daylight – their different wavelength sensitivities (long, medium, short) allow perception in the red, green and blue regions of the electromagnetic spectrum.

ORB – Object Request Broker, part of a distributed system that matches requests for remote procedure calls from clients to appropriate servers, providing a degree of location transparency between components.

ORCID – Open Researcher and Contributor Identifier, a method of identifying people, primarily in the context of academic authorship.

ORF – Open Reading Frame, the part of a nucleotide sequence (spanning a start codon through to a stop codon) that represents a putative or known gene; ORFs are commonly used to assist gene finding, long ORFs helping to identify candidate coding regions in DNA sequences.

organ – part of the body of a multicellular organism that's specialised for one or more vital functions.

organelle – a discrete component of a cell that's specialised for the performance of one or more vital functions, usually bounded within a lipid bilayer.

orthologues – homologous proteins that usually perform the same function in different species (*i.e.*, they arise following speciation events).

OS – Operating System, an essential component of the system software of a computer, the OS is a software suite that manages computer hardware resources and provides common services for computer programs.

ovine – of, or relating to, a sheep.

OWL – Web Ontology Language, a widely-used family of ontology languages.

P

P – the IUPAC single-letter abbreviation for proline.

P2P – Peer to Peer, an architecture for a distributed system in which all components behave as equals and co-operate to solve an overall problem.

pairwise alignment – the process of, or the product of the process of, comparing (aligning) two linear sequences of proteins or nucleic acids, usually necessitating the insertion of gaps to bring equivalent parts of the sequences into vertical register; pairwise alignment algorithms underpin database search programs, such as BLAST and FASTA.

PAM matrix – a matrix used to score protein sequence alignments, whose scores represent the probabilities for all possible amino acid substitutions; each matrix in the PAM series is based on the original 1 PAM (Point Accepted Mutation) matrix, which gives substitution probabilities for sequences in which there has been one point mutation per 100 amino acids.

PANTHER – a database that exploits HMMs to classify protein sequences into families and subfamilies.

paralogues – homologous proteins that usually perform different but nevertheless related functions in the same species (*i.e.*, they arise following gene-duplication events).

parallel execution – the simultaneous execution of more than one task; this requires the tasks to be independent of each other's output.

parathyroid hormone – or parathyrin, a polypeptide hormone secreted by the parathyroid glands, which acts in concert with calcitonin to regulate calcium concentration in the blood.

pathogenic process – a process that has the capacity to cause disease.

pattern – see consensus expression.

pattern recognition – the process of identifying characteristic patterns of conservation in biomolecular sequences and structures using some form of pre-calculated or trained signature or discriminator (*e.g.*, a consensus expression, fingerprint, profile or HMM); or the process of discovering unknown patterns in large data-sets (*e.g.*, using principal-component-analysis, neural-network or partial-least-squares methods).

Pauling, Linus – a chemist and biochemist, known principally for his research into the nature of the chemical bond, for which he was awarded the Nobel Prize in Chemistry in 1954; Pauling also won the Nobel Peace Prize in 1962.

PDB – Protein Data Bank, a database of 3D structural data of biological macromolecules (including proteins, nucleic acids, carbohydrates and protein-nucleic acid complexes), largely derived by means of X-ray diffraction, NMR spectroscopy or electron microscopy, and maintained by the RCSB; the PDB is overseen by the wwPDB.

PDBe – Protein Data Bank Europe, the European repository of biological macromolecular structures, maintained by EMBL-EBI; PDBe is overseen by the wwPDB.

PDBj – Protein Data Bank Japan, the Japanese repository of biological macromolecular structures, maintained by the National Bioscience Database Center of Japan and Osaka University; PDBj is overseen by the wwPDB.

PDF – Portable Document Format, a file format that specifies document layout, including text, fonts, graphics and so on, in a way that's software/hardware and operating-system independent.

PDZ domain – a structural domain found in signalling proteins that helps to anchor TM proteins to the cytoskeleton and hold signalling complexes together by binding to specific, short C-terminal regions of their target proteins.

Peer to Peer (P2P) – see P2P.

pepsin – a digestive protease enzyme, formed from pepsinogen; found in the gastric juice of vertebrates, it acts to cleave peptide bonds preferably linked to aromatic amino acids, tryptophan, tyrosine and phenlyalanine.

peptide – a compound containing two or more amino acid residues linked via peptide bonds, formed between the carboxyl group of one amino acid and the amino group of the next, with the elimination of water.

peptide bond – a covalent chemical bond (-C(=O)NH) formed between the carboxyl group of one molecular entity and the amino group of the next, with the elimination of water; the peptide bond has partial double-bond character, rendering it planar – polypeptides are chains of amino acid residues joined by peptide bonds, the torsional rotations about which lend proteins their characteristic secondary and tertiary structures.

Perl – a general-purpose scripting language with in-built regular-expression functionality.

personalised therapy – see personalised medicine.

personalised medicine – a proposed customised approach to health-care in which decisions and practices are tailored to individual patients, according to their genetic, proteomic and metabolomic profiles.

Perutz, Max – a molecular biologist known for his pioneering work in determining the first structures of proteins (notably, haemoglobin) at atomic resolution, for which he was jointly awarded the Nobel Prize in Chemistry with John Kendrew in 1962.

Pfam – a database that exploits HMMs to classify protein sequences into families and domains.

pH – a measure of the activity of solvated hydrogen ions; the pH scale, which relates to an agreed standard set of solutions, allows relative measurement of solution pH: pure water has a pH of ~7 at 25°C; solutions with pH <7 are acidic; those with pH >7 are basic.

pharmacology – the study of the characteristics, biological interactions and biological effects on normal or abnormal biochemical function of drugs or chemicals, and their modes of action.

Phe – the IUPAC three-letter abbreviation for phenylalanine.

phenotype – the full complement of observable functional and structural characteristics of an organism as determined by the interaction of its genotype with the environment in which it exists.

phenylalanine – 2-amino-3-phenylpropanoic acid, a chiral amino acid found in proteins.

pheromone – a substance that, when released by an organism, stimulates social responses in other organisms (usually of the same or similar species).

phosphate – the trivalent anion, PO_4^{3-}, derived from phosphoric acid.

photosynthesis – the synthesis by plants, algae and bacteria of organic compounds (particularly carbohydrates) from carbon dioxide using light energy derived from the sun.

Phred base-calling – the process of identifying a nucleotide base, from a fluorescence trace generated by an automated DNA sequencer, by means of the Phred base-calling software.

phylogenetic tree – a graphical diagram, in the form of a branching tree, that represents inferred evolutionary relationships among organisms, species or other entities according to similarities and differences in their physical or genetic characteristics; trees are usually constructed from multiple sequence alignments.

physicochemical property – a property that relates to the physical chemistry of a molecule, such as its mass, volume, pH, pI, hydropathy, surface charge, *etc.*

pI – the pH of a solution in which a molecular entity has zero mobility in an electric field; hence, at which it has zero net charge.

Pingala – an Indian mathematician who presented the first known description of a binary numeral system in his Sanskrit treatise called Chandahsastra.

PIR – Protein Information Resource, an integrated bioinformatics resource that supports genomic and proteomic research, including databases such as the PIR-PSD and PIRSF.

PIR-Protein Sequence Database (PIR–PSD) – see PIR–PSD.

PIR–PSD – PIR–Protein Sequence Database, a database of protein sequences that grew out of the *Atlas of Protein Sequence and Structure*, maintained collaboratively by PIR-International, a consortium including the NBRF, MIPS and JIPID.

PIRSF – a database that exploits clustering and HMM-based approaches to classify protein sequences into families and subfamilies.

point mutation – a mutation that involves the exchange of a single nucleotide base for another within DNA or RNA.

poly(A) tail – a sequence of (40–250) adenine bases located at the 3′ terminus of eukaryotic mRNA; polyadenylation is part of the process that's responsible for the production of mature mRNA.

polypeptide – a peptide molecule containing >10–20 amino acid residues joined by peptide bonds; proteins may contain one or more polypeptide chains (as many proteins contain only a single polypeptide chain, the term is sometimes used as a synonym for protein).

porcine – of, or relating to, a pig.

Portable Document Format (PDF) – see PDF.

Position-Specific Scoring Matrix (PSSM) – see PSSM.

position-specific weight matrix – see Position-Specific Scoring Matrix.

Post-Translational Modification (PTM) – see PTM.

PMC – PubMed Central, an archive of free, full-text biomedical literature hosted by the National Intitutes of Health's National Library of Medicine.

PRALINE – an iterative, consistency-based multiple protein sequence alignment tool based on a progressive alignment algorithm.

PRANK – a probabilistic, phylogeny aware multiple sequence alignment tool based on a progressive alignment algorithm.

precision – see selectivity.

primary structure – the first order of complexity of structural organisation in macro-molecules or biopolymers, referring to the sequence of monomeric units without regard to their spatial arrangement: for nucleic acids, it is the sequence of nucleotide bases; for proteins, it is the sequence of amino acid residues.

prime number – a natural number greater than 1 that can only be divided by 1 and itself without leaving a remainder.

PRINTS – a database that uses a fine-grained fingerprint (multiple-motif) approach to classify protein sequences into families and subfamilies.

prion – a membrane protein encoded in the mammalian genome that is rich in α-helices; in its misfolded state, it switches to a conformation that comprises β-sheets, which aggregate as rod-like deposits in the brain and other neural tissues; prions mediate a number of neurodegenerative diseases, and are believed to 'propagate' via this mis-folded state, causing tissue damage and cell death when and where it accumulates.

Pro – the IUPAC three-letter abbreviation for proline.

ProbCons – a probabilistic, consistency-based multiple protein sequence alignment tool based on a progressive alignment algorithm.

processing – in software engineering, the manipulation of modelled data according to the rules of the model; this might be simple mathematical processing, or powered by complex logic.

ProDom – a database that uses a recursive PSI–BLAST approach to classify protein sequences into domains.

profile – a form of PSSM, a table of position-specific amino acid weights and gap costs used to encapsulate the degree of conservation in a sequence alignment, or part of an alignment (such as in a motif or domain); the scores allow quantification of the degree of similarity between a query or database sequence and the aligned sequence profile.

Profiles database – see PROSITE.

program – a set of instructions for controlling the behaviour of a computer, particularly to allow a computer to manipulate data in a specified way, or to express an algorithm in a precise way.

programming language – a technical language in which computer programs are written: different languages are used for different applications – *e.g.,* there are low-level languages (like machine code), higher level languages (like Fortran and C), high-level scripting languages (like Perl and Python), and object-oriented languages (like Java and C++).

prokaryote – an organism whose genetic material is not bounded by a nuclear envelope within the cell (*i.e.,* prokaryotic cells lack a nucleus): bacteria, cyanobacteria and archaea are prokaryotes.

proline – pyrrolidine-2-carboxylic acid, a cyclic, chiral amino acid found in proteins.

promoter – a DNA sequence positioned 5′ to a gene that denotes the initiation site for transcription of that gene.

ProRule – a database of manually-derived rules that's used to augment the PROSITE database, increasing the potency of its discriminators by providing supplementary information about structurally or functionally important amino acids.

PROSITE – a database that exploits both consensus expressions and profiles to classify protein sequences into families and domains; the database also includes expressions designed to characterise non-family-specific functional sites.

prostaglandin – a member of a group of lipid compounds, derived enzymatically from fatty acids, that mediates strong physiological effects, such as the regulation of contraction and relaxation of smooth muscle tissue.

prostanoid – a class of compounds that includes prostaglandins and thromboxanes, which play important physiological roles as mediators of inflammatory reactions, regulators of muscle contraction, and so on.

protein – organic compound that contains one or more polypeptide chains, linear polymers of amino acids linked by peptide bonds; existing in globular or membrane-bound forms, proteins participate in virtually all cellular processes.

Protein Data Bank (PDB) – see PDB.

Protein Data Bank Europe (PDBe) – see PDBe.

Protein Data Bank Japan (PDBj) – see PDBj.

protein family – or gene family, a set of proteins (or gene products) that share a common evolutionary history (*i.e.*, are homologous), and typically exhibit similar (or related) sequences, 3D structures and functions.

protein family database – usually, a database that classifies proteins into protein or gene families, and domain families, using one or more types of diagnostic signature (*e.g.*, such as consensus expressions, fingerprints, profiles and HMMs).

protein fold – the 3D conformation of a polypeptide chain or protein; it is often used to denote a protein structural building-block or domain – commonly occurring domains have been termed super-folds.

protein folding – the exploratory process by which a linear polypeptide chain adopts its 3D conformation as it's synthesised at the ribosome.

Protein Information Resource (PIR) – see PIR.

protein moonlighting – the phenomenon by which a single protein may have multiple context-dependent functions or roles within an organism.

protein sequence – the primary structure of a polypeptide chain or protein, the linear sequence of amino acid residues, which runs from the N terminus (containing a free amino group) to the C terminus (containing a free carboxyl group).

protein sequencing – the process by which the amino acid sequence of a protein is determined, using techniques such as mass spectrometry or Edman degradation.

Protein Structure Initiative (PSI) – see PSI.

proteomics – the large-scale experimental analysis of the protein complement of an organism or system, including its post-translational modifications.

provenance – meta-data relating to the ownership, source and originator of a data-set, including how it has been used and modified.

pseudocode – informally written instructions that describe the workings of an algorithm.

PSI – Protein Structure Initiative, a major federal, university and industry effort that commenced in 2000, which aimed to reduce the cost and decrease the time taken to determine protein 3D structures by employing high-throughput techniques.

PSI–BLAST – an implementation of BLAST that creates a profile from sequences returned above a specific scoring-threshold from an initial BLAST search, which is used for subsequent iterative database searches; a kind of hybrid pairwise-alignment/profile approach, it has greater sensitivity than BLAST alone.

PSSM – Position-Specific Scoring Matrix, a scoring table that encapsulates the residue information within sets of aligned sequences – the derived scores may be a simple measure of the observed residue frequencies at particular positions in the alignment, or may result from more complex mathematical treatments, denoting, for example, the log-odds scores for finding a particular amino acid in a sequence at a particular position.

psychology – the study of mental functions and their role in individual and social behaviours, aiming to understand the physiological and neurobiological processes that underpin them.

PTM – Post-Translational Modification, the chemical modification of a protein following its translation from mRNA: such modifications include phosphorylation (addition of a phosphate group), glycosylation (addition of a glycosyl group), myristoylation (attachment of myristate), amidation (amide bond formation), glycation (sugar attachment), *etc.*

PubMed – an online database and information-retrieval system that provides access to the MEDLINE database of references and abstracts of life-science and biomedical articles.

PubMed Central (PMC) – see PMC.

punched card – a card containing digital information as designated by the presence or absence of holes in defined locations; in the 19th century, such cards were used to control textile looms; in the 20th century, they were used by digital computers to input programs and data.

purine – a heterocyclic, aromatic, organic compound containing fused imidazole and pyrimidine rings; adenine and guanine are purines, which constitute one of the two classes of nucleotide base found in DNA and RNA.

p-value – in a database search, the probability of a match occurring with a score greater than or equal to that of the retrieved match, relative to the expected distribution of scores from comparing random sequences of the same length and composition as the query to the database (the closer the value to 0, the more significant the match).

P vs NP problem – an unsolved problem in the theory of computation that relates the effectively computable (P) to the effectively verifiable (NP).

pyramid – a large masonry structure, with outer triangular faces rising from a square or polygonal base and converging to a single point at the apex; the Egyptian pyramids were built as tombs for the country's Pharaohs and their consorts.

pyrimidine – a six-membered heterocyclic, aromatic, organic compound containing nitrogen atoms at ring positions 1 and 3; cytosine, thymine and uracil are pyrimidines, which constitute one of the two classes of nucleotide base found in DNA and RNA.

Python – a popular interpreted object-oriented programming language used in a wide range of applications.

Q

Q – the IUPAC single-letter abbreviation for glutamine.

qualified identifier – an identifier that has a specific scope by virtue of being associated with a given namespace.

quaternary structure – the fourth order of complexity of structural organisation in multi-subunit macromolecules, referring to the arrangement in space of the individual subunits, and the ensemble of the inter-subunit contacts and interactions (irrespective of their internal geometry): *e.g.*, such as the arrangement of polypeptide chains in a multi-subunit protein molecule.

R

R – the IUPAC single-letter abbreviation for argininine.

random access – also known as direct access, the ability to directly access elements in a sequence (or values from memory) given their coordinates.

RCSB – Research Collaboratory for Structural Biology, a research group hosted at Rutgers, and home of the Protein Data Bank (PDB).

RDF – Resource Description Framework, a set of specifications for modelling information that is implemented as Web-based resources.

reasoner – a computer program used to check the validity of an ontology, or to infer new axioms from given ones.

recall – see sensitivity.

reductionist – an approach to understanding complex entities by reducing them to simpler, more fundamental and hence more tractable components.

regular expression – a regular language that allows us to match patterns of text, including repetitive and optional matching, but not recursive matching.

relational database – a database in which data and their attributes are stored in formally described tables from which information can be readily retrieved; they are the predominant choice for data storage.

release factor – a protein that participates in the release of a newly formed polypeptide chain from a ribosome, recognising the stop codon in an mRNA and halting translation.

Remote Procedure Call (RPC) – see RPC.

Research Collaboratory for Structural Biology (RCSB) – see RCSB.

resolution – the extent to which an imaging system can distinguish closely juxtaposed objects as separate entities; hence, the ability of the imaging system to discern detail in the entity being imaged.

resolving – in the context of computer science, the act of finding an object referred to by an identifier. In the case of opaque identifiers, this requires the object/identifier pairings to be looked up in some form of database; for non-opaque identifiers this can be achieved algorithmically.

Resource Description Framework (RDF) – see RDF.

reusability – in the context of computer science, the degree to which a callable piece of code – such as a function, or a Web service – can be used in multiple contexts, thereby reducing the amount of new code required to solve a given problem; reusability is one of the 12 (original) properties of a Research Object.

rhodopsin – a light-sensitive visual pigment found in the rod-shaped photoreceptor cells of the retinas of most vertebrates; it is an achromatic receptor, mediating vision in dim light, and has a characteristic 7TM architecture – rhodopsin is a member of the rhodopsin-like GPCR superfamily.

rhodopsin-like receptor – name given to a widespread superfamily of membrane proteins that includes hormone, neurotransmitter and light receptors; it is the largest group belonging to the GPCR clan.

ribonuclease – a nuclease enzyme that catalyses the degradation of RNA by cleaving phosphodiester bonds.

ribonucleic acid (RNA) – see RNA.

ribose – a simple sugar, or monosaccharide, with formula $C_5H_{10}O_5$, that constitutes the backbone of RNA.

ribosome – a large intracellular macromolecule, formed from protein and RNA, that mediates protein biosynthesis; ribosomes are molecular 'machines' that orchestrate the translation of mRNA via tRNAs charged with their cognate amino acids.

ribosomal RNA (rRNA) – see rRNA.

RNA – ribonucleic acid, one of two main types of nucleic acid, consisting of a long, unbranched macromolecule formed from ribonucleotides, the 3′-phosphate groups of all but the last being joined in 3′,5′-phosphodiester linkage to the 5′-hydroxyl group of the next; most RNA molecules are single-stranded – the principal RNA types are messenger RNA (mRNA), ribosomal RNA (rRNA) and transfer RNA (tRNA).

RNA polymerase – a ubiquitous enzyme, found in all organisms (including some viruses) that synthesises RNA from a DNA or RNA template; the process of generating RNA from DNA is termed transcription.

Roberts, Richard – a biochemist and molecular biologist known for his work on the discovery of introns and the mechanism of RNA splicing, for which he was jointly awarded the Nobel Prize in Physiology or Medicine with Phillip Sharp in 1993.

Rockefeller University – a private university, based in New York City in the USA, that largely hosts research in the biological and medical sciences.

rod photoreceptor – a light-sensitive, rod-shaped structure of the vertebrate retina that contains a photopigment (rhodopsin) responsible for vision in dim light.

RPC – Remote Procedure Call, the notion of calling functions (procedures) that are to be executed somewhere other than on the local machine.

rRNA – ribosomal RNA, the RNA constituent of ribosomes; rRNAs form two subunits, large and small, which each contain RNAs of different sizes.

rubredoxin – a low-molecular-weight, iron-containing protein found in sulphur-metabolising bacteria and archaea; small and highly conserved, it participates in electron transfer in biological systems.

Ruby – an interpreted, object-oriented programming language that has support for security built directly into the language, allowing data from untrusted sources to be tainted and hence precluded from processing by certain sensitive system functions.

Rutgers – the State University of New Jersey, the largest higher education institution in New Jersey, USA.

S

S – the IUPAC single-letter abbreviation for serine.

saccharide – see sugar.

Saccharomyces cerevisiae – baker's or budding yeast, a species of yeast exploited widely in baking, brewing and wine-making, and used extensively as a model eukaryotic organism in genetics, and in molecular and cell biology.

Saccharomyces Genome Database (SGD) – see SGD.

salt – a compound formed by the reaction of an acid and a base.

San Diego Supercomputer Center (SDSC) – see SDSC.

Sanger, Fred – biochemist, known for his work on the structure of proteins (notably that of insulin) and for determining the base sequences of nucleic acids, for each of which contributions he was awarded Nobel Prizes in Chemistry, in 1958 and 1980 respectively.

Sanger method – the dideoxy chain-termination method for DNA sequencing, a technique that allowed long stretches of DNA to be sequenced rapidly and accurately; initially, the method was used to sequence human mitochondrial DNA and bacteriophage λ, and went on to be used to sequence the complete human genome.

scientific visualisation – an interdisciplinary branch of computer science concerned with visualisation of 3D phenomena, focusing on realistic renderings of volumes, surfaces, illumination sources, and so on, to allow scientists to better understand and illustrate their data.

SCOP – Structural Classification of Proteins, a protein-fold classification database maintained at the Laboratory of Molecular Biology, Cambridge, UK; it is a largely manual classification of protein domains based on observed similarities of their structures and sequences.

scoring matrix – see Position-Specific Scoring Matrix.

Screen-Oriented Multiple Alignment Procedure (SOMAP) – see SOMAP.

SDSC – San Diego Supercomputer Center, a research unit of the University of California, San Diego, with a mission to develop and use technology to advance science; it is part of the RCSB.

sea lamprey – a parasitic lamprey that inhabits the Atlantic coasts of Europe and North America, the western Mediterranean Sea, and the Great Lakes; it attaches itself to the skin of its prey, and scrapes away the tissues while secreting substances to prevent the prey's blood from clotting.

SEAVIEW – a multi-platform, graphical user interface for multiple sequence alignment and molecular phylogeny.

secondary structure – the second order of complexity of structural organisation in macromolecules, referring to the local, regular organisation of the molecular backbone stabilised by hydrogen bonds: in proteins, common forms of secondary structure include α-helices, β-sheets and β-turns; in nucleic acids, familiar forms of secondary structure are found in the helical stem (resulting from base-pairing interactions) and in the loop structures of the tRNA clover-leaf.

Second World War – World War II, a global war that raged between 1939 and 1945, the most far-reaching war in history, involving most of the world's nations.

secretin – an intestinal hormone that regulates duodenal secretions and water homeostasis.

secretin-like receptor – name given to a superfamily of membrane proteins that includes receptors for secretin, calcitonin and parathyroid hormone peptides; secretin-like receptors belong to the GPCR clan.

seed alignment – a preliminary alignment of representative sequences used to initiate an iterative database search; the seed aims to represent the full breadth of variation exhibited by a particular protein or domain family.

selectivity – or precision, the proportion of all retrieved matches that belong to the true data-set, as given by the formula (number of true positives)/(number of true positives + number of false positives).

semantic integration – the process of integrating and organising information from heterogeneous sources using ontological rules to formalise the concepts they embody and the relationships between them.

semantics – in the context of computer science, the process of associating specific meaning with given symbols or identifiers.

sensitivity – or recall, the proportion of the true data-set that is retrieved, as given by the formula (number of true positives)/(number of true positives + number of false negatives).

sequence alignment – see multiple alignment, pairwise alignment.

sequence analysis – the process, or processes (including, for example, sequence alignment and pattern recognition), by which biomolecular sequences (protein, DNA, RNA) are studied in order to shed light on their structural, functional and/or evolutionary features.

sequence pattern – see consensus expression.

Sequence Read Archive (SRA) – see SRA.

sequential execution – carrying out tasks sequentially, one after the other, not moving on to the next task until the previous task is complete; this might be because the tasks rely on the results of their predecessors, or because only one task can be executed at any one time.

Ser – the IUPAC single-letter abbreviation for serine.

serine – 2-amino-3-hydroxypropanoic acid, a chiral amino acid found in proteins.

serine protease – an enzyme that uses a catalytic triad, comprising serine, histidine and aspartic acid, to cleave peptide bonds in proteins.

server – a component in a distributed system that accepts and executes functionality at the request of a client.

SGD – Saccharomyces Genome Database, a repository concerned with the molecular biology and genetics of the baker's or budding yeast, *Saccharomyces cerevisiae*, providing access to its complete genome, its genes and their products.

SH2 domain – a structural domain found in a variety of intracellular signalling proteins that facilitates binding to other proteins by recognising specific, phosphorylated tyrosine residues in its binding partners.

SH3 domain – a structural domain found in a variety of intracellular signalling proteins that mediates assembly of specific protein complexes, usually by binding to proline-rich peptides in its binding partners.

Sharp, Phillip – a geneticist and molecular biologist known for his work on the discovery of introns and the mechanism of RNA splicing, for which he was jointly awarded the Nobel Prize in Physiology or Medicine with Richard Roberts in 1993.

Shine–Dalgarno sequence – a specific ribosomal binding site in mRNA that mediates the recruitment of ribosomes to initiate protein synthesis.

SIB – Swiss Institute of Bioinformatics, a not-for-profit foundation that federates bioinformatics activities across Switzerland, dedicated to the provision of bioinformatics resources to the life-science research community, based in Geneva, Lausanne, Basel, Bern and Zurich, Switzerland.

side-chain – a chemical constituent attached to the α-carbon of an amino acid; side-chains are the unique components of amino acid molecules.

side-effect – a secondary effect of a medicine or drug; it may be either an adverse reaction or an unintended beneficial outcome.

signal peptide – historically, this term denoted the usually N-terminal (but sometimes C-terminal) 20-30 hydrophobic residues of secretory proteins, subsequently cleaved from the nascent protein; the term is now also used more generally to denote any stretch of amino acids in newly synthesised proteins that either directs proteins to particular organelles (the Golgi apparatus, endoplasmic reticulum, *etc.*) in eukaryotic cells, or from the cytoplasm to the periplasmic space of prokaryotic cells.

signal transduction – the process by which an extracellular signal (whether chemical, electrical or mechanical) is transduced into a cellular response (*e.g.*, the interaction of a hormone with a cell-surface GPCR, leading to signal amplification via downstream biochemical cascades within the cell).

silicon chip – an electronic circuit made by means of lithography.

Simple Knowledge Organisation System (SKOS) – see SKOS.

Simple Modular Architecture Research Tool (SMART) – see SMART.

Simple Object Access Protocol (SOAP) – see SOAP.

SKOS – Simple Knowledge Organisation System, a method used for representing thesauri or taxonomies.

small molecule – in biochemistry, a low-molecular-weight organic compound (not a polymer); in pharmacology, a molecule that binds with high affinity to a biopolymer, thereby altering its activity or function.

SMART – Simple Modular Architecture Research Tool, a repository for identification and analysis of protein domains, maintained at EMBL, Heidelberg, Germany.

SOAP – originally, Simple Object Access Protocol, but now just a name in its own right, SOAP specifies a method of exchanging data between components in a Web service environment.

social networking – in computing, a system or platform (usually Web-based) that represents users and their interactions, often based around a particular interest or activity.

sociology – a social science concerned with the study of society, using empirical investigation and critical analysis to understand human social activity.

software – a set of programs, procedures and algorithms that tell a computer what tasks to perform and how to perform them.

solar system – the sun, its planetary system and all the other objects gravitationally bound in orbit around it.

SOMAP – Screen-Oriented Multiple Alignment Procedure, a versatile tool for interactive editing of multiple sequence alignments, the first to provide a 'menus and windows' interface for VT100 terminals.

space complexity – how an algorithm behaves in terms of the space it requires as a function of the amount of data it must process.

species – broadly, a taxonomic category, a fundamental unit of biological classification, consisting of a group of closely related individuals that are capable of interbreeding to produce fertile offspring.

specificity – or precision, a measure of the proportion of true-negatives that are correctly identified as such, as given by the formula (number of true negatives)/(number of true negatives + number of false positives).

sperm whale – a marine mammal, a toothed whale with the largest brain of any animal (its name derives from the white, waxy substance found in the animal's head cavities, which it uses to control buoyancy).

splice form – or splice variant, a protein isoform that arises from translation of an mRNA following the connection of exons in a non-canonical form – most commonly, by skipping one or more of the available exons, leading to a shorter-than-normal protein product.

spliceosome – a macromolecular complex, formed from protein and small nuclear RNAs, that's assembled during splicing of the mRNA primary transcript to remove introns.

splice site – the sequence of bases at the termini of introns that specifies the point of splicing; the consensus sequence for the 5′ splice site is GU, while the consensus for the 3′ splice cite is AG.

splicing – in eukaryotes, the process, catalysed by the spliceosome, by which introns are removed from the primary RNA transcript and the exons are joined to form an mRNA suitable for translation.

SRA – Sequence Read Archive, a database of short reads generated from high-throughput sequencing projects, maintained collaboratively by members of the INSDC.

Src-like kinase – a member of the Src family of protein tyrosine kinases (enzymes that transfer phosphate groups from ATP to cellular proteins); Src-like kinases demonstrate a wide variety of functions, including roles in signal transduction pathways.

5S rRNA – a class of rRNA found in the large ribosomal subunits of prokaryotes and eukaryotes.

start codon – the first codon of an mRNA transcript to be translated by the ribosome, most commonly AUG – in eukaryotes, it encodes methionine; in prokaryotes, it encodes N-formyl-methionine.

Star Trek – a science fiction TV series that charted the intergalactic fortunes of the captain (James T. Kirk) and crew of a 23rd-century exploration vessel, the Starship Enterprise.

statelessness – a stateless system or function does not retain any state between invocations, so that previous history does not change future behaviour.

statistics – broadly, the study of the collection, analysis and interpretation of numerical data.

stop codon – an mRNA codon (which may be UAG, UAA or UGA) that specifically signifies the end of translation; unlike the start codon, it doesn't also specify an amino acid residue.

Structural Classification Of Proteins (SCOP) – see SCOP.

subtilisin – a broad-specificity bacterial serine protease that hydrolyses peptide bonds.

sugar – a generalised name for carbohydrate; sugars exist as monsaccharides, disaccharides, oligosaccharides and polysaccharides.

supercomputer – a (usually custom-built) computer that boasts computational speed, memory and storage space close to the current limits of what is technologically possible.

SUPERFAMILY – a database that exploits HMMs to classify protein sequences into structural domains, based on the SCOP superfamily level.

super-fold – a ubiquitous fold found in non-homologous proteins, such as the globin fold, the Jelly roll, the β-trefoil, the β-barrel.

super-secondary structure – or, sometimes, super-fold or folding motif, the arrangement of protein secondary structures into discrete folded units (*e.g.*, β-barrels, β–α–β-units, Greek keys, helix-turn-helix motifs, four-helix bundles), which may form components of larger, tertiary structures.

Swiss Institute of Bioinformatics (SIB) – see SIB.

Swiss-Prot – a non-redundant database that contains (largely) manually-annotated protein sequences, with annotation culled from the scientific literature and curator-evaluated sequence analysis.

syndrome – a set of concurrent symptoms, signs, phenomena or characteristics that are indicative of a particular pathological state.

syntax – the rules that govern the structure of sentences or statements in natural or computer languages.

systems biology – an interdisciplinary field of biology that adopts holistic rather than reductionist approaches, embracing complex interactions within complete biological systems.

T

T – the IUPAC single-letter abbreviation for threonine, or the pyrimidine base thymine.

TAIR – The Arabidopsis Information Resource, host of the database of genetic and molecular biology data for the model higher plant, *Arabidopsis thaliana*.

taste receptor – a type of receptor that mediates the sensation of taste; taste receptors are members of the GPCR clan.

taxonomy – a controlled vocabulary with particular semantics associated with the terms, arranged in a hierarchy.

T-Coffee – Tree-based Consistency Objective Function For alignment Evaluation, a multiple sequence alignment tool based on a progressive alignment algorithm.

template strand – or non-coding strand, in the context of DNA transcription, the strand of DNA that's used as a template for RNA synthesis; it has a base sequence that is complementary to the RNA transcript (but with thymine, not uracil); its complement is the non-template or coding strand.

tertiary structure – the third order of complexity of structural organisation in macromolecules, referring to the arrangement in space of the molecule as a result of its intramolecular contacts and interactions – in multi-subunit macromolecules, such as certain proteins (*e.g.*, haemoglobin), the tertiary structure refers to the 3D fold of each individual polypeptide chain.

text mining – broadly, the process by which specific information is retrieved and extracted computationally from unstructured, natural-language text.

The Arabidopsis Information Resource (TAIR) – see TAIR.

The Atomic City – a suspense film from 1952, set in Los Alamos, New Mexico, involving a terrorist plot to acquire the formula for the hydrogen bomb.

theory of computation – the branch of mathematics and computer science that deals with algorithmic problem-solving.

third-generation sequencing (3Gen) – see 3Gen.

Thr – the IUPAC three-letter abbreviation for threonine.

threonine – (2*R**,3*S**)-2-amino-3-hydroxybutanoic acid, a chiral amino acid found in proteins.

thrombin – a serine protease formed from prothrombin that plays a key role in blood coagulation.

thromboxane – a member of a group of lipid compounds that plays a role in clot formation.

thymine – 2,4-dihydroxy-5-methylpyrimidine, a pyrimidine derivative that's one of the five principal bases of nucleic acids.

TIGRFAMs – a collection of diagnostic protein family signatures that uses HMMs to identify functionally-related proteins.

time complexity – how an algorithm behaves in terms of the time it takes as a function of the amount of data it must process.

tinker – one who adapts and adjusts something in the course of making repairs or improvements.

TM domain – transmembrane domain, generally, an α-helical span of around 20 amino acids that crosses a membrane; transmembrane domains are usually hydrophobic in character.

topology – in the CATH database, the fold in the core of a protein domain, taking into account the overall shape and connectivity of its secondary structures.

torsion angle – see dihedral angle.

toxicity – the degree to which a substance or agent is poisonous or harmful to parts of an organism (cells, organs, *etc.*) or to a whole organism; toxic effects are generally dose-dependent.

transcription – the process of converting or transcribing the chemical code of one nucleic acid molecule into the complementary code of another, such as in the synthesis of

RNA from a DNA template, or DNA from an RNA template; in transcription of DNA to RNA, the RNA complement incorporates uracil in all locations where thymine would have occurred in a DNA complement.

TRANScription FACtor database (TRANSFAC) – see TRANSFAC.

transcriptomics – the study of transcriptomes, and in particular how patterns of transcription are influenced by development, disease, environmental factors, *etc.*

TRANSFAC – TRANScription FACtor database, an annotated database of eukaryotic transcription factors, including their genomic binding sites and DNA-binding profiles.

transfer RNA (tRNA) – see tRNA.

transistor – a basic building-block of modern electronic devices; transistors are semiconductor devices that are used to amplify and switch electronic signals and electrical power.

translation – the process of converting or translating the chemical code of nucleic acid molecules (*i.e.*, the specific sequence of codons in mRNA) into the chemical code of polypeptides (*i.e.*, the corresponding sequence of amino acids in a protein) in the third stage of protein biosynthesis.

Transmembrane domain (TM domain) – see TM domain.

Travelling Salesman Problem (TSP) – see TSP.

Tree-based Consistency Objective Function For alignment Evaluation (T-Coffee) – see T-Coffee.

TrEMBL – a database that contains computer-annotated protein sequences created by translation of annotated coding sequences in EMBL-Bank; TrEMBL was created to provide swifter access to genomic data than could be achieved using manual annotation methods.

tRNA – transfer RNA, a small adaptor RNA molecule that mediates the synthesis of proteins, by carrying specific amino acids to the appropriate codon specified by an mRNA template.

Trp – the IUPAC three-letter abbreviation for tryptophan.

true negative – in the context of database searching, an entity that does not belong to the true data-set.

true-positive match – in the context of database searching, a match to the conditions of a search that is a true member of the data-set.

trypsin – a digestive enzyme, a serine protease, formed from inactive trypsinogen; found in the pancreatic juice of vertebrates, it acts in the small intestine to hydrolyse protein.

tryptophan – 2-amino-3-(1-H-indol-3-yl)propanoic acid, a chiral amino acid found in proteins.

TSP – Travelling Salesman Problem, the problem faced by a travelling salesman who wishes to efficiently visit each of a set of cities connected by roads of different lengths; the TSP aims to find the shortest path around the network that visits every single node.

Tubby protein – a protein involved in cell signalling, common to multi-cellular eukaryotes, mutations in which have been shown to give rise to delayed-onset obesity.

Twilight Zone – a region of sequence identity (~15% to 25%) in which alignment scores are not statistically significant; in the absence of statistically significant similarity, homology inferences depend on collateral structural and/or functional information.

Tyr – the IUPAC three-letter abbreviation for tyrosine.

tyrosine – 2-amino-3-(4-hydroxyphenyl)propanoic acid, a chiral amino acid found in proteins.

U

ungulate – broadly, a hoofed animal.

Unicode – a comprehensive scheme for encoding text from most international languages and writing systems.

Uniform Resource Identifier (URI) – see URI.

Uniform Resource Locator (URL) – see URL.

Uniform Resource Name (URN) – see URN.

UniMES – UniProt Metagenomic and Environmental Sequences, a database of metagenomic and environmental data.

UniParc – UniProt Archive, a comprehensive, non-redundant database containing sequences from all the main, publicly available protein sequence databases.

UniProt – a comprehensive resource for protein sequences and their annotation.

UniProt Archive (UniParc) – see UniParc.

UniProt Consortium – the collaborating organisations responsible for the maintenance of UniProt: the EBI, SIB and PIR.

UniProtKB – UniProt Knowledgebase, a protein sequence database comprising the (mostly) manually annotated UniProtKB/Swiss-Prot and the fully automatically annotated UniProtKB/TrEMBL databases.

UniProt Knowledgebase (UniProtKB) – see UniProtKB.

UniProt Metagenomic and Environmental Sequences (UniMES) – see UniMES.

UniProt Reference Clusters (UniRef) – see UniRef.

UniRef – UniProt Reference Clusters, a set of three databases (UniRef100, UniRef90 and UniRef50) containing sets of protein sequences from UniProtKB and selected UniParc records clustered at different levels of sequence identity; clustering reduces database size and enables faster sequence searches.

unitary matrix – or identity matrix, a scoring matrix in which all elements score 0 except for those on the diagonal, which score 1 – hence, all identical (self-self) matches carry equal weight.

Universal Protein Resource (UniProt) – see UniProt.

untranslated region (UTR) – see UTR.

uracil – 2,4-dioxopyrimidine, a pyrimidine base that occurs in RNA but not DNA.

URI – Uniform Resource Identifier, a string of characters used to identify a name of a resource, often used in the context of Linked Data or the Web.

URL – Uniform Resource Locator, a form of URI that gives the address of a resource on the Web.

URN – Uniform Resource Name, a form of URI used to identify resources without implying a mechanism to find the resource.

US Department of Defense – the Executive Department of the Government of the USA, responsible for coordinating and over-seeing all government agencies and functions concerned with national security and the US armed forces.

user – the human agent who makes use of a computer system.

Utopia Documents – a PDF reader that uses semantic technologies to connect the static content of PDF documents to dynamic online content.

UTR – untranslated region, part of an mRNA sequence that resides either side of a coding sequence, *i.e.* at its 5′ and 3′ termini.

V

V – the IUPAC single-letter abbreviation for valine.

Val – the IUPAC three-letter abbreviation for valine.

valine – 2-amino-3-methylbutanoic acid, a chiral amino acid found in proteins.

valve – in electronics, a device that controls electric current through a vacuum in a sealed container, often made of glass.

van der Waals radius – half of the inter-nuclear distance between atoms at equilibrium, when long-range attractive forces and short-range repulsive forces are balanced.

variable – a name used to represent a value in a computer program; variables usually assume different values throughout the lifetime of a program, and are used to keep track of the current data state of an algorithm.

Venn diagram – a diagram, usually comprising a number of overlapping circles, showing all logical relations between a finite collection of items or elements.

vertebrate – an animal that possesses a vertebral column (backbone).

virtual machine – in computer science, either an emulator for one computer architecture that runs on another, or the software that translates the commands of a given programming language into the machine code required by the computer's hardware.

virus – a non-cellular infective agent that can reproduce only in a host cell; typically smaller than bacteria, viruses can infect all types of organism, from animals and plants to bacteria and archaea – the infective particle (virion) contains a nucleic acid core (DNA or RNA) surrounded by a protein capsid, sometimes coated with a lipid envelope.

Vitis vinifera – the common grape vine, a species of *Vitis* native to the Mediterranean, central Europe and southwestern Asia.

vomeronasal receptor – a type of receptor that mediates the detection of pheromones; vomeronasal receptos are members of the GPCR clan.

W

W – the IUPAC single-letter abbreviation for tryptophan.

water – a chemical substance that contains one oxygen and two covalently connected hydrogen atoms, H_2O; it is vital for all known forms of life.

Watson–Crick base-pairing – the canonical pairing of DNA bases in which adenine pairs with thymine, and guanine pairs with cytosine.

Watson, James – molecular biologist and geneticist known for proposing the double-helix structure of DNA, for which he was jointly awarded the Nobel Prize in Physiology or Medicine with Francis Crick and Maurice Wilkins in 1962.

Web – see WWW.

Web Ontology Language (OWL) – see OWL.

Web service – the server part of a client/server architecture that uses the infrastructure of the Web to exchange data or request the execution of functions.

Web Service Description Language (WSDL) – see WSDL.

Wellcome Trust Sanger Institute (WTSI) – see WTSI.

western Arabic numerals – the ten digits 0,1,2,3,4,5,6,7,8,9 commonly used to represent numbers.

WGS – Whole Genome Shotgun, a sequencing strategy for whole genomes, in which DNA is randomly broken into segments that are sufficiently small to be sequenced; several rounds of fragmentation and sequencing give rise to multiple overlapping sequence 'reads', which can be assembled computationally into a continuous sequence.

Whole Genome Shotgun (WGS) – see WGS.

Wilkins, Maurice – molecular biologist and physicist known for his work on the structure of DNA, for which he was jointly awarded the Nobel Prize in Physiology or Medicine with Francis Crick and James Watson in 1962.

wireless network – a network that connects computers without the use of cables to make physical connections between them; communication is mediated via radio waves.

wobble – the name given to the tendency of the third base in the anticodon of tRNA to bind with less strict specificity than dictated by canonical Watson-Crick base-pairing.

world-wide Protein Data Bank (wwPDB) – see wwPDB.

World-Wide Web (WWW) – see WWW.

WormBase – a database of *Caenorhabditis* genes and genomes, and the primary repository for genetic and molecular data for the model organism *C. elegans*.

WSDL – Web Service Description Language, an Interface Definition Language (IDL) for Web services.

WTSI – Wellcome Trust Sanger Institute, a genomics and genetics research institute, established as a large-scale DNA-sequencing centre to participate in the Human Genome Project, based on the Wellcome Trust Genome Campus, Hinxton, UK.

wwPDB – the world-wide Protein Data Bank, an international collaboration of macromolecular-structure-data providers whose mission is to maintain a single archive of macromolecular structure data; its partners are the RCSB PDB, PDBe, PDBj and the Biological Magnetic Resonance Data Bank (BMRB).

WWW – the system of interlinked documents, data, images, videos, and so on, accessed via the Internet with browser software.

X

xenology – relatedness that arises as a result of gene transfer between two organisms.

XML – eXtensible Markup Language, a textual format for representing arbitrary data structures for storage or transmission.

X-ray crystallography – a method for deducing atomic arrangements within crystals using X-ray diffraction; commonly used to determine the structures of small molecules, proteins and nucleic acids.

X-ray diffraction – a technique that exploits the diffraction patterns formed when X-rays pass through crystals, or other regular molecular arrays, to measure inter-atomic distances and to determine the 3D arrangement of atoms (or molecules) in a material; used extensively in the determination of the structures of proteins and nucleic acids.

Y

Y – the IUPAC single-letter abbreviation for tyrosine.

Z

Zuckerkandl, Emile – a biologist considered to be one of the pioneers of molecular evolution, and best known for his work with Linus Pauling introducing the concept of the molecular clock.

Quiz Answers

Chapter 1

1 Who first introduced the term bioinformatics?
   C Paulien Hogeweg

2 Who first sequenced a protein?
   A Fred Sanger

3 How long did the determination of the sequence of insulin take?
   D Ten years

4 Which was the first enzyme whose amino acid sequence was determined?
   B Ribonuclease

5 Which was the first protein whose structure was determined?
   C Myoglobin

6 Which of the following statements is true?
   C The first collection of protein sequences was the *Atlas of Protein Sequence and Structure*.

7 Who was responsible for the invention of the Difference Engine?
   B Charles Babbage

8 What is the smallest number that could be represented using the Babylonian counting scheme?
   A One

9 Binary representation was first conceived by:
   B Pingala

10 Which of the following statements is true?
   C There is a gap between the ability of computers to manipulate binary numbers and our desire to use them to manipulate concepts.

Bioinformatics challenges at the interface of biology and computer science: Mind the Gap. First Edition.
Teresa K. Attwood, Stephen R. Pettifer and David Thorne. Published 2016 © 2016 by John Wiley and Sons, Ltd.
Companion website: www.wiley.com/go/attwood/bioinformatics

Chapter 2

1 Which of the following bases are found in RNA but not DNA?
   C Uracil

2 3′ and 5′ describe the termini of which of the following molecular sequences?
   B DNA
   C RNA

3 Which of the following contain sulphur atoms?
   C Methionine

4 Which of the following are not valid amino acid sequences?
   A ATTWOOD
   C THORNE

5 Gln-Trp-Tyr-Lys-Glu-Asn-Asp corresponds to which of the following amino acid sequences?
   D QWYKEND

6 Which two of the following DNA sequences (written in the 5′ to 3′ direction) are complementary to one another?
   1) CTTACGG   **2)** TCGGATT   **3)** GAATGCC   **4)** ACCGTAA
   D 1 and 3

7 Which of the following statements about the standard genetic code is correct?
   C There are three possible stop codons.

8 Which of the following statements about translation is correct?
   A Translation is initiated at a start codon.

9 The primary structure of a protein is:
   C the linear sequence of amino acids in a polypeptide chain.

10 The spliceosome is the molecular machine responsible for:
   C RNA processing.

Chapter 3

1 Which of the following databases house nucleotide sequences?
   B ENA

2 Which of the following databases house diagnostic protein signatures?
   B PRINTS
   C InterPro

3 The PDB, PDBe and PDBj are sources of:
   C macromolecular structure data.

4 UniRef100, UniRef90 and UniRef50 are sources of:
   B protein sequences clustered at different levels of sequence identity.

5 Which of the following statements are correct?
   A Swiss-Prot initially took data from Dayhoff's *Atlas*.
   B Swiss-Prot was inspired by the format of entries in the EMBL data library.
   C Swiss-Prot was built on data in the PIR-PSD.

6 Which of the following statements are correct?
   B None of the above.

7 Signature databases were developed to:
   B help analyse and functionally annotate uncharacterised sequences.
   C classify sequences into evolutionarily related families.

8 Which of the following statements are correct?
   C The sequence/structure deficit refers to over-abundance of protein sequences relative to the number of known protein structures.

9 Which of the following statements are correct?
   B NGS technologies make fully manual annotation strategies unrealistic.

10 Which of the following statements are correct?
   B The goal of 3Gen sequencing technology is to make whole-genome sequencing cheaper and faster.
   C The goal of 3Gen sequencing technology is to make sequencing technology feasible as a diagnostic biomedical and clinical tool.

Chapter 4

1 Which of the following statements are correct?
   A Annotation helps computers to process sequence data.
   B Annotation can be free-text notes.
   C Annotation must be structured in order to be machine-accessible.
   D Annotation helps humans to understand sequence data.

2 A commonly used database flat-file format is:
   C EMBL

3 Which of the following statements are correct?
   C None of the above.

4 The database tag, FT is:
   D the Feature-Table tag describing notable characteristics or features of sequences.

5 Which of the following statements are correct?
   A Unitary matrices are sparse.
   B Unitary matrices score all identical matches equally.
   C The diagnostic power of unitary matrices is poor.
   D Unitary matrices score all non-identical matches equally.

6 Which of the following statements are correct?
   D None of the above.

7   Which of the following statements are false?
    A   A false-negative match is a match returned by a database search in error.
    C   A false-negative match is a true sequence matched by a database search.

8   Which of the following are alignment tools?
    A   FASTA
    B   PRANK
    C   CINEMA
    D   BLAST

9   Which of the following statements are false?
    A   Consensus expressions or patterns are probabilistic models.
    B   HMMs are more selective than profiles.
    C   Fingerprints use PAM 250 scoring.
    D   Profile pattern-matching is binary.

10 Which of the following statements are correct?
    A   InterPro is used for sequence annotation.
    C   InterPro synchronises and updates its component diagnostic methods with
        respect to UniProtKB.
    D   InterPro is diagnostically more powerful than its component methods.

Chapter 5

1   A gene may be considered to be:
    A   a heritable unit corresponding to an observable phenotype.
    B   a packet of genetic information that encodes RNA.
    C   a packet of genetic information that encodes a protein.
    D   a packet of genetic information that encodes multiple proteins.

2   Which of the following statements about sequence database entries is true?
    D   None of the above.

3   Which alignment tool uses evolutionary information to place gaps?
    D   PRANK

4   Which of the following amino acid colour-coding is based on a 'standard'
    property metaphor?
    C   R = Blue

5   Which of these statements is true?
    B   Structure-based alignments use a framework for comparison that differs from
        sequence-based alignments.

6   Which of the following belong to the GRAFS class of receptors?
    C   Glutamate, Rhodopsin, Adhesion, Frizzled, Secretin

7   A protein family may be considered to be a group of sequences that:
    A   share more than 50 per cent similarity.
    B   share the same function.
    C   have the same fold.
    D   share less than 30 per cent identity.

8   Homology is the acquisition of common features as a result of:
    A   the evolutionary process of divergence from a common ancestor.

9   Anfinsen's dogma states that:
    C   the native structure of small globular proteins is determined by the amino acid
        sequence alone.

10 The three ontologies within GO define:
    B   biological process, cellular component, molecular function.

Chapter 6

1   The Halting Problem:
    C   states that it is impossible to tell whether certain algorithms can complete in
        finite time.

2   The bubble sort algorithm:
    A   is polynomial in time, but linear in space.

3   A vector has an access time of:
    C   O(1)

4   A binary tree:
    D   has logarithmic search time.

5   Functional programming languages:
    B   allow only stateless functions.

6   Moore's law:
    C   predicts the period in which the number of components on a chip will
        double.

7   The Travelling Salesman Problem:
    C   is not effectively computable.

8   Java is:
    B   an imperative object-orientated language.

9   Operator short-circuiting:
    C   allows the calculation of an expression to terminate as soon as a certain
        outcome is known.

10 Embarrassingly parallel algorithms:
   D carry out the same task on large amounts of independent data and so are amenable to divide-and-conquer techniques.

Chapter 7

1 An opaque identifier:
   D can't be interpreted by a human or machine without dereferencing it.

2 LSIDs are:
   C both a URN and a URI.

3 XML is:
   A a framework for encoding structured information in text files.

4 SPARQL is used for:
   A extracting content from a graph of RDF triples.

5 A controlled vocabulary:
   C is an agreed set of symbols used in a particular context.

6 A thesaurus defines:
   B relationships between terms.

7 A taxonomy defines:
   B a hierarchical relationship between concepts.

8 Linked Data can be represented as:
   D any of the above.

9 Which of the following is not one of the fallacies of distributed computing?
   B CPU performance is unbounded.

10 HTTP is stateless because:
   A the protocol itself contains no concept of state.

Chapter 8

1 Which of the following statements is true?
   D None of the above.

2 A 'structured abstract':
   C attempts to capture the main content of an article in machine-readable form.

3 In scholarly communication, 'hedging' means:
   B 'softening' claims made in scientific articles.

4 MeSH is:
   D a controlled vocabulary of biological concepts.

5 A Research Object is:
    B a collection of data and metadata intended to encapsulate a scientific idea.

6 Digital resources are sensitive to:
    D all of the above.

7 PubMed central is:
    B a collection of Open Access full-text articles.

8 A DOI can be used to identify:
    D all of the above.

9 'Claim amplification' is the tendency for:
    C ambiguous claims to become concrete facts by repeated citation.

10 As article search engines become more 'semantic', they will tend to return:
    D impossible to tell.

Problems Answers

Chapter 1

1 The single-letter code for amino acids; the PAM scoring matrices.

2 GenBank. 606 sequence entries in December 1982; 97,084 sequence entries in December 1992; >188 million in November 2015.

3 The gap between what we can learn and what is known. It's disturbing because it means we can no longer be experts, even in supposedly narrow fields, because the growing volumes of published information have become impossible for individuals to assimilate.

Chapter 2

1 The genome of *S. alaskensis* encodes 3,165 genes. The genome size of *V. vinifera* is 490Mb. There's likely to be a discrepancy, because the relationship between genome size and number of genes isn't linear in eukaryotes, largely because they can contain significant amounts of sequence that are non-coding.

   This answer requires Table 2.4 to be turned into a graph, and the result to be compared with the online lists of sequenced bacterial genomes[1] and eukaryotic genomes[2].

2 It's harder to deduce the protein products of eukaryotic gene sequences than prokaryotic gene sequences because most eukaryotic coding sequences are interrupted by introns. This is a hard problem to tackle computationally because the signals at intron–exon boundaries are very small and variable, because introns may be extremely large, because the numbers of introns per gene vary – it's therefore impossible to identify coding sequences with the same fidelity as is achieved naturally. When the process goes wrong, gene structures are likely to be built incorrectly, for example by missing exons, by including incorrectly terminated exons, by erroneously placing exons in non-coding regions, *etc.*

---

[1] http://en.wikipedia.org/wiki/List_of_sequenced_prokaryotic_genomes
[2] http://en.wikipedia.org/wiki/List_of_sequenced_eukaryotic_genomes

Bioinformatics challenges at the interface of biology and computer science: Mind the Gap. First Edition.
Teresa K. Attwood, Stephen R. Pettifer and David Thorne. Published 2016 © 2016 by John Wiley and Sons, Ltd.
Companion website: www.wiley.com/go/attwood/bioinformatics

3 The protein sequence of bacteriorhodopsin from *H. salinarium* is:

```
MLELLPTAVEGVSQAQITGRPEWIWLALGTALMGLGTLYFLVKGMGVSDPDAKKFYAITTLVPAIAFT
MYLSMLLGYGLTMVPFGGEQNPIYWARYADWLFTTPLLLLDLALLVDADQGTILALVGADGIMIGTGL
VGALTKVYSYRFVWWAISTAAMLYILYVLFFGFTSKAESMRPEVASTFKVLRNVTVVLWSAYPVVWLI
GSEGAGIVPLNIETLLFMVLDVSAKVGFGLILLRSRAIFGEAEAPEPSAGDGAAATSD
```

The translation of the third exon of human rhodopsin is:

```
YIPEGLQCSCGIDYYTLKPEVNNESFVIYMFVVHFTIPMIIIFFCYGQLVFTVKE
```

And the fifth:

```
FRNCMLTTICCGKNPLGDDEASATVSKTETSQVAPA
```

Automating the process of DNA translation is complicated by the need to identify likely ORFs, which, in prokaryotes, are normally the longest pieces of sequence uninterrupted by stop codons (although, sometimes, ORFs may overlap, and sometimes the shortest may be the correct one). Next is the difficulty of knowing which of the six possible reading frames is the correct one (noting that coding sequences may occur in both forward and reverse frames).

For eukaryotes, the situation is more challenging, because coding sequences may be interrupted by non-coding sequences (introns). Hence, we need to be able to correctly recognise the signals that denote intron–exon boundaries (*e.g.*, GU at the 5' end and AG at the 3' end) – this is hard computationally because the signals at intron–exon boundaries are extremely small, and introns may be very large and very far apart. Deducing the correct reading frame, the correct number of introns and exons, their specific boundaries, and sewing them up into contiguous coding sequences is thus a highly error-prone computational process.

Other issues include the use of the appropriate genetic code, as the 'standard' code isn't quite universal, and the quality of the starting sequence – sequencing errors can lead to incorrectly assigned or missed start and/or stop codons, leading to truncation of a gene, extension of a gene, or missing genes.

To help reduce possible errors, additional evidence that may be used includes intrinsic sequence features, such as the GC-content characteristic of coding sequences, or translation start and stop sites (ribosome binding sites, promoter regions, *etc.*). Most reliable of all is to search a sequence database to see whether the ORF encodes a known protein.

4 According to the boxes, Trp – amphipathic; Phe – hydrophobic; His, Arg and Lys – hydrophilic. There are some discrepancies, *e.g.* for Pro and Cys: Box 2.2 suggests that both Pro and Cys are amphipathic, but in Table 2.3 in Box 2.3, Pro has hydrophilic values and Cys has hydrophobic values. According to the table, Ile appears to be the most hydrophobic amino acid, just ahead of Val; and Arg appears to be the most hydrophilic amino acid, ahead of Asp – these results are summarised in Table 2.8.

The hydropathic rankings suggest that Phe is very marginally more hydrophobic compared to Ile; and that Arg is (arguably) marginally more hydrophilic than Lys. The rankings show Pro as mostly occupying the hydrophilic half of the ranking, except in the Zimmerman scale; while Cys tends to be in the hydrophobic half of the ranking, except in the von Heijne scale. Similarly, the rankings show Lys as mostly occupying the hydrophilic half of the ranking, except in the Zimmerman scale; while

Table 2.8

Summary of results from Box 2.2 and Box 2.3 for each amino acid listed in the problem.

	Hydrophobic	Hydrophilic	Amphipathic
Trp	Most rankings (except Zimmerman); B2.3Hyd2	Zimmerman, B2.3Hyd1	B2.2
Phe	All rankings; B2.2; B2.3Hyd1, B2.3Hyd2		
Pro	Zimmerman	All rankings (except Zimmerman); B2.3Hyd1, B2.3Hyd2	B2.2
His		B2.2; B2.3Hyd1, Box2.3Hyd2	Rankings show half hydrophobic, half hydrophilic
Cys	All rankings (except von Heijne); B2.3Hyd1, B2.3Hyd2	von Heijne	B2.2
Arg		All rankings; B2.2; B2.3Hyd1, B2.3Hyd2	
Lys	Zimmerman	All rankings (except Zimmerman); B2.2; B2.3Hyd1, B2.3Hyd2	

Trp largely falls in the hydrophobic side, except in the Zimmerman scale. His, on the other hand, appears to be ambivalent, occurring equally in both sides of the ranking. The discrepancies result from the overlapping properties of the amino acids (the fact that they have both hydrophobic and hydrophilic substituents), and the ways in which the properties are measured experimentally – different methods give different results, because they are measuring different properties.

Chapter 3

1 AVYSIVCVVGLIGNYLVM doesn't match because it has an I at position 12, but the expression only allows L, V or F. Changing the I to L, V or F gives the following sequences: AVYSIVCVVGLLGNYLVM, AVYSIVCVVGLVGNYLVM and AVYSIVCVVGLFGNYLVM; each of these would match the expression.

The query most resembles the third fingerprint, because the size of the motifs and the distances between them are most like this one, and aren't compatible with the others. Two motifs (3 and 4) are missing.

2 CATH and SCOP are fold-classification databases. Their estimates of the number of unique folds differ because the databases use different methods to classify protein folds: CATH is primarily domain-based, and uses greater degrees of automation than SCOP, which takes a more family-centric view of protein space; their perspectives of the protein fold landscape are therefore different.

The total number of protein structures available from the PDB is the redundant figure of all protein structures submitted to the database (including multiple examples of commonly studied proteins, like haemoglobin, myoglobin, *etc.*). Clustering the database using different levels of sequence identity gets us closer to the number of unique protein families in the database, but even divergent sequences can have similar folds. Comparing the contents against the fold-classification databases reduces the number further, because SCOP and CATH use structure-comparison rather than sequence-comparison approaches (and structure is more conserved than sequence).

3 Non-identical composite databases tend to be larger than non-redundant composite databases because they remove only identical sequences, and retain those that are highly similar (*i.e.*, that are near identical – sequences, for example, with only one amino acid difference, such as a leading methionine).

UniProt and NRDB are widely used composite protein sequence databases: strictly, NRDB is non-identical, while UniProt is non-redundant – specifically, it provides UniRef: in UniRef90 and UniRef50, no pair of sequences in the representative set has >90 per cent or >50 per cent mutual sequence identity; in UniRef100, identical sequences and their sub-fragments are reduced to a single entry.

Both databases include errors: these can be sequencing errors, human errors or systematic computer errors. NRDB is particularly error-prone because it is made in a rather simplistic way: hence, multiple copies of the same protein may be retained as a result of polymorphisms or sequencing errors; errors that have been corrected in Swiss-Prot are re-introduced when sequences are re-translated back from nucleotide sequences in GenBank; and numerous sequences are duplicates of fragments.

4 InterPro is an integrated protein family database. The main diagnostic methods are based on the use of single motifs (*e.g.*, consensus expressions, which underpin PROSITE), multiple motifs (*e.g.*, fingerprints, which underpin PRINTS) or complete domains (*e.g.*, profiles, which underpin the Profile library, Pfam, PANTHER, *etc.*).

Some of the integration challenges faced in creating InterPro include: the different perspectives on protein families held by each of its sources (whether motif-based, domain-based, *etc.*), which give different views on family membership; the volume of data contained in some of the sources (*e.g.*, the size of PANTHER renders rapid manual processing impossible); the complexity of data contained in some of the sources (*e.g.*, structure-based data in Gene3D); the transient nature of automatically-generated sequence clusters (*e.g.*, as found in ProDom, which can make it hard to assign stable accession numbers to entries between successive database releases).

Chapter 4

1 The translations in the five other reading frames of the sequence are below. Several apparent ORFs are visible; however, most are so short that they are unlikely to encode peptides. Checking each of these in turn:

```
5'3' Frame 2
GATVKSATTASRQEPTSDTQKVRTVECRNDHEFFVRFEWStopHACTHRLHRGCFDASRPLNQQAAGI
YSRCGPNGGVRYFGMetVWNPRVGSVSDGSSVLNSVTSVPYStopLGRRVTHISSLGTVACWSYCQQQ
WRGYRRPRSPDVRSGSGStopRSVRRStopWDSGRSISSStopKGWASRTQMetQRNSTPSRRSSQPS
RSRCTSRCCWGMetASQWYRSVGSRTPSTGRGTLTGCSPRRCCCStopTSRCSLTRIRERSLRSSVPT
ASStopSGPAWSAHStopRRSTRTASCGGRSAPQRCCTSCTCCSSGSPRRPKACAPRSHPRSKYCVTL
PLCCGPRIPSCGStopSAAKVRESCRStopTSRRCCSWCLTStopARRSASGSSSCAVVRSSAKPKRR
SRPPATARPRPATDRTRRTAPQPARLCSTTHDESPTRSCT
```

Here, there are two short ORFs: neither encodes a known peptide.

```
5'3' Frame 3
VQPStopSPPRPRHDRSRPATPRRCERLSAATITSFSCASSGNTRARIDFTAGVSTPAGRStopTSR
QRAFTAAVAQMetVGCAILVWFGIRVSAPCLTVHRFStopIPSRAYHTDWVVELHTYPRStopVLLH
VGVIANSSGGGIAGPDHRTSGVDLASARYGANGTRDALFPRERDGRLGPRCKEILRHHDARPSHRVH
DVPLDAAGVWPHNGTVRWGAEPHLLGAVRStopLAVHHAAVVVRPRVARStopRGSGNDPCARRCRR
HHDRDRPGRRTDEGLLVPLRVVGDQHRSDAVHPVRAVLRVHLEGRKHAPRGRIHVQSTAStopRYRC
VVVRVSRRVADRQRRCGNRAAEHRDAAVHGAStopREREGRLRAHPPAQSCDLRRSRSAGAVRRRRR
GRDQRLIAHAGQPHNRRGCVQRHTMetSPPLGLVL
```

Here, there are two shorter ORFs, neither of which encodes a known peptide.

```
3'5' Frame 1
EYKTEWGTHRVSLNTAAPVVGLSCVCDQSLVAAAPSPADGSGASASPKIARLRRRMetSPKPTFALT
SSTMetNSSVSMetFSGTIPAPSLPISHTTGYADHNTTVTLRSTLNVDATSGRMetLSAFEVNPKNS
TYRMetYSIAAVLIAHHTKRYEStopTFVSAPTRPVPIMetMetPSAPTSARIVPStopSASTSNAR
SNNNSGVVNSQSAYRAQStopMetGFCSPPNGTIVRPYPSSIERYIVNAMetAGTSVVMetAStopN
FFASGSETPIPFTRKStopSVPSPISAVPSASQIHSGRPVIWACDTPSTAVGNNSNMetQQYLTRIC
VStopLYDPISMetVRSStopRNLEPMetNRQTRSRHADSKPYQNSAPHHLGHSGCKCPLPAGSTAG
WRRNTRGEVDACTRVTTRSARKTRDRCGTQPFAPSGCRWSAPVVTRSWRTSRLH
```

Here, there is one long and four short ORFs – none encodes a known peptide.

```
3'5' Frame 2
STRPSGGLIVCRStopTQPRRLWGCPACAISRWSRPRRRRRTAPALRLRRRSHDCAGGStopARSRP
SRSRQAPStopTAASRCSAARFPHLRCRSATRRDTRTTTQRStopRYAVLStopTWMetRPRGACFR
PSRStopTRRTARTGCTASLRCStopSPTTRSGTSRPSSVRRPGRSRSStopCRRHRRAQGSFPDPR
QRATRGLTTTAAWStopTASQRTAPSRWGSAPHRTVPLStopGHTPAASRGTSStopTRWLGRASSt
opWRRISLHLGPRRPSLSRGNRASRVPLAPYRALARSTPDVRStopSGPAIPPPLLLAITPTCNSTS
topRGYVCNSTTQSVWYARDGIStopNRStopTVRHGADTRIPNHTKIAHPTIWATAAVNARCLLVQ
RPAGVETPAVKSMetRARVLPLEAHEKLVIVAALNRSHLLGVAGRLLSStopRGRGGLHGCT
```

Here, there are two shorter ORFs, neither of which encodes a known peptide.

```
3'5' Frame 3
VQDRVGDSSCVVEHSRAGCGAVLRVRSVAGRGRAVAGGRLRRFGFAEDRTTAQEDEPEADLRAHVKH
HEQQRLDVQRHDSRTFAADQPHDGIRGPQHNGNVTQYFERGCDLGAHAFGLRGEPEEQHVQDVQHRC
GADRPPHEAVRVDLRQCADQAGPDHDAVGTDERKDRSLIRVNEQREVStopQQQRRGEQPVSVPRPV
DGVLLPTERYHCEAIPQQHREVHRERDGWDERRDGVEFLCIWVRDAHPFHEEIERPESHStopRRTE
RStopPDPLRTSGDLGLRYPLHCCWQStopLQHATVPNEDMetCVTLRPNQYGTLVTEFRTDEPSDT
EPTRGFQTIPKStopRTPPFGPQRLStopMetPAACWFNGRLASKHPRStopSRCVHACYHSKRTKN
SStopSLRHSTVRTFWVSLVGSCRDAVVADFTVAP
```

Here, there are two shorter ORFs: neither encodes a known peptide.

2   The sequence depicted in Figure 4.10 is from version 83 of EMBL, with entry ID HSAGL1, accession number V00488[3], and sequence version V00488.1.

3   The left-hand hydropathy profile shows seven peaks, one of which is just above the strict cut-off and five are between this and the loose cut-off. The profile is suggestive of a TM protein, but the lack of confidence in the majority of peaks urges caution in the interpretation. The right-hand profile is much clearer – seven peaks sit strongly above the strict cut-off. The right-hand protein is hence likely to be a membrane protein with seven TM domains, while the left-hand protein is probably a globular

---

[3] http://www.ebi.ac.uk/cgi-bin/sva/sva.pl?index=25&view=323349557

protein with seven significant hydrophobic structures within its core (presumably buried helices).

4   A BLAST search with this sequence reveals several proteins with titles suggesting that this is a GPCR. Specifically, UniProtKB states that the sequence is 'probable G protein-coupled receptor 153'. However, the results from the protein family databases are inconclusive: InterPro says that the sequence is a rhodopsin-like GPCR; specifically, a member of the GPCR 153 subfamily of the GPCR 153/162 family. Exploring this result, we see that the PROSITE profile and the Pfam HMM diagnose the sequence as a rhodopsin-like GPCR. The PROSITE pattern, however, doesn't make a match. Furthermore, the fingerprints matched by the sequence are family GPR153/162, and subfamily GPR153; none of the PRINTS rhodopsin-like GPCR superfamily fingerprints is matched.

From this result, several things are evident. First, the specific 'GPCR' annotation derives from the HMM and the profile, which we know to be very sensitive discriminators. This diagnosis isn't supported by the PROSITE pattern or by the PRINTS fingerprint. Given the number of rhodopsin-like GPCR fingerprints available (there are ~250 fingerprints for rhodopsin-like GPCRs in PRINTS), it's curious that the sequence matches only its own fingerprint, and that of its parent family (GPR153/162). A hydropathy profile suggests that the sequence is likely to be a TM protein (with 7 TM domains), but the fingerprint profile indicates that it may not in fact be a rhodopsin-like GPCR (not all 7TM proteins are GPCRs) - see Figure 4.38.

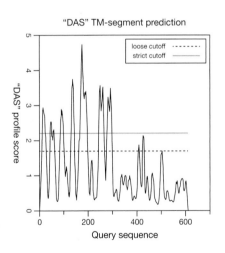
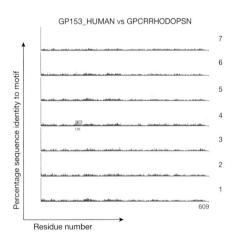

Figure 4.38
7TM profile (left panel), and rhodopsin-like GPCR fingerprint plot (right panel).

Chapter 5

1   Looking at the history for CAH2_HUMAN in December 2015, we find that there are 202 sequence versions (by the time you read this, there are likely to be several more). Part of the answer to this question would involve inspecting each of these entries and noting the number of unique DE lines. But the answer isn't that straightforward, because the history file appears to start at Swiss-Prot 9.0 in 1988, wheareas the sequence entered Swiss-Prot from PIR in July 1986. This means that there are at

least eight Swiss-Prot versions, and at least one PIR version not included in this history that we'd have to reconstruct by other means. One clue might be to determine how many different PIR and Swiss-Prot AC#s the sequence has had (at least four in Swiss-Prot (P00918, B2R7G8, Q6FI12 and Q96ET9) and eight in PIR (A27175; A23202; A92194; A92147; I37214; I51863; I51871; A01141)); and we might be able to find out more by looking in UniParc. This is not a trivial exercise.

2   What the published alignment failed to show correctly was that the proteins are characterised by seven small hydrophobic regions; these map to seven inner helices, each of which contains a GxxG motif, which together comprise an inner α-helical toroid structure (Hint: Illingworth, C. J. R., Parkes, K. E., Snell, C. R., Mullineaux, P. M. and Reynolds, C. A. (2008) Criteria for confirming sequence periodicity identified by Fourier transform analysis: application to GCR2, a candidate plant GPCR? *Biophys. Chem.*, **133**, 28–35). The misaligned features are the three GxxG motifs in the N-terminal portion of the alignment.

   As we might expect, the alignments created using MUSCLE, T-Coffee and Multalin are all different: the alignment created by Multalin, like ClustalW, misses the first three GxxG motifs; while those created using MUSCLE and T-Coffee come closest to preserving the seven GxxG motifs, getting six of them right (both get the second wrong). We should avoid considering either of these the 'best' alignment, unless we strictly define our frame of reference: the MUSCLE and T-Coffee alignments have done the best, if the frame of reference is to align the GxxG sequence motifs; but, as we would expect, none is faithful to the 3D structure (*i.e.*, all contain insertions in the α-helical regions). The MUS-CLE alignment is shown in Figure 5.15, with the correctly aligned GxxG motifs high-lighted in green boxes, and the incorrectly aligned GxxG motif denoted by the red box.

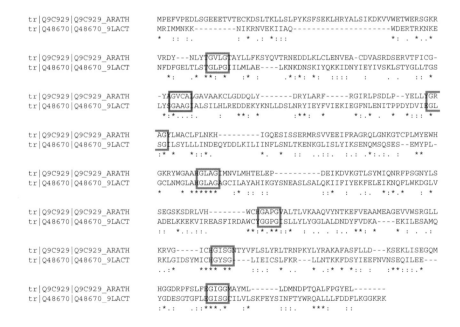

Figure 5.15

MUSCLE alignment highlighting the incorrectly aligned (red box) and correctly aligned GxxG motifs (green boxes).

3 The annotation of Q23293_CAEEL could have changed for several reasons: i) TrEMBL's analysis pipeline might have changed, in terms of the tools it uses for annotation, or of the parameters and priorities given to the tools in the pipeline; ii) the underlying sequence database will have changed (*i.e.*, it will be substantially bigger than it was in 1996), and other sequences are now more similar; and iii) the source data (from ENA) may have changed (*e.g.*, because the *C. elegans* genome has been revised). Any one or all of these will have influenced the annotation of this protein during the 20 years or so since it was deposited in TrEMBL.

The DMSR annotation doesn't obviously derive from the protein family databases, which all seem to suggest that the sequence may be a member of the GPCR clan (although there is some ambiguity about the precise family to which it might belong). 'DMSR' seems to have made its first appearance in the UniProtKB/TrEMBL description line in March 2012 (version 76). Within that entry, the DMSR annotation can be tracked via ENA's Sequence Version Archive (SVA): sequence FO081573, version 4[4], Release 111, 5 January 2012, identifies the gene as ZC404.11, protein product ZC404.11; four days later, however, version 5[5] identifies the gene as dmsr-11, protein product DMSR-11 (two other genes within this entry also bear the dmsr annotation). At this point, it isn't clear what DMSR-11 means, or why the gene and product names changed. If we follow the link to the WormBase entry[6](WormBase:WBGene00022605) we get closer to the annotation source – here, the gene is identified as 'DroMyoSuppressin Receptor related, dmsr-11'. Exploring the Gene class link[7] reveals that the designating lab was that of Oliver Hobert, of Columbia University, but not *how* the annotation was derived. We're left to deduce that the gene must have been inferred to be homologous to the *Drosophila* myosuppressin receptor following a BLAST search.

This exercise reveals some of the ambiguity and volatility of computationally-derived annotations; it also highlights the fact that different analytical processes are used to annotate DNA and protein sequences, and that these processes aren't synchronised. Consequently, we've had to work very hard to find out what DMSR means! Note that in December 2015, the detective work has a different starting point, as the EMBL/ENA link is via BX284605 rather than FO081573. Within entry BX284605, DMSR-11 is one of a dozen genes bearing the DroMyoSuppressin Receptor related annotation.

4 A BLAST search of sequence Q23293[8], using default parameters returns numerous putative uncharacterised sequences, and proteins with names like DMSR or CBG towards the top of the list. On subsequent pages are occasional matches to proteins with more meaningful annotations, like Class A rhodopsin-like GPCR, myosuppressin receptor, neuropeptide receptor, and so on. Selecting half-a-dozen of them and aligning them with CINEMA, gives a result like the one shown in Figure 5.16.

---

[4] http://www.ebi.ac.uk/cgi-bin/sva/sva.pl?index=5&view=1403324248
[5] http://www.ebi.ac.uk/cgi-bin/sva/sva.pl?index=4&view=1407952378
[6] http://www.wormbase.org/species/c_elegans/gene/WBGene00022605#04-9e-10
[7] http://www.wormbase.org/resources/gene_class/dmsr#01-10
[8] http://www.uniprot.org/uniprot/q23293

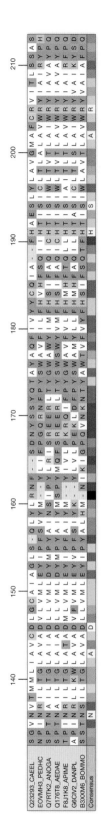

Figure 5.16

Alignment of selected sequences from a BLAST search of UniProtKB/TrEMBL.

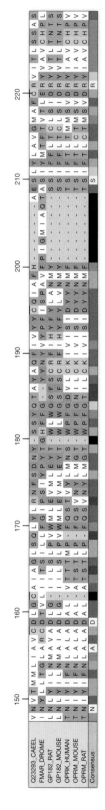

Figure 5.17

Alignment of selected sequences from a BLAST search of UniProtKB/Swiss-Prot.

Re-running BLAST just on the Swiss-Prot component of UniProtKB gives a very different answer: in December 2015, the output shows top matches to FMRF amide receptor, to GPCR 182 and to a range of μ- and κ-type opioid receptors (by the time you run the BLAST search, the results might look slightly different again). Aligning half-a-dozen of these gives a result that looks like Figure 5.17.

There are similarities between the results, but there are also significant differences. Clearly, TrEMBL is much larger than Swiss-Prot; the TrEMBL search thus revealed sequences that were more similar than the Swiss-Prot search had access to. Most of the retrieved matches from Swiss-Prot were μ- and κ-type opioid receptors, just as they must have been when Q23293 was first deposited in TrEMBL; but it's clear that this sequence doesn't share a number of key features (*e.g.*, the DRY motif). Similarly, the aligned sequences from the TrEMBL search were all annotated as receptors of some sort; but again, Q23293 doesn't share various features (*e.g.*, the WRY motif). In both alignments, Q23293 appears to be an outlier – it shares similarities with both sets of sequences, but it isn't the same as either. A fingerprint scan[9] of the sequence hits the rhodopsin-like GPCR superfamily (the graphical output depicted in Figure 5.18 shows a weak partial match); but no specific family or subfamily is matched. Thus, taken together, the results suggest that Q23293 could be a receptor, but that it's a functional GPCR isn't conclusive.

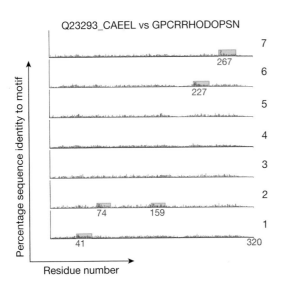

Figure 5.18

A fingerprint scan of Q23293.

Chapter 6

1 This algorithm takes a similar approach to that of bubble sort, with each pass iterating through the entire list in order to guarantee a single correctly-placed value. Because of that, it requires the same number of potential comparisons as a naïve bubble sort, giving it a time complexity of $O(n^2)$. Unlike bubble sort, however, which we saw was amenable to optimisation by progressively comparing one fewer value

---

[9] http://www.bioinf.manchester.ac.uk/cgi-bin/dbbrowser/fingerPRINTScan/FPScan_fam.cgi

each pass, this algorithm forces us to continue comparing values even after they've been replaced with -1, making it less efficient than the optimised bubble sort.

Because values are not swapped in place, but moved to a separate list, the space required to perform the algorithm is not that of a single item (used during the swap operation); rather, an entirely new list is required of equal length to the input. This gives it a space complexity not of bubble sort's constant $O(1)$, but a linear $O(n)$.

2   Given the number of entries in UniProtKB, some form of associative array, with fast key-based access, would be most efficient. Hash maps decide where to place an item by *hashing* its key – turning it into a near-unique position within the map. Assuming the hashing algorithm (the method, in this case, for turning an accession number into a position in the data structure) was quick enough, this would provide almost immediate access to amino acid sequences, with little or no traversal of the structure required. Hashing algorithms take into account the size of the container, so increasing the size of the hash map to cater for new records would require all the items to be rehashed and the map to be rebuilt – a very time-consuming process; thankfully the problem stated that downtime owing to updates was acceptable, so this disadvantage does not apply.

3   As hash maps need to be completely rebuilt from scratch when they're resized, using a hash map for such a fluid set of data would require regular time-consuming periods of restructuring that would adversely affect performance. With that in mind, a data structure that can quickly and dynamically restructure itself on inserts would be preferable, such as one that uses a red-black tree to organise its keys. This would marginally slow down access to individual sequences owing to tree traversal, and would require more time in general to insert a new sequence (owing to the tree restructuring itself), but would entirely circumvent the need to rebuild the entire tree.

4   In general, XML (a *context-free language*) cannot be parsed with a regular expression (designed for parsing *regular languages*); the differences between the two types of language simply make the general case of this impossible. One example of something allowed in XML, but that cannot be matched with a regular expression, is arbitrarily nested elements: the very core of what makes XML useful prevents regular expressions from parsing it[10].

That said, if the XML documents you were trying to parse were simple enough and were known never to diverge from a certain template, then it would be possible to use regular expressions to correctly match and parse small fragments of them. We could, for example, match simple opening element tags if there were no attributes that included the closing tag character ('>'). We might even be able to parse out the value of a particular attribute, so long as we could guarantee the ordering of attributes that we would come across, so as not to confuse the regular expression.

So, if we can guarantee the structure of the XML we will be parsing, or if the fragments we wish to match are simple enough, then regular expressions can probably

---

[10]The general inapplicability of regular expressions to XML and HTML causes much consternation among those without computer science training, and much frustration among those with computer science training. See: http://stackoverflow.com/questions/1732348/regex-match-open-tags-except-xhtml-self-contained-tags/1732454#1732454

serve us well. The moment the matching becomes more complex, or the XML is no longer under our control, regular expressions fail to be useful, and a proper XML parser should be used.

5 The BLAST algorithm exhibits a number of qualities that make it difficult to implement simply in generic languages. The heavy use of matrix calculations, the large (and ever-increasing) data-sets over which it runs, and the repetitive nature of the algorithm make the task difficult to do quickly. We therefore need to choose a language that will allow us to optimise as much of the algorithm as possible, and that means relatively low-level access to processing hardware.

Matrix calculations can be done in an optimised way on some processors, but only if the language has access to such methods, which discounts most scripting and interpreted languages. In order to have very tight control over exactly what memory is used, the ability to work around precise memory usage is also needed. This leaves only the low-level compiled languages such as C and C++ – *i.e.*, languages that don't hide any (or much) complexity from the user.

There is, however, a readily available class of processors that excel in massively repetitive tasks, including such things as matrix multiplication: Graphical Processing Units (GPUs). These processors, designed primarily for dealing with 2D and 3D graphics tasks, happen to provide just the right kind of tools to implement the BLAST algorithm. Programming them with languages such as NVIDIA's CUDA, or the Open Computing Language (OpenCL), is not a simple task, but, done correctly, can speed up the execution of the BLAST algorithm by an order of magnitude or more.

Chapter 7

1 The MD5 ('Message Digest Version 5') hashing algorithm takes a variable-length sequence of 512-bit blocks and, using some fairly hard-core maths, generates a 128-bit 'signature' (that's often represented as a 32-digit hexadecimal number). Assuming that we're encoding amino acids as 8-bit characters, that means 64 amino acids per block, with some padding if necessary to make up the block length. The intention of the algorithm was that no two different input sequences should give the same result, so the signature could be used as a reliable and compact checksum for chunks of data of any size.

For example, if we had a sequence for bovine rhodopsin (P02699) that consisted of the characters

```
MNGTEGPNFYVPFSNKTGVVRSPFEAPQYYLAEPWQFSMLAAYMFLLIML
GFPINFLTLYVTVQHKKLRTPLNYILLNLAVADLFMVFGGFTTTLYTSLH
GYFVFGPTGCNLEGFFATLGGEIALWSLVVLAIERYVVVCKPMSNFRFGE
NHAIMGVAFTWVMALACAAPPLVGWSRYIPEGMQCSCGIDYYTPHEETNN
ESFVIYMFVVHFIIPLIVIFFCYGQLVFTVKEAAAQQQESATTQKAEKEV
TRMVIIMVIAFLICWLPYAGVAFYIFTHQGSDFGPIFMTIPAFFAKTSAV
YNPVIYIMMNKQFRNCMVTTLCCGKNPLGDDEASTTVSKTETSQVAPA
```

in a text file containing nothing but these characters (*i.e.*, no line breaks or other extraneous whitespace), and ran the MD5 algorithm over this, we would get the result 2277dd9e488b82fc17d0e1908fb33f8a, represented in hexadecimal, back as a signature. While there's no way of reversing this process to turn the signature back into the sequence, the 32-digit hex number could be seen as a handy shorthand for its source sequence.

On the face of it, this might seem like a neat way of generating a unique identifier for the amino acid sequences we've been asked to manage in this scenario. But there are a number of pitfalls.

The first is that a number of flaws in the MD5 algorithm means that it's nowhere near as 'collision resistant' as was first thought, and that it's actually quite straight-forward to artificially construct two input sequences that 'collide' to give the same output. For security applications, this is disastrous, but even in situations where there's little chance of malicious manipulation, it means that using MD5 is a risk that's not worth taking. Other, more recent, hashing algorithms (such as SHA-256) are much more resistant to collisions, and therefore more suitable replacements for MD5.

Whereas the first pitfall is one of balancing probabilities – the risk of an accidental collision or the consequences of malicious manipulation – the second concern is much more fundamental, and hinges on the idea that hashing a sequence to generate an identi-fier breaks the principle of opacity discussed in Section 7.3.2. Although 2277dd9e488b-82fc17d0e1908fb33f8a appears opaque to humans, in that there's no way of figuring out what sequence it refers to just by inspecting the hexadecimal signature, using this mechanism creates a strong semantic link between the sequence and the identifier (as the latter was generated from the former). This is almost always a bad thing to do. For example, consider a situation where an error is later found in the amino acid sequence, and the record needs to be updated. Even if we do the right thing and maintain the prov-enance trail so that it's computationally possible to see that an update to the original sequence has happened (rather than by silently just fixing the sequence *in situ*), we end up with a semantically untidy situation, as the hashed signature is now inconsistent with the sequence of bytes that generated it in the first place. We could say that this inconsist-ency doesn't really matter if we want to *treat* the identifier as being semantically opaque, but in that case, why bother with the hashing at all? As dull as it might sound, an integer incremented for each record, and suitably qualified with a namespace would have done the job far better and with fewer chances of confusion.

2   The slight strangeness that remains in the format of http-style URIs comes from the inclusion of the domain name component. Domain names, such as www.uniprot.org, when read left to right, start off being quite specific and get more general (the machine called 'www', in a domain called 'uniprot', which is a kind of 'organisation'). URIs, on the other hand, get more specific when read left to right (this is a kind of http URI, hosted by this particular domain, referring to this set of resources, and specifically to this particular one called P02699). Although this may rankle with computer-science purists, in practice, this oddity doesn't cause any real problems.

Chapter 8

1   It's really hard to tell how many times '*Defrosting the Digital Library*' has been cited – and any numbers that you've found should probably be taken with a large pinch of salt. In February 2015, for example, Google Scholar suggested 159 citations, whereas Scopus gave 59; it's very hard to tell which of these is the more accurate,

as the corpus of documents being analysed by both systems isn't public knowledge, and neither are the details of the algorithms. Google Scholar probably has access to a bigger but scruffier corpus; Scopus to a smaller but better defined one; but really all you can do is use these numbers as some kind of indicator of… well, something. Probably. If you look at the 'article-level metrics' on PLoS, you'll be able to get an indication of how many times the article has been *downloaded* as a PDF, or *visited* as a Web-page. What that means in terms of how many times the article has *actually been read by a human* is anybody's guess.

2   EDEM1 is Endoplasmic Reticulum (ER) Degradation Enhancing α-Mannosidase-like protein 1; ERAD is the ER-Associated Degradation system. EDEM1 is a chaperone for rod opsin, and its expression can be used both to promote correct folding *and* enhanced degradation of mutant proteins in the ER. Retinitis pigmentosa (RP) is a neurodegenerative, misfolding disease caused by mutations in rod opsin – as EDEM1 is able to promote correct folding and enhanced degradation of mutant proteins in the ER, this suggests that it may be able to combat protein-misfolding diseases. The most common mutation that causes RP in North America is a Pro to His (P -> H) mutation in RP4. It has been suggested that EDEM1 may clear terminally misfolded P23H rod opsin, allowing other chaperones to interact with the folding-competent mutant rod opsin and to assist in productive folding; on the other hand, enhanced degradation of mutant rod opsin could reduce aggregation, and hence allow productive folding to occur in the absence of aggregation seeds and proteasome inhibition. Almost none of this information is captured in the UniProtKB entry. The UniProt curators could have made it easier to access this information had they given specific provenance trails for the facts cited, rather than lists of articles from which users are forced to reconstruct their steps.

3   Amongst the listed 'Open Access' papers, notice that the copyright for Roberts *et al.* resides with the authors and not the journal's publishers; here, the authors have paid for their article to be made freely available under the terms of the Creative Commons Attribution Licence, which permits unrestricted use, distribution and reproduction in any medium, provided the original work is properly cited. Therefore, as long as you give credit for having used the article, you can legally text-mine this work. The attribution part of the CC-BY licence is really important though: if your text-mining algorithm does find some crucial fact in this paper that had hitherto been overlooked, you'd really have to give the authors proper credit for their work or you'd be in breach of the licence.

   Chen *et al.'s* paper is a little more complex. It too is published under a Creative Commons licence, but this time with the 'Attribution-NonCommercial-NoDerivative' licence. Like CC-BY, this means that you have to give credit to the authors for using their work, but you cannot use it if you intend to get a 'commercial advantage' or receive any monetary compensation for doing so. The 'NoDerivatives' clause means that you can't 'remix, transform or build upon' the material either. So, although this means that the paper is free to read with your eyes and brain, and you can text-mine it if you're not doing so for commercial purposes, you can't do anything meaningful with any results you create!

   In our own paper, Pettifer *et al.*, the situation is even muddier, as the downloadable article doesn't say anything about the licence under which it's made available:

it's free to download and read, but all it tells you is that the copyright resides with the authors. It was our intention to make this a fully open article that you could do anything with, but we stumbled at the last hurdle in terms of getting the legal licence in place. If you do happen to find the cure for cancer in the article though, we'll happily share the glory!

4　CITO provides lots of different citation types, allowing you to 'agree with', 'cite as evidence', 'correct' or even 'deride' other works. In this question, we have implied that CITO is an 'authoritative description or definition of the subject under discussion', so in CITO's own terms, we used a 'cites as authority' citation.

5　The somewhat tortuous exemplar sentence from Box 8.3 is 'Treatment of primary cells from newly diagnosed CML patients in chronic phase as well as BCR-ABL+ cell lines with imatinib increased IRF-8 transcription'. What we think this means is perhaps more clearly written as 'Treatment with *imatinib increases IRF-8 transcription* in *primary cells* from *newly diagnosed chronic-phase* CML *patients*, and also in *BCR-ABL+ cell lines*', where the most important concepts are in italics. A very crude first pass at representing this in something approaching a 'machine readable' state would be to break the sentence into the kind of structure shown in Figure 8.11, from which it should be fairly straightforward to see how identifiers or terms from a controlled vocabulary or ontology could be assigned to various different components. In the case of 'imatinib', for example, we've already ended up with a node in the graph that contains a single unique concept – in this case, a drug that's already catalogued in numerous places (*e.g.*, it has the ChemSpider identifier 5101, and in ChEMBL it's CHEMBL941). At this very crude level, some of the other nodes are trickier, and we'd likely want to break things down further: for example, to encode the fact that, while IRF-8 is a gene, we're talking about the process of transcribing that gene here.

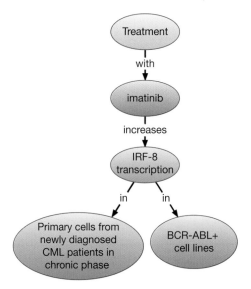

Figure 8.11

Crude parse tree for first exemplar sentence from Box 8.3.

It turns out, then, that, while it's possible for us to make a reasonable attempt to understand what the text means, provided we mentally rearrange the sentence, the task is much harder for a computer, suggesting that the machine's interpretation may not be entirely safe. As hinted above, the authors, editors and/or publishers could have made this more machine accessible had they supplied database identifiers for the main concepts (the drug, the gene, *etc.*).

Index

Bioinformatics challenges at the interface of biology and computer science: Mind the Gap. First Edition.
Teresa K. Attwood, Stephen R. Pettifer and David Thorne. Published 2016 © 2016 by John Wiley and Sons, Ltd.
Companion website: www.wiley.com/go/attwood/bioinformatics